Research for Development

More information about this series at http://www.springer.com/series/13084

Joaquim A.O. Barros · Liberato Ferrara
Enzo Martinelli
Editors

Recent Advances on Green Concrete for Structural Purposes

The Contribution of the EU-FP7 Project EnCoRe

Springer

Editors
Joaquim A.O. Barros
Department of Civil Engineering
University of Minho
Guimarães
Portugal

Enzo Martinelli
Department of Civil Engineering
University of Salerno
Fisciano, Salerno
Italy

Liberato Ferrara
Civil and Environmental Engineering
Politecnico di Milano
Milan
Italy

ISSN 2198-7300 ISSN 2198-7319 (electronic)
Research for Development
ISBN 978-3-319-56795-2 ISBN 978-3-319-56797-6 (eBook)
DOI 10.1007/978-3-319-56797-6

Library of Congress Control Number: 2017937297

Printed on acid-free paper

This Springer imprint is published by Springer Nature
The registered company is Springer International Publishing AG
The registered company address is: Gewerbestrasse 11, 6330 Cham, Switzerland

Preface

This book aims at summarising the results obtained in the Research Project entitled "Environmentally-friendly solutions for Concrete with Recycled and natural components" (EnCoRe, FP7-PEOPLE-IRSES-2011, Project ID: 295283), funded by the European Union as part of the 7th Framework Programme. The Project, whose activities have been developed during the three years 2012–2014, gathered three European Beneficiary Institutions (namely, *Università degli Studi di Salerno* and *Politecnico di Milano*, from Italy, and *Universidade do Minho*, from Portugal) and three non-European Partners (namely, *Universidad de Buenos Aires* and *Universidad Nacional de Tucuman*, from Argentina, and *Universidade do Rio de Janeiro*, from Brazil).

As stated by the title, EnCoRe was intended at investigating the physical and mechanical behaviour of cementitious composites made out of recycled and natural constituents. In fact, this is a subject of current relevance in both building technology and structural engineering, as a result of the growing interest to make the construction industry "greener". Specifically, the three following classes of materials have been considered:

1. Concrete with recycled aggregates and partial replacement of Ordinary Portland Cement (OPC);
2. Concrete reinforced with recycled fibers;
3. Cementitious composites internally reinforced with natural fibers or textiles.

However, this book is structured into two main parts. Part 1 covers the behaviour of a material belonging to the first one of the aforementioned classes and often referred to as Recycled Aggregate Concrete (RAC), as it is made with Recycled Concrete Aggregates (RCAs). Further insights are also reported on the effect of replacing OPC with Fly Ash (FA), a by-product of carbon-fed power plants, which is characterised by marked pozzolanic properties. Part 2 summarises the results obtained on cementitious composites internally reinforced with either recycled or natural fibers. Specifically, Recycled Steel Fibers (RSFs) obtained from post-consumed pneumatic tyres, and Natural Fibers (NFs) produced from tropical plants, such as sisal, are considered in this section.

The two main parts of this book consist of a number of *sections* treating the relevant knowledge derived from the research carried out as part of EnCoRe, which was also complemented with the available information in the performed literature review. The organization and contents of the sections aims to provide information from the technology up to the design. In fact, they summarise the empirical evidence about the physical and mechanical behaviour of the materials under consideration, and outline the theoretical models and the numerical procedures that can be formulated and implemented for simulating the behaviour of these materials. Moreover, a first attempt to propose a consistent conceptual formulation capable to make "predictable" their mechanical properties is also reported.

Although enhancing sustainability of cementitious composites is the fundamental motivation of this study, no consideration is reported within the book about the Life Cycle Assessment (LCA) of the materials addressed by the EnCoRe project. However, since the Project has contributed to an advance of knowledge on the mechanical behaviour of the aforementioned "environmentally-friendly" materials (and, hence, on their potential to be employed in field applications), its results can be used as input data by environmental scientists eventually interested in performing LCA on the materials investigated as part of the EnCoRe Project.

Finally, the book editors wish to acknowledge the tremendous contribution given by the members of the research groups belonging to the six Institutions that took part in EnCoRe: their names are listed at the beginning of each book section with the twofold aim to highlight the role played by the each researcher and the cooperation developed between the concerned research groups.

Guimarães, Portugal — Joaquim A.O. Barros
Milan, Italy — Liberato Ferrara
Salerno, Italy — Enzo Martinelli

Acknowledgements

This book mainly derives from the project "Environmentally-friendly solutions for Concrete with Recycled and natural components" (FP7-PEOPLE-2011-IRSES, n.295283): both Editors and Authors take this opportunity to gratefully acknowledge the financial contribution of the European Union.

Moreover, further Institutions and Funding Agencies have contributed to the activities herein reported:

- Joaquim A.O. Barros wishes to acknowledge the support provided by FCT through the grant SFRH/BSAB/114302/2016;
- Liberato Ferrara wishes to acknowledge the support of Politecnico di Milano through the Young Researchers Grant 2011–2013;
- Enzo Martinelli wishes to acknowledge the support of the University of Salerno through the FARB 2013 Grant.

Contents

Part I
Recycled Aggregate Concrete

Chapter 1
State of Knowledge on Green Concrete with Recycled Aggregates and Cement Replacement

Enzo Martinelli, Eduardus A.B. Koenders and Marco Pepe

Abstract Since the construction industry is characterized by a huge demand for both energy and raw materials, it is fully concerned by the need for enhancing sustainability, which is certainly the main challenge for all industrial sectors in the twenty-first century. Therefore, several solutions are nowadays under investigation to reduce the environmental impact of concrete production. They often consist of partially replacing the ordinary concrete constituents with recycled ones, in view of the objective of reducing both the demand of raw materials and the amount of waste to be disposed in landfills. The most recent advances in this field are summarized in this chapter, which is intended at drawing the line of the current state of knowledge on "sustainable" structural concrete.

Concrete is the most widely used construction material and, hence, the reduction of the environmental impact induced by its production processes is a relevant and timely challenge for modern science and technology.

As a matter of fact, the production of concrete is characterised by a considerable demand for energy and raw materials and results in significant emission of Greenhouse Gases (GHG). Specifically, the cement production industry is deemed responsible for about 5% of the total CO_2 emissions, whereas the whole concrete production leads to almost double this share (Moya et al. 2010).

Moreover, the construction of new buildings, as well as the maintenance and/or demolition of existing ones, is responsible for the production of large amount of waste, commonly referred to as Construction & Demolition Waste (CDW), which generally require environment-sensitive and expensive disposal procedures (Moll et al. 2005).

Therefore, recycling these waste to replace part of ordinary aggregates (OA) is a straightforward and rational solution to produce more sustainable and

E. Martinelli (✉) · M. Pepe
University of Salerno, Fisciano, Italy
e-mail: e.martinelli@unisa.it

E.A.B. Koenders
TU Darmstadt, Darmstadt, Germany

J.A.O. Barros et al. (eds.), *Recent Advances on Green Concrete for Structural Purposes*, Research for Development, DOI 10.1007/978-3-319-56797-6_1

environmentally-friendly concrete, also for structural applications (USDT 2004). Furthermore, replacing part of the Portland cement needed in ordinary concrete mixtures with alternative binders, often obtained by recycling industrial by-products, is a viable option for reducing the emissions of GHG due to the production of concrete (Lothenbach et al. 2011). These and other possibilities are currently under investigation for enhancing sustainability in the concrete industry by reducing the environmental impact due to producing and supplying the aforementioned ordinary constituents and the water always needed in concrete mixtures (Sandrolini and Franzoni 2001).

This chapter is intended at providing an overview of the current state of knowledge about the physical and mechanical properties of a wide class of materials, often referred to as "green concretes" (fib 2013).

First of all, Sect. 1.1 analyses the most promising solutions for producing and supplying alternative constituents whose use in concrete production can result in a significant reduction of the environmental impact. Specifically, this close examination is subdivided into three main parts dealing with recycled aggregates, alternative binders and further solutions for the other concrete constituents. The main physical properties, also connected to the most industrially feasible processing solutions, are examined in these subsections that are intended at describing the main differences between these "sustainable" concrete constituents and the ordinary ones, in terms of both physical and mechanical properties.

Although several options have been explored for producing concrete and other cementitious composites with recycled constituents, the following sections focus on the ones intended at obtaining structural concrete by employing Recycled Concrete Aggregates (RCAs), possibly in conjunction with industrial by-products, such as Fly-Ash (FA) or Silica Fume (SF). Therefore, Sect. 1.2 outlines the most recent findings about the fresh-state behaviour of concrete made with RCAs, generally referred to as Recycled Aggregate Concrete (RAC), and the influence of the possible use of the aforementioned mineral additions. Moreover, the current state of knowledge on the physical and mechanical properties of RAC and their correlation with the relevant engineering properties of the constituents is outlined. Unveiling these correlations is a key step for making predictable the resulting physical characteristics and mechanical properties of these "green" concretes. However, no general well-established theory has been formulated and validated so far. In this respect, the current codes and guidelines generally provide strict limitations on the use of recycled constituents: a wide overview of these codes and guidelines is proposed in Sect. 1.4.

Moreover, Sect. 1.5 proposes an overview of the main contributions about the environmental implications of both producing concrete with the aforementioned components and quantifying the possible beneficial effects on sustainability.

It is worth highlighting that the following sections focus on the most recent advances in this broad field of research and, hence, it refers to theoretical and experimental contribution appeared in the last decade in the international scientific literature. For a further discussion about the first studies on the mechanical characterisation of cementitious composites made out by replacing part of the ordinary constituents with recycled ones, the Reader may refer to well-established texts (Hansen and Narud 1983; Hendriks et al. 2005).

1.1 Sustainable Concrete Constituents

Sustainable concrete constituents can be obtained by recycling various classes of waste and by-products (Pacheco-Torgal et al. 2013). Since aggregates and binders are generally the main ingredients in any concrete mixtures, the following subsections only deal with the most recent advanced in classifying, producing, processing and employing these two main constituents for producing structural concrete. Although some studies address the effect of recycling water (CCAA 2007), this aspect is not covered in the following subsections.

It is worth highlighting that the definitions adopted hereinafter are inspired to the classical "terminology" adopted by Hansen (1986a, b): therefore, the notation adopted by other authors is sometimes modified for being consistent with the aforementioned work.

1.1.1 Recycled Aggregates

Recycled aggregates can be produced by using various types of waste, often deriving from Construction and Demolition Waste (CDW), but also obtained through other types of waste that are not strictly connected with the construction section and the concrete production (Kuosa 2012). Other sources of waste, not only belonging to the class of CDW, that may be recycled and employed as aggregate in concrete are listed and discussed into details by de Brito and Saikia (2013). Further classifications are available both in national pre-standard regulation documents (Kreijger 1981) and in the international scientific literature (Butler et al. 2014). Other proposals deals with the use of such aggregates for specific purposes (Zhu et al. 2011; Tebaldi et al. 2012). However, the present discussion is restricted to CDW and the expression "Recycled Aggregates" (RAs) identifies aggregates produced by crushing and processing any kinds of CDW. Specifically, Recycled Concrete Aggregates (RCAs) are those obtained by selecting, crushing and processing concrete members coming from different sources, such as the demolition of existing buildings or the recovery of residuals in pre-cast concrete factories (Pedro et al. 2014) and unused concrete returned to plant (Ferrari et al. 2014).

Although RCAs are the most relevant option for the subject under discussion in this book, RAs can be further classified by considering their original materials and, hence, the following main classes can be recognised within CDW:

- Recycled Masonry Aggregates (RMAs), obtained by crushing masonry bricks (Corinaldesi 2012);

- Recycled Ceramic Aggregates (RCerAs), obtained by crushing waste ceramic tiles and sanitary ware (Alves et al. 2014);
- Recycled Glass Aggregates (RGAs), mainly intended at replacing the fine fraction of natural concrete aggregates (Adaway and Wang 2015; Mardani-Aghabaglou et al. 2015).

Moreover, RAs derived by industrial activities that are connected with construction and demolition (such as the extraction and transformation of marble stones) are also considered in the scientific literature as a viable source of recycled aggregate for concrete (Corinaldesi et al. 2010).

However, demolition and processing generally imply that the various types of aggregates get mixed and, hence, some classifications adopted within the scientific literature take into account the possibility that recycled aggregates do not belong to a unique waste source (Yang et al. 2011). To this extent, the following classification was recently proposed (Agrela et al. 2011):

- Concrete Recycled Aggregate (CRAs) in which concrete content is at least 90% (in weight);
- Mixed Recycled Aggregate (MixRAs) in which the ceramic content ranges between 10 and 30%;
- Ceramic Recycled Aggregate (CerRAs), containing more than 30% of ceramic particles.

Since concrete and ceramic (sometimes using the latter term in a broader sense including also masonry) are the main sources of waste produced in construction and demolition of buildings, a further classification criterion intended at a visual selection of these two main waste streams can be based on their colors and, hence, defines the two following "fraction" (Toledo Filho et al. 2013):

- the *grey* fraction, consisting of particles mainly made of structural concrete (and, in a minor portion, mortar) debris
- the *red* fraction, including clay bricks and ceramic-based (i.e. tiles) materials.

Classifications based on the original source of RAs are certainly useful, as they can drive the selection and separation processes, either during demolition or in dedicated recycling plants (Mas et al. 2012). However, a *performance related approach* (WRAP 2007), intended at classifying recycled aggregates in terms of their relevant engineering properties, would be more useful for the design of concrete mixtures employing these sustainable constituents. Experimental evidences highlighted that the actual content of *red* particles controls the main engineering properties of aggregate (Agrela et al. 2011). However, the mere classification by origin does not lead to CDW aggregates with homogeneous properties: as a matter of fact, water absorption measured in concrete aggregates, albeit collected in the same geographic area, can be extremely variable (Angulo et al. 2010).

Therefore, a more accurate performance-based classification, based on the tests usually carried out for concrete aggregates (CEN 2013) and intended at determining the main geometric (such as grain size distribution and shape parameters), physical (such as density and water absorption) and chemical properties (such as total sulfur, acid-soluble sulfate and chlorides contents) is needed (McNeil and Kang 2013; Rodríguez-Robles et al. 2014). For instance, the sink–float technique is a reasonably feasible tool for separating CDW in different density classes (Angulo et al. 2010).

Based on the results of these tests, a close correlation between water absorption (WA) and oven-dried density (ODD) was recently unveiled by analyzing the data regarding 589 aggregates of different types, sizes, origins, and sourced drawn from 116 publications. The statistical processing of these data and, particularly, a regression analysis carried on the two aforementioned quantities led to the following polynomial relationship (Silva et al. 2014):

$$\mathrm{WA} = A_3 \cdot \mathrm{ODD}^3 + A_2 \cdot \mathrm{ODD}^2 + A_1 \cdot \mathrm{ODD} + \mathrm{A}_0 \tag{1.1}$$

where WA is expressed in percentage and ODD in kg/m^3, and the constants assume the following values: $A_3 = 2.9373\ 10^{-9}$, $A_2 = -9.4014\ 10^{-6}$, $A_1 = 1.8977\ 10^{-2}$ and $A_0 = 65.745$. Silva et al. (2014) also reported further statistical information intended at describing the distribution of the experimental results with respect to the relationship (1.1); particularly, a coefficient of determination $R^2 = 0.878$ was estimated.

The correlation described by relationship (1.1) is the basis for a classification proposed by the same Authors for CDW waste in view of their use as RCAs (Silva et al. 2014): four classes, denoted with A, B, C and D, ranging from the better to the worse one, were defined.

A similar idea is presented herein, around the definition of a Quality Index QI of RCAs:

$$\mathrm{QI}^{\mathrm{RCA}} = \frac{\mathrm{WA}^{\mathrm{OA}}}{\mathrm{WA}^{\mathrm{RCA}}} \tag{1.2}$$

where $\mathrm{WA}^{\mathrm{OA}}$ is the water absorption of ordinary aggregate that is going to be replaced, and $\mathrm{WA}^{\mathrm{RCA}}$ is the one of the recycled aggregate under consideration. In principle, the QI should be defined for each relevant size range considered in the mixture.

Both criteria and any other classification approach based on either of the aforementioned quantities is based on the physical observation that, especially in RCA (Pepe et al. 2014a, b), but also in other types of RAs (Corinaldesi and Moriconi 2009a; García-González et al. 2014), porosity is mainly related to attached mortar (AM). Since AM is one of the key parameters affecting the relevant properties of concrete at both fresh and hardened states (Duan and Poon, 2014), it should be carefully considered in rational mix-design rules for RAC (Fathifazl et al. 2010). As a matter of fact, these quantities can also be controlled by means of various techniques intended at "cleaning" the outer part of AM and reducing the particle of crushed concrete as close as desired to the original properties of the original OA. In fact, the "autogenous cleaning" process, based on treating crushed concrete particles in a rotating mill for a certain time, is a viable solution for reducing the amount of AM and, hence, controlling all the related engineering properties of RCAs, such as density, water absorption and porosity (Pepe et al. 2014b). As it can be expected, the processing procedure plays a key role on the resulting quality of RCAs and, hence, of RAC. In fact, experimental results demonstrated that RCAs processed by coupling a primary plus secondary crushing (PSC), using a jaw crusher followed by a hammer mill, performed significantly better than the case in which a simple primary crushing process is executed (Pedro et al. 2014).

The results show that the PSC process leads to higher performance, especially in terms of durability. The experimental evidence highlights that higher contents of AM with scarce presence of natural aggregates characterizes the smaller size fractions of RAC (Evangelista et al. 2015). This is the reason why, as it will be reported in Chap. 2, the use of RCA for replacing the fine fraction of is not generally allowed for producing structural concrete (NTC 2008). However, several studies have demonstrated that the weaknesses induced by the higher porosity of RCA can be somehow balanced by means of mineral additions often derived by recycling environmentally harmful industrial by-products, such as FA and SF. Particularly, the former, when employed in partial replacement of the finer fraction of aggregates, has been shown to be able to enhance workability and strength of RAC (Kou et al. 2008; Lima et al. 2013). Moreover, the addition of fly ash has been also very effective in reducing carbonation and chloride ion penetration depths in concrete, even in RAC (Corinaldesi and Moriconi 2009b). The aforementioned by-products are also characterised by pozzolanic properties (Wang et al. 2013; Dilbas et al. 2014) and, hence, they can be employed in partial substitution of Portland cement, in view of an even more sustainable structural concrete.

Finally, RCA have been proved to be suited also for producing High-Performance concrete: however, the quality (i.e. the mechanical properties) of the original concrete plays a decisive role in limiting the maximum strength of RAC (Gonzalez and Etxeberria 2014).

1.1.2 Alternative Binders

Portland Cement (PC) is an essential constituent of concrete as it has been produced and utilized so far. As already said, the production of PC is deemed responsible for a significant share of the global GHG emissions. Therefore, reducing the amount of PC needed for producing structural concrete of a given *quality* (in terms of resulting physical and mechanical properties) would result in a straightforward reduction in the environmental impact of the concrete industry and the construction sector as a whole.

The hydration reaction of PC developing in concrete mixtures during the phase generally referred to as "setting" and "hardening" are characterized by a number of coupled chemical processes whose kinetics is determined by both the nature of the process and the state of the system at that instant. A macroscopic interpretation of these reactions based upon combining two well-established physical tools, such as the heat transfer theory and the Arrhenius principle, is capable of reproducing the influence of thermal boundary conditions on the resulting rate of hydration and the time development of the reaction heat and temperature throughout concrete (Azenha 2009; Ventura-Gouveia 2011; Martinelli et al. 2013).

However, a more fundamental analysis of the hydration reaction highlights six main chemical processes (generally referred to as: *dissolution/dissociation, diffusion, growth, nucleation, complexation, adsorption*) that may develop either in series or parallel, the latter case resulting in a further complication of this complex chemical system (Bullard et al. 2011).

The growing interest for developing more sustainable concretes and cementitious composite materials is leading to considering more and more the possible employment of secondary mineral additions, generally referred to as *supplementary cementitious materials* (SCMs, Lothenbach et al. 2011) and often originating as by-products of other industrial activities. Therefore, a more complete knowledge of the fundamental mechanisms of hydration is needed to provide a rational basis for selecting the most effective constituents and designing more sustainable concrete mixtures. Furthermore, PC is not the ideal binder for all construction applications, as it suffers from durability problems in particularly aggressive environments. Several alternative binders have been available for almost as long as Portland cement but, despite this, they have not been yet extensively used (Lollini et al. 2014).

The most promising alternative binders currently available can be classified as follows (Juenger et al. 2011):

- *Calcium aluminate cements* featuring rapid strength development and good durability in high sulfate environments;
- *Calcium sulfoaluminate cements* characterized by low CO_2 emissions and energy demand, but with several unknown aspects on the time development of mechanical properties and long-term durability;
- *Alkali-activated binders* often obtained by recycling from waste materials and industrial by-product and, hence, exposed to the natural variability about physical compositions and chemical properties of these waste and by-products;
- *Supersulfated cements* almost entirely made from waste materials, coupled with low heat production and good durability in aggressive environments such as seawater.

Based on the above short descriptions, it is clear that the last two classes have potential to be employed in "green concrete" for reducing the demand of PC. As regards the alkali-activated binders (Pacheco-Torgal et al. 2008), the following five categories can be introduced and considered in the following classification (Shi et al. 2011):

- *Alkali-activated slag-based cements*, including blast furnace slag cement, phosphorus slag cement, blast furnace slag-fly ash cement, blast furnace slag-steel slag cement, blast furnace slag-MgO cement, blast furnace slag-based multiple component cement

- *Alkali-activated pozzolan cements*, including fly ash cement, natural pozzolan cement, metakaolin cement, soda lime glass cement;
- *Alkali-activated lime-pozzolan/slag cements*, lime–natural pozzolan cement, lime–fly ash cement, lime–metakaolin cement, lime–blast furnace slag cement;
- *Alkali-activated calcium aluminate blended cement*, including combinations of calcium aluminate cement (CAC) with metakaolin, pozzolan and fly-ash;
- *Alkali-activated Portland blended cement (hybrid cements)*, including Portland blast furnace slag cement, Portland phosphorus slag cement, Portland Fly ash cement, Portland blast furnace slag–steel slag cement, Portland blast furnace slag–fly ash cement, multiple components blended cements.

The use of Fly Ash and Silica Fume in partial replacement of cement has been also employed in junction with RCA (Corinaldesi and Moriconi 2009b; Mahmoud et al. 2013; Lima et al., 2013).

More recent solutions for a partial replacement of cement in concrete were developed by considering the ash obtained by burning municipal solid waste of agricultural waste. Three relevant examples of these emerging supplemental cementing materials are reported below:

- *Rice-Husk Ash* (*RHA*);
- *Sugar Cane Bagasse* (*SCB*) *ash*;
- *Municipal Solid Waste Incinerator* (*MSWI*) *ash.*

Rice husk is an agricultural residue that accounts for 20% of the 649.7 million tons of rice produced annually worldwide. The chemical composition of rice husk is found to vary from one sample to another due to the differences in the type of paddy, crop year, climate and geographical conditions. Burning the husk under controlled temperature below 800 °C can produce ash with silica mainly in amorphous form Ghassan and Hilmi (2010). The performance of RHA as a supplemental cementing materials was even investigated in Ultra-High Performance Concrete (UHPC) and the obtained results highlighted that RHA acts both as highly pozzolanic admixture and internal curing agent in UHPC (Van et al. 2014).

SCB ash is a by-product of the sugar/ethanol agro-industry abundantly available in some regions of the world and has cementitious properties indicating that it can be used together with cement (Fairbairn et al. 2010).

MSWI ashes have several applications. As regards the applications of relevance in cement and concrete industry, on the one hand, the addition of MSWI ash for clinker production has been demonstrated to shorten the setting time and decrease workability. On the other hand, experimental results demonstrate that addition of up to 50% treated MSWI fly ash does not significantly affect the mechanical properties (Lam et al. 2010).

Finally, nanotechnology offers further options for reducing the amount of cement needed in concrete. Specifically, incorporating colloidal Nano-Silica in concrete with 100% coarse RCAs led to similar results, in terms of mechanical properties, with respect to a reference concrete mixture (Mukharjee and Barai 2015).

1.2 Fresh-State Behaviour

The peculiar features characterising RCAs have a significant influence in affecting the properties of concrete at the fresh-state. Particularly, they generally affect workability due to the two main reasons explained below:

- the higher porosity and water absorption capacity of RCAs have an interaction on the water actually available in the mixture: initially dry RCAs result in a reduction of the free water and, hence, a reduction of workability, whereas, part of the water absorbed in initially saturated ones can have the opposite effect (Pepe et al. 2014a);
- the higher irregularity and roughness of RCA particles with respect to ordinary aggregates generally leads to reduced workability (Safiuddin et al. 2011).

On the contrary, the addition of FA to RAC mixtures generally enhances workability: this is mainly due to the higher regularity and fineness of FA particles, which contribute to reducing friction interactions among aggregates in both normal (Lima et al. 2013) and self-compacting (Revathi et al. 2013) RAC.

Although the slump test (CEN 2009) is the most common methodology for quantifying workability in fresh concrete mixture, a systematic investigation carried out on concrete mixtures characterised by variable aggregate replacement ratio and water compensation rate demonstrated that the VeBe time and flow table tests are more suitable to determine workability of RAC. In fact, these testing techniques result are more capable to detecting the influence of relevant quantities, such as free water content and aggregates' shape (Leite et al. 2013). As a general observation, as it is easy to expect that, by increasing the water compensation rate results in enhancing workability (as a result of the well-known Lyse's rule) and reducing compressive strength at the hardened state (as a result of the well-known Abrams' law).

Therefore, chemical admixtures are required to guarantee the target workability without inducing detrimental effects on the resulting mechanical properties of RAC. Particularly, a new generation of superplasticizers containing some copolymer polycarboxylate makes it possible to significantly improve the fluidity of the RAC (Braymand et al. 2015). To a certain extent this superplasticizers can also compensate the detrimental effect induced by replacing the fine fraction of aggregate with recycled sand (Pereira et al. 2012).

An alternative option, intended at reducing the interaction of RCAs with free water, is based on tailored polymer-based treatments which consist in soaking these aggregates in polymer solutions and developing a water-repellent film on the aggregate outer surface. A recent study compared the effect of various polymers and variable concentrations of the aforementioned solutions on both the fresh and hardened state properties of RAC (Spaeth and Tegguer 2013).

Finally, due to the same reasons affecting workability, the presence of RCA has been proved to play a role in the initiation and development of the hydration reaction of concrete at early age: the aggregate replacement ratio modifies the reaction kinetics (Koenders et al. 2014) and the initial moisture condition of aggregates modifies the initial rate of reaction and the time development of cement hydration (Pepe et al. 2014a).

1.3 Hardened-State Behaviour

The properties of aggregates play a significant role also on the resulting behaviour of RAC at the hardened state, both in terms of mechanical response and physical durability-related parameters.

The present review of the most recent advances on this topic focuses on RAC made from RCAs, as it is proved to be the best suited option for producing "green" structural concretes. However, it is worth mentioning that studies on the mechanical characterisation of concrete with coarse RMAs or MixRAs (see Sect. 1.1) are available in the literature (Gomes and de Brito 2009). They demonstrate that limited replacement of natural aggregates with the aforementioned mixed recycled ones (lower than 25% in weight) result in concrete characterised by sufficient strength and durability for housing construction and almost the same physical properties with respect to the reference mixtures (Medina et al. 2014). Moreover, studies on the use of RGAs are also available: they are often intended at investigating the conditions of occurrence and the solutions for suppressing the Alkali-Silica Reaction (ASR) which can be harmful for the material durability (Rajabipour et al. 2010). The resulting behaviour of RAC made from mixed glass/concrete RAs has been also investigated for understanding the concurrent influence of the lighter glass particles and the more porous crushed concrete ones (Mardani-Aghabaglou et al. 2015). Furthermore, although the focus of the present review is on RAC for structural purposes, RCA seems particularly suited for some specific non structural applications, such as the production of pervious concrete (Chen et al. 2012; Güneyisi et al. 2016), as they are mainly made of coarse particle and should guarantee a significant water porosity (Sriravindrarajah et al. 2012). In fact, replacing the fine fraction of aggregates with fine RCA results in several detrimental effects on the resulting mechanical performance of RAC, among which a significant increase in shrinkage and creep deformation (Cartuxo et al. 2015).

Since the compressive strength f_c (generally determined after 28 days of curing) is the main design parameter for structural concrete, several studies on RAC focus on determining the role of RCA on the resulting value of this property. Although these are often limited to an empirical observations of the role of RCA on f_c, scrutinising the cement reactions developing in RAC unveils the fundamental influences of the main engineering parameters of both aggregates (i.e. either open porosity or water absorption capacity, along with their initial moisture condition at mixing) and mixtures (i.e. the water/cement ratio) on the resulting hydration kinetics (Koenders et al. 2014). However, it was demonstrated that RAC is affected by curing conditions (either laboratory conditions, external environment or wet chamber) roughly in the same way as ordinary concrete (Fonseca et al. 2011).

Between the mere empirical observation and these fundamental modelling approach, various quantitative relationships have been recently proposed for expressing the correlation between f_c depending on the main parameters describing the mix composition. These relationship are often formulated in terms of the $f_{c,RAC}/f_{c,RC}$ ratio between the compressive $f_{c,RAC}$ of RAC and the one, denoted $f_{c,RC}$, of the reference concrete (RC) mixture made with only ordinary aggregates and the same size grading of RAC (de Brito and Alves 2010):

$$\frac{f_{c,RAC}}{f_{c,RC}} = 1 - 2.619 \cdot \left(1 - \frac{D_{RAC}}{D_{RC}}\right) \tag{1.3}$$

where D_{RAC} and D_{RC} denote the *weighted density* of RAC and RC, respectively, determined across the various size fraction of aggregates included in the concrete mixtures under consideration. The relationship (1.3), determined on seven different series of experimental tests (mainly carried out on mixtures of RCAs) was calibrated for weighted density ratios ranging between 0 and 0.10 and exhibited a reasonably good correlation with the aforementioned experimental results (expressed by a coefficient of determination $R^2 = 0.7615$). The same authors (de Brito and Alves 2010) proposed similar relationships intended at expressing the correlation between the same weighted density ratio and other concrete properties, such as compressive strength at 7 days, splitting and flexural tensile strength, modulus of elasticity, abrasion resistance, shrinkage, water absorption, carbonation penetration and chloride penetration: the analytical expressions of these relationships are omitted herein and the interested Reader can refer to the cited paper for further details.

The knowledge achieved so far about the influence of RCAs on the resulting strength of RAC led to the formulation of generalised mix-design rules for the latter (Fathifazl et al. 2009; Yehia et al. 2015): a physically-based conceptual proposal based on the findings of a recent Ph.D. thesis (Pepe 2015) will be presented in Chap. 6. The structural scale response of RAC was investigated under both permanent and variable loads. On the one hand, the results obtained on RAC with 100% of coarse RCAs reveal that the ratio between strength determined at low loading rate and the standard one in compression and in tension is similar for RAC and ordinary concrete (González-Fonteboa et al. 2012). On the other hand, beam-to-column joints made of RAC with 30% limited replacement ratio (30% of the coarse aggregates) and subjected to cyclic-actions highlighted the suitability of using RAC in seismic zones (Corinaldesi et al. 2011).

Moreover, an empirical correlation was proposed for expressing the time evolution of strength (Malesev et al. 2010)

$$f_c(t) = \frac{a \cdot t}{t + b}, \tag{1.4}$$

where a and b are two constants whose values can be derived through best-fitting of the available experimental result: as it is clear, the parameter a represents the asymptotic value $f_{c,\infty}$ of f_c, whereas b (dimensionally a time quantity) controls the initial rate of growth of compressive strength. The relationship (1.4) will be considered and calibrated in the following Chap. 3 for the concrete mixtures reported therein.

Due to the peculiar properties of RAC and their interaction with the other mixture constituents, the resulting correlation between the compressive strength f_c and the other mechanical parameters cannot generally be expressed by means of the same analytical expressions adopted in Codes (CEN 2005) and Guidelines (fib 2013) for ordinary structural concrete. As regards the Young's Modulus E_c, the following correlation between its mean value and the cubic compressive strength of concrete $f_{c,cube}$ proposed (Corinaldesi 2010):

$$E_{cm} = 18800 \cdot \sqrt[3]{\frac{0.83 \cdot f_{c,cube}}{10}}. \tag{1.5}$$

This relationship was calibrated on the results obtained from RAC specimens with five different water/cement ratios (ranging from 0.40 to 0.60) and replacement ratio of 30% of the coarse aggregates. The experimental results highlighted that compressive strength was almost unaffected by RCAs, whereas RAC exhibited lower static elastic modulus: this reduction was around 10% and justifies the reduction of the coefficient adopted in Eq. (1.5) with respect to the one currently in use for ordinary concrete (CEN 2005; fib 2013).

Moreover, analytical relationships were also proposed for generalising the Sargin curve, adopted by the aforementioned Codes and Guidelines, to take into account the increase in axial deformability observed in RAC. Specifically, three coefficients α_c^{rec}, β_{cu}^{rec} and φ_{cm}^{rec} were calibrated by González-Fonteboa et al. (2011) for modifying the values of ε_{c2}, ε_{cu} and E_{cm} defining the stress-strain curve for ordinary structural concrete (CEN 2005; fib 2013):

$$\alpha_c^{rec} = 0.0021 \cdot RR_{CRCA} + 1, \tag{1.6}$$

$$\beta_c^{rec} = 0.0022 \cdot RR_{CRCA} + 1, \tag{1.7}$$

$$\varphi_{cm}^{rec} = -0.0020 \cdot RR_{CRCA} + 1, \tag{1.8}$$

where RR_{CRCA} is the percentage of coarse RCA employed in the concrete mixture. The relationship (1.8) confirms that the reduction in E_{cm} expected for RAC with replacement ratio lower than 30% foresees a reduction in strength lower than 10%, as already found by the aforementioned Author (Corinaldesi 2010). Moreover, similar conclusions can be achieved by means of alternative analytical expressions proposed for the stress-strain curve of RAC under uniaxial compression (Wardeh et al. 2015). A complete constitutive formulation capable of simulating the response and predicting the failure mode of members made of RAC and subjected to tri-axial stress states (Folino and Xargay 2014) would also lead to the same conclusions. Further mechanical properties of interest for reinforced concrete structures, such as bond between RAC and deformed steel rebars, were recently investigated for understanding the influence of RCA and the bond-slip relationships generally adopted for ordinary concrete were recalibrated for RAC characterised by either normal (Prince and Singh 2014a) or high strength (Prince and Singh 2014b). Moreover, the widely adopted non-destructive testing techniques, such as those based on ultrasonic velocity (Rao et al. 2011) and acoustic emissions (Kencanawati et al. 2013), were recalibrated for RAC.

The mechanical response under sustained loads and, particularly, the development of creep and shrinkage phenomena are other features of relevance in RAC for structural purposes. Experimental observations confirm that the evolution of these phenomena in RAC is fairly similar to the one of ordinary structural concrete, although a certain influence of the aggregate replacement ratio can be detected. An experimental study available in the literature shows creep increase of around 50% and shrinkage increase of about 70% in RAC specimens with 100% replacement of coarse aggregate tested in uniaxial compression (Domingo et al. 2010). However, the same study highlighted much lower differences for RAC specimens with lower replacement ratio, the one made from 20% of recycled aggregate being almost unaffected by the effect of RCA on creep and shrinkage. For this reason, it is generally accepted that no formal changes are needed to the general models available in Codes and Guidelines for ordinary concrete. Fathifazl and Razaqpur (2013) suggested only to introduce a new coefficient for increasing the basic predictions based on those models. As regards the response under tensile stresses, a uniaxial restrained shrinkage cracking test was executed to investigate the tensile creep properties caused by the restraint of drying shrinkage of RAC: the results highlighted that the tensile creep of RAC caused by the restraint of shrinkage was about 20–30% higher than that of the corresponding RC (Seo and Lee 2015).

Finally, durability is the other main aspect of concern in RAC. Correlation between compressive strength and durability-related properties, mainly controlling carbonation and chloride ingress, are available in the literature. Specifically, the following relationship between chloride migration coefficient D_{nssm} and compressive strength f_c was recently proposed (Silva et al., 2015a, b):

$$D_{nssm} = 47.618 \cdot e^{-0.024 \cdot f_c}. \tag{1.9}$$

Besides the common belief, the presence of fine recycled aggregate (FRA), that is technically feasible for low replacement ratios (e.g. <30%), does not lead to any reduction in durability (Evangelista and de Brito 2010). Conversely, the possible "contamination" of RAC due to the presence of chlorides or sulphates plays a significant role on mechanical and durability-related properties of RAC (Debieb et al. 2010).

All the aspects mentioned in this section will be further developed in the following sections of this chapter. For other issues, not addressed in this book, the Reader may refer to Behera et al. (2014).

1.4 Codes and Guidelines

Several codes, regulations and guidelines dealing with the use of recycled aggregates in concrete are available worldwide. Particularly, the most comprehensive documents concern countries, such as Hong Kong and North European countries, where waste disposal represents a crucial problem due to morphological and environmental conditions, or industrial countries, which have promoted a new urban development process (e.g. Germany during the last few decades). This section proposes an overview of the most significant documents that are currently in force in various regions of the world.

1.4.1 Europe

RILEM, among the first concerned Institutions, proposed some specifications for concrete with recycled coarse aggregates, while the use of recycled sand was not recommended (RILEM 1994). Therefore, these recommendations deal with Recycled Coarse Aggregate Concrete (RCAC) and suggest a classification of this material according to material composition: RCAC Type I is mainly constituted by crushed brick, RCAC Type II is made of crushed concrete, while RCAC Type III is recycled material from concrete and brick mixture containing up to 50% brick. Focusing the attention on RCAC Type II, it is allowed a total replacement of natural coarse aggregates with recycled ones in concretes up to class C50/60.

Similarly, the structural code currently in force in Italy (NTC 2008) only allows the use of recycled aggregates for replacing the coarse fraction of aggregates in new concrete production. A total replacement of natural aggregates, by recycled ones made of CDW, it is allowed only for concrete produced for non-structural applications. In the case of structural concrete, the maximum allowed percentage of RCAs is strongly limited for the usual compressive strength targets related to structural elements.

In Germany, recycled aggregates are classified into four types, depending on the material composition (DIN 4226-100 2002). Particularly, Type 1 and Type 2 derive both from demolition of concrete structures, but the minimum content of concrete plus natural aggregate should be at least 90 and 70% by mass, respectively; the other part may consist of clinker and calcium silicate bricks. Type 3 shall contain more than 80% of dense bricks, and it generally is obtained from pure brick masonry demolition. Finally, Type 4 is a mixture of all mineral building materials without strict specification of the constituents. In terms of applications and mechanical requirements, the best reference is the "Guideline of the German Committee for Reinforced Concrete (DAfStb 1998)". This document specifies that only aggregates with equivalent size bigger than 2 mm belonging to Type 1 or Type 2 can be used in producing structural concrete; moreover, it proposes correlations between replacement percentage and mechanical performance of recycled aggregate concrete.

In the United Kingdom, the BS 8500-2 (2006) provides general requirements for coarse recycled aggregate. In accordance with the use of recycled concrete aggregate in new concrete production, a maximum of 20% replacement of coarse aggregate is allowed and the corresponding compressive strength is limited between 20 and 40 MPa. Moreover, it is specified that RAC can be used for unreinforced members, internal elements or external elements not exposed to chlorides or subject to de-icing salts. RAC also cannot be used in foundations or paving elements. Finally, it is useful to note that no provisions are given in BS 8500-2 for the use of fine recycled aggregates, but their use is not it precluded in principle.

The Spanish Code on Structural Concrete EHE-08 (2008), in annex 15 "*Recommendations for using recycled concrete*", specifies that the use of coarse recycled concrete aggregates is allowed in structural concrete for replacing up to 20% (by weight) of the total amount of coarse aggregates. However, this is only allowed for concrete with cylindrical compressive strength up to 40 MPa and the recycled aggregates should be characterised by a water absorption capacity lower than 7%.

The same limitation in terms of replacement ratio is provided by the French standard NF EN 206-1/CN (2012) for concrete classes up to C35/45 to be employed in the exposure classes XC1, XC2, XC3, XC4 or XF1, provided that the origin of demolished/deconstructed concrete is traceable.

1.4.2 United States, Hong Kong and Australia

The American Concrete Institute (ACI) highlights that possible sources of RAs can be identified in concrete pavements, structures, sidewalks, curbs and gutters that when are removed can be used in concrete production. Particularly, ACI E-701 (2007) specifies that new concrete mixtures can contain both fine and coarse recycled aggregate. Although up to 100% of the coarse aggregates can be made of recycled materials, the percentage of fine aggregate replacement is usually limited to 10–20%.

The Buildings Department of Hong Kong proposed one of the most detailed Guidelines about the use of recycled concrete aggregates (HKBD 2009). These Technical Guidelines specify that concrete with 100% of recycled coarse aggregates shall only be used for non-structural works. Both 100 and 20% recycled coarse aggregates analysed in these guidelines, shall be produced by crushing old concrete.

The Cement Concrete & Aggregates Australia (CCAA), that is the main national body in Australia (representing the interests of six billion dollar a year heavy construction materials industry), recently published an interesting document reporting the current knowledge about the use of recycled aggregates in new concrete production (CCAA 2008). The Commonwealth Scientific and Industrial Research Organisation (CSIRO 2002) and the Standards of the Concrete Institute of Australia are the most important references for this document. Five types of recycled aggregates are identified and classified: Recycled Concrete Aggregate

(RCA), Recycled Concrete and Masonry (RCM), Reclaimed Aggregate (RA), Reclaimed Asphalt Pavement (RAP) and Reclaimed Asphalt Aggregate (RAA). As CSIRO reported, Class 1A RCA (which is a good quality RCA with no more than 0.5% brick content) has the potential for being used in a wide range of applications. Applications include partial replacement of ordinary material in concrete production for non-structural components, such as kerbs and gutters. Although CSIRO (2002) emphasizes that the current field experience with the use of RCAs for structural applications is scarce, it contributes to clarify this matter and defines two different grades of RAC both made by using Class 1A RCAs:

- Grade 1 RAC, characterized by a maximum 30% replacement ratio with Class 1A coarse RCAs, has a maximum specified compressive strength limit of 40 MPa;
- Grade 2 RAC, made of up to 100% Class 1A coarse RCAs recycled, has a maximum specified compressive strength limit of 25 MPa.

1.4.3 Some Remarks About Existing Regulations and Standards

As the aforementioned recommendations show, the use of RCAs is intended mostly to replace the coarse fraction of ordinary aggregates. In fact, recycled fine aggregate from concrete exhibit deleterious characteristics that might affect performance and workability of recycled concrete, especially if the concrete mixture is not accurately designed.

The requirements that aggregate shall meet in order to be used as RCA, seem to be almost the same for the documents proposed from different institutions: they are outlined in Table 1.1. Although it is required that RCAs mainly derive from demolished concrete, a certain limited content of other "alien" materials, such as metals, plastics, clay lumps and glass, is allowed. In this respect, only ACI 555R-01 (2001), among the current regulations, provides clear indications about a selective process of demolition intended at safeguarding the "purity" of RCAs. Conversely, the processing procedure implemented for producing recycled aggregates, is generally designed by operator companies, according to their specific practices.

A total replacement of coarse natural aggregates is allowed only for non-structural concrete, due to the decrease of compressive strength that generally occurs considering the recycled material source.

Moreover, the use of recycled coarse concrete aggregate is still limited in structural applications, as several international standards define an upper limit between 20 and 30% for their replacement ratio (in volume). Table 1.2 summarizes the main requirements and limitations provided by the regulations and guidelines considered herein.

Table 1.1 Recycled concrete aggregate requirements: synoptic overview

Code/guideline	Material source	Maximum content of fine (%)	Maximum content of "alien" materials
Italian Ministry of Infrastructure and Transportation (NTC 2008)	Building demolition (for non structural concrete) and concrete demolition (for structural concrete)	–	–
RILEM (1994)	–	5	1%
DAfStB (1998)	Demolished concrete structures	–	≤0.2% (Type 1)
			≤0.5% (Type 2)
British Standard Institution (BS 8500-2 2006)	Crushing hard concrete	5	1%
Building Department Hong Kong Government (HKBD 2009)	Crushing old concrete	4	1%
American Concrete Institute (ACI E-701 2007)	Removed pavements, structures, sidewalks, curbs, and gutter	–	2 kg/m^3
Cement Concrete & Aggregates Australia (CCAA 2008)	Demolition waste of at least 95% concrete	–	–

Note Metals, plastics, clay lumps and glass are considered as "alien" materials

Finally, selective demolition is generally needed in order to obtain materials that may be easily turned into RAC with limited need for further screening processes and decontamination procedures (HKBD 2004).

1.5 Insights into Concrete Sustainability

According to its original and most cited definition, "Sustainable development" should "meet the needs of the present without compromising the ability of future generations to meet their own needs" (WCED 1987). Therefore, no further speculation, that would be out of the scope of this work, is actually needed for understanding that the construction industry is fully concerned by the challenge of making its processes "sustainable". In fact, the building construction sector and the production of cement and concrete is responsible for a significant share of the global emissions and raw material demand (van den Heede and De Belie 2012). Particularly, 40% of anthropogenic GHG global emissions and 40% of raw materials are attributed to building construction sector, whereas the global annual production of concrete is going to approach 25 gigatonnes, namely 3.8 t per person (Gursel et al. 2014). Moreover, the production of CDW requires more and more

Table 1.2 Main characteristics of recycled concrete: synoptic overview

Country (guideline)	Application	Replaceable aggregate fraction	Maximum replacement percentage (%)	Maximum cylindrical compressive strength (28 days) (MPa)
Italy (NTC 2008)	Non-structural	Coarse	100	8
	Structural	Coarse	30	30
			60	20
RILEM (1994)	Not specified	Coarse	100	50
Germany (DAfStB 1998)	Structural	Coarse	35	25
			25	35
UK (BS 8500-2 2006)	Not specified	Coarse	20	40
Spain (EHE-08 2008)	Structural	Coarse	20	40
France (NF EN 206-1/CN 2012)	Structural	Coarse	20	35
Hong Kong (HKBD 2009)	Non-structural	Coarse	100	20
	Structural	Coarse	20	30
USA (ACI E-701 2007)	Not specified	Coarse fine	100	Not specified
			20	
Australia (CCAA 2008)	Grade 2 concrete	Coarse	100	25
	Grade 1 concrete	Coarse	30	40

landfilling capacity and this, especially in some countries, implies significant impact on the environment (Kien et al. 2013). As a matter of fact, cost-benefit analyses reveal positive results when evaluating CDW recycling solutions and highlight that the landfill charge is a key factor in determining those results and the achievement of a breakeven post after the initial investment (Yaun et al. 2011).

Therefore, any action capable of even slightly reducing both GHG emissions and raw material demand results in a significant global effect on the environment, due to the abovementioned huge figures. In this light, recycling CDW for transforming them into sustainable “second raw materials” is the solution to answer the above requirements and achieve a higher sustainability for the construction industry. Besides the general understanding of recycling solutions for producing “green concrete”, quantitative assessment methodologies capable of quantifying and comparing alternative solutions for sustainable concrete are needed for approaching this problem in a rational way. Particularly, the six quantification methodologies listed below can be generally recognised in the studies available in the literature, some of which being also possibly combined in some specific cases (Wu et al. 2014):

- site visit (SV) method, based on direct or indirect surveys carried out at the construction or demolition sites by duly skilled and trained personnel;
- generation rate calculation (GRC) method, intended at determining the waste generation rate for a particular activity unit (i.e. kg/m^2, and m^3/m^2) by means of alternative approaches (such as per capita multiplier, financial value extrapolation and area-based calculation) determined on similar situations analysed in the past;
- lifetime analysis (LA) method, mainly implemented for demolition waste, and based on the principle of material mass balance when turning buildings into demolition rubbles;
- classification system accumulation (CSA) method, based on GRC method with a further classification system providing a tool for determining the contribution of a given material;
- variables modelling (VM) method, consisting in a simulation of the produced amount of CDW taking into account the variables controlling the production of waste, such as economic indicators, construction areas, on-site working conditions;
- other particular methods, such as the assumption of a given percentage of waste generation or other global parameters (the generation of CDW estimated on the amount of annual cement production).

As a proof of concept, the economic viability of building a construction and demolition waste recycling plant in Portugal was analysed by Coelho and de Brito (2013a). According to the analysis proposed in that study, the break-even point is around 2 years and, hence, in spite of the significant initial investment, the construction of this kind of plant can be a profitable investment. Moreover, since the factors affecting this result can be significantly variable even in the short period, a sensitivity analysis was carried out for investigating the influence of various relevant parameters, such as CDW generation rate and landfilling charges and rejected materials. In the worst scenario the return on investment was eight years and, hence, still fairly acceptable (Coelho and de Brito 2013b). A more general "environmental analysis" and the corresponding sensitivity investigation were carried out by the same authors by using primary energy consumption and $CO_{2,eq}$ emission impact factors as environmental impact performance indicators (Coelho and de Brito 2013c, d).

References

ACI 555R-01 (2001) Removal and reuse of hardened concrete. American Concrete Institute

ACI E-701 (2007) Aggregates for concrete. ACI Education Bulletin E1-07, Materials for Concrete Construction

Adaway M, Wang Y (2015) Recycled glass as a partial replacement for fine aggregate in structural concrete-effects on compressive strength. Special issue: Electron J Struct Eng 14(1). Available online at http://www.ejse.org/Archives/Fulltext/2015-1/2015-1-11.pdf. Accessed on 1 Sept 2015

Agrela F, Sánchez de Juan M, Ayuso J, Geraldes VL, Jiménez JR (2011) Limiting properties in the characterisation of mixed recycled aggregates for use in the manufacture of concrete. Constr Build Mater 25(10):3950–3955

Alves AV, Vieira TF, de Brito J, Correia R (2014) Mechanical properties of structural concrete with fine recycled ceramic aggregates. Constr Build Mater 64:103–113

Angulo SC, Carrijo PM, Figueiredo AD, Chaves AP, John VM (2010) On the classification of mixed construction and demolition waste aggregate by porosity and its impact on the mechanical performance of concrete. Mater Struct 43(4):519–528

Azenha MAD (2009) Numerical simulation of the structural behaviour of concrete since its early ages. Ph.D. thesis. Faculty of Engineering of University of Porto

Behera M, Bhattacharyya SK, Minocha AK, Deoliya R, Maiti S (2014) Recycled aggregate from C&D waste & its use in concrete—a breakthrough towards sustainability in construction sector: a review. Constr Build Mater 68:501–516

Braymand S, François P, Feugeas F, Fond C (2015) Rheological properties of recycled aggregate concrete using superplasticizers. J Civil Eng Archit 9(2015):591–597

BS 8500-2 (2006) Concrete. Complementary British Standard to BS EN 206-1. Specification for constituent materials and concrete. British Standards Institution, London, UK

Bullard JW, Jennings HM, Livingston RA, Nonat A, Scherer GW, Schweitzer JS, Scrivener KL, Thomas JJ (2011) Mechanisms of cement hydration. Cem Concr Res 41(12):1208–1223

Butler LJ, West JS, Tighe SL (2014) Towards the classification of recycled concrete aggregates: influence of fundamental aggregate properties on recycled concrete performance. J Sustain Cem Based Mater 3(2):140–163

Cartuxo F, de Brito J, Evangelista L, Jiménez JR, Ledesma EF (2015) Rheological behaviour of concrete made with fine recycled concrete aggregates—influence of the superplasticizer. Constr Build Mater 89:36–47

CCAA (2007) Use of recycled water in concrete production, cement concrete & aggregates Australia. Available online at http://www.ccaa.com.au. Accessed on 1 Sept 2015

CCAA (2008) Use of recycled aggregates in construction. Cement Concrete Aggregates Australia

CEN (2005) Eurocode 2, Design of concrete structures, part 1–1: General rules and rules for buildings, EN 1992-1-1. European Committee for Standardization, Brussels, BE

CEN (2009) EN 12350-2:2009 Testing fresh concrete. Slump-test. European Committee for Standardization, Brussels, BE

CEN (2013) EN 12620:2013 Aggregates for concrete. European Committee for Standardization, Brussels, BE

Chen Y, Wang KJ, Liang D (2012) Mechanical properties of pervious cement concrete. J Cent South Univ 19(11):3329–3334

Coelho A, de Brito J (2013a) Economic viability analysis of a construction and demolition waste recycling plant in Portugal—part I: location, materials, technology and economic analysis. J Clean Prod 39:338–352

Coelho A, de Brito J (2013b) Economic viability analysis of a construction and demolition waste recycling plant in Portugal—part II: economic sensitivity analysis. J Clean Prod 39:329–337

Coelho A, de Brito J (2013c) Environmental analysis of a construction and demolition waste recycling plant in Portugal—Part I: Energy consumption and CO_2 emissions. Waste Manage 33(5):1258–1267

Coelho A, de Brito J (2013d) Environmental analysis of a construction and demolition waste recycling plant in Portugal—Part II: Environmental sensitivity analysis. Waste Manage 33(1):147–161

Corinaldesi V (2010) Mechanical and elastic behaviour of concretes made of recycled concrete coarse aggregates. Constr Build Mater 24:1616–1620

Corinaldesi V (2012) Environmentally-friendly bedding mortars for repair of historical buildings. Constr Build Mater 35:778–784

Corinaldesi V, Moriconi G (2009a) Behaviour of cementitious mortars containing different kinds of recycled aggregate. Construction and Building Materials 23(1):289–294

Corinaldesi V, Moriconi G (2009b) Influence of mineral additions on the performance of 100% recycled aggregate concrete. Constr Build Mater 23(8):2869–2876

Corinaldesi V, Moriconi G, Naik TR (2010) Characterization of marble powder for its use in mortar and concrete. Constr Build Mater 24(1):113–117

Corinaldesi V, Letelier V, Moriconi G (2011) Behaviour of beam–column joints made of recycled-aggregate concrete under cyclic loading. Constr Build Mater 25(2011):1877–1882

CSIRO (2002) Guide to the use of recycled concrete and masonry materials (HB 155). Commonwealth Scientific and Industrial Research Organisation. Standards Australia

DAfStb (1998) Guideline of the German committee for reinforced concrete. Deutscher Ausschuss für Stahlbeton, Germany

de Brito J, Alves F (2010) Concrete with recycled aggregates: the Portuguese experimental research. Mater Struct 43(1):35–51

de Brito J, Saikia N (2013) Recycled aggregate in concrete—use of industrial, construction and demolition waste. Springer London, p 445

Debieb F, Courard L, Kenai S, Degeimbre R (2010) Mechanical and durability properties of concrete using contaminated recycled aggregates. Cement Concr Compos 32(6):421–426

Dilbas H, Şimşek M, Çakır Ö (2014) An investigation on mechanical and physical properties of recycled aggregate concrete (RAC) with and without silica fume. Constr Build Mater 55:29–37

DIN 4226-100 (2002) Aggregates for concrete and mortar—Part 100: Recycled aggregates. Deutsches Institut für Normung, Germany

Domingo A, Lázaro C, Gayarre FL, Serrano MA, López-Colina C (2010) Long term deformations by creep and shrinkage in recycled aggregate concrete. Mater Struct 43(8):1147–1160

Duan ZH, Poon CS (2014) Properties of recycled aggregate concrete made with recycled aggregates with different amounts of old adhered mortars. Mater Des 58:19–29

EHE-08 (2008) Code on structural concrete, permanent commission of the concrete. Spanish Ministry of Public Works, Madrid, Spain

Evangelista L, de Brito J (2010) Durability performance of concrete made with fine recycled concrete aggregates. Cement Concr Compos 32(1):9–14

Evangelista L, Guedes M, de Brito J, Ferro AC, Pereira MF (2015) Physical, chemical and mineralogical properties of fine recycled aggregates made from concrete waste. Constr Build Mater 86:178–188

Fairbairn EM, Americano BB, Cordeiro GC, Paula TP, Toledo Filho RD, Silvoso MM (2010) Cement replacement by sugar cane bagasse ash: CO_2 emissions reduction and potential for carbon credits. J Environ Manage 91(9):1864–1871

Fathifazl G, Razaqpur AG (2013) Creep rheological models for recycled aggregate concrete. ACI Mater J 110(2):115–126

Fathifazl G, Abbas A, Razaqpur AG, Isgor OB, Fournier B, Foo S (2009) New mix design method for recycled aggregate concrete. J Mater Civ Eng ASCE 21(10):601–611

Fathifazl G, Razaqpur AG, Isgor OB, Abbas A, Fournier B, Foo S (2010) Proportioning concrete mixtures with recycled concrete aggregate: a novel method. Int Mag ACI 32(3):37–43

Ferrari G, Miyamoto M, Ferrari A (2014) New sustainable technology for recycling returned concrete. Constr Build Mater 67:353–359

fib (2013) Model code for concrete structures 2010. Fédération international du beton. Ernst Sohn Verlag, Berlin, Germany

Folino P, Xargay H (2014) Recycled aggregate concrete—mechanical behavior under uniaxial and triaxial compression. Constr Build Mater 56:21–31

Fonseca N, de Brito J, Evangelista L (2011) The influence of curing conditions on the mechanical performance of concrete made with recycled concrete waste. Cement Concr Compos 33 (6):637–643

García-González J, Rodríguez-Robles D, Juan-Valdés A, Morán-del Pozo JM, Guerra-Romero MI (2014) Pre-saturation technique of the recycled aggregates: solution to the water absorption drawback in the recycled concrete manufacture. Materials 7:6224–6236

Ghassan AH, Hilmi BM (2010) Study on properties of rice husk ash and its use as cement replacement material. Mater Res 13(2):185–190

Gomes M, de Brito J (2009) Structural concrete with incorporation of coarse recycled concrete and ceramic aggregates: durability performance. Mater Struct 42(5):663–675

Gonzalez A, Etxeberria M (2014) Experimental analysis of properties of high performance recycled aggregate concrete. Constr Build Mater 52:227–235

González-Fonteboa B, Martínez-Abella F, Carro López D, Seara-Paz S (2011) Stress–strain relationship in axial compression for concrete using recycled saturated coarse aggregate. Constr Build Mater 25(5):2335–2342

González-Fonteboa B, Martínez-Abella F, Herrador MF, Seara-Paz S (2012) Structural recycled concrete: behaviour under low loading rate. Constr Build Mater 28:111–116

Güneyisi E, Gesoğlu M, Kareem Q, İpek S (2016) Effect of different substitution of natural aggregate by recycled aggregate on performance characteristics of pervious concrete. Mater Struct 49(1–2):521–536

Gursel AP, Masanet E, Horvath A, Stadel A (2014) Life-cycle inventory analysis of concrete production: a critical review. Cement Concrete Comp 51:38–48

Hansen TC (1986a) Physical structure of hardened cement paste. A classical approach. Mater Struct 19(6):423–436

Hansen TC (1986b) Recycled aggregates and recycled aggregate concrete second state-of-the-art report developments 1945–1985. Mater Struct 19(3):201–246

Hansen TC, Narud H (1983) Strength of recycled concrete made from crushed concrete coarse aggregate. ACI Concr Int 5(1):79–83

Hendriks CF, Janssen GMT, Vázquez E (2005) Use of recycled materials—final report of RILEM TC 198-URM, RILEM Publications SARL

HKBD (2004) Code of practice for demolition of buildings. Hong Kong Buildings Department

HKBD (2009) Use of recycled aggregates in concrete. Hong Kong Buildings Department

Juenger MCG, Winnefeld F, Provis JL, Ideker JH (2011) Advances in alternative cementitious binders. Cement Concr Res 41:1232–1243

Kencanawati NN, Iizasa S, Shigeishi M (2013) Fracture process and reliability of concrete made from high grade recycled aggregate using acoustic emission technique under compression. Mater Struct 46(9):1441–1448

Kien TT, Thanh LT, Lu PV (2013) Recycling construction demolition waste in the world and in Vietnam. In: The international conference on sustainable built environment for now and the future, Hanoi (VN), 26–27 Mar 2013, pp 247–256

Koenders EA, Pepe M, Martinelli E (2014) Compressive strength and hydration processes of concrete with recycled aggregates. Cem Concr Res 56:203–212

Kou SC, Poon CS, Chan D (2008) Influence of fly ash as a cement addition on the hardened properties of recycled aggregate concrete. Mater Struct 41(7):1191–1201

Kreijger PC (1981). Classification of recycled aggregate, proposed in the Netherlands. In: Adhesion problems in the recycling of concrete. Springer US, pp 309–310

Kuosa H (2012), Reuse of recycled aggregates and other C&D wastes, Project 72480—NeReMa: Report VTT-R-05984-12. Available online at http://www.vtt.fi/inf/julkaisut/muut/2012/VTT-R-05984-12.pdf. Accessed on 1 July 2015

Lam CHK, Ip AWM, Barford JP, McKay G (2010) Use of incineration MSW ash: a review. Sustainability 2:1943–1968

Leite MB, Figueire do Filho JGL, Lima PRL (2013) Workability study of concretes made with recycled mortar aggregate. Mater Struct 46:1765–1778

Lima C, Caggiano A, Faella C, Martinelli E, Pepe M, Realfonzo R (2013) Physical properties and mechanical behaviour of concrete made with recycled aggregates and fly ash". Constr Build Mater 47:547–559

Lollini F, Redaelli E, Bertolini L (2014) Effects of portland cement replacement with limestone on the properties of hardened concrete. Cement Concr Compos 46:32–40

Lothenbach B, Scrivener K, Hooton RD (2011) Supplementary cementitious materials. Cem Concr Res 41(12):1244–1256

Mahmoud E, Ibrahim A, El-Chabib H, Patibandla VC (2013) Self-consolidating concrete incorporating high volume of fly ash, slag, and recycled asphalt pavement. Int J Concr Struct Mater 7(2):155–163

Malesev M, Radonjanin V, Marincovic S (2010) Recycled concrete as aggregate for structural concrete production. Sustainability 2:1204–1225

Mardani-Aghabaglou A, Tuyan M, Ramyar K (2015) Mechanical and durability performance of concrete incorporating fine recycled concrete and glass aggregates. Mater Struct 48(8):2629–2640

Martinelli E, Koenders EA, Caggiano A (2013) A numerical recipe for modelling hydration and heat flow in hardening concrete. Cement Concr Compos 40:48–58

Mas B, Cladera A, del Olmo T, Pitarch F (2012) Influence of the amount of mixed recycled aggregates on the properties of concrete for non-structural use. Constr Build Mater 27(1):612–622

McNeil K, Kang THK (2013) Recycled concrete aggregates: a review. Int J Concr Struct Mater 7 (1):61–69

Medina C, Zhu W, Howind T, Sánchez de Rojas MI, Frías M (2014) Influence of mixed recycled aggregate on the physical e mechanical properties of recycled concrete. J Clean Prod 68:216–225

Moll S, Bringezu S, Schütz H, (2005) Resource use in European countries—material flows and resource management. Wuppertal Institute for Climate, Environment and Energy, Wuppertal, DE

Moya JA, Pardo N, Mercier A (2010) Energy efficiency and CO_2 emissions: prospective scenarios for the cement industry. JRC Scientific and Technical Report

Mukharjee BB, Barai SV (2015) Influence of Nano-Silica on the properties of recycled aggregate concrete. Constr Build Mater 55:29–37

NF EN 206-1/CF (2012) Béton—Partie 1: spécification, performance, production et conformité—Complément national à la norme NF EN 206-1

NTC (2008) DM 14.01. 2008: Norme Tecniche per le Costruzioni. Italian Ministry of Infrastructures and Transportation (in Italian)

Pacheco-Torgal F, Castro-Gomes J, Jalali S (2008) Alkali-activated binders: a review—part 1. Historical background, terminology, reaction mechanisms and hydration products. Constr Build Mater 22, 1305–1314

Pacheco-Torgal F, Jalali S, Labrincha J, John VM (eds) (2013) Eco-efficient concrete. Woodhead publishing series in civil and structural engineering, 1st edn. Woodhead Publishing Limited, p 582

Pedro D, de Brito J, Evangelista L (2014) Influence of the use of recycled concrete aggregates from different sources on structural concrete. Constr Build Mater 71:141–151

Pepe M (2015) A conceptual model for designing recycled aggregate concrete for structural applications. Springer theses series. Springer

Pepe M, Koenders EAB, Faella C, Martinelli E (2014a) Structural concrete made with recycled aggregates: Hydration process and compressive strength models. Mech Res Commun 58:139–145

Pepe M, Toledo Filho RD, Koenders EAB, Martinelli E (2014b) Alternative processing procedures for recycled aggregates in structural concrete. Constr Build Mater 69:124–132

Pereira P, Evangelista L, de Brito J (2012) The effect of superplasticisers on the workability and compressive strength of concrete made with fine recycled concrete aggregates. Constr Build Mater 28(1):722–729

Prince MJR, Singh B (2014a) Bond behaviour between recycled aggregate concrete and deformed steel bars. Mater Struct 47(3):503–516
Prince MJR, Singh B (2014b) Bond strength of deformed steel bars in high-strength recycled aggregate concrete. Mater Struct. doi:10.1617/s11527-014-0452-y
Rajabipour F, Maraghechi H, Fischer G (2010) Investigating the Alkali-Silica reaction of recycled glass aggregates in concrete materials. J Mater Civ Eng 22(12):1201–1208
Rao MC, Bhattacharyya SK, Barai SV (2011) Influence of field recycled coarse aggregate on properties of concrete. Mater Struct 44(1):205–220
Revathi R, Selvi RS, Velin SS (2013) Investigations on fresh and hardened properties of recycled aggregate self compacting concrete. J Inst Eng India Ser A 94(3):179–185
RILEM (1994) Specifications for concrete with recycled aggregates, RILEM TC 121-DRG, RILEM Publications SARL
Rodríguez-Robles D, García-González J, Juan-Valdés A, Morán-del Pozo MJ, Guerra-Romero MI (2014) Quality assessment of mixed and ceramic recycled aggregates from construction and demolition wastes in the concrete manufacture according to the Spanish standard. Materials 7:5843–5857
Safiuddin M, Alengaram UJ, Salam MA, Jumaat MZ, Jaafar FF, Saad HB (2011) Properties of high-workability concrete with recycled concrete aggregate. Mater Res 14(2):248–255
Sandrolini F, Franzoni E (2001) Waste wash water recycling in ready-mixed concrete plants. Cem Concr Res 31(3):485–489
Seo TS, Lee MS (2015) Experimental study on tensile creep of coarse recycled aggregate concrete. Int J Concr Struct Mater 9(3):337–343
Shi C, Fernández Jiménez A, Palomo A (2011) New cements for the 21st century: the pursuit of an alternative to Portland cement. Cement Concr Res 41:750–763
Silva RV, de Brito J, Dhir RK (2014) Properties and composition of recycled aggregates from construction and demolition waste suitable for concrete production. Constr Build Mater 65:201–217
Silva RV, de Brito J, Neves R, Dhird R (2015a) Prediction of chloride ion penetration of recycled aggregate concrete. Mater Res 18(2):427–440
Silva RV, Neves R, de Brito J, Dhir RK (2015b) Carbonation behaviour of recycled aggregate concrete. Cement Concr Compos 62:22–32
Spaeth V, Tegguer AD (2013) Improvement of recycled concrete aggregate properties. Int J Sustain Built Environ 2:143–152
Sriravindrarajah R, Huai Wang ND, Wen Ervin LJ (2012) Mix design for pervious recycled aggregate concrete. Int J Concr Struct Mater 6(4):239–246
Tebaldi G, Dave E, Marsac P, Muraya P, Hugener M, Pasetto M, Graziani A, Grilli A, Marradi A, Wendling L, Gaudefroy V, Jenkins K, Loizos A, Bocci M (2012) Classication of recycled asphalt (RA) material. In: Proceedings of the 2nd International Symposium on Asphalt Pavement and Environment, 1–3 Oct 2012, in Fortaleza (BR). Available online at https://hal.archives-ouvertes.fr/hal-00849456/document. Accessed on 10 June 2015
Toledo Filho RD, Koenders EAB, Pepe M, Chagas Cordeiro G, Fairbairn E, Martinelli E (2013) Rio 2016 sustainable construction commitments lead to new developments in recycled aggregate concrete. Proc Inst Civ Eng Civ Eng 166(6), 28–35
USDT (2004) Transportation applications of recycled concrete aggregate—FHWA state of the practice national review 2004; U.S. Department of Transportation Federal Highway Administration: Washington, DC, USA
Van den Heede P, De Belie N (2012) Environmental impact and life cycle assessment (LCA) of traditional and 'green' concretes: Literature review and theoretical calculations. Cement Concr Compos 34(4):431–442
Van VTA, Rößler C, Bui DD, Ludwig HM (2014) Rice husk ash as both pozzolanic admixture and internal curing agent in ultra-high performance concrete. Cement Concr Compos 53:270–278
Ventura-Gouveia A (2011) Constitutive models for the material nonlinear analysis of concrete structures including time dependent effects. Ph.D. thesis. University of Minho

Wang H, Wang J, Sun X, Jin W (2013) Improving performance of recycled aggregate concrete with superfine pozzolanic powders. J Cent South Univ 20:3715−3722

Wardeh G, Ghorbel E, Gomart H (2015) Mix design and properties of recycled aggregate concretes: applicability of Eurocode 2. Int J Concr Struct Mater 9(1):1–20

WCED (1987) Our common future. World Commission on Environment and Development. Oxford University Press, Oxford, p 43

WRAP (2007) Performance related approach to use of recycled aggregates, aggregate research programme. Project code: AGG0074, Waste & Resources Action Programme. Available online at http://www.sustainableaggregates.com/library/docs/wrap/L0344_AGG0074_Dundee.pdf. Accessed on 1 Sept 2015

Wu Z, Yu ATW, Shen L, Liu G (2014) Quantifying construction and demolition waste: an analytical review. Waste Manag 34:1683–1692

Yang J, Du Q, Bao Y (2011) Concrete with recycled concrete aggregate and crushed clay bricks. Constr Build Mater 25(4):1935–1945

Yuan F, Shen LY, Li QM (2011) Emergy analysis of the recycling options for construction and demolition waste. Waste Manage 31(12):2503–2511

Yehia S, Helal K, Abusharkh A, Zaher A, Istaitiyeh H (2015) Strength and durability evaluation of recycled aggregate concrete. Int J Concr Struct Mater 9(2):219–239

Zhu JQ, Wu SP, Zhong JJ, Wang DM (2011) Classification of coarse recycled aggregate used in asphalt concrete. Appl Mech Mater 71–78:1025–1030

Chapter 2
Concrete with Recycled Aggregates: Experimental Investigations

Carmine Lima, Marco Pepe, Ciro Faella and Enzo Martinelli

Abstract The mechanical behaviour of Recycled Aggregate Concrete (RAC) is investigated by reporting the main results of experimental tests intended at understanding the influence of Recycled Concrete Aggregates (RCAs) on the resulting mechanical properties of concrete. The focus is placed on the higher porosity of RCAs and their higher water absorption capacity. Consequently, the role of the initial moisture conditions of RCAs at mixing is also unveiled and its consequences on both the hydration reaction and the time evolution of compressive strength are highlighted. The influence of processing procedures intended at reducing the aforementioned porosity is also discussed.

The experimental activities carried out on Recycled Aggregate Concretes have been aimed at understanding the role played by recycled aggregates in affecting the relevant properties of structural concrete. Specifically, several experimental campaigns have been performed with the scope of investigating the influence on the concrete performance of the following parameters:

- Processing procedures for RCAs;
- Initial moisture condition of coarse aggregates;
- Aggregate replacement ratio;
- Nominal water-to-cement ratio.

C. Lima · M. Pepe · C. Faella · E. Martinelli (✉)
University of Salerno, Fisciano, Italy
e-mail: e.martinelli@unisa.it

J.A.O. Barros et al. (eds.), *Recent Advances on Green Concrete for Structural Purposes*, Research for Development, DOI 10.1007/978-3-319-56797-6_2

2.1 The Influence of Processing Procedures for RCAs in RAC

The first experimental campaign described in this chapter has been mainly aimed at investigating the influence of processing procedures on the physical and mechanical performance of the resulting recycled aggregate concrete.

2.1.1 *Materials and Methods*

All the mixtures have been produced by using "high initial strength Portland cement", indicated as CP V ARI RS, according to the National Brazilian Standard NBR 5733:1991, characterised by a specific mass of 3100 kg/m^3. The grain size distribution of the Portland cement have shown that 95% of its particles are smaller than 50 μm, and 50% of the particles are smaller than 15 μm. Moreover, the polycarboxilate superplasticizer Glenium 51 was used for workability control, which is characterised by a specific mass of 1.07 kg/l and a solid concentration content of 30% (www.basf-cc.com.br).

Both natural and recycled aggregates were employed in this experimental campaign and, for natural aggregates, common crushed limestone were classified in three size classes according to the Brazilian standard (NBR 7211 2009). Meanwhile, the recycled concrete aggregates were obtained after the demolition of the hospital Clementino Fraga Filho in Rio de Janeiro (BR). The recycled aggregates were selected and analysed by the university laboratory for construction materials LABEST (PEC/COPPE—UFRJ Rio de Janeiro), and were processed as described by Pepe et al. (2014). More specifically, both NCA and RCA were

Table 2.1 Influence of processing procedures for RCAs in RCA, natural and recycled aggregates

Class size	Natural		Recycled		Recycled—CL	
	A (%)	γ (kg/m^3)	A (%)	γ (kg/m^3)	A (%)	γ (kg/m^3)
Sand (d < 4.75 mm)	1.40	2668	–	–	–	–
C1 (4.75 mm < d < 9.5 mm)	3.39	2547	11.94	1946	5.56	2261
C2 (9.5 mm < d < 19 mm)	1.28	2634	4.94	2268	4.09	2328

Table 2.2 Mixture compositions

Mix	CEM	w	ads.w	SP	w/c	Natural			Recycled		Recycled - CL	
						Sand	C1	C2	C1	C2	C1	C2
	(kg/m^3)					(kg/m^3)						
REF	300	160	31.4	4.02	0.53	952.6	439.9	470.2	–	–	–	–
RAC			71.8			950.4	–	–	346.6	404.0	–	–
RAC CL			49.9			951.3	–	–	–	–	403.1	415.1

employed with the aim to analyse the effect of processing procedures on the mechanical properties of the resulting recycled aggregate concrete (Table 2.1).

Based on the results obtained so far, crushed concrete debris were processed to recycled aggregates, and three different mixtures were designed to analyse the different behaviour of concrete made with original (uncleaned) and cleaned recycled aggregates. Table 2.2 report the mix compositions of the produced concretes.

Specifically, a reference mixture, indicated as REF, was prepared with all natural components. Moreover, two additional mixtures, referred to as RAC and RAC CL, were designed with 50% (by volume) of the natural aggregates were replaced with recycled ones, with and without 15 min of autogenous cleaning, respectively (Pepe et al. 2014).

All batches were produced with 300 kg/m^3 cement and a w/c ratio of 0.53. To compensate the water absorption of both dry recycled and natural aggregates, additional water was poured during mixing while taking into account the water absorption tests (Table 2.1).

A slump test was performed and concrete was cast (in three steps on a vibrating table for expelling the entrapped air) in steel cylinders. After one day, specimens were demoulded and the concrete was placed in a water curing room (21 °C) up to the designated times for testing the compressive strength, elastic modulus and tensile splitting strength.

2.1.2 Results and Analysis

This subsection reports a summary of the experimental results obtained from RAC samples produced as described in the previous subsection. As mentioned, three different mixtures were designed for analysing the effect of alternative processing procedure for RCAs on the mechanical properties of concretes made with them.

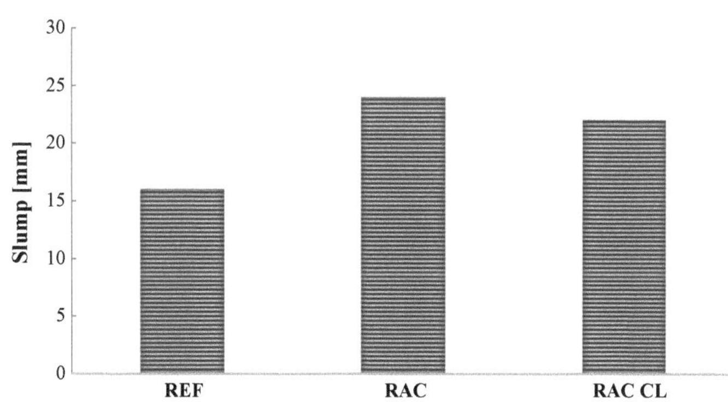

Fig. 2.1 Slump test results

The fresh concrete properties were investigated through slump tests (EN 12350-2:2009). Results are reported in Fig. 2.1 and clearly highlight the effect autogenous cleaning has on the mix performance in the fresh state.

First of all, it is worth mentioning that a higher slump value was observed for mixtures with uncleaned recycled aggregates (RAC) with respect to the corresponding reference mix (REF). This largely depends on the fact that the same amount of superplasticizer was added for all mixtures, whereas the absorption compensation water was added on the bases of water absorption tests carried out on aggregates after 24 h of absorption time. In fact, recycled aggregates within the concrete mix cannot absorb such an amount of water in a short period time, which is equal to the mixing time. Since the mixing process only takes 10–15 min, which is much less than the absorption time used in the absorption tests (24 h), the remaining part of the water just modifies (increases) the water content in the mix (and the w/c ratio), leading to an increase in workability (and a decrease in compressive strength). Thus, such a higher amount of total mixing water available in the RAC mix, led to slump values significantly higher than those corresponding to the reference mix. Further investigations are needed to better understand the role of added water in RACs and, possibly, to achieve a sound definition of such a key parameter. The same effect is observed for the mix with cleaned aggregates (RAC CL), whose lower water absorption capacity required a lower amount of added compensation water and, led to lower slump values.

The effectiveness of autogenous cleaning of RCAs was also evaluated by conducting several tests intended at determining compressive strength, elastic modulus and tensile splitting strength. The compressive strength of the three mixtures was determined at an age of 2, 7, 14, 28 and 60 days (five tests for each curing age per

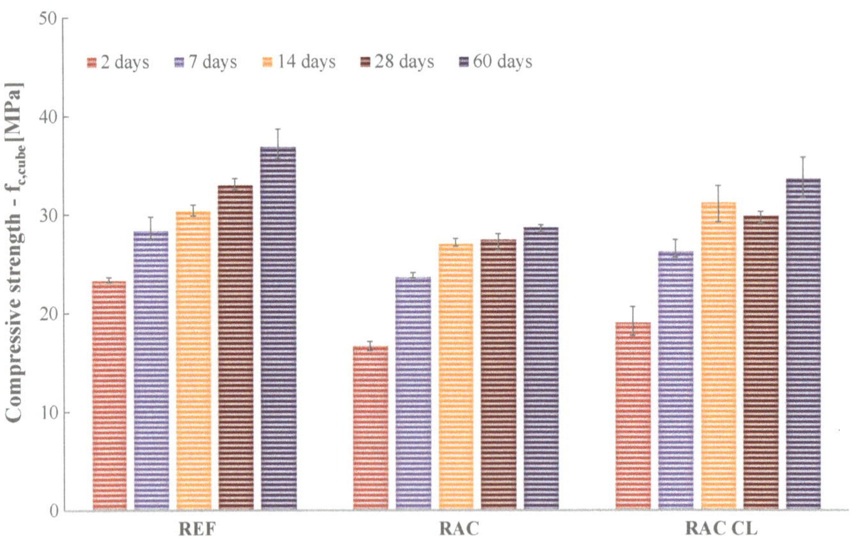

Fig. 2.2 Compressive strength results

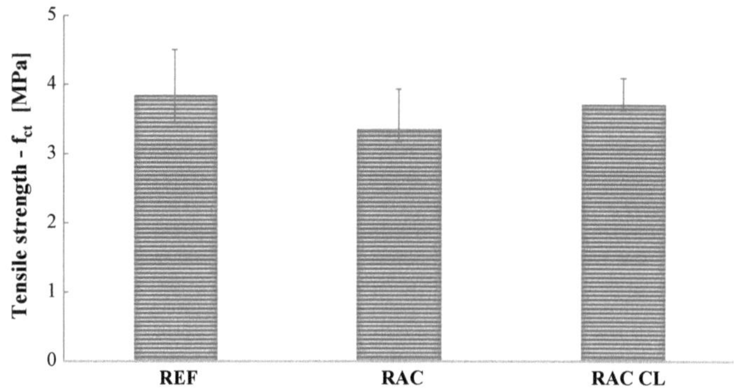

Fig. 2.3 Tensile strength

mixture) according to the NBR 5739 (2007) on cylindrical specimens with a nominal diameter of 100 mm and a height of 200 mm.

Figure 2.2 provides the results of the compressive strength evolution. Samples of the REF mix show an average 28-day compressive strength of 33 MPa and, because of ongoing hydration, a compressive strength of 37 MPa was reached after 60 days of curing. As can be observed from the results presented in Fig. 2.2, the compressive strength of RAC concrete was reduced by about 20%. This is mainly due to a higher amount of water added for compensating absorption that increased the effective value of the w/c ratio. This effect and the change in the affective amount of free water was already observed when discussing the workability effects. However, the beneficial effect of autogenous cleaning clearly emerges when analysing the compressive strength results obtained for RAC CL, where a significantly smaller reduction was measured (8.9%).

The Brazilian tensile splitting strength was determined from cylindrical specimens (ASTM C496/C496M 2011) after 28 days of water curing (three specimens for each batch). The results show that the use of uncleaned recycled aggregates reduces the tensile splitting strength by about 13%, while autogenous cleaning has led to both higher tensile strength and a reduction in scatter (less than 4%) of the results obtained from the three tests performed on RAC CL specimens (Fig. 2.3).

2.2 Influence of the Initial Moisture Condition of RCAs

As recycled aggregates are characterised by water absorption capacity higher than natural ones, the initial moisture condition of RCA has an influence on the total amount of free water available in the mixture. Hence, since free water plays a key role on the evolution of the physical and mechanical properties of both hardened and fresh concrete, initial moisture conditions of RCAs are expected to have an influence on such properties.

For this reason, the experimental campaign described hereafter has investigated the changes on mechanical and physical performance of RACs by considering the possible variation of the initial moisture condition for RCAs.

2.2.1 Materials and Methods

Four different concrete mixtures with 30% recycled aggregate replacement have been considered in the present study with emphasis on the relationship between the compressive strength development and the corresponding time evolution of the degree of hydration. Two key parameters, i.e. the water to cement ratio and the initial moisture condition of the recycled aggregates, respectively, have been considered in this campaign and their impact on the final concrete quality has been investigated. Specifically, their influence on both the hydration reaction and the evolution of the compressive strength of concretes made with recycled aggregates has been investigated. Therefore, two mixtures are designed with a different nominal water/cement ratio, i.e. 0.45 and 0.60. Moreover, for the recycled aggregates two different initial moisture conditions have been adopted, according to the two following definitions:

- Dry condition (DRY): the coarse aggregates have been dried for 24 h in an oven with a constant temperature of 100 °C;
- Saturated condition (SAT): the coarse aggregates have been saturated for 24 h in water and, before mixing, their surface has been dried with a cloth.

In fact, natural aggregates used in ordinary concretes have been generally characterised by a low water absorption capacity and their corresponding portion of "absorbed water" can easily be accounted for in the concrete mix design. On the contrary, a higher water absorption capacity of RCAs clearly depends on their production process. Particularly, internal damage and cracks due to demolition and crushing, results in a non-negligible influence of RCAs water absorption capacity on the concrete mix performance, in both the fresh state (in terms of actual workability and rheological properties) and the hardened state (in terms of mechanical properties). The processing of the RCAs, as considered in this study, has led to the following ranges of the grain sizes (Table 2.3):

Table 2.3 Influence of the initial moisture condition of RCAs, natural and recycled aggregates

Class size	Natural		Recycled	
	A (%)	γ (kg/m^3)	A (%)	γ (kg/m^3)
Sand (d < 4.75 mm)	1.2	2690	–	–
C1 (4.75 mm < d < 9.5 mm)	0.5	2690	6.0	2231
C2 (9.5 mm < d < 19 mm)	0.4	2690	3.0	2231
C3 (19 mm < d < 9.5 mm)	0.3	2690	1.8	2231

Table 2.4 Influence of the initial moisture condition of RCAs, mixture composition

Mix	CEM	w	ads. W	w/c	Natural			Recycled	
					Sand	C1	C2	C2	C3
	(kg/m^3)				(kg/m^3)				
0.45DRY	410	185	0	0.45	760	130	300	100	400
0.45SAT			21						
0.60DRY	310		0	0.60	850				
0.60SAT			21						

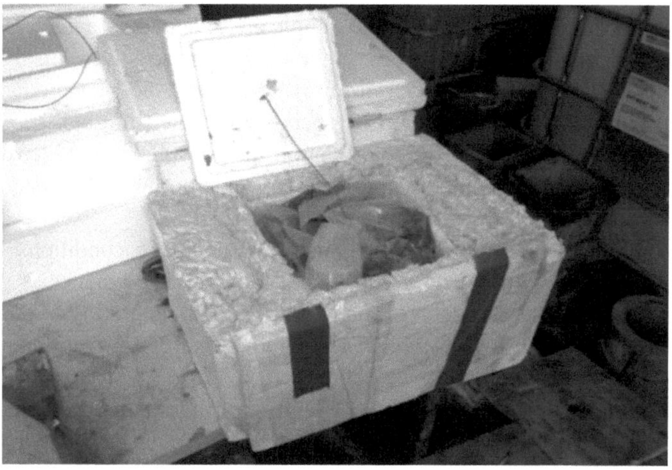

Fig. 2.4 Insulated mould inducing semi-adiabatic boundary conditions on the curing concrete sample

- Sand, nominal size smaller than 4.75 mm;
- C1, nominal size 4.75–9.5 mm;
- C2, nominal size 9.5–19 mm;
- C3, nominal size 19–31.5 mm.

Table 2.4 describes the actual composition of the four concrete mixtures considered and differentiates between the two aforementioned values of the nominal water to cement ratios and the two initial moisture conditions.

The amount of RCAs has been kept constant to 30% of the total amount of aggregates, with a total replacement of the coarse fraction, a partial replacement of the finer fraction, and no replacement of sand.

The volume of water absorbed by the saturated aggregates is not included in the calculation of the w/c ratio. The absorbed volumes are estimated by considering the amount of the various aggregate fractions and their respective water absorption capacity, which was determined on both the natural and recycled aggregates.

Table 2.4 reports the amounts of water absorbed by recycled aggregate in saturated conditions, apart from the regular mixing water.

Finally, a common Portland cement, type CEM I 42.5 R (EN 197-1 2011), was used as a binder in all concrete mixes. For all the investigated concrete mixtures, the hydration process was monitored by measuring the temperature evolution in the centre of a concrete cube during the first seven days after casting. To this end, a cubic sample of each concrete mix was cured in semi-adiabatic conditions with the aim of measuring the temperature evolution and to use these results as input data in simulation model capable to calculating the hydration process (Martinelli et al. 2013). For this purpose, this concrete sample was cast within an insulated mould with an edge size of 150 mm (Fig. 2.4).

Four out of the six faces of the cubic sample are bounded by a thick layer (about 100 mm) of insulating material, whereas the other two faces were insulated with a significantly thinner layer (of about 40 mm). Therefore, the heat produced by the hydration reaction was supposed to be mainly dissipated through the two faces bordering with the thinner layers of insulation material. Since these two faces are placed opposite from each other (namely, the top and bottom of the system depicted in Fig. 2.4) a 1D heat flow was supposed to occur.

The evolving exothermic reaction under semi-adiabatic conditions leads to a temperature increase within the cube, which also affects the kinetics of the cement hydration reaction. The evolution of the hydration temperature inside the cube was monitored with respect to the ambient temperature using two thermocouple wires (Fig. 2.5).

Moreover, apart from the one cured within the insulated mould, ten other concrete samples for each mix were cast and cured in a water bath under isothermal conditions at a temperature T_R of 22 °C. Couples of these samples were tested in compression (Fig. 2.6) after 2, 3, 7, 14 and 28 days of curing, with the aim of determining the average strength at the aforementioned curing ages.

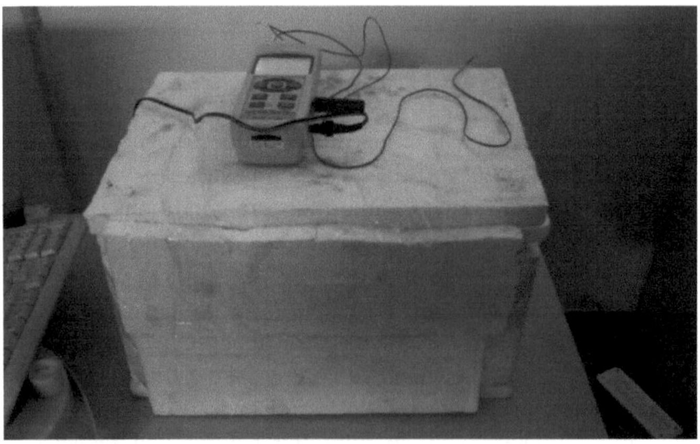

Fig. 2.5 Temperature measurement inside insulated mould

Fig. 2.6 Compression test

Fig. 2.7 Time evolution of temperature on curing concrete samples (w/c = 0.45)

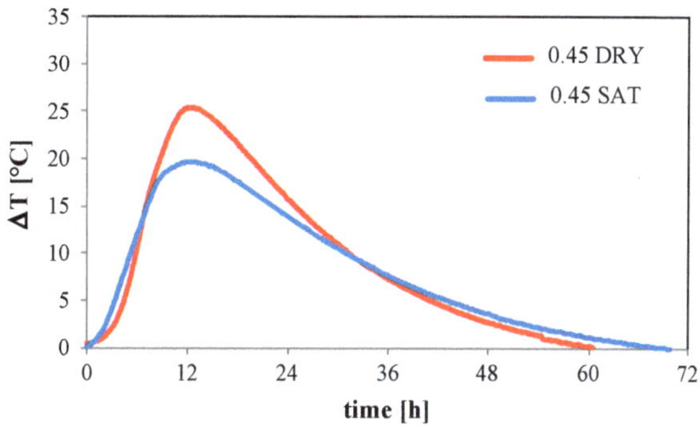

2.2.2 *Results and Analysis*

The time evolution of temperature, measured as described above, allows scrutinizing the actual influence of the mix ingredients and moisture conditions of the recycled aggregates on the resulting cement hydration reaction. Figure 2.7 shows the temperature development measured in the two concrete samples with a w/c = 0.45. It can be observed that the sample with dry aggregates reaches the highest peak value for the temperature as well as a slightly higher rate of the hydration reaction for the ascending branch. This could likely be attributed to the relatively lower amount of water in this mix with dry aggregates. In fact, the role of the w/c ratio on the hydration process is well known and this observation is basically in line with this well-established knowledge (van Breugel 1991). On the contrary, the sample with saturated recycled aggregates exhibited a slightly longer reaction period, which could likely be attributed to the higher amount of available water, and a slower temperature decay in the post-peak branch.

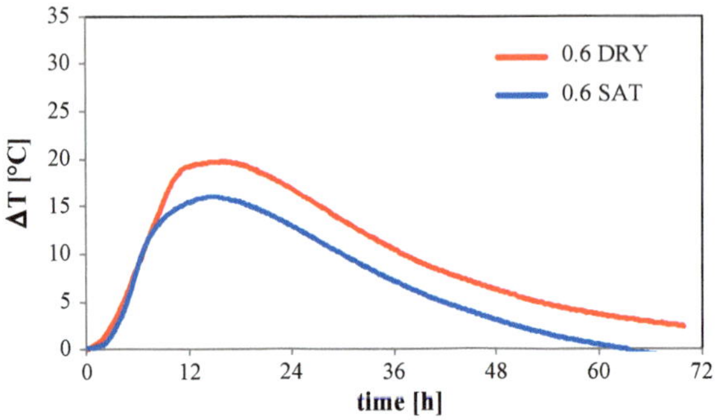

Fig. 2.8 Time evolution of temperature on curing concrete samples (w/c = 0.60)

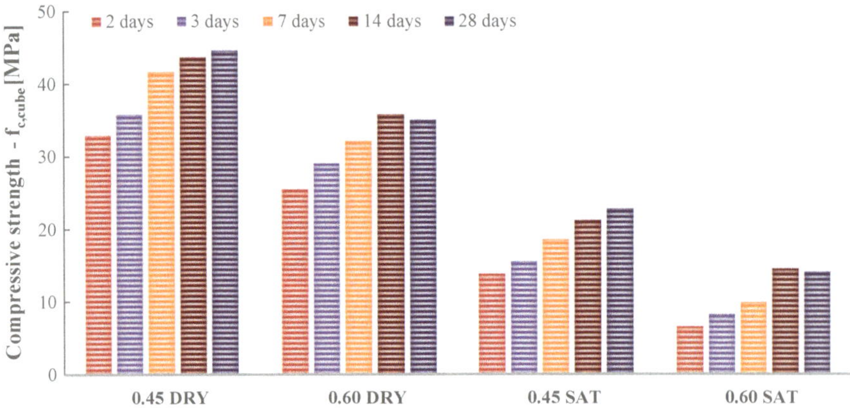

Fig. 2.9 Time evolution of the average compressive strength on concrete samples

Similar considerations hold for the two concrete mixtures with a w/c of 0.60 (Fig. 2.8). Higher peak temperatures were reached in the concrete made with dry aggregates, but in this case the post-peak temperature decay observed in the two samples showed similar rates.

For the compressive strength evolution, the influence of the two parameters under consideration, i.e. the w/c ratio and the moisture condition of the recycled aggregates, on the hydration reaction of the four mixtures (Table 2.4) is expected to have a clear effect on the time evolution of the compressive strength.

Figure 2.9 shows the results of the compressive strength development in terms of average cube compressive strength $f_{c,cube}$, according to the procedure described in the previous subsection. The results confirm that the initial moisture content of the recycled aggregates plays an important role in the development of the compressive strength of concrete Fig. 2.9. In this respect, dry aggregates lead to higher values of $f_{c,cube}$ over the investigated time span of 28 days for both w/c ratios considered in this study. Moreover, the compressive strength of specimens denoted as 0.60DRY (namely, those with w/c = 0.6 and dry aggregates) resulted in significantly higher than the corresponding mixtures produced with initially saturated aggregates (namely, 0.60SAT). For the corresponding tests performed for the w/c ratio of 0.45 with recycled aggregates under dry and saturated conditions this reduction is about 50%. Particularly, both mixtures made with saturated aggregates were characterised by a very low compressive strength, apparently as a result of the higher total amount of water (mixing water + absorbed water) which potentially led to a higher value of the actual (local) w/c ratio, depending on the possible release of the absorbed water from the aggregates into the mix.

The correlation between the hydration processes and the resulting compressive strength can be figured out by calculating the evolution of the degree of hydration from the temperature curves as presented in Figs. 2.7 and 2.8 and by comparing the results with the strength measurements reported in Fig. 2.9. A deeper investigation towards the quantitative relationship between these hydration-related measurements and the compressive strength is proposed in the following Sect. 2.3.

2.3 Influence of the Aggregate Replacement and Water to Cement Ratios

Once having assessed the key role of the initial moisture condition of RCA on the resulting properties of RACs, it is important to understand how this parameter affects the nominal value of the water-to cement ratio. With this aim, the experimental campaign described in this section has considered the variation of the initial moisture condition by combining the possible variation of the aggregate replacement ratio and the water to cement ratio.

2.3.1 Materials and Methods

All mixtures were produced by using a high initial strength cement, denoted as CEM I 52.5 R according to EN 197-1 (2011). Meanwhile, three different size ranges were considered for the aggregates: sand (nominal diameter smaller than

Table 2.5 Influence of the aggregate replacement and water to cement ratios: natural and recycled aggregates properties

Class size	Natural		Recycled	
	A (%)	γ (kg/m^3)	A (%)	γ (kg/m^3)
Sand (d < 4.75 mm)	1.20	2690	–	–
C1 (4.75 mm < d < 9.5 mm)	0.50	2690	8.70	2127
C2 (9.5 mm < d < 19 mm)	0.40	2690	6.60	2290

Table 2.6 Influence of the aggregate replacement and water to cement ratios, mixture composition

Mix	CEM	w	w/c	Natural			Recycled	
				Sand	C1	C2	C1	C2
	(kg/m³)			(kg/m³)				
0.50NATDRY	344	172	0.50	742	554	554	–	–
0.50NATSAT					557	557	–	–
0.50RAC30DRY					554	–	–	474
0.50RAC30SAT					557	–	–	505
0.50RAC60DRY					–	–	440	474
0.50RAC60SAT					–	–	478	505
0.40RAC60DRY	430		0.40	712	–	–	422	454
0.40RAC60SAT					–	–	458	484
0.60RAC60DRY	287		0.60	762	–	–	452	487
0.60RAC60SAT					–	–	491	519

4.75 mm) and coarse aggregates (C1 and C2 classes already defined in the previous sections). The main physical properties of both natural and recycled aggregates employed in this study, are reported in Table 2.5, meanwhile the mixtures composition are reported in Table 2.6.

It is worth highlighting that natural sand obtained from crushing limestone rocks has been employed in this study. In fact, no recycled sand is employed, as it would have been too porous and, then, it would have had a significantly detrimental effect on the resulting concrete properties (Lima et al. 2013). Conversely, a combination of both NAs and RCAs has been employed as coarse aggregates. Ten different concrete mixtures have been produced in order to investigate the influence of the three following parameters:

- (nominal) value of the water-to-cement ratio: 0.40, 0.50 and 0.60;
- RCAs-to-NAs replacement ratio ranging from 0 to 30 to 60% relative to the total volume of fine and coarse aggregates (i.e. considering also the natural sand);
- initial moisture condition of the coarse aggregates (Koenders et al. 2014): oven-dried assured by heating the aggregates for 24 h at a temperature of 100 ± 5 °C (DRY), and saturated surface dry, obtained by submerging the aggregates in water for 24 h (SAT).

Figure 2.10 shows a schematic synopsis of the tests, while Table 2.6 describes into detail the mix compositions. In all mixtures, the fine fractions (i.e., sand) represented 40% (by volume) of the total amount of aggregates, while the remaining 60% is equally divided into two fractions of coarse aggregates (i.e., C1 and C2).

The first two reference mixtures described in Table 2.6 were obtained with only natural aggregates (NAT in the mixture labels), a nominal water-cement ratio of 0.50 (also mentioned in the labels) and two alternative initial moisture conditions of the coarse aggregates (i.e., DRY and SAT).

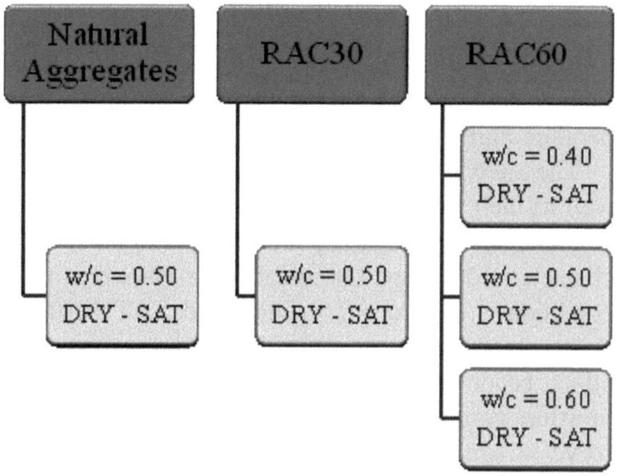

Fig. 2.10 Experimental campaign

Therefore, a RCAs-to-NAs coarse aggregate replacement of 30% was obtained via a complete replacement of the coarse fraction (namely class C2) with the corresponding ones made of RCAs, whereas 60% was achieved by replacing both fractions C1 and C2 of the NAs with the corresponding fractions of RCAs. Hence, the two mixtures denoted as 0.50RAC30DRY and 0.50RAC30SAT were derived from 0.50NATDRY and 0.50NATSAT, by fully replacing the C2 fraction with an equal volume of RACs. Similarly, the two mixtures, denoted as 0.50RAC60DRY and 0.50RAC60SAT, were obtained from the reference ones by replacing all the coarse aggregates (i.e. classes C1 and C2).

Finally, the last four rows refer to the composition of the two mixtures obtained by either reducing to 0.40 or raising to 0.60 the nominal w/c ratio. For each mixture, nine compressive strength tests were performed on cubic specimens according EN 12390-3:2009 after 1, 3 and 28 days of curing in a water bath under isothermal conditions at a temperature 20 ± 2 °C. Moreover, the time development of temperature was measured in the centre of the cubes, prepared for all concrete mixtures and cured in semi-adiabatic conditions so that an indirect monitoring of the hydration process could be registered and applied to the numerical procedure as recently proposed in the scientific literature (Martinelli et al. 2013). However, the concrete specimens tested in compression have been cured in isothermal conditions, at room temperature. However, due to the small size of such specimens and due to the quick dissipation of the reaction heat to the environment, no significant temperature enhancement could be measured. Therefore, a special insulated mould was designed to measure the evolution of the hydration temperature inside a cube such that a significant temperature enhancement was monitored with respect to the room temperature, using two thermocouple wires (as already described in the previous section).

2.3.2 Results and Analysis

Figures 2.11 and 2.12 show the time evolution of compressive strength and temperature during the first 28 days of hardening.

The analysis of these results clearly show that the use of RCAs in concrete, as well as their initial moisture condition, affects the resulting concrete performance.

First, when comparing the compressive strength results of the concrete mixes characterised by equal nominal water-to-cement ratio and the same initial moisture condition (i.e., w/c = 0.5, DRY and SAT) it turns out that with increasing the aggregate replacement ratio, the 28 day compressive strength is decreasing. This effect becomes even more pronounced whenever employing saturated initial moisture conditions (SAT). This result can be explained by the higher porosity of recycled aggregates. In fact, when a SAT condition is employed, the aggregates tend to release accumulated water into the mixture and consequently to change the

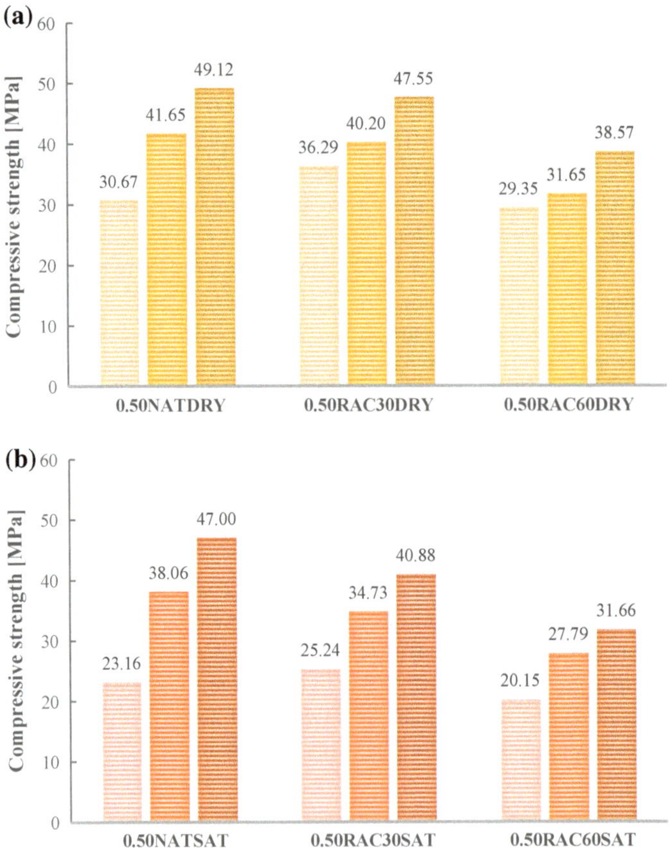

Fig. 2.11 Time-evolution of compressive strength

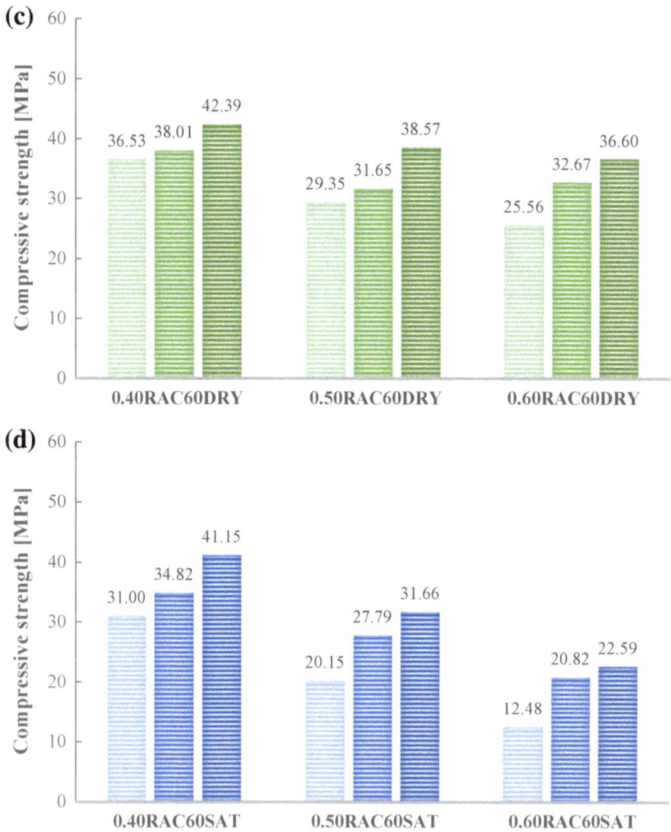

Fig. 2.11 (continued)

water-to-cement ratio. On the contrary, when DRY conditions are adopted, the aggregates tend to absorb part of the mixing water and, then, reduce the actual water-to-cement ratio. This phenomenon is less evident when natural aggregates are used because of their lower absorption capacity and, consequently, the amount of water, potentially released or absorbed into a mixture is negligible compared to the total amount of mixing water. On the other hand, analysis of the compressive strength results of those mixtures characterised by the same aggregate replacement ratio (i.e., RAC60), but with different nominal water-to-cement ratios (i.e., moving from 0.40 to 0.50 to 0.60), highlights the key impact of the w/c ratio on the final strength of concrete, as already known from literature. It is worth mentioning that also in this case, the initial moisture condition plays an important role. In fact, the gap (in terms of compressive strength at 28 days) between DRY and SAT mixtures also increases with increasing the nominal water-to-cement ratio.

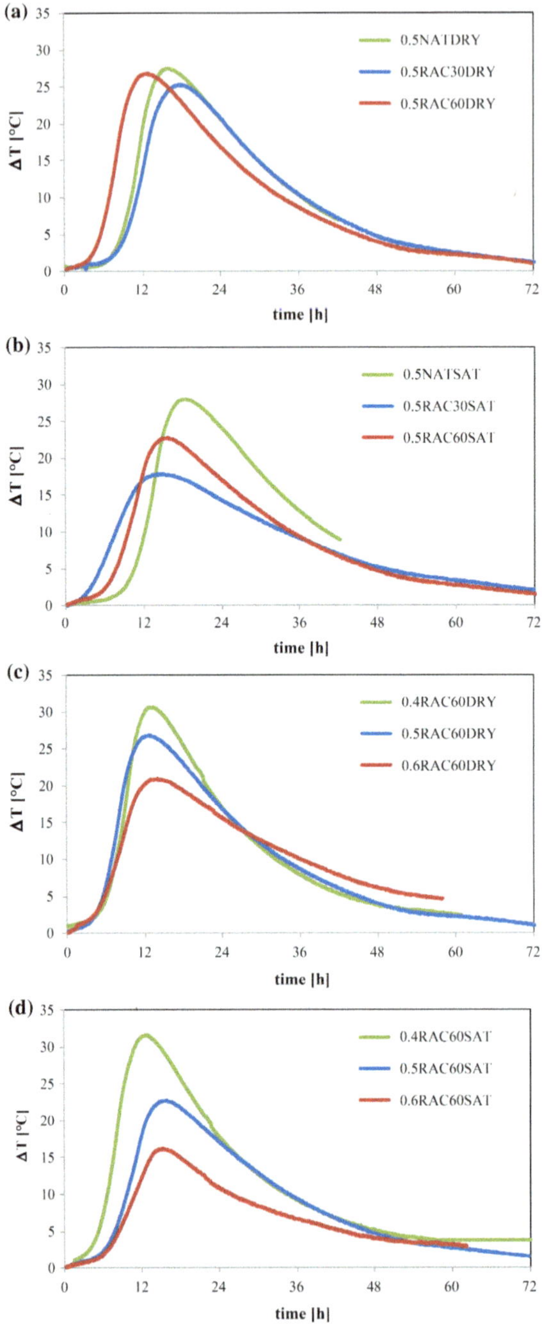

Fig. 2.12 Temperature-time evolution

A more fundamental analysis can be performed by considering the differences of the temperature measurements in time, as reported in Fig. 2.12. Experimental results highlight that a lower w/c ratio results in a faster acceleration of temperature. A similar effect was observed by enhancing aggregate replacement ratio or in case of DRY aggregates.

Based on the above considerations and the analysis of results, it is clear that the initial moisture condition and the aggregate porosity tend to modify the nominal water-to-cement ratio and, hence, an effective value, taking into account both the initial moisture condition and the aggregate porosity, can be defined as follows:

$$\left(\frac{w}{c}\right)_{eff} = \frac{w}{c} + \frac{w_{add}}{c} - \delta \cdot \left(\sum_{i=1}^{n} \frac{p_i \cdot P_i}{c}\right) \tag{2.1}$$

where w/c is the nominal water-to-cement ratio, w_{add} is the extra water added to the mix from the (partially or totally) soaked the aggregates, p_i and P_i represent the absorption capacity and the weight in the mixture of the ith aggregate fraction, and δ is a parameter that takes into account the initial moisture condition of the aggregates and is zero in SAT conditions and 0.5 in DRY ones. The calibration $\delta = 0.5$ for DRY condition is that during mixing and casting it turned out that the aggregates are able to absorb an amount of water equivalent to 50% of their 24 h capacity.

2.4 Concluding Remarks

The experimental results summarised in this section highlight that:

- the presence of RCAs in the concrete mixtures modifies their fundamental properties, as even the cement hydration reaction is affected by replacing ordinary aggregates with recycled ones;
- RCAs are significantly more porous than NAs and this is the main reason behind the difference in the observed behaviour of ordinary concrete and RACs;
- moreover, the initial moisture condition of aggregates at mixing plays a significant role, as it modify the actual amount of water (and, hence, the water-cement ratio) available for the hydration process;
- cleaning of RCAs may be an option to modify porosity (and, hence, water absorption capacity) in RCAs and, hence, reduce the difference between RCAs and NAs.

Finally, three main parameters are supposed to control the behaviour of RACs (namely, water-cement ratio, aggregate replacement ratio and initial moisture conditions): their influence may be condensed by defining an "effective" water-cement ratio that can be employed in a generalised Abrams' law capable to predict compressive strength of RAC, as will be explained in Chap. 6.

References

ASTM C496/C496M (2011) Standard test method for splitting tensile strength of cylindrical concrete specimens. Am Soc Test Mater

EN 12350-2 (2009) Testing fresh concrete—part 2: slump-test. CEN, European Committee for Standardization

EN 12390-3 (2009) Testing hardened concrete—part 3: compressive strength of test specimens. CEN, European Committee for Standardization

EN 197-1 (2011) Cement—part 1: composition, specifications and conformity criteria for common cements. CEN, European Committee for Standardization

Koenders EA, Pepe M, Martinelli E (2014) Compressive strength and hydration processes of concrete with recycled aggregates. Cem Concr Res 56:203–212

Lima C, Caggiano A, Faella C, Martinelli E, Pepe M, Realfonzo R (2013) Physical properties and mechanical behaviour of concrete made with recycled aggregates and fly ash. Constr Build Mater 47:547–559

Martinelli E, Koenders EA, Caggiano A (2013) A numerical recipe for modelling hydration and heat flow in hardening concrete. Cement Concr Compos 40:48–58

NBR 5733 (1991) High early strength Portland cement—specification. ABNT, Associação Brasileira de Normas Técnicas

NBR 5739 (2007) Concrete—compression test of cylindric specimens—method of test. ABNT, Associação Brasileira de Normas Técnicas

NBR 7211 (2009) Aggregates for concrete—specification. ABNT, Associação Brasileira de Normas Técnicas

Pepe M, Toledo Filho RD, Koenders EAB, Martinelli E (2014) Alternative processing procedures for recycled aggregates in structural concrete. Constr Build Mater 69:124–132

van Breugel K (1991) Simulation of hydration and formation of structure in hardening cement-based materials. Ph.D. thesis, Delft University of Technology, The Netherlands

Chapter 3
Cement Replacement: Experimental Results for Concrete with Recycled Aggregates and Fly-Ash

Carmine Lima, Ciro Faella, Marco Pepe and Enzo Martinelli

Abstract The experimental activity reported in this chapter was aimed at enhancing the knowledge about the mechanical behaviour and durability of concretes made with Recycled Concrete Aggregates (RCAs) and coal Fly Ash (FA) and their possible use for structural purposes. To this end, starting from a reference concrete composition, twelve mixtures were designed by replacing part of the ordinary constituents (i.e. cement, sand and coarse aggregates) with the FA and RCAs. The time evolution of the compressive strength, as well as the splitting strength, were measured with the aim to monitor the mechanical performance, whereas the durability performance was scrutinised by measuring water permeability, carbonation depth and chloride-ions ingress. The obtained results unveil the influence of both RCAs and FA on the resulting concrete performance and highlight that their combined use can lead to a synergistic effect in terms of the relevant physical and mechanical properties of structural concrete.

This section presents the results obtained in an experimental study on the performances of "green concretes", i.e. of concrete made by combining the use of both an alternative binder (to partially replace cement) and recycled concrete aggregates (up to the total replacement of natural ones) possibly revealing the consequences of their combined use. Particularly, several NAC and RAC mixtures were produced with different combinations of recycled aggregates derived from old existing concrete elements (RCAs) and coal Fly Ashes (FAs). They were tested at both the Laboratory of Material Testing and Structures (LMS) of the University of Salerno (UniSA) and the Laboratory of the company General Admixtures SpA, in Ponzano Veneto, which supported the experimental activities reported in this section. The behaviour of the concrete mixtures under consideration was investigated both fresh and hardened state (Lima et al. 2013).

Since some properties of RCAs (i.e. their surface shape, sizing and water absorption capacity) are rather different with respect to the corresponding ones

C. Lima · C. Faella · M. Pepe · E. Martinelli (✉)
Department of Civil Engineering, University of Salerno, Fisciano, Italy
e-mail: e.martinelli@unisa.it

J.A.O. Barros et al. (eds.), *Recent Advances on Green Concrete for Structural Purposes*, Research for Development, DOI 10.1007/978-3-319-56797-6_3

generally characterising natural aggregates, the mix-design approach was carefully revised and adapted to better achieve the target properties at both the fresh- and the hardened-states.

The experimental campaign was aimed at investigating the mechanical and physical properties of the produced mixtures such as the compressive and tensile strength, the bond behaviour, the water permeability, the chloride penetration and the carbonation resistance as indexes of their durability performance (Faella et al. 2016).

Finally, the key durability-related issues and the mechanical properties of RAC samples, also controlled by the physical and chemical nature of recycled aggregates and fly-ash, were also investigated through the above-mentioned experimental tests.

3.1 The Experimental Campaign

Thirteen different concrete mixes were prepared and batches of samples of such mixes were tested to measure the relevant physical and mechanical properties and, in some case, monitor their time evolution during the setting and hardening phases. The aforementioned mixes were conceived as follows:

- one reference mix with cement and natural aggregates;
- four mixes with FA as a cement replacement up to the maximum allowed replacement ratio allowed by the current European regulations (EN 450 2005), and different aggregates replacement ratio;
- four mixes with FA as both cement replacement (up to the maximum replacement ratio allowed by the current European regulations) and filler (in partial substitution the fine aggregates) and different aggregates replacement ratio;
- four mixes with FA as a cement replacement beyond the maximum allowed replacement ratio allowed by the current European regulations, and different aggregates replacement ratio.

The compositions of the concrete mixes described above are fully detailed in the following subsections.

3.1.1 Materials

Crushed limestone was used as natural aggregate (EN12620 2002), while RCAs used in concrete mixes made with recycled concrete were obtained by crushing the rubbles derived from both demolition of existing concrete structures and waste materials resulting from the concrete industry (EN 13242 2008). An Italian company certified according to EN 13242 (2008) supplied them. The mix composition and the mechanical properties of the original concrete were unknown.

An accurate screening and the mechanical characterization of aggregates deriving from recycling processes was preliminarily performed in order to obtain

Fig. 3.1 Sieving operations

Fig. 3.2 Sieved aggregates used in concrete mixtures

high quality recycled materials. Then, the crushed concrete particles were selected, cleaned and sieved at the LMS of UniSA.

The grain size distribution of the particles to be used as aggregates in RAC specimens was established through a preliminary sieving process. Figure 3.1 shows the operation of both screening and sieving phases on RCAs.

Both natural (common crushed limestone) and recycled aggregates, whose nominal maximum diameter was 31.5 mm, were subdivided in four size fractions (Fig. 3.2):

- N3, with nominal diameters ranging between 20 and 31.5 mm;
- N2, with nominal diameters ranging between 10 and 20 mm;
- N1, with nominal diameters ranging between 2 and 10 mm;
- sand, with nominal maximum diameters smaller than 2 mm.

Table 3.1 Specific gravity and water absorption (%) at 24 h of both natural and recycled aggregates

Type	Specific gravity (kg/m^3)	Water absorption (%)			
		Sand (0–2 mm)	N1 (2–10 mm)	N2 (10–20 mm)	N3 (20–31.5 mm)
Natural	2690	1.20	0.70	0.50	0.30
Recycled	2369	12.20	6.00	3.00	1.80

Table 3.2 Chemical composition and physical properties of cement and fly ash

	CEM I 42.5 R	Fly ash
CaO (%)	64.06	2.30
SiO_2 (%)	18.90	46.90
Al_2O_3 (%)	4.90	28.50
Fe_2O_3 (%)	3.66	6.22
SO_3 (%)	2.92	0.04
MgO (%)	0.82	1.23
Loss to ignition (%)	3.18	6.20
Specific gravity (kg/m^3)	3110	2100

The recycled concrete aggregate (RCA) consisted of a combination of pieces of uncracked and fractured stone with mortar adhering to stone itself. Figure 3.2 shows the different aggregates divided in the four size fractions.

The aggregate fractions (sand, fine and coarse natural and recycled aggregates) were mixed according to the Bolomey curve (Bolomey 1947), with the aim of reproducing an optimal grain size distribution.

The water absorption capacity (at 24 h) of both natural and recycled aggregates was measured according to ASTM C127-15 (2015) standard for coarse aggregates and ASTM C128-15 (2015) code for fine ones. Table 3.1 reports the relevant results and confirms that recycled aggregates absorbed much more water than natural ones. The specific gravity was also measured. Since RCAs were more porous than natural aggregates they result in lower specific gravity. This peculiarity is a well-known feature of RCAs (de Juan and Gutierrez 2009).

Portland cement type CEM I 42.5 R in accordance with EN 197-1 (2001) was used for producing concrete mixtures. It contains more than 95% of clinker in order to study the effects of only cement and FA without any additions. Cement has a specific weight equal to 3110 kg/m^3. The fly ash employed for partially replacing Portland cement and sand was classified as Class F coal FAs according to ASTM C618 (2015) and it was also compliant with both EN 450 (2005) and EN 12620 (2002). It was produced in an Italian thermo-electrical power plants. Table 3.2 shows the key physical and chemical properties of both cement and fly ash employed in this experimental campaign.

Finally, an acrylic-based super-plasticizer was also used in order to control workability of the fresh concrete without changing the water to cement (w/c) ratio.

Table 3.3 Mixture proportions (in kg/m^3) of mixes

Mix	CEM I 42.5R (kg/m^3)	Fly ash (kg/m^3)	FA/C	RCA (%)	w (l/m^3)	Absorption water w_{add} (l/m^3)	Natural aggregates (kg/m^3)				RCA (kg/m^3)			
							N3	N2	N1	sand	N3	N2	N1	Sand
N	280	–	0	0	150	14.02	505	470	165	830	–	–	–	–
LN	250	80	0.32	0	150	14.02	505	470	165	750	–	–	–	–
LR30	250	80	0.32	30	150	21.86	–	408	165	750	445	55	–	–
LR60	250	80	0.32	60	150	38.16	–	–	–	750	445	415	145	–
LR100	250	80	0.32	100	150	109.65	–	–	–	–	445	415	145	660
MN	250	220	0.88	0	150	11.28	545	490	170	500	–	–	–	–
MR30	250	220	0.88	30	150	17.85	35	490	170	500	450	–	–	–
MR60	250	220	0.88	60	150	28.99	–	–	185	500	455	450	–	–
MR100	250	220	0.88	100	150	98.84	–	–	–	–	400	375	130	595
HN	200	255	1.27	0	150	11.28	545	490	170	500	–	–	–	–
HR30	200	255	1.27	30	150	17.85	35	490	170	500	450	–	–	–
HR60	200	255	1.27	60	150	28.68	–	–	175	500	455	445	–	–
HR100	200	255	1.27	100	150	98.84	–	–	–	–	400	375	130	595

3.1.2 Concrete Mixtures Proportions

A total number of thirteen different mixes was made by changing the substitution ratio of natural aggregates with RCAs and adding FA. At the same time, the amount of cement and fine aggregates was reduced. Table 3.3 reports the mix proportions of the investigated mixtures.

The first column of the table reports a label denoting the mixture and provides key information about the type of aggregates and the fly ash content. Particularly, the natural aggregate type is defined by the letter "N"; conversely, with "R" the presence of a given percentage of recycled aggregate is indicated; this percentage is provided through a number following the "R" symbol, i.e. 30, 60 or 100. If any, a first letter denotes the content of fly ash ("L" for low content, "M" for medium and "H" for high). As an example, "LN" refers to a mix containing only natural aggregates and the lowest content of fly ash among those considered (i.e. 80 kg/m^3); label "LR60", instead, denotes a mix characterised by the same fly ash content, whereas the 60% of natural aggregates is replaced by an equivalent volume amount of recycled ones.

The reference mixture was the "N" designed with only natural aggregates according to EN 206-1 (2006) for an exposure class XC2 in strength class C25/30. The minimum amount of cement *c* (equal to 280 kg/m^3), suggested by European Standards (EN 206-1 2006), was used for the reference mix.

Other twelve mixes were grouped by taking into account both the increasing amount of FA used in substitution of the cement and the substitution of natural aggregate with recycled one. Mixes "LN", "LRxx", "MN" and "MRxx", in which "xx" represents the percentage of substitution of natural aggregate with RCA, were made by reducing the cement amount of 30 kg/m^3 according to EN 206-1 (2006) limitation:

$$\Delta c = k \cdot (c - 200) \cong 32\,\mathrm{kg/m^3} \tag{3.1}$$

in which k coefficient is a sort of "cementing efficiency index" of fly ash, and the following "k-value" can be used for a concrete containing CEM I conforming to EN 197-1 (2001):

- k = 0.2 in case of CEM-I 32.5;
- k = 0.4 in case of CEM-I 42.5 and higher.

The reduction in cement was compensated by adding FA. According to EN 206-1 (2006) provisions, a maximum of 33% by weight of FA with respect to the cement used in the mixes (250 kg/m^3) can be considered as binder. On the basis of the mentioned limitation, mixes "LN" and "LRxx" were produced by adding 80 kg/m^3 of FA which represents about the 33% of the minimum amount of cement allowed by EN 206-1 (2006). However, taking into account the different specific gravity of FA and cement, fine aggregates were reduced in order to balance the volumes (Table 3.3).

The mixes "MN" and "MRxx" were made by adding 220 kg/m^3 of FA. According EN 206-1 (2006), about 80 kg/m^3 can be taken into account as binder, while other 140 kg/m^3 of FA should be considered as fine aggregate (Table 3.3).

The mixes of the third group (namely "HN" and "HRxx") were made by reducing the content of cement beyond the limit given by Eq. (1.1) and further increasing the amount of FA. Indeed, 200 kg/m^3 of cement and 255 kg/m^3 of FA were used.

Other information reported in Table 3.3 are:

- the mass per unit volume of cement (CEM I 42.5R) and fly ash (FA) with their ratio (FA/C);
- the percentage of recycled aggregates used in substitution of natural ones (RCA);
- the free water contents per unit volume (w) and the water surpluses needed for the water absorption of aggregates (w_{add});
- the quantities per unit volume of both natural and recycled aggregates for each sieve fraction.

All mixes were produced with 150 l/m^3 of free water and further water was added in order to consider the water absorption capacity (at 24 h) of aggregates (Table 3.1). The additional water content has been estimated by assuming that aggregates were initially dry and considering the actual amount of NA and RCA for each sieve fraction. In particular, the water surpluses needed in mixes for taking into account the above mentioned additional water are reported in Table 3.3 in the column titled "Absorption water (w_{add})". The control mixture only contains NAs and Portland cement and features a total water content of about 164 kg/m^3; therefore, the water–cement ratio ($w_{tot}/C = 164.02/280 = 0.59$) is within the range given by the European standard (EN 206-1 2006).

According to the aforementioned European Standard, the minimum cement content C_{min} may be reduced if fly ash is added in the mixture, so that the following relationship is satisfied:

$$FA/C < 0.33 \text{ (by mass)} \tag{3.2}$$

According to the EN 206-1 (2006), if a greater amount of fly ash is used, the excess cannot be taken into account for the calculation of the water-to-binder ratio:

$$\frac{w}{b} = \frac{w}{C + k \cdot FA} \tag{3.3}$$

Two different water–binder ratios have been evaluated and reported in Table 3.4 by considering the total amount of FA computing the total binder content "b_0":

$$\frac{w_{tot}}{b_0} = \frac{w + w_{add}}{C + k \cdot FA_{tot}} \tag{3.4}$$

Table 3.4 Water-binder ratios of mixes

Mix	w_{tot}/b_0^a	w_{tot}/b^b
N	0.59	0.59
LN	0.58	0.58
LR30	0.61	0.61
LR60	0.67	0.67
LR100	0.92	0.92
MN	0.48	0.57
MR30	0.50	0.59
MR60	0.53	0.63
MR100	0.74	0.88
HN	0.53	0.71
HR30	0.56	0.74
HR60	0.59	0.79
HR100	0.82	1.10

[a]Evaluating $b_0 = C + k\ FA_{tot}$: the total content of FA is considered
[b]Evaluating $b = C + k\ FA$: a maximum amount of FA equal to the 33% of cement is considered

and limiting the amount of FA as request by the European Standard:

$$\frac{w_{tot}}{b} = \frac{w + w_{add}}{C + k \cdot FA} \tag{3.5}$$

Table 3.4 highlights that not all the mixes comply with the limit value of the water–binder ratio given by EN 206-1 for the exposure class XC2. The four mixes identified by L as first letter of the label are fully in agreement with the code provisions: in fact, according to Eq. (1.1) the cement content was reduced to C = 250 kg/m^3 and, correspondingly, 80 kg/m^3 of FA were added. The group M of four mixes were produced by using 220 kg/m^3 of FA and 250 kg/m^3 of cement.

Therefore, these mixes contain about 140 kg/m^3 of FA that cannot be considered as binder and the content of fine aggregates (i.e., the sand) was adjusted for volume balance. The third group of four mixes (first letter H) was obtained by further reducing the content of cement up to 200 kg/m^3, i.e. well beyond the limits allowed by Eq. (1.1). In this group, in order to keep the total volume of binder equal to the one of the second group of mixes 255 kg/m^3 of coal FA were used.

3.1.3 *Test Matrix*

For each type of concrete, eighteen 150 × 150 × 150 mm^3 cube samples were cast in polyurethane forms (EN 12390-3 2002). Moreover, two 150 mm diameter cylindrical samples with height equal to 300 mm and one 102 mm × 300 mm cylindrical sample were prepared; this last was cut in order to obtain four

102 mm × 50 mm cylinder useful for chloride diffusion tests. Other four 150 × 150 × 150 mm^3 cube specimens were cast incorporating ø10 deformed steel bar made of B450C steel (NTC 2008) in order to perform pull out tests. The bond length of steel bars embedded in concrete length was 50 mm, according to (RILEM 7-II-28 1994).

In particular, the following tests were planned for each concrete mixture:

- 14 compressive strength tests on 150 × 150 × 150 mm^3 cube specimens (EN 12390-4 2000) at 2, 7, 28, 60, 90 and 365 days of curing (2 tests at 2 days, 1 at 7 days, 6 at 28 days, 2 at 60 days, 1 at 90 days and 2 at 365 days);
- 2 splitting strength tests on 150 mm × 300 mm cylindrical samples (EN 12390-6 2000) at 28 days, to evaluate the concrete tensile strength;
- 4 pull out tests on cubic specimens including a ø10 steel bar (RILEM 7-II-28 1994) at 28 days of curing;
- 2 water permeability tests on 150 × 150 × 150 mm^3 cubes (EN 12390-8 2000) at 90 and 365 days;
- 2 carbonation depth tests on 150 × 150 × 150 mm^3 cubes at 90 and 365 days of curing;
- 4 cylinders 102 mm × 50 mm prepared from the 102 mm × 300 mm cylinder in order to perform chloride penetration tests according ASTM C1202-12 (2012).

The entire test matrix of the whole proposed experimental programme is shown in Table 3.5. It reports information about the specimen geometry and tests made for investigating both mechanical and durability properties.

Therefore, 260 and 104 tests were performed in total for investigating mechanical and durability aspects, respectively: 182 compressive strength tests, 26 splitting tests, 52 pull out tests, 26 water permeability tests, 26 carbonation tests and 52 rapid chloride-ion penetrability tests.

The performance of composite materials is controlled by both the properties of its constituents and the interactions between them. In concrete, aggregate and cement paste, as well as the interfacial zone between them, affect both the mechanical behaviour and the resulting permeability (Mather 2004). Moreover, concrete durability is significantly affected by the latter: as it is well known, concretes with lower permeability show better resistance against chemical attacks and are also more capable to resist freeze-thaw cycles (Mather 2004).

However, concrete permeability is controlled by the so-called "open" porosity and microcracks possibly developed within the concrete matrix may significantly contribute to increasing permeability. Hence, concrete permeability depends on the porosity of the paste, on the aggregate and, consequently, on the paste to aggregate ratio, that define the interface porosity (Chia and Zhang 2002).

The whole experimental programme developed and herein summarised was aimed at investigating the mechanical and physical properties of hardened samples of the aforementioned concrete mixtures made with a variable amount and fraction of natural and recycled aggregates.

Table 3.5 Experimental programme

Type of test and geometry of specimens		Days	Number of tests
Compressive Cube: 150 × 150 × 150 mm^3		2	2
		7	1
		28	6
		60	2
		90	1
		365	2
Tensile Cylinder: ϕ150 mm, h = 300 mm	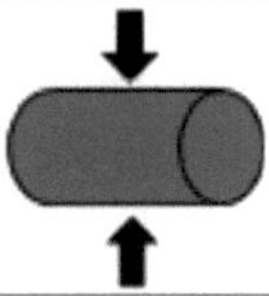	28	2
Pull-out Cube: 150 × 150 × 150 mm^3 ϕ10 steel bar	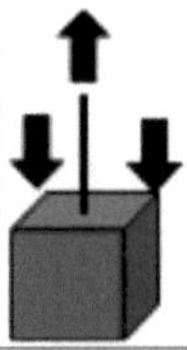	28	4
Permeability Cube: 150 × 150 × 150 mm^3	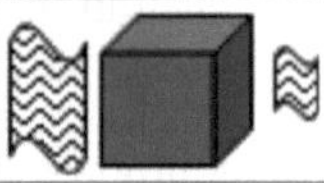	90	1
		365	1
Carbonation depth Cube: 150 × 150 × 150 mm^3		90	1
		365	1
Chloride diffusion Cylinder: ϕ100 mm, h = 50 mm		90	2
		365	2

3.1.4 *Casting and Curing Processes*

Concretes described in the previous section were prepared by using a small mixer available in the Laboratory of Materials testing and Structures of the University of Salerno.

Both coarse and fine aggregates were first blended with the absorption water; subsequently, cement and fly ash were added and dry mixed with the aggregates. Finally, the free water and super-plasticizer was added and the mixing operation continued for about 10 min (Fig. 3.3).

Then, slump tests were performed (Fig. 3.4) to evaluate the workability of the various concrete mixes. Moreover, the density of fresh concrete was measured. Finally, concretes were cast in cubic and cylindrical polyurethane moulds, and duly

Fig. 3.3 Mixing operation

Fig. 3.4 Slump test

vibrated (Fig. 3.5). Steel bars were accurately included in cubic samples made for performing pull out tests (Fig. 3.6).

After 36 h the concrete samples have been removed from the moulds and 28 days cured at 22 °C with about 100% (under water) (EN 12390-3 2002).

After 28 days of curing, specimens were kept at environmental condition inside the laboratory where the temperature range was between 15 °C (in cold season) and 30 °C (in hot season) with a relative humidity ranging between 40 and 65%.

3.2 Experimental Results

Recycled aggregates in concrete significantly affect the mechanical properties and durability of concrete at both fresh and hardened state. In the following the main results in terms of both states are outlined and discussed.

Fig. 3.5 Casting and vibration of cubic specimens

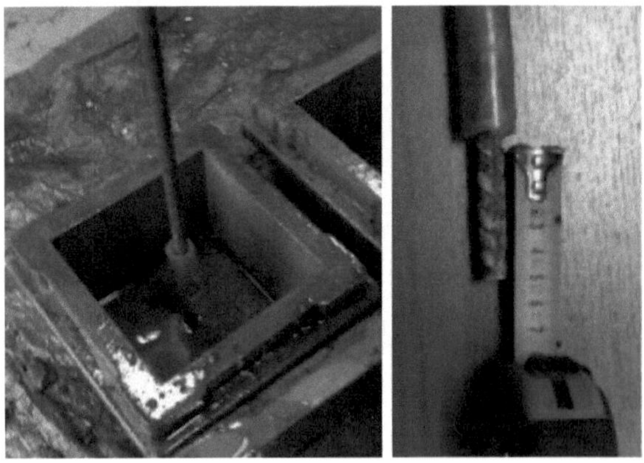

Fig. 3.6 Casting of samples for pull out tests

A special emphasis is reserved in the following on the possible synergic effects induced by combining RCA and FA in the concrete mixes.

3.2.1 *Workability*

Slump tests were performed in order to assess the concrete workability. The reference concrete "N" made without RCA and FA was designed in order to obtain mixes class S4 (EN 206-1 2006) in which slump values range between 160 and 210 mm. A small quantity of a super-plasticizer (0.22% of the total volume) was added in that mix. Larger amounts of the same super-plasticizer were added in the other mixes trying to achieve comparable slumps.

All mixes achieved good workability and were readily consolidated under laboratory conditions. However, the substitution of natural aggregates with recycled

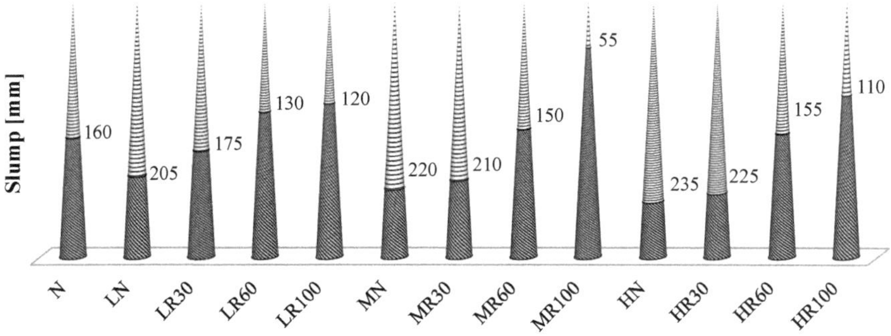

Fig. 3.7 Results of the slump tests

ones led to a loss in workability due to the rougher surface texture of crushed particles and more irregular form compared with natural stones. Fly ash gave the most significant improvement in workability and required a shorter vibration time. This was reflected in the slump measurements presented in Fig. 3.7. The results of the slump tests performed were also influenced by the amount of acrylic super-plasticizer adjoined to the mixes. Concrete mixtures were produced with variable amounts of super-plasticizer in order to obtain good workability and, consequently, highly compacted specimens.

The results reported in Fig. 3.7 show the loss in workability observed for mixes made with 100% of recycled aggregates. However, in the same mixes a greater amount of super-plasticizer was used (e.g., for mixes "MR100" and "HR100").

Definitely, the percentage increase of recycled aggregates has been detrimental in terms of workability.

3.2.2 Density

As it is well known, the mass density of concrete depends on the amount and density of coarse and fine aggregates, the amount of entrained air (and entrapped air) and on the water content. "Ordinary" structural concrete is generally characterised by a mass density around 2400 kg/m^3 and, as a matter of fact, lower values of this quantity results from higher water contents, which, in turn, lead to lower mechanical strength. In the case of RAC, density values also depend on the content of coarse and fine recycled aggregate.

Table 3.6 shows the theoretical value of the fresh concrete density calculated for each mix.

All specimens were weighted with the aim of evaluating the specific gravity of concrete mixes under investigation. 182 specimens have been weighted at different days: 26 samples at 2, 13 at 7, 78 at 28, 26 at 60, and 13 at 90 days and 26 at 365 days: Fig. 3.8 reports the average values of their measured densities compared

Table 3.6 Concrete densities evaluated from the mix proportion

Mix	NA (%)	RCA (%)	CEM I 42.5R (kg/m^3)	Fly ash (kg/m^3)	w_{tot} (kg/m^3)	NA (kg/m^3)	RCA (kg/m^3)	D (kg/m^3)
N	100	0	280	–	164.02	1970	0	2414
LN	100	0	250	80	164.02	1890	0	2384
LR30	70	30	250	80	171.86	1323	500	2325
LR60	40	60	250	80	188.16	750	1005	2273
LR100	0	100	250	80	259.65	0	1665	2255
MN	100	0	250	220	161.28	1705	0	2336
MR30	70	30	250	220	167.85	1195	450	2283
MR60	40	60	250	220	178.99	685	905	2239
MR100	0	100	250	220	248.84	0	1500	2219
HN	100	0	200	255	161.28	1705	0	2321
HR30	70	30	200	255	167.85	1195	450	2268
HR60	40	60	200	255	178.68	675	900	2209
HR100	0	100	200	255	248.84	0	1500	2204

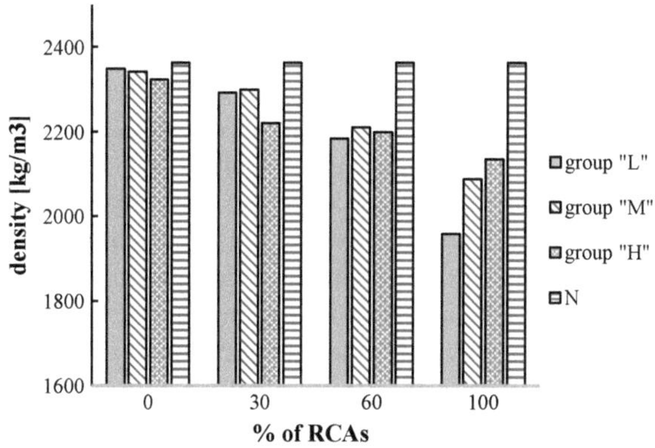

Fig. 3.8 Density of the analysed concrete mixes

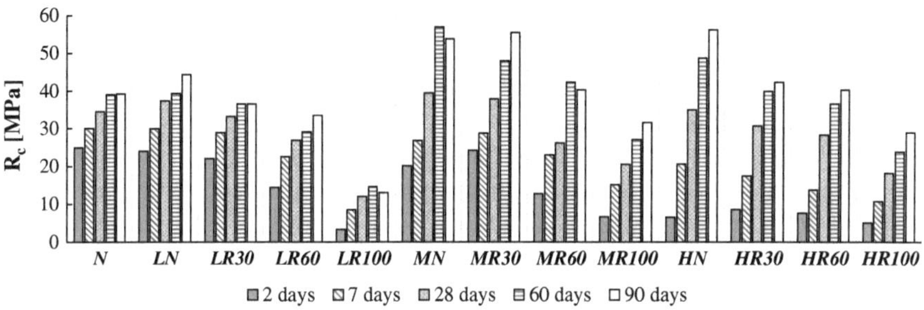

Fig. 3.9 Cube compressive strength of concrete for each curing time

with the theoretical ones evaluated in Table 3.6. It can be observed that the density of RAC was always lower than the one of concrete made with natural aggregates; this result was in agreement with the observation that recycled aggregates were characterised by higher percentages of internal pores.

Both Table 3.6 and Fig. 3.8 indicate that the RCA content affects the concrete density. In particular, the higher the percentage of substitution of natural aggregates with recycled ones, the lower the mean value of the concrete density; such a reduction in density is almost negligible in the case of concretes with only 30% of RCAs, whereas, the scatters between the minimum and maximum value increase with the amount of RCAs.

3.2.3 Compressive Strength

Figure 3.9 reports the mean values of the compressive strength evaluated by testing several cubic samples made of the 13 concrete mixes under consideration; tests have been performed at different curing times, ranging between 2 and 365 days.

The experimental results demonstrated that the replacement of natural aggregate with recycled ones significantly affects the compressive strength of concrete for each curing time. In particular, the compressive strength decreases for high amount of RCA used in substitution of natural aggregates. Nevertheless, the concrete strength of a RAC mix can be significantly improved by partially substituting the fine fraction of natural aggregate with fly ash. With reference to the case of "MR100" mix, the improvement of the concrete compressive strength due to fly ash contribution was considerable if compared with the mix "LR100". Compressive strength of mixes made with low amount of FA and natural aggregates was similar to the one achieved for the reference mixture (i.e. "N" and "LN"). However, in the above listed case, the 80 kg/m^3 of FA introduced as partial substitution of binder can be considered to be equivalent to the 30 kg/m^3 of cement type CEMI 42.5R removed. Furthermore, the substitution of 30% of natural aggregate with RCA and the use of low amount of FA (i.e. mix "LR30") resulted in concrete with compressive strength similar to the reference mixture ("N"). A significant reduction of the compressive strength was observed, in comparison with the reference mix (Fig. 3.9), when the amount of RCA used in substitution of natural aggregates increases (i.e. "LR60" and "LR100").

Figure 3.9 outlines that increasing amounts of FA could always mitigate the negative effects of RCA on the compressive strength. The replacement of 30% of NAs with RCAs together with the addition of a low amount of FA (namely, the mix "LR30") has led to a concrete having compressive strength comparable to that of the mixture taken as reference (i.e. the mix "N"). In fact, regardless the curing times, a difference not larger than 10% between the mean strength values of these two types of mixtures has been observed.

The concrete strength of mixtures with high content of RCA can be improved by partially substituting the finest portion of aggregates with FA (compare the strengths of RAC100% samples, i.e. mixes "LR100", "MR100" and "HR100"). Actually, compressive strengths of "MR60" mix reached higher values than those achieved for "LR60" mix. Finally, mixes "HN" and "HR30-60-100" were made with very low amount of cement. The same behaviour observed for the mixes discussed above can be observed. However, as it was expected, the high reduction in cement results in low compressive strength at short curing times (i.e. 2, 7 and 28 days). For longer curing times, mixes "HN", "HR60" and "HR100" develop compressive strengths similar to the ones achieved for mixes "MN", "MR60" and "MR100", respectively.

Figures 3.10, 3.11 and 3.12 highlight the time evolution of the compressive strength of the concrete mixtures under investigation. Particularly, they report the experimental data determined at 2, 7, 28, 60, 90 and 365 days by reporting the average value measured for each curing age. The analysis of the results points out the key roles of both FAs and RCAs on the resulting time evolution of the compressive strength.

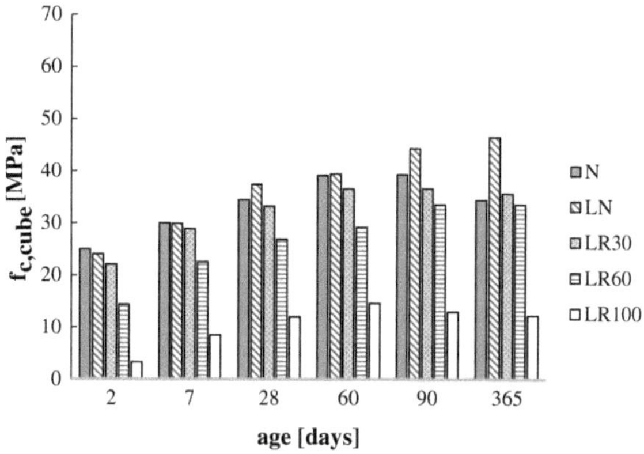

Fig. 3.10 Compressive strength of NAC and RAC with low content of FA

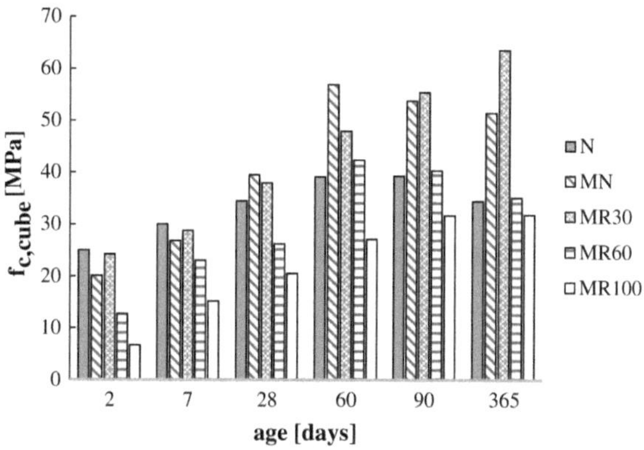

Fig. 3.11 Compressive strength of NAC and RAC with medium con-tent of FA

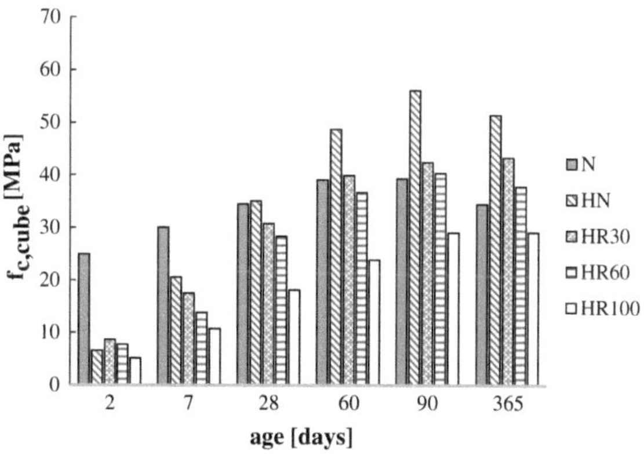

Fig. 3.12 Compressive strength of NAC and RAC with high content of FA

Natural aggregate concrete specimens made with FA in substitution of a part of Portland cement (i.e. "LN") have lower strengths than the reference ones ("N") at 2 and 7 days of curing, but they reach similar strength values at 28 days, probably due to the different times of hydration of the fly ash against the cement reaction (Fig. 3.10).

Mixes with high amount of FA (i.e. "MN" and "HN") highlight that the pozzolanic effects of fly ash enhance the compressive strength of concrete after 28 days of curing. Indeed, after 7 days of curing the compressive strength of mix "MN" was lower than the one achieved for mixes "N" and "LN", while at 28 days of curing they results in similar values. However, for curing times longer than 60 days, the compressive strength of mix "MN" was higher than ones developed in mixes without or with low amount of FA (Fig. 3.11).

Particularly, after 60 days of curing, samples made of "MN" and "HN" mixtures have exhibited average values of the compressive strength about 40% higher than those resulting from tests on samples made of "N" and "LN" mixtures (Figs. 3.11 and 3.12). This means that:

- the amount of fly ash exceeding the 33% C (3.2), did not behave as inert as it has produced a non-negligible increase of strength, though at longer curing times;
- even reducing the cement content well beyond the limits allowed by the European Standard, it is still possible to reach optimal levels of the compressive strength after 60 days (i.e. "HN and HRxx" mixes).

3.2.4 Tensile Strength

The tensile strength of the examined concrete were determined via splitting tests on cylindrical samples at 28 days. Two splitting tensile tests for each mix have been performed and the mean values f_{ctm} are depicted in Fig. 3.13. The mean values of the tensile strength are reported as a function of the RCA introduced in the concrete mixes.

Specimens only made with natural aggregate concrete NAs have shown a splitting strength higher than RACs. As for the compression, a progressive strength reduction has been observed in splitting tests by increasing the percentage of recycled aggregates in the mix.

Fig. 3.13 Tensile strength of concrete mixes

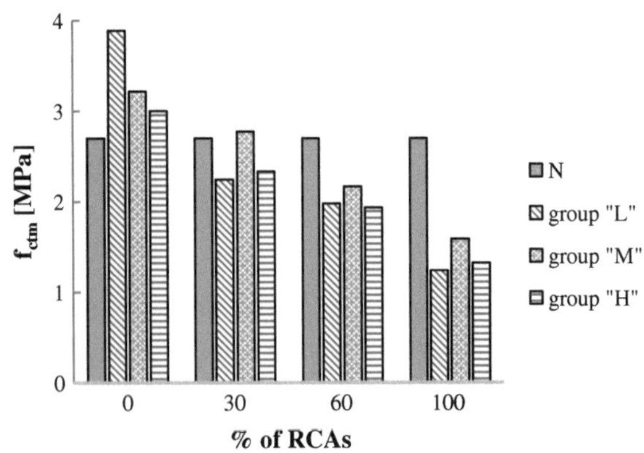

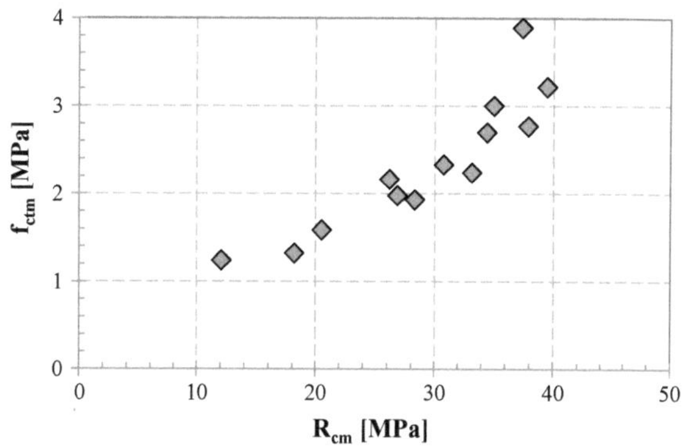

Fig. 3.14 Comparison between tensile and compressive strengths at 28 days of curing

The results reported in Fig. 3.13 outline that only mixes with less than 30% of substitution of natural aggregates with recycled ones developed values of tensile strength close to the ones of the reference concrete "N". Specimens having the 60% or 100% of RCAs content have shown a very strong reduction in the tensile strength at 28 days.

Finally, Fig. 3.14 relates the tensile and compressive strengths at 28 days of curing for the investigated concrete mixes. The quasi-linear correlation, observed in Fig. 3.14, outlined as the RCA affected both compressive and tensile strengths, in the same percentage.

Several contributions available in the scientific literature point out that the compressive and splitting tensile strengths of a RAC are mainly affected by the quality of the recycled aggregate, rather than by its quantity (Malesev et al. 2010; Tavakoli and Soroushian 1996). However, the experimental results of the herein reported investigation do not allow focusing on this topic, since only one type of RCA has been used in producing concretes. Nevertheless, the obtained results are in agreement with other previous studies, which already observed that the compressive and tensile strengths reduces in mixtures with higher RCA content (Thomas et al. 2013) (and this is the reason why the aggregate replacement ratio is often limited to approximately 30%).

3.2.5 *Bond-Slip*

Fifty-two 150 mm-edge cubes (i.e. four for each of the 13 mixes) were cast for each concrete mix, according to the geometric standards whose layout is depicted in Fig. 3.15. Steel rebars having diameter equal to 10 mm and anchorage length of 50 mm were considered. Direct pull-out tests were carried out at 28 days of curing.

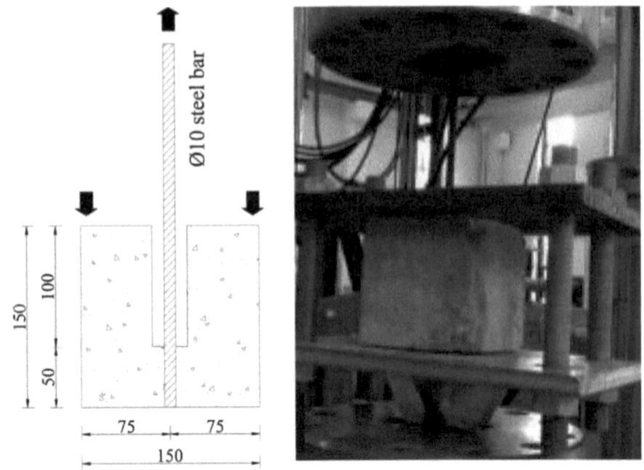

Fig. 3.15 Sketch of the pull-out specimen and picture of the test

Table 3.7 Bond-slip strengths

Mix	f_{bm} (MPa)
N	25.90
LN	17.10
LR30	19.10
LR60	14.40
LR100	11.50
MN	17.60
MR30	21.90
MR60	22.80
MR100	13.40
HN	12.20
HR30	19.70
HR60	17.70
HR100	6.50

The bond strength values have been calculated assuming a uniform stress distribution along the length L_b:

$$f_b = \frac{P}{\pi \cdot D_b \cdot L_b} \tag{3.6}$$

where P is the applied pull-out load and D_b the bar diameter.

Table 3.7 reports the average values of the bond-slip strength f_{bm} achieved on the basis of four specimens for each mix. The mixes made with recycled aggregates always developed bond-slip strengths lower than ones achieved by the reference mix (i.e. "N"). Moreover, the use of FA significantly improves bond-slip performance.

However, further experimental tests on bond-slip behaviour are required because the test set-up, depicted in Fig. 3.15 and suggested by RILEM 7-II-28 (1994), did not result in a realistic stress state due to the undesirable confined compression which take places between the surrounding concrete and steel bar (ACI 408R 2003). A more realistic test set-up in this topic is represented by the pull-out bending tests (RILEM 1982; Mazaheripour et al. 2013).

3.2.6 *Permeability*

Permeability is defined as the property that governs the rate of flow of a fluid into a porous solid. The overall permeability of concrete to water is a function of the permeability of the paste, the permeability and gradation of the aggregate, and the relative paste to aggregate proportion. Water permeability increases with the water–cement ratio and with the RCA content (Thomas et al. 2013). The durability of concrete is significantly affected by its permeability. Increased water tightness improves concrete resistance to re-saturation, sulphate and other chemical attack, and chloride ion penetration. Permeability also affects the destructiveness of saturated freezing.

Permeability tests performed according to EN 12390-8 (2000) were aimed to evaluate the average height of water penetration (Water Penetration Depth WPD) in cube specimens under a pressure of 5 bar at 90 and 365 days of curing age. After 72 h cubes were split in two parts and the wet profile was measured.

Twenty-six permeability tests were performed at 90 and 365 days (13 + 13 tests) according to aforementioned procedure: Fig. 3.16 shows the experimental setup and two specimens after test.

The average depth of the water penetration, h_m, was evaluated as the ratio between the area of the wet surface and the width of the cubic sample: of course, higher values of h_m correspond to higher permeability and, hence, concretes that are supposed to be more easily affected by degradation phenomena. Figure 3.17 shows

Fig. 3.16 Some concrete specimens during and after the permeability tests

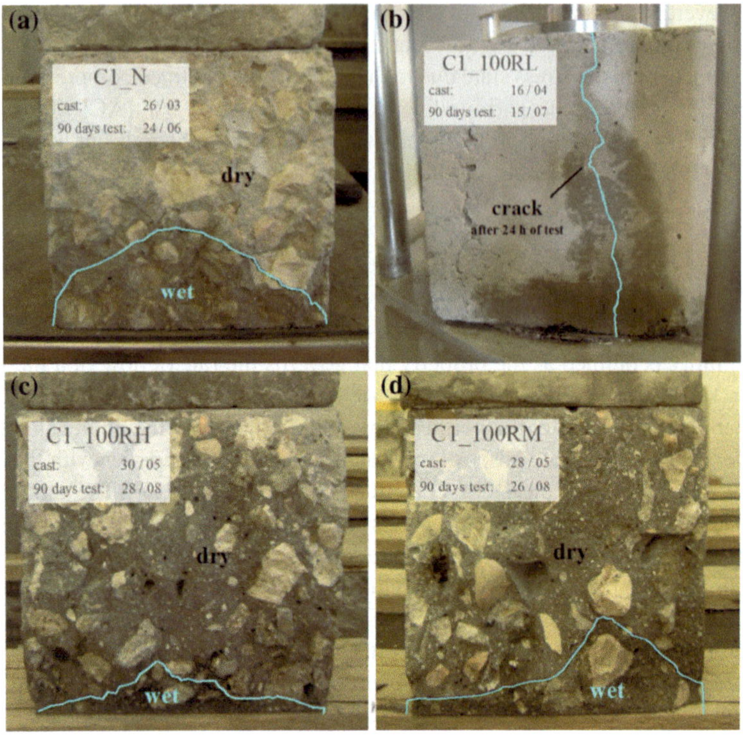

Fig. 3.17 Wet profiles derived by permeability tests

the contour lines describing of the wet surface resulting at the end of some of these permeability tests.

The permeability test on the mix "LR100" made with 100% of RCA and low amount of FA returned very negative values at both 90 and 365 days. Furthermore, the tests were interrupted after about 24 h because due to the presence of some microcracks, subsequently enlarged by the action of the water under pressure, the water completely penetrated through specimens (Fig. 3.17b).

Figure 3.18 summarises the results of permeability tests in terms of average depth of water penetration h_m, evaluated as the ratio between the wet area and the cube width (150 mm). Higher values of h_m denote lower durability of concrete as it results more porous and thus more subject to degradation. As expected, the use of RCA in concrete mixes affects durability mainly due to the higher percentage of internal pores. However, almost all the concrete mixtures produced with RCAs and FA show a lower water permeability capacity in comparison with the "ordinary" control mixture *N*, both at 90 and 365 of curing (the only exception are the mixtures *LN* and *LR60* that, at 90 days, show almost the same permeability of the reference mixture). The use of FA improves the durability properties of concrete. In particular, a higher amount of FA led to a significant reduction in the average height of water penetration.

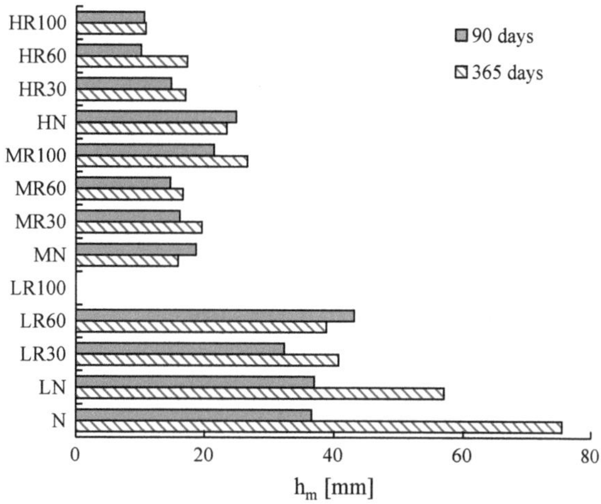

Fig. 3.18 Results of permeability test

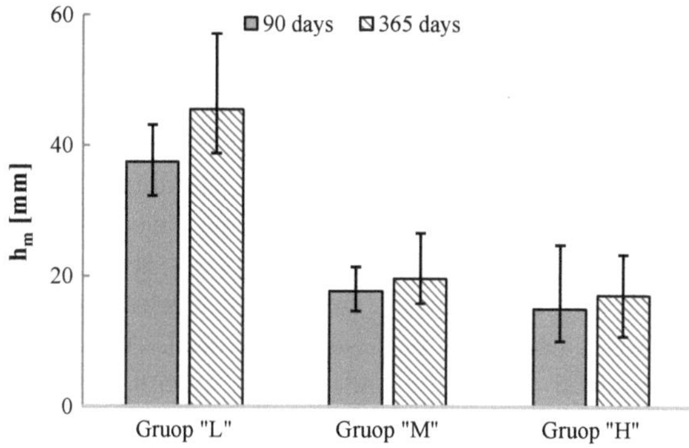

Fig. 3.19 The influence of RCAs and FAs on water penetration depth for concrete

The role of FA clearly emerges from these experimental observations. It influences the microstructure of paste and enhances the water tightness of the concrete, also counterbalancing the expected increase in porosity and, hence, water permeability due the presence of RCAs. As a matter of principle, the lower mean value of the nominal diameter characterising the grain size distribution of fly ash compared with cement particles, leads to a reduction of voids inside the concrete microstructure, which directly affects the water permeability more than the aggregate replacement. Therefore, concrete mixes containing the highest amounts of FA are characterised by the lowest water permeability.

As regards RCAs, it is well known that they are generally more porous than natural aggregates (de Juan and Gutierrez 2009) but nevertheless, despite some studies have demonstrated that the concrete permeability increases by substituting amounts of NAs with RCAs, Fig. 3.19 show that in this research the influence of recycled aggregates is negligible. In fact, Fig. 3.19 reports the average value of water permeability measured, both at 90 and 365 days, for each group mixtures characterised by the same total amount of FA (i.e., L, M and H). Moreover, the error lines represent the scatter of results obtained on mixtures characterised by the same total amount of binders, but different amount of RCAs. Since this scatter is reasonably limited, one can argue that the resulting water permeability is relatively more affected by the content of FA (that, in fact, leads in a more compact concrete paste) than RCAs.

3.2.7 Resistance to Chlorides

Chloride ion penetration is one of the major problems that affect the durability of reinforced concrete structures. Although chloride ions do not directly cause important damage in concrete, they contribute to the corrosion of steel rebars embedded in the structures. Moreover, there are interactions between chloride ions and hydration products of the cement, which may contribute to the development of damage under frost conditions.

The durability of the concrete mixes exposed to chlorides was investigated. The primary initial interest was to evaluate the durability of concrete mixes with RCA and FA in seaside applications. The resistance to chloride ion penetration was determined through Rapid Chloride Penetration tests (RCPT) according to ASTM C1202-12 (2012). To this end, cylindrical specimens (102 mm diameter and 50 mm high) were cast (Fig. 3.20b). Twenty-six specimens (two per mix) were tested at 90 days and other twenty-six at 365 days of curing for a total of 52 concrete cylinders. The amount of electric current passed through the cylinder during a period of 6 h was monitored. A potential difference of 60 V dc was maintained across the ends of the specimens, one of which was immersed in a sodium chloride solution, the other in a sodium hydroxide solution (Fig. 3.20c).

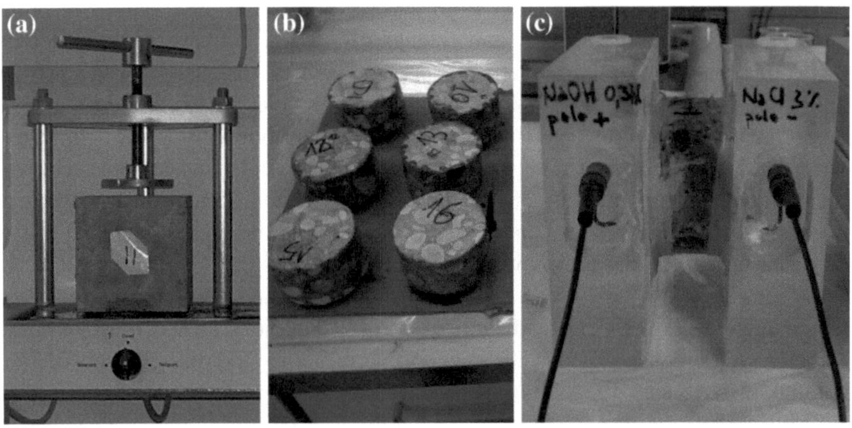

Fig. 3.20 Specimens for investigating the durability properties of the considered concretes

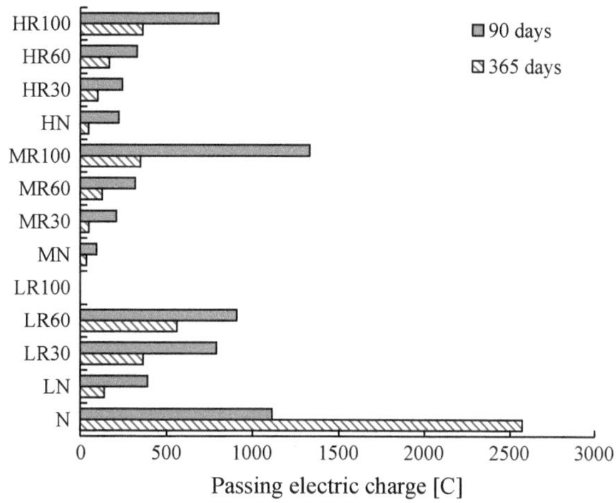

Fig. 3.21 Results of chloride diffusion tests

Table 3.8 Correlation between passing electric charge and chloride diffusion according to ASTM C 1202-12 (2005)

Electric charge (C)	Diffusion of chloride
>4000	High
2000–4000	Moderate
1000–2000	Low
100–1000	Very low
<1000	Negligible

The amount of electric current passing through the cylinder during a period of 6 h was monitored. Generally, a very low diffusive capacity of chlorides characterized the more compact concretes.

Figure 3.21 shows the electric charge (in Coulombs) measured for the mixes under investigation. Thus, when a low electricity charge passes through the specimen, a higher resistance to chlorides characterises the concrete under consideration. However, test results can also be examined by considering the qualitative correlation between the passing electrical charge and the diffusion of chlorides as reported in Table 3.8. Fly ash used in partial substitution of cement led to enhancing the resistance to chloride diffusion in concrete. This positive trend can easily be observed in Fig. 3.21 by comparing the results of mixes “LN”, “MN” and “HN” with the ones of the reference mix “N”. Moreover, the use of RCA in concrete provide low resistance to the chloride diffusion (i.e., mix “LR100” in Table 3.3).

Furthermore, all the concrete mixtures containing FA show resistances to the chloride-ion penetration at 365 days remarkably lower than those achieved at 90 days. This can be attributed to the aforementioned delayed reaction of FA that contributes to make concrete mortar less permeable to chlorides.

Finally, the obtained experimental results clearly indicate that the water permeability data cannot be used to predict the concrete resistance to the chloride-ion penetration and vice versa, especially when RCAs are used in concrete. As already evidenced in the technical literature, this discrepancy may be related to the different transport mechanisms involved in the water permeability test and in the chloride penetration tests (Chia and Zhang 2002).

3.2.8 Carbonation Depth

Carbonation is a major cause of deterioration in concrete structures: although it is not a detrimental phenomenon for the concrete itself, it is recognised for triggering corrosion in steel reinforcement (Bertolini et al. 2013).

The carbonation rate is controlled by the increase in CO_2 concentration inside the concrete pores which, in turn, is influenced by the relative humidity of concrete. As a general trend, within 1–2 days the surface of fresh concrete reacts with CO_2 and, then, the process penetrates deeper into the concrete at a rate proportional to the square root of time; after a year carbonation typically affects depths from about 1 mm (concrete of low permeability made with a low w/b ratio) up to 5 mm or more (porous and permeable concrete made characterised by a high w/b) (Pacheco Torgal et al. 2012).

Although the carbonation is often considered less serious than the chloride-ions penetration, it is much more widespread since it involves carbon dioxide in the air. Furthermore, carbonation coupled with chloride-ion penetration reduces the durability of concretes exposed to marine environment (Mather 2004).

The most widely used method for assessing carbonation depth consists in spraying concrete broken faces with a phenolphthalein indicator solution (Pacheco Torgal et al. 2012). In this study carbonation tests were performed on standard concrete cubes (150 mm): two specimens per mixture were cast and cured in controlled conditions (in water at 20 ± 2 °C) for 28 days. Subsequently, the cubes were placed outside at environmental conditions and tested at 90 and 365 days.

At the end of curing age, specimens were split into two parts and a phenolphthalein solution (the indicator was obtained by dissolving 1% phenolphthalein in isopropyl alcohol) was applied on the broken surfaces. The "affected depths"—i.e. the distances of the coloured parts from the external surface—were readily shown and were easily measured (Fig. 3.22).

Finally, the carbonation depth D was evaluated as follows:

$$D = \frac{A_1 + A_2 + B_1 + B_2 + C_1 + C_2 + D_1 + D_2}{8} \quad (3.7)$$

The pictures collected in Figs. 3.23, 3.24 and 3.25 highlight the resulting carbonation depth D of the tested samples. Particularly, they show the cross-sections of specimens made with low, medium and high content of FA, respectively.

Fig. 3.22 Carbonation test

Figure 3.26 shows the obtained carbonation depths, determined by means of Eq. (3.7). It is worth highlighting that the large variability in the constituents of the thirteen mixtures (i.e. percentage of RCAs; amount of FA; water content-cement ratio; amount of Portland cement) does not allow to easily identify the influence of some of them in influencing the carbonation level. As expected Fig. 3.26 confirms that carbonation is higher for concrete specimens tested after 365 than 90 days of curing. All the mixtures made with RCAs and FAs were generally affected by a higher value of carbonation depth with respect to the control mixture *N*. In addition, the presence of RCAs seems to play a role in the carbonation resistance.

Moreover, as well know from the literature, the carbonation depth D can be correlated with time through the following equation (Silva et al. 2015a, b)

$$D = k_c \bullet \sqrt{t} \tag{3.8}$$

where k_c is the carbonation factor and t is the air exposure time (expressed in days). The carbonation factor was determined for all the produced mixtures and, for instance, Fig. 3.27 reports the calibration determined for some of the employed mixtures, whereas Table 3.9 reports the carbonation factor calculated for each mixture.

Finally, the emerging correlation between the carbonation factor and the amount of FA and RCA confirms the above statement (Fig. 3.28).

Fig. 3.23 Natural carbonation at 90 and 365 days of concrete samples containing low amounts of FA

Fig. 3.24 Natural carbonation at 90 and 365 days of concrete samples containing medium amounts of FA

Fig. 3.25 Natural carbonation at 90 and 365 days of concrete samples containing highest amounts of FA

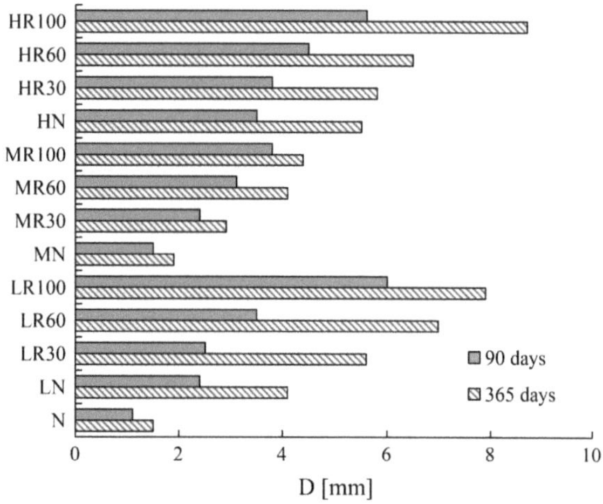

Fig. 3.26 Carbonation depths of all concrete specimens

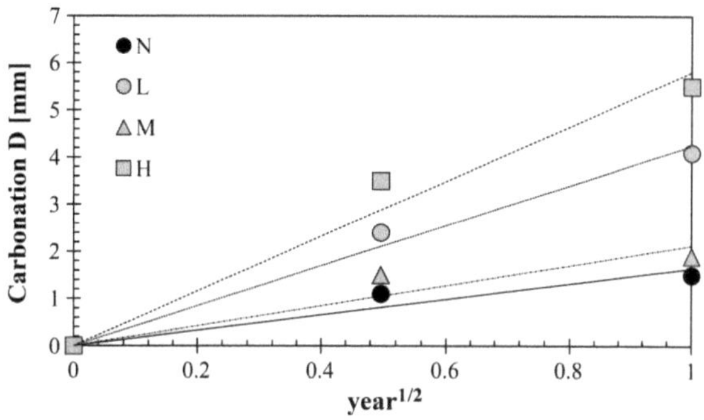

Fig. 3.27 Calibration of the carbonation coefficient for the concrete mixtures

Table 3.9 Carbonation factors

Mix	D (mm)		K_C	R^2
	90 days	365 days		
N	1.1	1.5	1.64	0.92
LN	2.4	4.1	4.25	0.99
LR30	2.5	5.6	5.49	1.00
LR60	3.5	7.0	7.01	1.00
LR100	6.0	7.9	8.73	0.90
MN	1.5	1.9	2.22	0.88
MR30	2.4	2.9	3.28	0.85
MR60	3.1	4.1	4.52	0.90
MR100	3.8	4.4	5.04	0.82
HN	3.5	5.5	5.81	0.97
HR30	3.8	5.8	6.17	0.96
HR60	4.5	6.5	7.01	0.94
HR100	5.6	8.7	9.21	0.97

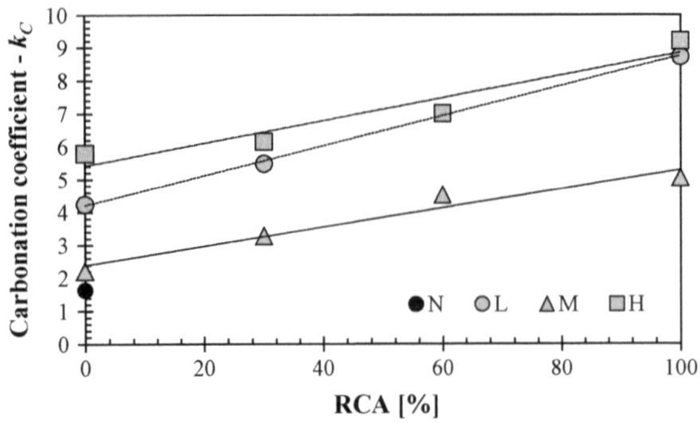

Fig. 3.28 Influence of RCAs and FAs on the carbonation coefficient for concrete

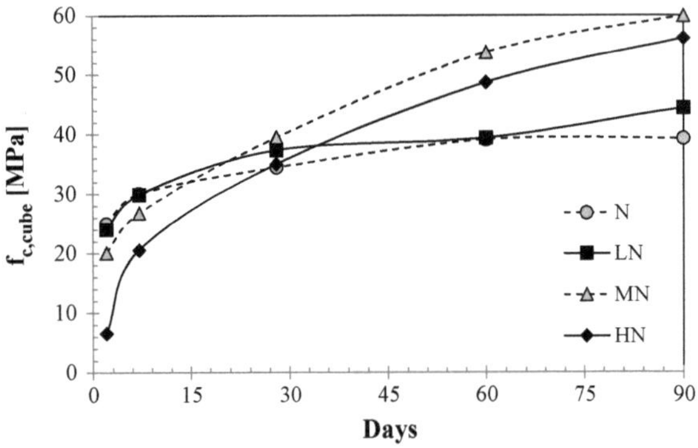

Fig. 3.29 Effects of fly ash on the development of compressive strength

3.2.9 Analysis of Results and Evolution of the Compressive Strength

Figure 3.29 shows the time evolution of the average cube compressive strength. The comparison was carried out for mixes made with increasing amounts of FA in substitution of Portland cement and without recycled aggregates. The experimental results on mixes "N", "LN", "MN" and "HN" demonstrated as concrete with the addition of FA attained higher compressive strengths than the referenced mixes (e.g. "N") for longer curing time (generally longer than 28 days). Moreover, the reduction of cement content in mixes with FA led to a loss in compressive strength for curing times shorter than 28 days. The effect of FA outlined above should be carefully taken into account in engineering applications.

The prediction of the time evolution of the average cubic compressive strength $f_{c,cube}$ of concrete is a relevant issue for RACs. An analytical formulation recently proposed for this purpose was developed by Malesev et al. (2010):

$$f_{c,cube} = \frac{a \cdot t}{b + t} \tag{3.9}$$

Table 3.10 Values of the a and b coefficients and of R^2

MIX	a	b	R^2
N	39.70	2.42	0.862
LN	41.47	1.79	0.880
LR30	36.33	1.42	0.959
LR60	31.57	2.63	0.934
LR100	14.86	5.82	0.971
MN	58.88	7.05	0.870
MR30	50.59	3.63	0.786
MR60	41.69	6.20	0.854
MR100	32.05	9.30	0.943
HN	65.86	19.80	0.983
HR30	47.29	12.17	0.991
HR60	46.73	16.32	0.993
HR100	31.98	16.46	0.970

in which *a* and *b* represent two coefficients that should be calibrated on the basis of compressive experimental tests and *t* is the curing age. The values of the coefficients *a* and *b* derived for the mixes under investigation are listed in Table 3.10 in which also the R^2, values representing the correlation index between the experimental and the theoretical values of the compressive strength, are reported. Indeed, values of R^2 close to the unit represent good agreement between the experimental evidence and the values evaluated with the analytical formulation provided by Eq. (1.9).

The graphics plotted in Fig. 3.30 show the evolution of the cube compressive strength with the curing age of the concrete mixes under investigation compared with the values achieved through the analytical formulation (1.9) and the coefficients outlined in Table 3.10. In particular, Fig. 3.30 depict the results obtained for mixes made without, with 30, 60 and 100% of recycled aggregates. It was clearly stated that the theoretical formulation with the coefficients *a* and *b*, evaluated for each mix, was able to well approximate the experimental evidence.

Figure 3.31 highlights the trend of the parameter *a* (on the vertical axis) evaluated for concrete mixes under investigation against the amount of RCA (on the horizontal axis) used as substitution of natural aggregate. In particular, the horizontal axis reports the mixes made with 30, 60 and 100% of recycled aggregates in which “y” should be substitute with “L”, “M” and “H” for taking back concrete mixes produced with low, medium and high amount of FA (Table 3.3), respectively.

Furthermore, with the same criteria adopted in Figs. 3.31 and 3.32 shows the trend of the parameter *b* (on the vertical axis) with respect to the increasing amount of FA (on the horizontal axis) used in concrete mixes. As expected, a higher amount of FA results in higher values of *b*. As a matter of fact, by observing the structure of Eq. (1.9), high values of *b* involve low compressive strengths at short curing age.

Finally, it can be stated that the coefficient *a* affects the value of the maximum compressive strength, while *b* controls the gradient of the first branch of the

Fig. 3.30 Theoretical evaluation of the compressive strength with the curing age

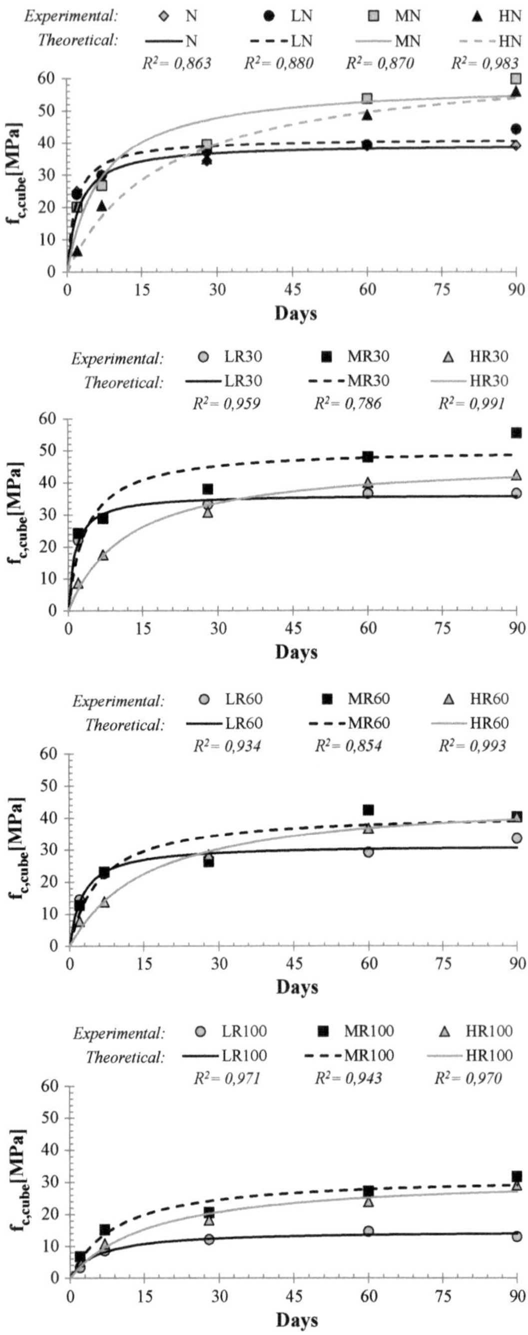

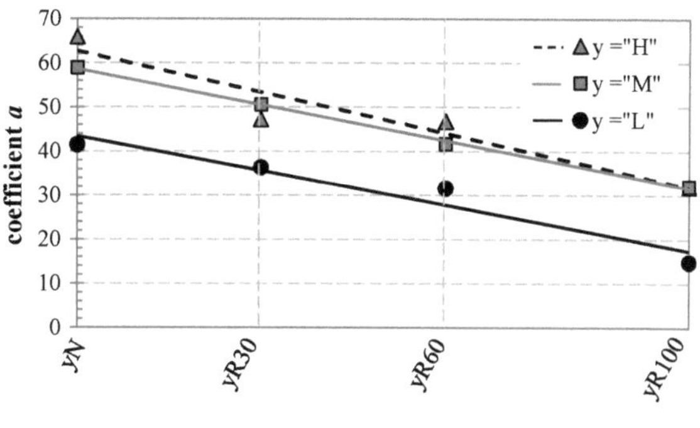

Fig. 3.31 State of the coefficient a in comparison with the amount of RCA in concrete mixes

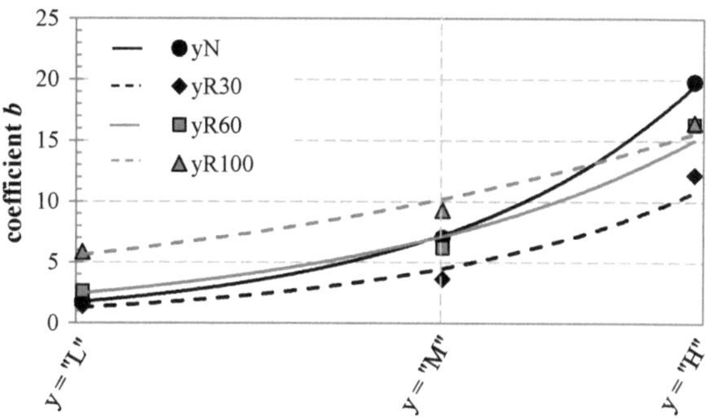

Fig. 3.32 State of the coefficient b against the amount of FA in concrete mixes

relationship relating the curing age and the compressive strength. Moreover, high values of *b* well represent concrete mixes with high amounts of FA, used in substitution of cement as they develop low compressive strengths at short curing time.

3.3 Concluding Remarks

The behaviour of concrete with both recycled aggregates (up to the total replacement of natural ones) and fly ash (mainly adopted as a filler) was investigated in the reported experimental campaign.

As a matter of fact, both in term of mechanical and durability performance, the mixtures produced with even high amount of RCAs and FAs show acceptable results for structural application requirements. The analysis of the results highlighted that the weaker mechanical and durability behaviour of concrete made by RACs, generally induced by the high porosity of the recycled aggregates, can be correlated with the higher amount of water/binder ratio that derive from the standard procedures adopted for structural concrete production. The water permeability of concrete is mainly governed by the paste characteristics than the aggregate ones;

moreover, the presence of FA in concrete reduces the permeability of the composite and, in addition, the presence of high volume of FA mitigates the possible detrimental effects due to the use of RCAs.

The chloride ions penetration is affected by the presence of RCA in concrete, but, at the same time, the use of FA enhances the resistance to chloride-ion penetration. The carbonation depth is mainly controlled by the paste characteristics: it is higher for higher values of the effective water-to-binder ratio defined in this study.

Also concrete samples obtained by reducing the cement content (and, correspondingly adding fly ash) far beyond the limits allowed by the current European Standards have been investigated. Although these concretes have shown lower mechanical and physical properties, a possible widening of such limits is still possible and should be implemented in the future versions of European Standards, as a further driver towards the production of more sustainable structural concretes.

References

ACI 408R-03 (2003) Bond and development of straight reinforcing bars in tension. American Concrete Institute

ASTM C1202-12 (2012) Standard test method for electrical indication of concrete's ability to resist chloride ion penetration. ASTM International, West Conshohocken, PA. www.astm.org

ASTM C127-15 (2015) Standard test method for relative density (specific gravity) and absorption of coarse aggregate. ASTM International, West Conshohocken, PA. www.astm.org

ASTM C128-15 (2015) Standard test method for relative density (specific gravity) and absorption of fine aggregate. ASTM International, West Conshohocken, PA. www.astm.org

ASTM C618-15 (2015) Standard specification for coal fly ash and raw or calcined natural pozzolan for use in concrete. ASTM International, West Conshohocken, PA. www.astm.org

Bertolini L, Elsener B, Pedeferri P, Redaelli E, Polder RB (2013) Corrosion of steel in concrete: prevention, diagnosis, repair. John Wiley & Sons

Bolomey J (1947) The grading of aggregate and its influence on the characteristics of concrete, Rev Matér Constr Trav Publ, 147–149

Chia KS, Zhang MH (2002) Water permeability and chloride penetrability of high-strength lightweight aggregate concrete. Cem Concr Res 32:639–645

de Juan MS, Gutiérrez PA (2009) Study on the influence of attached mortar content on the properties of recycled concrete aggregate. Constr Build Mater 23(2):872–877

EN 12390-3 (2002) Testing hardened concrete—part 3: compressive strength of test european committee for standardization, Brussels, BE

EN 12390-4 (2000) Testing hardened concrete—part 4: compressive strength specification for testing machines. European Committee for Standardization, Brussels, BE

EN 12390-8 (2000) Testing hardened concrete—part 8: depth of penetration of water under pressure. European Committee for Standardization, Brussels, BE

EN 12390-6 (2000) Testing hardened concrete—part 6: tensile splitting strength of test specimens. European Committee for Standardization, Brussels, BE

EN 12620 (2002) Aggregates for concrete. European Committee for Standardization, Brussels, BE

EN 13242 (2008) Aggregates for unbound and hydraulically bound materials for use in civil engineering work and road construction. European Committee for Standardization, Brussels, BE

EN 197-1 (2001) Cement: composition, specifications and conformity criteria for common cements. European Committee for Standardization, Brussels, BE

EN 206-1 (2006) Concrete—part 1: Specification, performance, production and conformity. European Committee for Standardization, Brussels, BE

EN 450 (2005) Fly ash for concrete. Definition, specifications and conformity criteria. European Committee for Standardization, Brussels, BE

Faella C, Lima C, Martinelli E, Pepe M, Realfonzo R (2016) Mechanical and durability performance of sustainable structural concretes: an experimental study. Cement Concr Compos 71:85–96

Lima C, Caggiano A, Faella C, Martinelli E, Pepe M, Realfonzo R (2013) Physical properties and mechanical behaviour of concrete made with recycled aggregates and fly ash. Constr Build Mater 47:547–559

Malesev M, Radonjanin V, Marincovic S (2010) Recycled concrete as aggregate for structural concrete production. Sustainability 2:1204–1225

Mather B (2004) Concrete durability. Cement Concr Compos 26(1):3–4

Mazaheripour H, Barros JAO, Sena-Cruz JM, Pepe M, Martinelli E (2013) Experimental study on bond performance of GFRP bars in self-compacting steel fiber reinforced concrete. Compos Struct 95(2013):202–212

NTC (2008) DM 14.01. 2008: Norme Tecniche per le Costruzioni. Italian Ministry of Infrastructures and Transportation (in Italian)

Pacheco Torgal F, Miraldo S, Labrincha JA, De Brito J (2012) An overview on concrete carbonation in the context of eco-efficient construction: evaluation, use of SCMs and/or RAC. Constr Build Mater 2012(36):141–150

RILEM MR-4 recommendation (1982) RILEM recommendations for the testing and use of constructions materials

RILEM 7-II-28 recommendation (1994) Specification for concrete with recycled aggregates. Mater. Struct. 27(173):557–559

Silva RV, de Brito J, Neves R, Dhird R (2015a) Prediction of chloride ion penetration of recycled aggregate concrete. Mater Res 18(2):427–440

Silva RV, Neves R, de Brito J, Dhir RK (2015b) Carbonation behaviour of recycled aggregate concrete. Cement Concr Compos 62:22–32

Tavakoli M, Soroushian P (1996) Strengths of recycled aggregate concrete made using field-demolished concrete as aggregate. ACI Mater J 93(2) [March–April]

Thomas C, Setién J, Polanco JA, Alaejos P (2013) Juan MSD. Durability of recycled aggregate concrete. Constr Build Mater 40:1054–1065

Chapter 4
Insights into the Triaxial Behaviour of Recycled Aggregate Concrete

Paula Folino and Hernán Xargay

Abstract In this chapter, the results of an experimental programme aiming to investigate the incidence of different replacements ratios of natural coarse aggregates by recycled ones on the mechanical behaviour of Recycled Aggregate Concrete (RAC) under tri-axial compression stress states are presented and discussed. The experimental campaign performed at the University of Buenos Aires considers four concrete mixtures including a reference or parent concrete and RACs with 30, 60 and 100% of coarse aggregate replacements. Besides the aforementioned tri-axial compression tests, the results of splitting tensile and uniaxial compression experimental tests are also presented. It is showed that increasing levels of recycled aggregates lead to concretes more sensitive to confinement.

Concrete is usually characterised by a stress-strain relationship obtained from uni-axial compression tests. There is no doubt that the uni-axial compressive strength is an essential property of concrete, as it is sufficient to model the behaviour of 1D structural members (such as beams, beam-columns and struts) which are among the most common in reinforced concrete (RC) structures. Moreover, from a "practical" standpoint, it can be easily determined in laboratory tests and it is generally assumed as the key parameter to describe all the relevant mechanical properties of structural concrete (Folino et al. 2009; Folino and Etse 2011).

However, as a matter of fact, a significant wide class of RC members and structures are subjected to multi-axial stress states. Besides 2D and 3D members such as plates or block foundations, respectively, tri-axial stress states can be also activated in common 1D members, such as beam-columns as a result of the confinement effect induced by either steel stirrups or steel/FRP jackets.

Figure 4.1 shows the results obtained from a uni-axial compression test performed under displacements control on a cylinder specimen with diameter of 100 mm and height of 200 mm (Lu 2005). Particularly, it can be observed that the

P. Folino (✉) · H. Xargay
Universidad de Buenos Aires, Facultad de Ingeniería,
LMNI-INTECIN (UBA-CONICET), Buenos Aires, Argentina
e-mail: pfolino@fi.uba.ar

J.A.O. Barros et al. (eds.), *Recent Advances on Green Concrete for Structural Purposes*, Research for Development, DOI 10.1007/978-3-319-56797-6_4

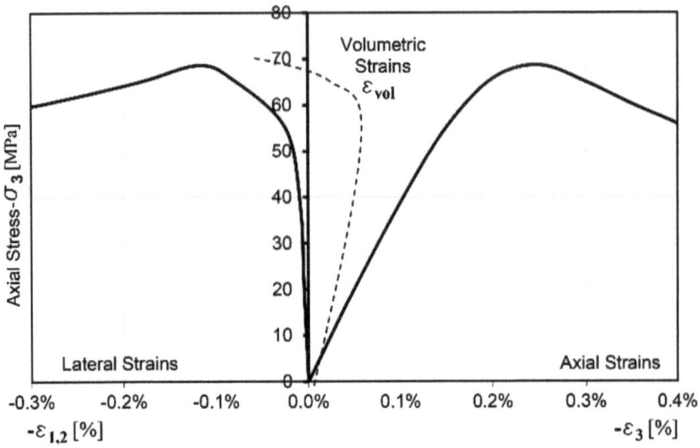

Fig. 4.1 Typical stress-strain relationship obtained from uniaxial compression test. Data extracted from (Lu 2005)

axial stress versus axial strains relationship presented on the right side of the figure, exhibits the three following behaviour zones

- a linear elastic zone when stresses are very low;
- a hardening zone characterised by plastic strains;
- a softening branch after the peak stress is exceeded.

On the left side of Fig. 4.1 the axial stress versus lateral strains relationship is presented. It can be observed that while a shortening occurs in axial direction, a dilation takes place in transverse direction. An important feature of concrete behaviour can be observed by analysing the evolution of the volumetric strains as defined by Eq. (4.1) and represented by a dashed line in Fig. 4.1:

$$\varepsilon_{vol} = \varepsilon_1 + \varepsilon_2 + \varepsilon_3 \tag{4.1}$$

being εi, with i = 1 to 3, the strain corresponding to principal axis i.

Up to a certain level of stress (called "critical stress"), volumetric strains are negative and increase in absolute value, meaning that volume decreases. A turning point can be observed on the dashed curve at such a critical stress, beyond which volumetric strains decrease in absolute value, even reaching positive values in some cases. This phenomenon is called dilatancy and mainly depends on concrete composition. It is strongly influenced by the coarse aggregate volume content (Shah and Chandra 1968), being higher for higher coarse aggregate contents. Furthermore, dilatancy is also influenced by concrete quality considering that high strength concretes present less dilatancy than that of normal strength concretes (van Mier 1997; Folino and Etse 2012).

When concrete is subjected to tri-axial compression, dilatancy is reduced or directly avoided, depending on the applied confining pressure. This implies that a greater stress is needed to cause material failure, with the consequence of an increment in the material strength. On the contrary, a small lateral tensile stress would reduce concrete strength. Regarding these features, concrete failure surface is complex to be fully determined, but necessary to predict failure and behaviour of concrete structures.

Figures 4.2 and 4.3 show typical results of triaxial compression tests performed on concrete cubic samples of 100 mm height (van Geel 1998). Figure 4.1 shows that not only peak stress increases with confinement but also ductility is larger as the confinement pressure increases. The softening branch is less pronounced for higher levels of confinement. Moreover, in Fig. 4.3, the evolution of the volumetric strains is plotted demonstrating that dilatancy tends to decrease with increasing confinement as it was above mentioned.

As a matter of fact, a state of compressive confinement applied to concrete members, results in a significant enhancement of their structural performance, especially in terms of ductility. Recent studies reported the properties of recycled aggregate concrete confined by outer tubes (state of passive confinement): e.g., steel tubes or fibre reinforced polymer (FRP) tubes (Xiao et al. 2012a, b).

Since the research on confined RAC is currently at an "early stage", there is a lack of experimental results related with the mechanical behaviour under tri-axial compression stress states of this concrete type and its possible correlations with the mix properties (i.e., in terms of aggregate replacement ratio). Tri-axial compression tests are relevant for the calibration of constitutive models. It is important to point out the relevance that the development of constitutive formulations that accurately predict the mechanical behaviour of structures built with new materials has for structural engineering in order to ensure adequate safety factors. This was the key motivation for the experimental programme developed at the Faculty of

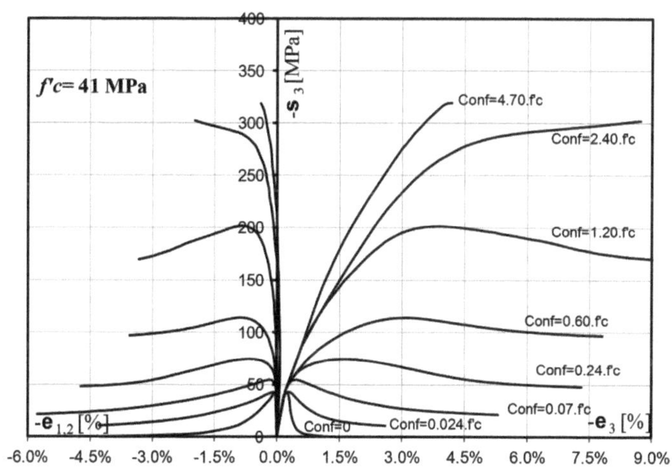

Fig. 4.2 Stress-strain relationships obtained from triaxial compression tests under different levels of confinement (van Geel 1998)

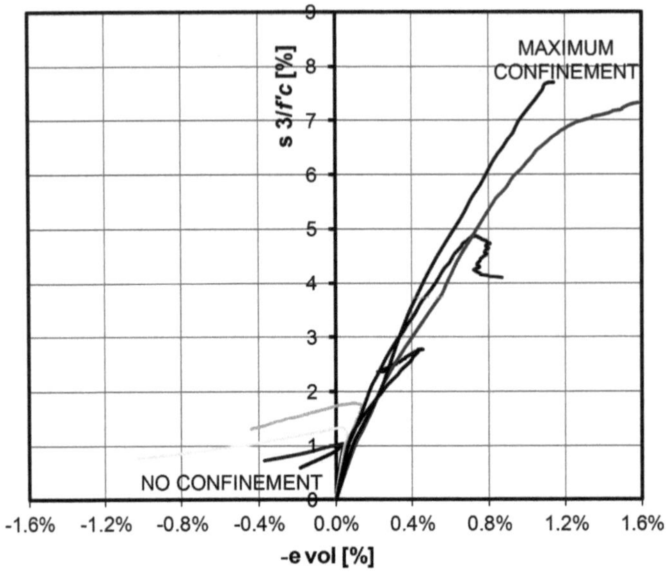

Fig. 4.3 Stress-volumetric strain relationships obtained from tri-axial compression tests under different levels of confinement (van Geel 1998)

Engineering of the University of Buenos Aires (FIUBA) with the purpose of analysing the mechanical behaviour of RAC under active tri-axial compression states and different levels of confinement.

In this section, the results of the FIUBA experimental program, including uni-axial compression, tri-axial compression and splitting tensile tests on RAC are presented. A standard concrete with a target mean compressive strength of 35 MPa was designed, produced and subjected to different experimental tests. Then, it was crushed and reused to produce RAC. Three different replacement percentages *R* of natural coarse aggregates by recycled ones were considered. Results related to the physical properties of recycled aggregates, fresh concrete properties, and mechanical behaviour are addressed and discussed.

4.1 The Experimental Campaign

The aim of the experimental campaign was to evaluate the degradation in main properties when recycled concrete aggregates are used. Four different concrete types were analysed, including a reference standard concrete and three RACs. Mix proportioning of RACs was not done considering a target compressive strength but maintaining almost the same mix proportioning of that of the standard concrete. The following concrete types were considered

- NAC: a reference concrete with natural coarse aggregates and a target 28 days mean compressive cylinder strength of 35 MPa.
- RAC30%: RAC with natural and recycled coarse aggregates (R = 30%)
- RAC60%: RAC with natural and recycled coarse aggregates (R = 60%)
- RAC100%: RAC with only recycled coarse aggregates (R = 100%)

being R the replacement percentage (by volume) of natural coarse aggregates by recycled ones.

4.1.1 Materials

Fine aggregates were constituted by a combination of siliceous river sand with a fineness modulus FM = 2.00 and granitic crushed sand with FM = 3.82. The resulting combination had a final FM = 2.84. No recycled fine aggregates were used.

High early-strength Portland cement—analogous to Type III according to ASTM C150/C150M-12 (2012) having a 28 days compressive strength of 50 MPa was used. A water-reducing additive (naphthalene-based) was incorporated during the mixing operations. No mineral admixtures were used.

Granitic crushed stone was used as natural coarse aggregate (see Fig. 4.4). Recycled coarse aggregates in RACs were obtained by the crushing of concrete specimens corresponding to the NAC that had already been tested. Therefore, in this research the recycled aggregates parent concrete properties were known. The crushing process was carried out at an external laboratory. Recycled aggregates passing sieve size 4.75 mm were discarded. Nominal maximum coarse aggregate size was Φ_{max} = 19 mm in all cases.

Fig. 4.4 Granitic natural coarse aggregates

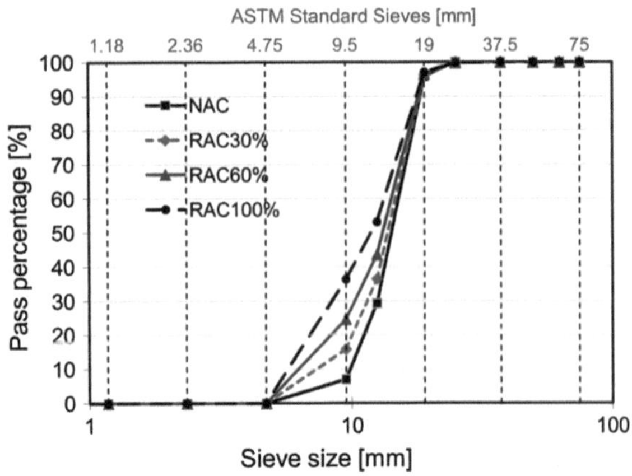

Fig. 4.5 Particle size distribution of coarse aggregates

Table 4.1 Water absorption and density of aggregates

	Absorption (%)	Density (kg/m^3)
Fine aggregates		
River sand	0.26	2620
Crushed sand	0.89	2630
Coarse aggregates		
Natural	0.31	2730
Recycled	2.69	2570

4.1.2 Physical Properties of Recycled Coarse Aggregates

The particle size distribution of coarse recycled aggregates obtained as result of the sieve analysis is presented in Fig. 4.5. Standard sieves according to ASTM C33/C33M-13 (2013) were used. It can be observed that although the recycled aggregates were obtained from the NAC, the resulting combinations of natural and recycled aggregates led to different particle size distributions, with decreasing values of FM for increasing replacement percentages *R*, being the final FM values: 6.97 for the NAC, 6.88 for the RAC30%, 6.79 for the RAC60% and 6.67 for the RAC100%. Considering that a smaller fineness modulus FM indicates a greater surface area of aggregates, concretes elaborated with the above detailed combinations of natural and recycled coarse aggregates are expected to be less workable than NAC, particularly regarding that the free water has not been modified in this experimental campaign. Furthermore, workability is also affected by the rougher surface of recycled aggregates in comparison of that of natural aggregates (Neville 1995).

The results of absorption properties and density of aggregates are presented in Table 4.1. Recycled coarse aggregates exhibited a lower density, and consequently a greater porosity than natural aggregates, coinciding with the general tendency observed in different research works.

Since standard methods for determining water absorption of coarse aggregates were found to be inappropriate in the case of recycled aggregates, as it is explained in the next paragraph, the absorption properties of these aggregates were determined by applying the hydrostatic weighing approach proposed in Djerbi (2012).

The first step to apply standard methods for determining water absorption is to obtain a saturated surface-dry (SSD) condition (ASTM C127-01 2001). For this purpose, after a soaking period of 24 h, aggregates are towel-dried to eliminate surface water. In the recycled concrete aggregates case, two problems arise. On the one hand, usually 24 h are not enough to saturate these aggregates, and on the other hand, the drying process using a towel not only removes the surface water but also some of the old cement paste that has been adhered to that surface. The hydrostatic weighing method that was applied solves both problems. The absorption properties of natural aggregates were determined by both standard and hydrostatic weighing methods, obtaining very similar results.

As expected, recycled aggregates absorbed more water (2.69%) than natural ones (0.31%). However, the obtained absorption results are smaller than other values in the literature (Ryu 2002; Padmini et al. 2009). Certainly, the absorption percentage can be considered as a measure of porosity and quality of recycled aggregates, and consequently, can be used as an important parameter in the prediction of RAC properties.

4.1.3 Mixture Proportioning and Mixing Procedure

The details of the different concrete mixes are summarized in Table 4.2. Mixture proportioning of the NAC was done in order to achieve a mean compressive cylinder strength of 35 MPa at 28 days.

Table 4.2 Concrete mixes

	NAC	RAC30%	RAC60%	RAC60%
Cement (c) (kg/m^3)	358.0	358.0	358.0	358.0
Free water (w) (kg/m^3)	178.3	178.3	178.3	178.3
Absorption water (w_{ab}) (kg/m^3)	7.8	14.6	21.6	31.0
Effective water (w_{eff}) (kg/m^3)	178.3	185.1	192.1	201.5
Fine aggregate (kg/m^3)	730.4	730.4	730.4	730.4
Natural coarse aggregate (kg/m^5)	1109.4	783.6	458.3	0.0
Recycled coarse aggregate (kg/m^3)	0.0	299.3	598.4	1020.0
Superplastizier (kg/m^3)	0.30	0.30	0.30	0.30
w/b[a] (–)	0.498	0.498	0.498	0.498
w_{eff}/b[b] (–)	0.498	0.517	0.537	0.563

[a]Water/binder ratio (being b the binder, i.e. cement plus mineral admixtures contents)
[b]Effective water/binder ratio

The criterion for the RACs proportioning was based on accepting almost the same material proportions of those of the NAC, with the following exceptions:

- a percentage R of natural coarse aggregates content was replaced by recycled aggregates, while the total coarse aggregate volume remained unchanged, and
- total water content was modified in order to consider the absorption properties of the different aggregate types.

All mixes were produced with 178.3 l/m^3 of free water (w), i.e. water available for chemical reactions, while the greater water absorption characterizing the recycled aggregates, was taken into account by adding "absorption water" (w_{ab}), which volume was evaluated in terms of the aggregates absorption properties.

Considering that in standard methods for concrete proportioning, the water content refers to the free water, and that aggregates are introduced saturated—and therefore, already including the extra water needed due to absorption properties—it can be observed that from the viewpoint of mix proportioning, all the mixes had the same water/cement ratio (w/c).

In the frame of this work, another water content called "effective water content" (w_{eff}) was defined as

$$w_{eff} = w + w_{ab} - w_{ab}^{NA} \tag{4.2}$$

where w_{ab}^{NA} is the absorption water corresponding to the natural aggregates (including fine and coarse aggregates). The effective water represents the sum of the free water and the extra water needed to saturate the cement paste adhered to recycled aggregates.

Observe that the mix proportioning criterion adopted for RACs did not have the aim of obtaining the same uniaxial compressive strength of that of the reference concrete, but to analyse the degradation in mechanical and physical properties when recycled coarse aggregates are used.

Concrete constituents were mixed in an electric laboratory mixer (pan type with vertical axis). Coarse and fine aggregates, in air dried condition, were first blended with 70% of the total water content. Then, cement was added, and finally, the remaining 30% of the total water content and the water-reducing additive were incorporated.

4.1.4 Fresh Concrete Test Results

The results of slump and density measured on fresh state for the different concretes are presented in Table 4.3. It can be observed that in this experimental campaign RACs showed lower workability and lower density in comparison with NAC (see Fig. 4.6). The observed tendency in workability is mainly due to the mixes design used in this work, the particle size distribution, shape and quality of aggregates, and the fact that recycled aggregates were not incorporated in SSD condition.

Table 4.3 Tests results on fresh concrete

Concrete	Slump (mm)	Unit weight (kg/m^3)
NAC	180	2420
RAC30%	170	2385
RAC60%	80	2382
RAC100%	55	2346

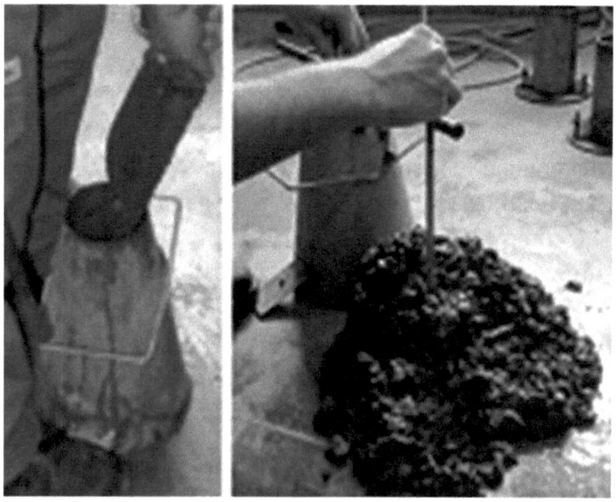

Fig. 4.6 Fresh state test for estimating the workability

Fig. 4.7 Specimens preparation at FIUBA

4.1.5 Curing, Casting, and Specimen's Preparation

Concrete was cast in cylindrical steel moulds, nominally 100 mm in diameter and 200 mm in height. After 24 h, the specimens were demolded and cured in moisture curing room with 100% relative humidity and at 20 °C temperature for 28 days (Fig. 4.7).

Samples to be tested under uniaxial compression were capped with a sulfur compound at both ends, about 10 h before being tested, while the preparation of specimens for splitting tensile and triaxial compression tests only consisted in polishing the top and bottom surfaces.

Besides, in samples for triaxial tests, the larger voids on their surface were filled with cement mortar in order to avoid the damage of the neoprene sleeve surrounding the sample during the test, which would produce the penetration of the confining fluid in the sample.

4.1.6 Equipment for Mechanical Tests

Mechanical tests were conducted in a 2000 kN closed-loop servo hydraulic GCTS rock testing machine with an axial stiffness of 3500 kN/mm and a stroke of 100 mm. A GCTS HTRX-070-L high pressure triaxial cell with a confining pressure capacity of 70 MPa was used for the triaxial compression tests (Fig. 4.8).

Fig. 4.8 General view of the GCTS compression frame (*left*) and the confining cell (*right*)

4.2 Mechanical Tests

4.2.1 Uniaxial Compression Tests (UC)

Uniaxial compression tests were performed according to ASTM C39-05 (2005). Instrumentation of the specimens can be observed in Fig. 4.9a, consisting on two Linear Variable Differential Transformers (LVDTs) and one circumferential LVDT (all with ±2.5 mm range), for measuring axial and circumferential displacements, respectively. The axial gage length adopted was 60 mm. Tests were done under displacement control with a rate of 0.002 mm/s.

Three samples were tested for each concrete type, at 28 days. Mean obtained results are summarized in Table 4.4. It can be observed that both the compressive strength and the Young's modulus decrease for increasing replacement ratio *R* of natural by recycled coarse aggregates, coinciding with other reports in the literature (Hernández and Fornasier 2005; Etxeberria et al. 2007; Yang et al. 2008; Casuccio et al. 2008; Padmini et al. 2009; Corinaldesi 2010; Lima et al. 2013).

As for the axial strain (ε_0) corresponding to the peak stress, despite the observed decrease in Young's modulus only slight differences were detected as can be also observed in Table 4.4, with increasing peak strains for increasing replacement ratio *R* (Fig. 4.10)

Fig. 4.9 Experimental setup: **a** uniaxial compression test, **b** splitting tensile test

Table 4.4 Uniaxial compression test results

Property	NAC	RAC30%	RAC60%	RAC100%
f_c' (MPa)	36.52	33.59	30.42	29.10
ε_0 (%)	1.84	1.64	1.73	2.05
E (MPa)	31,667	28,617	24,533	20,750

f_c' uniaxial compressive strength; ε_0 peak axial strain; E Young's modulus

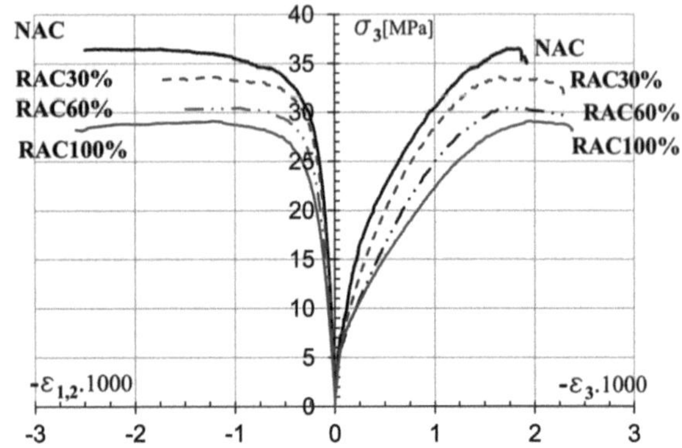

Fig. 4.10 Uniaxial compression: axial stress versus axial and lateral strains for concretes with different replacement ratio of natural by recycled coarse aggregates

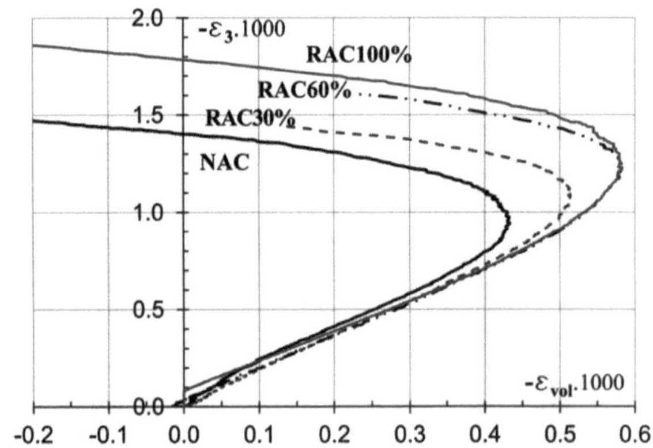

Fig. 4.11 Uniaxial compression: axial strain versus volumetric strains for concretes with different replacement ratio of natural by recycled coarse aggregates

Figure 4.12 shows axial stress versus axial and lateral strains curves for the different analysed concretes. In this figure it can be appreciated the decrease in strength and elastic properties above mentioned. Clearly, the addition of recycled aggregates involves a degradation of concrete quality. Nevertheless, the behaviour does not substantially differ from that of a natural concrete with the same uniaxial compressive strength, i.e., the plots for the RAC60% with a peak stress of 30 MPa is not different from that of an ordinary concrete with the same strength.

Figures 4.11 and 4.12 present the volumetric strains behaviour for the different concretes in this study, being the volumetric strain defined by Eq. (4.1). The first figure, presents the relations between axial and volumetric strains, while the second one, between axial stresses and volumetric strains.

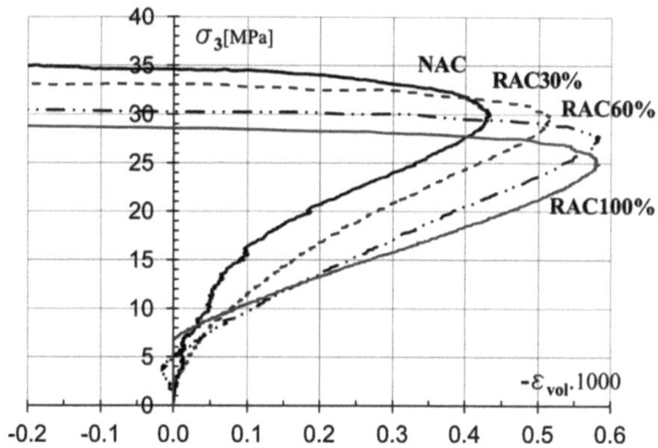

Fig. 4.12 Uniaxial compression: axial stress versus volumetric strains for concretes with different replacement ratio of natural by recycled coarse aggregates

It can be observed that the initial tangent to the curves in Fig. 4.12 representing the bulk modulus, decreases when *R* increases. The latter observation is analogous to what is observed in the comparison of natural aggregates concretes with different uniaxial compressive strengths where bulk modulus is lower for normal strength concretes than for high strength concretes.

In all cases, volume decreases till a certain point situated before reaching the peak stress. From that point, volume starts increasing, even turning positive the sign of volumetric strains. From the figures, it can be observed that the highest decrease in volume occurred for the RAC60% and RAC100%, and also, that the transition from volume compaction to dilation occurs closer to the peak stress for greater replacement percentages *R*.

Figures 4.13 and 4.14 show photographs of the resulting failure modes in uniaxial compression for RAC30% and RAC60%, respectively. It can be observed that a distributed damage occurred in all cases, being damage more evident in the case of the concrete with the lowest quality, i.e. RAC60%.

4.2.2 *Splitting Tensile Tests (ST)*

Splitting tests were performed according to ASTM C496/C496M-11 (2011). Three samples were tested for each concrete type at 30 days of being casted. Typical failure of cylindrical samples under splitting tension can be observed in Fig. 4.15. Obtained results corresponding to the mean tensile strength f'_{t-st} are presented in Table 4.5, with

$$f'_{t-st} = \frac{2P_{peak}}{\pi dL} \tag{4.3}$$

Fig. 4.13 RAC30%: failure modes in uniaxial compression

Fig. 4.14 RAC60%: failure modes in uniaxial compression

being P_{peak} the peak load, and d and L the sample diameter and length, respectively. A significant scatter was observed in results. For example, for RAC100%, the three obtained results were 3.56, 3.75 and 2.66 MPa, leading to a mean tensile strength of 3.32 MPa.

It can be observed that the decrease in strength detected in uniaxial compression tests for increasing R, is more evident than that of the splitting tensile strength. Only small differences in the mean values were obtained and not a clear tendency in this property could be detected for increasing R. This coincides with the general behaviour observed in previous research works related with concretes of different qualities where it was found that compressive strength is much more sensitive to concrete quality than tensile strength (Folino et al. 2009)

Fig. 4.15 Splitting tensile test: typical failure mode

Table 4.5 Splitting tensile test results

Property	NAC	RAC30%	RAC60%	RAC100%
$f_{t\text{-}st}$ (MPa)	4.04	3.87	3.90	3.32
$f_{t\text{-}st}/f_c$	0.111	0.115	0.128	0.114

$f_{t\text{-}st}$ tensile strength from the ST

4.2.3 *Triaxial Compression Tests (TC)*

Since there is a lack of available experimental data related with the tri-axial compression behaviour of RACs, the experimental program presented herein included tri-axial tests involving different levels of confinement, from low to high. For the RAC100%, a tri-axial test under very high confinement was also included. Samples were tested at an age between 35 and 50 days. Table 4.6 presents the different confinement pressures adopted in the performed tests.

Figure 4.16 shows the instrumentation of one specimen in the tri-axial compression tests. It can be observed that it is almost equal to that of the uniaxial compression tests as the available set of measuring devices is suitable to be used under pressure inside the tri-axial cell. The only difference is that in the TC case, the

Table 4.6 Confinement pressure in triaxial compression tests

Confinement (MPa)	NAC	RAC30%	RAC60%	RAC100%
4.5	X	X	X	X
15	X	X	X	X
21	X	X	X	X
40	–	–	–	X

Fig. 4.16 Experimental setup in the tri-axial compression test

concrete sample is placed in the tri-axial cell surrounded by a neoprene sleeve impermeable to the confinement fluid. The axial gage length adopted was 60 mm.

The loading path included two phases: (1) an hydrostatic phase, during which confining pressure was continuously increased with a rate of $\sim$0.35 MPa/s until the target confining pressure was reached, and (2) a deviatoric phase, under constant confining pressure, imposing a constant displacement rate of $\sim$0.002 mm/s. Figure 4.17 presents in a meridian view of the Haigh Westergaard stresses space the loading paths for different confinement pressures. Stress coordinates ξ, ρ, θ in that space are defined as

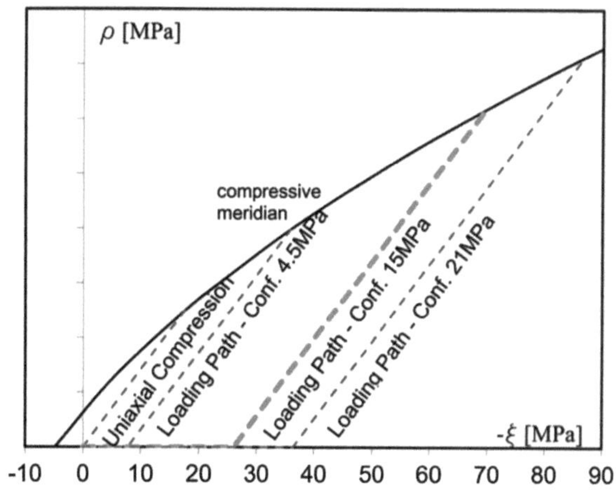

Fig. 4.17 Loading path in tri-axial compression tests

$$\xi = \frac{I_1}{\sqrt{3}}; \quad \rho = \sqrt{2J_2}; \quad \cos(3\theta) = \frac{3\sqrt{3}}{2}\frac{J_3}{J_2^{3/2}} \tag{4.4}$$

being I_1 the first invariant of the stress tensor, while J_2 and J_3 are the second and third invariants of the deviatoric stress tensor, respectively. Observe that in Fig. 4.17, only the compressive meridian is shown by setting $\theta = \pi/3$.

The results of peak stresses obtained in the different tri-axial tests are included in Table 4.7. Figure 4.18 shows axial stress versus axial and lateral strains curves for the RAC30% concrete under three different confinement levels corresponding to 4.5, 15 and 21 MPa, respectively. Analogously, Figs. 4.19 and 4.20 present the results corresponding to the RAC60% and RAC100% concretes and for different confinement levels. From the figures, it can be observed that confinement considerably improves compressive strength. It was demonstrated (Folino and Xargay 2014) that similarly to ordinary concrete, the obtained results lead to a curve compressive failure meridian following a quadratic variation with confinement and not a linear one. Moreover, that with increasing *R* concrete is slightly more sensitive to confinement. In other words, under the same confinement pressure, the increase in strength is greater for increasing *R* replacement percentages. This can be explained regarding that when *R* increases, concrete quality decreases, leading to a concrete with greater porosity, and consequently, more susceptible to confinement effects. Figure 4.21 presents the comparison of the increase in strength achieved in the different TC tests for all concrete types.

It must be pointed out that some problems occurred in the measuring of strains for the RAC60% case under 15 MPa, regarding that almost no lateral strains were captured.

Figures 4.22, 4.23 and 4.24 present the relations between axial and volumetric strains for the RAC30%, RAC60% and RAC100% concretes, respectively, under different confinement levels. It can be observed that for each one of the considered

Table 4.7 Comparative analysis: input data

σ_{I} (MPa)	σ_{II} (MPa)	σ_{III} (MPa)	$\xi = I_1\sqrt{3}$ (MPa)	$\rho = \sqrt{2}J_2$ (MPa)
NAC				
0.00	0.00	−36.52	−21.08	29.82
−4.50	−4.50	−62.00	−40.99	46.65
−15.00	−15.00	−04.05	−71.62	64.54
−21.00	−21.00	−118.48	92.65	79.59
RAC30%				
0.00	0.00	−33.59	−19.39	27.43
−4.50	−4.50	−59.80	−39.72	45.15
−15.00	−15.00	−88.00	−68.13	59.60
−21.00	−21.00	−108.29	−86.77	71.27
RAC60%				
0.00	0.00	−30.42	−17.56	24.84
−4.50	−4.50	−66.59	−43.59	50.62
−15.00	−15.00	−84.00	−65.82	56.34
−21.00	−21.00	−115.00	−90.64	76.75
RAC100%				
0.00	0.00	−29.10	−16.80	23.76
−4.50	−4.50	−75.99	−49.07	58.37
−15.00	−15.00	−87.10	−67.61	58.87
−21.00	−21.00	−102.74	−83.57	66.74
−40.00	−40.00	−144.90	−129.85	85.65

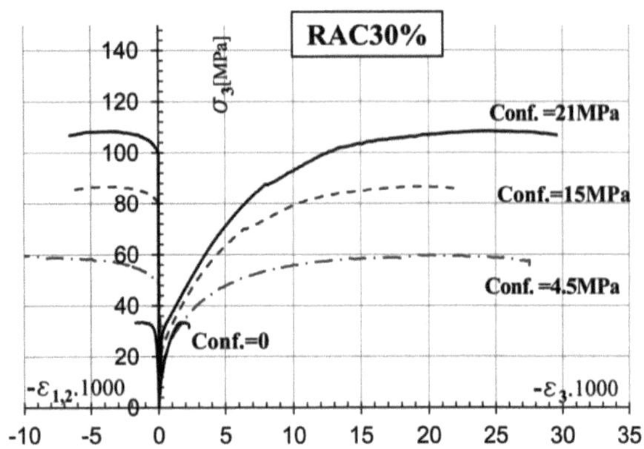

Fig. 4.18 RAC30% tri-axial compression: axial stress versus axial and lateral strains for 0, 4.5, 15 and 21 MPa confinement levels

concretes, the initial slopes are almost the same for all the confining pressures. Volume compaction considerably increases with increasing confinement. This is analogous to what is observed in the case of ordinary concretes (Lu 2005).

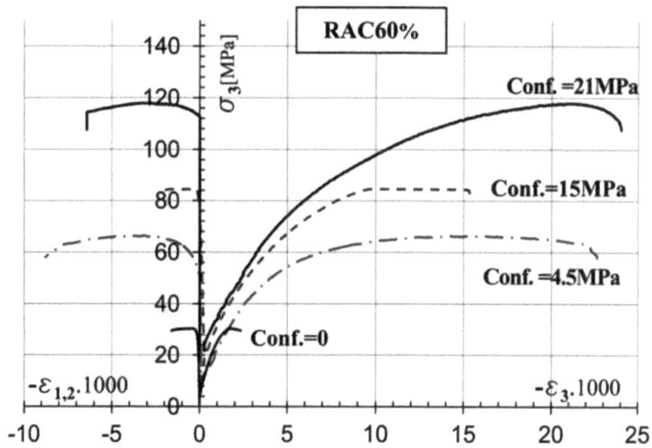

Fig. 4.19 RAC60% tri-axial compression: axial stress versus axial and lateral strains for 0, 4.5, 15 and 21 MPa confinement levels

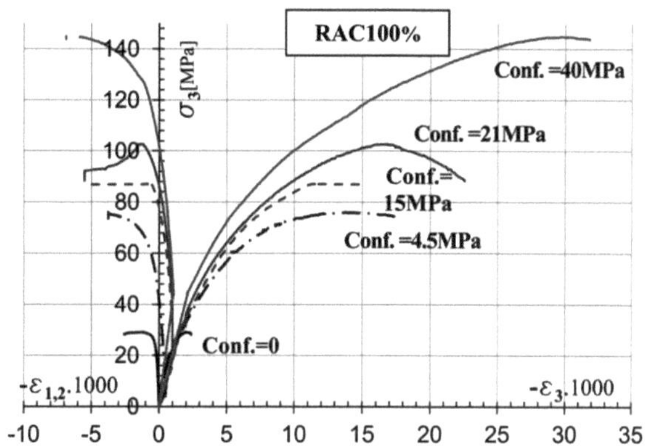

Fig. 4.20 RAC100% tri-axial compression: axial stress versus axial and lateral strains for 0, 4.5, 15, 21 and 40 MPa confinement levels

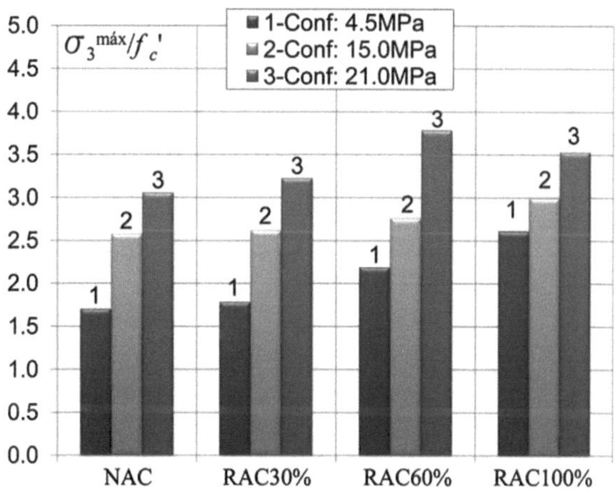

Fig. 4.21 RACs with different R: sensitivity to confinement

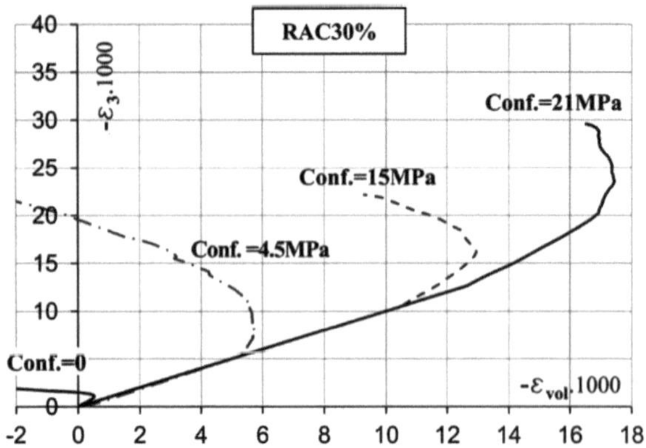

Fig. 4.22 RAC30% tri-axial compression: axial stress versus volumetric strains for 0, 4.5, 15 and 21 MPa confinement levels

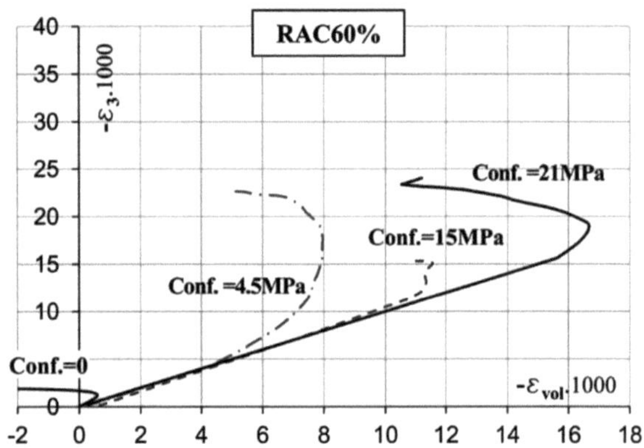

Fig. 4.23 RAC60% tri-axial compression: axial stress versus volumetric strains for 0, 4.5, 15 and 21 MPa confinement levels

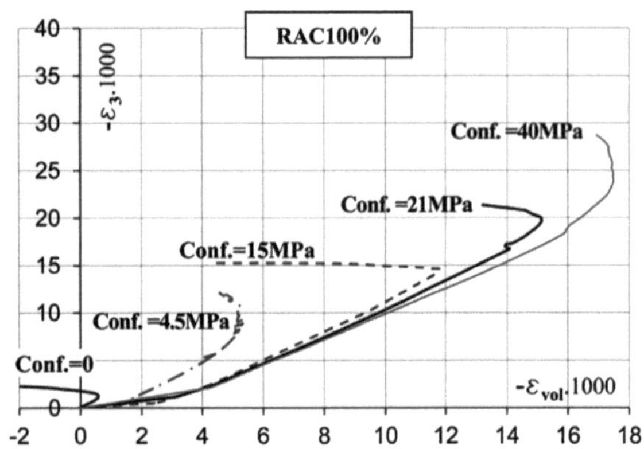

Fig. 4.24 RAC100% tri-axial compression: axial stress versus volumetric strains for 0, 4.5, 15, 21 and 40 MPa confinement levels

Fig. 4.25 RAC30%: failure modes in tri-axial compression for 4.5 MPa (*left*) and 15 MPa (*right*) confinement levels

Fig. 4.26 RAC60%: failure modes in tri-axial compression for 4.5 MPa (*left*) and 15 MPa (*right*) confinement levels

Figures 4.25 and 4.26 show photographs of the resulting failure modes in tri-axial compression for the RAC30% and RAC60%, respectively. In both cases, the photo on the left corresponds to 4.5 MPa of confinement, while the one on right side corresponds to 15 MPa.

Analogously, Fig. 4.27 presents three pictures corresponding to RAC100% under TC with confinement pressures of 4.5, 15 and 21 MPa, respectively.

Fig. 4.27 RAC100%: failure modes in tri-axial compression for 4.5 MPa (*left*), 15 MPa (*center*) and 21 MPa (*right*) confinement levels

It can be observed that a more distributed damage occurred in the 4.5 MPa confinement case, while for higher confinement levels, failure was mainly produced through a single inclined plane, for all concrete types. No substantial differences were observed between the failure modes of the different concretes.

4.3 Concluding Remarks

This section summarised the results of an experimental program intended at observing the tri-axial behaviour of RAC. A normal strength concrete with natural coarse aggregates was designed and used to produce cylindrical samples which, after being tested, were crushed and reused as coarse aggregates in the elaboration of RAC with replacement ratios of recycled aggregates of 30, 60 and 100%, respectively. All RACs in this experimental campaign present the same cement content, free water content and aggregates volumes. Concrete samples were submitted to uniaxial compression, splitting tensile and tri-axial compression tests.

In agreement with previous researches in the subject, it was found that recycled aggregates have a greater porosity than natural ones, and consequently, increasing replacement ratios lead to a worse workability and to a noticeable degradation in both uniaxial compression strength and elastic properties. In this investigation, it was found that volume compaction occurring in the first part of the uniaxial compression test was larger for greater replacement ratios. Also, that concretes with a higher replacement ratio of recycled aggregates are more sensitive to confinement, and present an increasing volume compaction. Both facts can be explained considering the greater porosity.

References

ASTM C127-01 (2001) Standard test method for density, relative density (specific gravity), and absorption of coarse aggregate. Am Soc Test Mater, USA

ASTM C150/C150M-12 (2012) Standard specification for Portland cement. Am Soc Test Mater

ASTM C33/C33M-13 (2013) Standard specification for concrete aggregates. Am Soc Test Mater

ASTM C39-05 (2005) Standard test method for compressive strength of cylindrical concrete specimens. Am Soc Test Mater, USA

ASTM C496/C496M-11 (2011) Standard test method for splitting tensile strength of cylindrical concrete specimens. Am Soc Test Mater

Casuccio M, Torrijos MC, Giaccio G, Zerbino R (2008) Failure mechanism of recycled aggregate concrete. Constr Build Mater 22:1500–1506

Corinaldesi V (2010) Mechanical and elastic behaviour of concretes made of recycled concrete coarse aggregates. Constr Build Mater 24:1616–1620

Djerbi Tegguer A (2012) Determining the water absorption of recycled aggregates utilizing hydrostatic weighing approach. Constr Build Mater 27(1):112–116

Etxeberria M, Vázquez E, Marí A, Barra M (2007) Influence of amount of recycled coarse aggregates and production process on properties of recycled aggregate concrete. Cem Concr Res 37:735–742

Folino P, Etse G (2011) Validation of the performance dependent failure criterion for concretes. ACI Mater J 108(3):261–269

Folino P, Etse G (2012) Performance dependent model for normal and high strength concretes. Int J Solids Struct 49(5):701–719

Folino P, Xargay H (2014) Recycled aggregate concrete—Mechanical behavior under uniaxial and triaxial compression. Constr Build Mater 56:21–31

Folino P, Etse G, Will A (2009) A performance dependent failure criterion for normal and high strength concretes. ASCE J Eng Mech 135(12):1393–1409

Hernández C, Fornasier G (2005) Caracterización de hormigones elaborados con agregado grueso reciclado (in Spanish). Hormigonar 7:6–13

Lima C, Caggiano A, Faella C, Martinelli E, Pepe M, Realfonzo R (2013) Physical properties and mechanical behaviour of concrete made with recycled aggregates and fly ash. Constr Build Mater 47:547–559

Lu X (2005) Uniaxial and triaxial behavior of high strength concrete with and without steel fibers. PhD thesis, New Jersey Institute of Technology, USA

Neville A (1995) Properties of concrete. Longman, UK

Padmini A, Ramamurthy K, Mathews M (2009) Influence of parent concrete on the properties of recycled aggregate concrete. Constr Build Mater 23:829–836

Ryu J (2002) An experimental study on the effect of recycled aggregate on concrete properties. Mag Concr Res 54(1):7–12

Shah SP, Chandra S (1968) Critical stress, volume change, and microcracking of concrete. ACI J Proc 65(9):770–780

van Geel E (1998) Concrete behavior in multiaxial compression. Ph.D. thesis, Technische Universiteit Eindhoven, The Netherlands

van Mier J (1997) Fracture processes of concrete. CRC Press, Boca Raton

Xiao J, Li W, Fan Y, Huang X (2012a) An overview of study on recycled aggregate concrete in China (1996–2011). Constr Build Mater 31:364–383
Xiao J, Li W, Sun Z, Shah S (2012b) Crack propagation in recycled aggregate concrete under uniaxial compressive loading. ACI Mater J 109(4):451–462
Yang K, Chung H, Ashour A (2008) Influence of type and replacement level of recycled aggregates on concrete properties. ACI Mater J 105(3):289–296

Chapter 5
Constitutive Formulations for Concrete with Recycled Aggregates

Antonio Caggiano, Guillermo Etse, Paula Folino, Marianela Ripani and Sonia Vrech

Abstract In this Chapter, a thermodynamically consistent gradient model is proposed for natural aggregate concrete and then, modified to take into account the addition of different contents of recycled aggregates and its influence on concrete mechanical response. A particular and simple form of gradient-based plasticity is considered, where the state variables are the only ones of non-local character. After describing the material formulation for natural and recycled aggregate concretes, the model calibration is performed with experimental data taken from literature. A comprehensive numerical analysis is presented, where the effects of the recycled aggregate content on the performance of concrete in pre and post-peak behavior are evaluated and discussed, for different stress states. Finally, the ability of the model to capture the variation of mechanical response of concrete with different recycled aggregate contents is demonstrated for different mechanical tests.

In the last years, an increasing motivation has emerged for producing concrete with recycled constituents, with the aim of reducing the negative environmental impacts of concrete production. This new trend makes imperative the formulation of theoretical models, capable of predicting the mechanical behaviour of materials and structures made with Recycled Aggregate Concrete (RAC) and characterizing the brittle or ductile failure modes of concrete at different loading regimes. It is well known that the failure behaviour of quasi-brittle materials such as concrete, strongly depends on both the governing stresses and the physico-chemical features of its micro- and meso-structure.

A. Caggiano · P. Folino (✉) · M. Ripani
Facultad de Ingeniería, Universidad de Buenos Aires,
LMNI-INTECIN (UBA-CONICET), Buenos Aires, Argentina
e-mail: pfolino@fi.uba.ar

G. Etse · S. Vrech
National University of Tucumán, Tucumán, Argentina

A. Caggiano · G. Etse · S. Vrech
CONICET, Buenos Aires, Argentina

J.A.O. Barros et al. (eds.), *Recent Advances on Green Concrete for Structural Purposes*, Research for Development, DOI 10.1007/978-3-319-56797-6_5

In the tensile regime, the material response under mechanical loading is highly brittle in nature, as the damage entirely localizes in one single surface of zero thickness crack, while the material outside the crack remains practically undamaged. Then, the failure process and the related cracking mechanisms exclusively depend on the fracture energy properties throughout such cracking surfaces.

Conversely, in the compressive regime, the ductility of concrete strongly increases with the confining pressure. In this case, the failure mechanism is characterised by the appearance of several microcracks developing in the normal direction to the local maximum principal stress and the occurrence of the material damage process in the zones located in between cracks or microcracks. Under increasing monotonic loading, both the microcracks and the portion of the material subjected to the degradation process coalesce in shear bands and aggregate size-dependent widths. In uniaxial compression, the shear band width approximates the maximum aggregate size for normal strength concretes and continuously increases for higher confinement levels. This highly variable failure behaviour of quasi-brittle concrete-like materials has been observed in several experimental researches published in the scientific literature (Hurlbut 1985; van Geel 1998; Sfer et al. 2002; Lu 2005). These works have shown that failure and mechanical behaviour are not only influenced by the loading scenario, but also by concrete "quality" (Folino and Etse 2011). The different failure mechanisms of concrete, either under tension or under uniaxial and triaxial compression tests, can be recognized also by assessing the final status and appearance of the tested specimens. In case of tests under uniaxial or biaxial tension, the volume change is almost zero. However, in uniaxial compression and, even more clearly, in triaxial compression with increasing confining pressure, the volume change is significant due to the more widespread distribution of microcracks and to the resulting enlargement of the damaged zone. Detailed descriptions of the mechanisms governing concrete strength decay can be found in Vile 1968; Petersson 1981; van Mier 1997. As a matter of fact, different theoretical frameworks need to be combined with the aim to accurately reproduce the entire spectrum of possible concrete failure modes when subjected to tensile or triaxial compression loading under low and high confinement levels, and, moreover, the continuous transition from brittle to ductile failure behaviour.

Within the framework of the Smeared Crack Approach (SCA), brittle materials require dissipative formulations based on fracture mechanics concepts (Bazant and Oh 1983; Etse and Willam 1994; Carpintieri et al. 1997; Duan et al. 2007). On the other hand, the simulation of the mechanical decay of ductile materials requires non-local theories that appropriately and objectively describe the development of (non-zero thickness) shear bands, like the non-local gradient theory, as proposed by Vardoulakis and Aifantis (1991), Sluys et al. (1993).

Some constitutive formulations have been proposed to predict the mechanical behavior of RAC based on empirical considerations and mainly related to the compressive stress-strain relationships (see a.o. Du et al. 2010). Another formulation based on microplane theory has been developed by Li et al. (2010).

In this work a thermodynamically consistent elastoplastic gradient-dependent material model is proposed to predict the mechanical response of RAC for all possible stress paths, considering variable amounts of recycled aggregates into the cementitious matrix. A reformulation of the so-called Leon-Drucker-Prager (LDP) model for Natural Aggregate Concrete (NAC) by Vrech and Etse (2009), based on non-local gradient and fracture energy-based concepts is proposed to account for the recycled aggregate content and its effect on the pre-peak stiffness, maximum strength capacity and post-peak regime, as well as on the inelastic dilatancy. An isotropic and local hardening formulation that turns non-local in the softening regime is also accounted. Particularly, the strength decohesion in post-peak regime is controlled by two independent mechanisms: (i) micro and macrocracking process, described by a fracture energy-based plasticity formulation according to Willam et al. (1985) and Etse and Willam (1994); and (ii) degradation of the continuum or material located in between cracks, formulated by means of the gradient-based non-local plasticity based on Vrech and Etse (2009). Two characteristic lengths are included, one related to the fracture energy released in the active cracks and the other related to the gradient-based formulation. To realistically reproduce the strong dependence of concrete failure modes on the acting confining pressure, both characteristic lengths are defined in terms of this variable. The effects of the recycled aggregates on the overall mechanical response behavior of RAC is taken into account by means of the so-called *concrete mixture recycling factor* which is introduced in the re-formulation of the hardening and softening laws, the maximum strength criterion and the non-associativity.

5.1 Thermodynamic Framework of Gradient Constitutive Theory

Following the thermodynamically consistent gradient material theory for small strain kinematics by Svedberg and Runesson (1997), the free energy density can be additively decomposed into three components, elastic, local and non-local plastic, as follows

$$\psi(\varepsilon^{e}, \kappa, \nabla\kappa) = \psi^{e}(\varepsilon^{e}) + \psi^{l}(\kappa) + \psi^{nl}(\nabla\kappa) \tag{5.1}$$

being ε^{e} the elastic strain tensor. The non-local effects are considered to be restricted to the state hardening/softening variable κ by means of $\nabla\kappa$, regarding an isotropic formulation. The constitutive equation is obtained from Coleman's relations as

$$\sigma = \rho\frac{\partial\psi}{\partial\varepsilon^{e}} \tag{5.2}$$

being σ the stress tensor and ρ the material density; while the local and non-local dissipative stresses are computed as

$$\mathrm{K}^{l}=-\rho\frac{\partial\psi}{\partial\kappa},\quad \mathrm{K}^{nl}=\nabla\left[\rho\frac{\partial\psi}{\partial(\nabla\kappa)}\right]. \tag{5.3}$$

In the realm of the flow theory of plasticity, the rates of plastic strain tensor and internal variable are defined as

$$\dot{\varepsilon}^{\mathrm{p}}=\dot{\lambda}\frac{\partial \mathrm{Q}}{\partial\sigma},\quad \dot{\kappa}=\dot{\lambda}\frac{\partial \mathrm{Q}}{\partial \mathrm{K}} \tag{5.4}$$

where Q is the dissipative potential, $\dot{\lambda}$ the rate of the plastic multiplier and the total dissipative stress $\mathrm{K}=\mathrm{K}^{\mathrm{l}}+\mathrm{K}^{\mathrm{nl}}$. To complete this formulation, the Kuhn-Tucker conditions are introduced

$$\dot{\lambda}\geq 0,\quad \mathrm{F}(\boldsymbol{\sigma},\mathrm{K})\leq 0,\quad \dot{\lambda}\mathrm{F}(\boldsymbol{\sigma},\mathrm{K})=0 \tag{5.5}$$

where F symbolizes the yield function.

5.2 Leon-Drucker-Prager Model for NAC

In this section the gradient and fracture energy-based Leon-Drucker-Prager model (LDP) for NAC proposed by Vrech and Etse (2009), is summarized. This constitutive formulation follows the original thermodynamically consistent gradient-regularized material theory for small strain kinematics by Svedberg and Runesson (1997). The maximum strength surface of the LDP model is obtained by combining the compressive meridian of the failure criterion by Leon (1935), with the Drucker-Prager circular deviatoric shape. The mathematical description of the LDP criterion for concrete is

$$F(\tau,\sigma)=\frac{3}{2}\tau^{2}+\mathrm{m}_{0}\left(\frac{\tau}{\sqrt{6}}+\sigma\right)-\mathrm{c}_{0}=0 \tag{5.6}$$

expressed in terms of the first and second Haigh-Westergaard stress coordinates normalized respect to the uniaxial compressive strength

$$\tau=\frac{\sqrt{2J_{2}}}{f_{c}},\quad \sigma=\frac{I_{1}}{3f_{c}} \tag{5.7}$$

being I_1 the first invariant of the stress tensor $\boldsymbol{\sigma}$ and J_2 the second invariant of the deviatoric stress tensor **S**. Cohesive and frictional parameters, c_0 and m_0, are calibrated in terms of f_c and f_t, the uniaxial tensile strength, resulting

$$c_0 = 1, \quad m_0 = \frac{3}{2}\frac{\left(f_c^2 - f_t^2\right)}{f_c f_t}. \tag{5.8}$$

Beyond the elastic regime, the evolution of the yield surface in the hardening and softening regimes are encompassed by one single equation as follows

$$F(\tau, \sigma, K^+, K^-) = \frac{3}{2}\tau^2 + m_0 K^+ \left(\frac{\tau}{\sqrt{6}} + \sigma\right) - K^+ K^- = 0 \tag{5.9}$$

being K^+ and K^- the pre- and post-peak dissipative stresses, varying according $K_0^+ \leq K^+ \leq 1$ and $1 \geq K^- \geq 0$. $K_0^+ = 0.1$ represents the initial value of the hardening dissipative stress. A non-associated plastic flow describes the inelastic behavior of concrete. The adopted plastic potential is based on a volumetric modification of the yield condition

$$Q(\tau, \sigma, K^+, K^-) = \frac{3}{2}\tau^2 + m_0 \left(\frac{\tau}{\sqrt{6}} + \eta\sigma\right) - K^+ K^- = 0 \tag{5.10}$$

being η the volumetric non-associativity degree, which varies between $0 \leq \eta \leq 1$. The extreme case when $\eta = 0$ corresponds to the isochoric plastic flow, while $\eta = 1$ results in associated plasticity.

5.3 The Extended LDP for Failure Analysis of RAC

The well-established theoretical and mechanical LDP framework has been extended to numerically reproduce the mechanical response and failure behavior of RAC under general load states. In the following and separately, the modifications in the yield condition, maximum strength criterion, post-peak regime and the description of the inelastic volumetric deformation due to the considered amount of recycled aggregates are described.

5.3.1 *Maximum Strength Surface and Yield Condition*

The maximum strength criterion of the Extended LDP follows from Eq. (5.6) as

$$F(\tau, \sigma) = \frac{3}{2}\tau^2 + m_{RAC}\left(\frac{\tau}{\sqrt{6}} + \sigma\right) - c_{RAC} = 0 \tag{5.11}$$

being the cohesive and frictional parameters for RAC

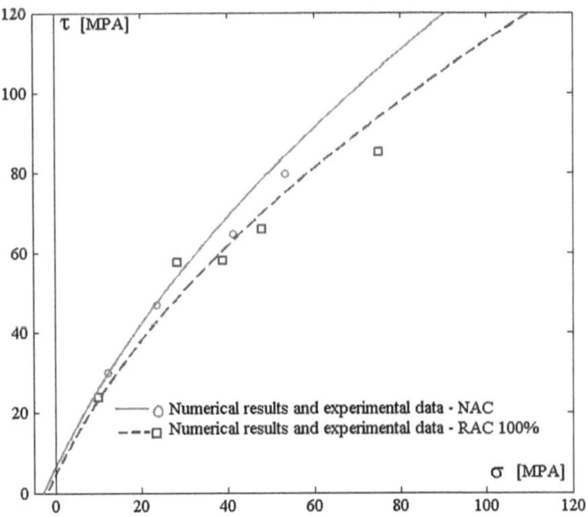

Fig. 5.1 Comparison of the LDP model extended to RAC against experimental data in compressive regime by Folino and Xargay (2014)

$$c_{RAC} = \alpha_{RAC}^2 c_0,\ m_{RAC} = \alpha_{RAC} m_0 \tag{5.12}$$

and the so-called *concrete mixture recycling factor*

$$\alpha_{RAC} = 1 - \alpha_{R1} RA^{\alpha_{R2}} = \frac{f_c^{RAC}}{f_c} \tag{5.13}$$

recycled aggregates. Thereby *RA* is the content of recycled aggregate, while α_{R1} α_{R2} are internal parameters to be calibrated from a set of uniaxial compression tests elaborated with a wide spectrum of different recycled aggregate contents. Figure 5.1 shows the variation of the maximum strength surface of the Extended LDP model with the aggregate content *RA*. In this case, the internal parameters $\alpha_{R1} = 0.00075$, $\alpha_{R2} = 1.25$ were adopted based on the set of experimental results by Folino and Xargay (2014). In the same form to the maximum strength criterion, the expression of the yield condition in hardening/softening regime of the Extended LDP model is modified from its original formulation in Eq. (5.9) to the following

$$F\left(\tau, \sigma, \alpha_{RAC}, K_{RAC}^{+}, K_{RAC}^{-}\right) = \frac{3}{2}\tau^2 + m_{RAC} K_{RAC}^{+}\left(\frac{\tau}{\sqrt{6}} + \sigma\right) - \alpha_{RAC} K_{RAC}^{+} K_{RAC}^{-} = 0 \tag{5.14}$$

5.3.2 Ductility in Pre-peak Regime

RAC is characterized by a reduced ductility in pre- and post-peak regimes. Actually, the degradation of stiffness in RAC as compared to NAC is much quicker

Fig. 5.2 Hardening dissipative stresses at variable confinement levels

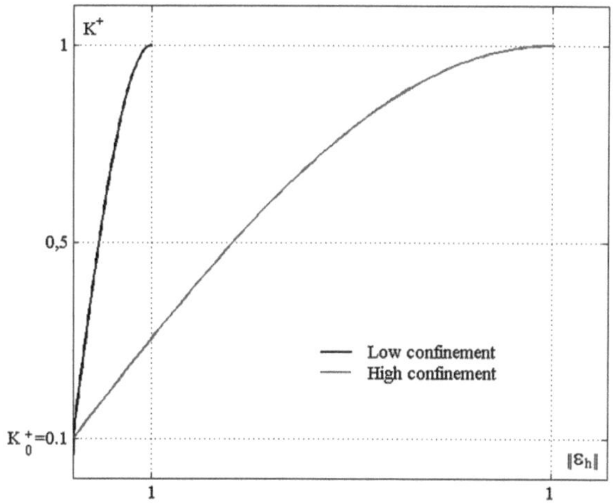

than that of peak strength. This effect even increases with the temperature. Different experimental results (see a.o. Li et al. 2012) indicate that the reduced ductility of RAC is mainly due to the poorer mechanical properties of the Interfacial Transition Zone (ITZ) which is an important phase influencing the overall material response. The reduction of ductility in the pre-peak regime of RAC is taken into account in the Extended LDP model through the redefinition of the hardening ductility parameter.

The hardening dissipative stresses are given by

$$\mathrm{K}^{+}_{RAC} = 0.1 + 0.9 \sin\left(\frac{\pi}{2} \frac{\|\varepsilon_h\|}{x_{RAC}}\right) \tag{5.15}$$

being ε_h the equivalent plastic strain in hardening and x_{RAC} the hardening ductility measure, defined not only as a function of the normalized confining pressure, but also of the concrete mixture recycling factor as

$$x_{RAC} = \alpha_{RAC} \mathrm{A_h} \exp(\mathrm{B_h} \sigma). \tag{5.16}$$

The coefficients $\mathrm{A_h}$ and $\mathrm{B_h}$ must be calibrated by triaxial compression tests under low and high confinement. The evolution of the hardening dissipative stresses with variable confinement levels, temperatures and RAC content is shown in Fig. 5.2. As can be observed in this figure, concrete ductility during the evolution of the hardening dissipative stresses in pre-peak regime decreases with the amount of RAC in the concrete mixture, while it increases with the applied confinement and temperature.

5.3.3 Ductility in Post-peak Regime

Concrete failure process in post-peak regime of the LDP model is defined in terms of two mechanisms in parallel: one is based on fracture energy while the other one consists on a non-local gradient formulation. Thus, K_{RAC}^{-} can be decomposed as

$$K_{RAC}^{-} = K_{RAC}^{f-} + K^{g-} \tag{5.17}$$

where K_{RAC}^{f-} is the fracture energy-based strength decay associated to further propagations of active cracks, and K^{g-} the gradient-based softening mechanism, which accounts for the strength reduction due to the material degradation in-between active cracks. In the LDP model formulation, both characteristic lengths, the fracture energy and the gradient one, are defined in terms of the confining pressure. The presence of recycled aggregates only affects the fracture energy-based strength degradation K_{RAC}^{f-}. This is because, the fracture energy is directly related to the mechanical properties of both aggregates and ITZ. Moreover, it is considered that the post-peak strength contribution related to the gradient based formulation, is not affected by the presence of recycled aggregates as this one is mainly related to the mechanical effects. The local or fracture-energy based dissipative stress is now defined as

$$K_{RAC}^{f-} = \exp\left[-5\|\langle\varepsilon_f\rangle\|\frac{h_f}{u_{RAC}}\right] \tag{5.18}$$

being ε_f the local fracture strain. The Mc Auley brackets indicate that only the tensile principal plastic strains contribute to energy density during the fracture process. Then, u_{RAC} represents the maximum crack opening displacement of RAC in pure mode I of failure, which is defined in terms of the concrete mixture recycling factor

$$u_{RAC} = \alpha_{RAC} u_r \tag{5.19}$$

with u_r the maximum crack opening displacement of the associated NAC, which is obtained when $\alpha_{RAC} \rightarrow 1$. The characteristic length h_f represents the distance between active microcracks. It decreases under increasing confinement according to

$$h_f = \frac{h_t}{R_f(\sigma)} \tag{5.20}$$

being h_t the maximum possible value of h_f corresponding to fracture in mode I and

$$R_f(\sigma) = \left\{ \begin{array}{ccc} 1 & if & \sigma \geq 0 \\ 50.75 + 49.75\sin\left(2\sigma - \frac{\pi}{2}\right) & if & -1.5 > \sigma > 0 \\ 100 & if & \sigma \geq 1.5 \end{array} \right\} \tag{5.21}$$

The pressure-dependent function R_f accounts for the ratio between the fracture energies G_{II}^f and G_I^f, which are considered as constants during each infinitesimal evolution of the cracking process.

The non-local dissipative stress in softening is defined as

$$\mathrm{K}^{g-}(\nabla\lambda) = -l_c^2 \mathrm{H}^g \nabla^2 \lambda \tag{5.22}$$

being H^g the non-local gradient modulus and l_c the width of the shear band in which the degradation of the material located in between active cracks ideally occurs. This geometrical measure, the so-called gradient-based characteristic length, is defined here in terms of the acting confinement, but independently of the recycled aggregate content, as

$$l_c(\sigma) = \left\{ \begin{array}{ccc} 0 & if & \sigma \geq 0 \\ A_g l_c \left[1 + \sin\left(B_g - \frac{\pi}{2}\right)\right] & if & -1.5 > \sigma > 0 \\ l_{cm} & if & \sigma \geq 1.5 \end{array} \right\} \tag{5.23}$$

being l_{cm} the maximum possible value of l_c, while A_g and B_g are internal parameters to be calibrated from sets of triaxial compression tests on concrete cylinders subjected to different confinements.

5.3.4 *Inelastic Volumetric Deformation and Plastic Potential*

In the low confinement regime, the inelastic lateral deformation of concrete significantly increases with the amount of recycled aggregate. This is also a consequence of the poorer mechanical properties of the ITZ. In the flow rule of plasticity, the theoretical framework for the formulation of this Extended LDP for RAC, the inelastic volumetric deformation is controlled by the degree of volumetric non-associativity, i.e. the terms of the plastic potential depending on the effective first invariant of the stress tensor. In this model formulation, the plastic potential is defined as

$$\begin{aligned} \mathrm{Q}\left(\tau, \sigma, \alpha_{RAC}, \eta_{RAC}, \mathrm{K}_{RAC}^+, \mathrm{K}_{RAC}^-\right) = & \frac{3}{2}\tau^2 + m_{RAC}\left(\frac{\tau}{\sqrt{6}} + \eta_{RAC}\sigma\right) \\ & - \alpha_{RAC}\mathrm{K}_{RAC}^+\mathrm{K}_{RAC}^- = 0 \end{aligned} \tag{5.24}$$

with the degree of volumetric non-associativity η_{RAC} formulated in terms of the concrete mixture recycling factor as

$$\eta_{RAC} = \eta_0 - (1 - \eta_0)\exp[\alpha_{RAC}(\sigma - \sigma_0) - 1] \tag{5.25}$$

Thereby, η_0 is the minimum value of the degree of volumetric non-associativity which is obtained on the edge of the plastic potential, where it intersects the hydrostatic axis. At this stress state the normalized form of the effective first stress coordinate approaches its maximum value $\sigma = \sigma_0$.

5.4 Numerical Analysis

In this section, the predictive capabilities of the proposed material model to reproduce the failure behavior of RAC are verified. Variable content of recycled aggregates are considered in the analyses. At the finite element (FE) level, the dual-mixed FE formulation proposed by Svedberg and Runesson (1997) is considered for the numerical analyses. Figure 5.3 shows the adopted FE discretization based on constant strain triangles (CST), and the boundary conditions. The last, correspond to inhomogeneous distributions of the stress and strain fields.

Three types of stress paths were considered: uniaxial tensile test, uniaxial compression test and triaxial compression test.

The particular geometry of the rectangular FE patch in Fig. 5.3 and its stress condition (axisymmetric or plane stress) were varied in the different numerical analyses according to the experimental boundary conditions. In all considered stress histories, incremental axial loads are applied under displacement control on the nodes along the specimen top edge, while the lateral confinement is applied under force control on the nodes along the external side of the FE mesh. The relevant internal parameters and material properties adopted in the numerical analyses are summarized in Tables 5.1 and 5.2. Thereby, are included all parameters of the hardening and softening (local and non-local) laws.

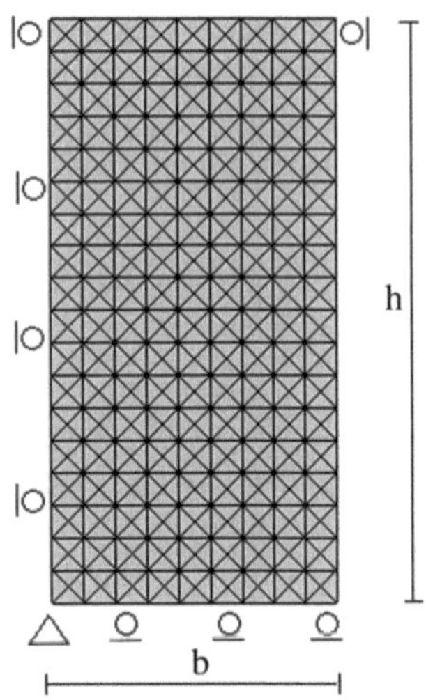

Fig. 5.3 FE discretization and inhomogeneous boundary conditions of a quarter of a concrete specimen (cylindrical or rectangular panel) considered in the numerical analyses

Table 5.1 Common model internal parameters considered in the numerical analyses

Parameter	Value
A_h ; B_h	−0.002; −0.13
A_g; B_g	0.2; 0.5
h_t	100 mm
u_r	0.127 mm
l_{cm}	50 mm
H^g	1 MPa
η_0	0.12

Table 5.2 Particular material properties and model parameters considered in the numerical analyses

	Akita et al. (2003)	Guo et al. (2014)	Folino and Xargay (2014)
f_c (MPa)	22.5	56.52	36.52
f_t (MPa)	2.5	6.28	4.04
E (MPa)	11	34.91	31.67
α_{R1}	0.00055	0.000375	0.00075
α_{R2}	1.05	1.2	1.25

5.4.1 *Uniaxial Tensile Test*

At first, the numerical predictions of the uniaxial experimental tests by Akita et al. (2003), performed on concrete prisms of 100 × 100 × 400 mm are considered. In this case, two FE patches like the one indicated in Fig. 5.3 were adopted, one on top of the other, to reproduce the 1 × 4 relation between specimen width and height, while plane stress conditions were considered. Moreover, the material strength of the 4 CST finite elements corresponding to the square on the extreme right hand side of the bottom row in the FE array was reduced in 10% to represent the notch considered in the experimental tests by Akita et al. (2003). Figure 5.4 shows the prediction of the proposed model to the uniaxial tensile tests for NAC and RAC with 100% recycled aggregate content. The comparison with the experimental results shows very good agreement regarding peak strength, post-peak behavior and residual strength, for both NAC and RAC cases.

5.4.2 *Uniaxial Compression Test*

In the following, model predictions of the uniaxial compressive behavior of RAC are evaluated. Thereby, the experimental results by Guo et al. (2014), are considered for NAC and RAC with 100% content of recycled aggregates. Axisymmetric

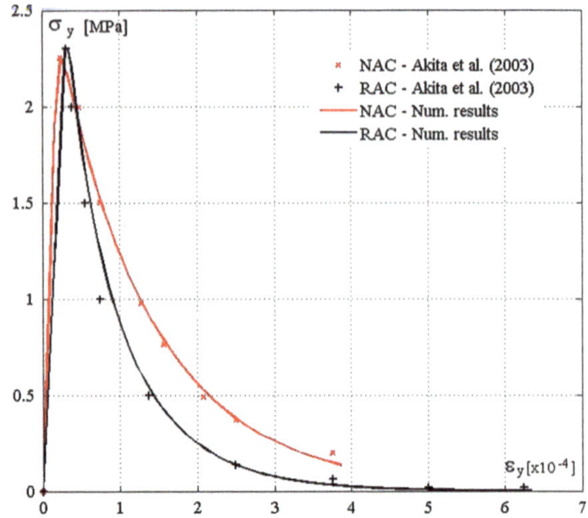

Fig. 5.4 Experimental results on uniaxial tensile test by Akita et al. (2003) for NAC and RAC with 100% recycled aggregate, and numerical predictions with the proposed model

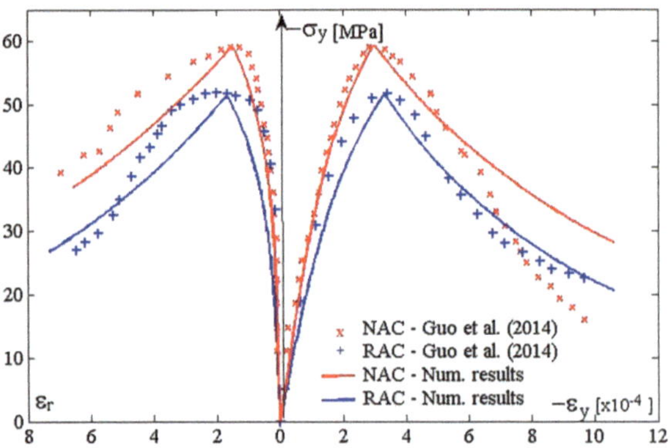

Fig. 5.5 Experimental results by Guo et al. (2014) of uniaxial compression tests on cylinders composed by NAC and RAC (100% recycled aggregate content), and numerical predictions with proposed constitutive model

stress conditions are considered to model the concrete cylindrical specimens of 150 × 300 mm, under non-homogeneous boundary conditions. The comparisons between experimental and numerical results in Fig. 5.5 confirm the predictive capabilities of the proposed model regarding pre- and post-peak behaviors, peak strength and inelastic deformation of RAC in the uniaxial compression test.

Next, the uniaxial compression experimental results by Folino and Xargay (2014) on concrete cylinders of 100 × 200 mm with different contents of recycled aggregates are considered. The comparison between numerical and experimental results in Fig. 5.6, for NAC and RAC with 30, 60 and 100% recycled aggregate contents show good agreement regarding both the pre-peak behavior and the peak strength for all different cases. The post-peak behaviors obtained by Folino and Xargay (2014) show no softening but very ductile responses.

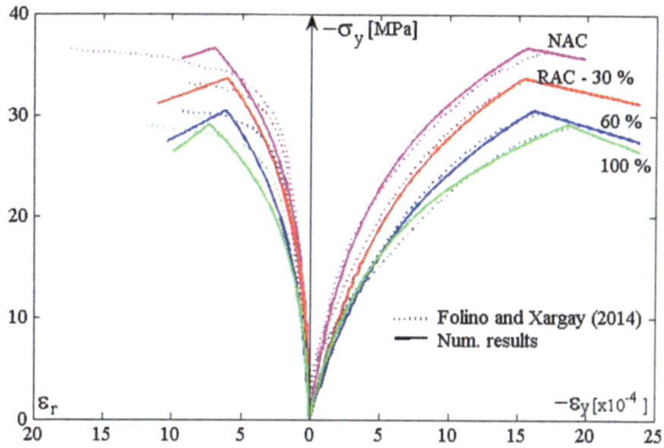

Fig. 5.6 Experimental results of uniaxial compression tests with NAC and RAC with different contents of recycled aggregates by Folino and Xargay (2014), and numerical predictions with the proposed constitutive model

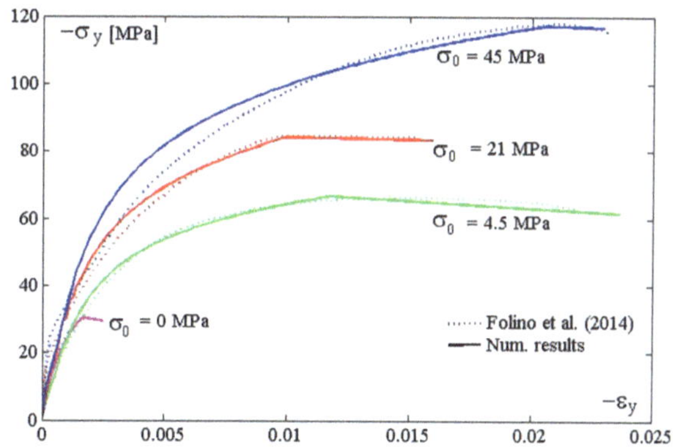

Fig. 5.7 Triaxial compression tests by Folino and Xargay (2014) for RAC with 60% recycled aggregate content and numerical predictions with the proposed mode

5.4.3 Triaxial Compression Test

Finally, the triaxial compression experimental results by Folino and Xargay (2014) on concrete cylinders of 100 × 200 mm are considered. They were performed with different content of recycled aggregates and different lateral confinements. Figure 5.7 illustrates the comparison between experimental and numerical results for the RAC specimens with 60% of recycled aggregate content. Once again, very good agreement between the numerical predictions with the proposed model and the experimental results for all different confinement pressures are obtained. It is very interesting to note that the model is able to reproduce the variation of ductility and peak strength of RAC with the applied confinement under triaxial stress conditions.

References

Akita H, Koide H, Ojima M (2003) Tensile behavior of a recycled concrete. In: Brandt AM, Ki VC, Marshall IH (eds) Proc. Inter. Symp. Britte Matrix Composites 7. Warsaw, pp 263–268

Bazant ZP, Oh B (1983) Crack band theory for fracture of concrete. RILEM Mater Struct 16: 155–177
Du T, Wang W, Liu Z, Lin H, Guo T (2010) The complete stress-strain curve of recycled aggregate concrete under uniaxial compression loading. J Wuhan Univ Technol Mater Sci 25 (5):862–865
Duan K, Hu K, Wittmann FH (2007) Size effect on specific fracture energy of concrete. Engrg Fract Mech 74(1–2):87–96
Etse G, Willam K (1994) Fracture energy-based constitutive theory for inelastic behavior of plain concrete. J Eng Mech 120(9):1983–2011
Folino P, Etse G (2011) Validation of the performance dependent failure criterion for concretes. ACI Mater J 108(3):261–269
Folino P, Xargay H (2014) Recycled aggregate concrete—mechanical behavior under uniaxial and triaxial compression. Constr Build Mater 56:21–31
Guo Y, Zhang J, Chen G, Xie Z (2014) Compressive behaviour of concrete structures incorporating recycled concrete aggregates, rubber crumb and reinforced with steel fibre, subjected to elevated temperatures. J Clean Prod 72:193–203
Hurlbut B (1985) Experimental and computational investigation of strain-softening in concrete. Master's thesis, University of Colorado, USA
Leon A (1935) Ueber das Mass der Anstrenung bei Beton. Ingenieur Archiv 4:421–431
Li J, Xiao J, Huang J (2010) Microplane model for recycled aggregate concrete. In: 2nd International Conference on waste engineering and management—ICWEM 2010
Li W, Xiao J, Sun Z, Kawashima S, Shah S (2012) Interfacial transition zones in recycled aggregate concrete with different mixing approaches. Constr Build Mater 35:1045–1055
Lu X (2005) Uniaxial and triaxial behavior of high strength concrete with and without steel fibers. PhD thesis, New Jersey Institute of Technology, USA
Petersson PE (1981) Crack growth and development of fracture zones in plain concrete and similar materials, report TVBM-1006, technical report. Lund Institute of Technology, Sweden
Sfer D, Gettu R, Carol I, Etse G (2002) Experimental study of the triaxial behavior of concrete. J Engrg Mech ASCE 128(2):156–163
Sluys LJ, de Borst R, Muelhaus M (1993) Wave propagation, localization and dispersion in a gradient-dependent medium. Int J Solids Struct 30:1153–1171
Svedberg T, Runesson K (1997) A thermodynamically consistent theory of gradient-regularized plasticity coupled to damage. Int J of Plast 13(6–7):669–696
van Geel E (1998) Concrete behavior in multiaxial compression. Ph.D. thesis, Technische Universiteit Eindhoven, The Netherlands
van Mier J (1997) Fracture processes of concrete. CRC Press, Boca Raton
Vardoulakis I, Aifantis EC (1991) A gradient flow theory of plasticity for granular materials. Acta Mech 87:197–217
Vile GWD (1968) The strength of concrete under short-term static biaxial stress. In: International Conference on the structure of concrete. Cement and Concrete Association, London, p 275
Vrech S, Etse G (2009) Gradient and fracture energy-based plasticity theory for quasi-brittle materials like concrete. Comp Meth App Mech Eng 199(1–4):136–147
Willam K, Hurbult B, Sture S (1985) Experimental constitutive aspects of concrete failure. In: US −Japan seminar on finite element analysis of reinforced concrete structures. ASCE-Special Publication (Meth Appl Mech Eng 199(1–4):136–147

Chapter 6
Generalised Mix Design Rules for Concrete with Recycled Aggregates

Eduardus A.B. Koenders, Enzo Martinelli, Marco Pepe and Romildo Dias Toledo Filho

Abstract A conceptual formulation for controlling the resulting mechanical properties of Recycled Aggregate Concretes (RACs) is proposed via a set of generalised mix-design rules intended at covering the specific features of Recycled Concrete Aggregates (RCAs). As a matter of fact, the RCAs are characterised by a higher porosity and water absorption capacity than ordinary aggregates and, thus, general mix-design rules for ordinary structural concrete cannot be applied to RACs as such. Therefore, the formulations proposed herein are intended at generalising those rules taking into account the key properties of RCAs, as they are possibly influenced by the alternative processing procedures, which can be applied when turning demolition debris into concrete aggregates. Particularly, these formulations aim at predicting both the final value and the time evolution of compressive strength of RACs depending on their production procedure and mixture composition. The proposed formulations are calibrated and validated on the results of various experimental campaigns covering the effect of several aspects and parameters, such as the processing procedures, the source for RCAs, the actual aggregate replacement ratio, the water-to-cement ratio, the water absorption capacity and the initial moisture condition of coarse recycled aggregates. Design charts of the proposed formulations show ease of the method as well as the potential of employing this rational design method for RAC.

E.A.B. Koenders (✉)
TU Darmstadt, Darmstadt, Germany
e-mail: koenders@wib.tu-darmstadt.de

E. Martinelli · M. Pepe
University of Salerno, Salerno, Italy

R.D.T. Filho
Federal University of Rio de Janero, Rio de Janero, Brazil

J.A.O. Barros et al. (eds.), *Recent Advances on Green Concrete for Structural Purposes*, Research for Development, DOI 10.1007/978-3-319-56797-6_6

The construction sector is among the most resource and energy demanding industrial activity on the planet. Reducing its impact askes for scientifically based solutions that enables a massive reuse of natural resources. The design of integrated recycling processes aimed at reducing waste from construction activities to produce new materials is of particular interest for lowering its environmental footprint and enhancing its sustainability. In fact, hardened concrete elements can be crushed and reused in the production of "green concretes" in which ordinary coarse aggregates are partially or totally replaced by Recycled Concrete Aggregates (RCAs); this practice also results in reducing the so-called Construction and Demolition Wastes (Mc Neil and Kang 2013). For considering RCAs as a credible sustainable alternative of natural aggregates (NAs), the compressive strength evolution of the resulting concrete (generally referred to as Recycled Aggregate Concrete, RAC) has to be predictable as it is in the case of ordinary structural concrete. However, the definition of accurate relationships capable of predicting the relevant properties of concrete made out of these constituents is still considered an open issue. The main reason of this is the lack of knowledge on the actual composition and the resulting engineering properties of RCAs (Behera et al. 2014). In fact, RCAs can be considered as a two phase composite consisting of "original" natural coarse aggregates and Attached Mortar (AM), made of sand, hydration products and fractions of un-hydrated cement. The Attached Mortar content is also the part that is responsible for the increased adsorption capacity of RCAs due to its relatively higher porosity in comparison with NAs.

The rational mix design approach proposed in this study, for designing, controlling and predicting the compressive strength evolution of RAC is mainly based on identifying a key parameter that is characterising the RCA properties and, consequently, the resulting concrete performance. As a matter of fact, RCA porosity was recognised as the main parameter controlling the physical and mechanical properties of RCAs, and the associated water transport phenomena occurring between cement and aggregates during setting and hardening of RAC (Pepe 2015). In fact, the role of free water inside a RAC mixture becomes fundamental as it determines the time evolution of the cement paste strength and bond strength between aggregates and cement (ITZs). In addition, the presence of a porous medium (i.e. RCAs) in RAC results in a possible weak link for the resulting concrete composite. Based on this approach, the authors propose a series of simplified formulations taking into account the porosity of RCA and how it affects the resulting properties for both RCAs and RACs.

6.1 Experimental Results and Modelling Techniques

This section describes the materials, techniques, methods and theoretical formulations used for both investigating the physical properties of RCAs and unveiling their influence on the resulting concrete performance (Pepe 2015).

6.1.1 Experimental Characterisation of RCAs

The RCAs under investigation were collected from various sources, such as from demolished concrete structures and from crushed concrete samples already tested in laboratory conditions. For the coarse aggregates three different size ranges were considered, i.e. class 1, 2 and 3, representing a nominal diameter ranging between 4.75–9.5, 9.5–19 and 19–31 mm, respectively. Moreover, with the aim of better understanding the influence alternative processing procedure have on aggregate properties, an autogenous cleaning process was conceived and performed (Pepe et al. 2014).

Several experimental tests were performed both on single RCA particles and on representative samples taken from RAC fractions. Relevant physical properties of RCAs were determined, such as, evaluation of the AM content by performing CT-scan analysis (ASTM E1570 2011), water absorption capacity at 24 h, particle density analysis, and open porosity by Mercury Intrusion Porosimetry (ASTM D4404 2010). An example of a CT scanned RCA is presented in Fig. 6.1. This experimental technique is used for characterising the internal morphology and porosity of RCAs and for determining the actual volume percentages of phases present inside RCA particles. The bright areas represent the "old" natural aggregates, the dark grey parts the AM, and the black spots are representing pockets of closed porosity.

The relationship between particle density and porosity is determined for RCAs received from various sources. In Fig. 6.2, results are provided for different percentages of AM where it can be seen that for higher amounts of AM a higher porosity and, consequently, a lower particle density was measured.

Finally, based on the analysis conducted on the experimental results obtained from this study, some analytical correlations between the actual AM content and the physical properties of RCAs, such as water absorption (that is strictly related to porosity) and particle density, are proposed for RAC in the next section.

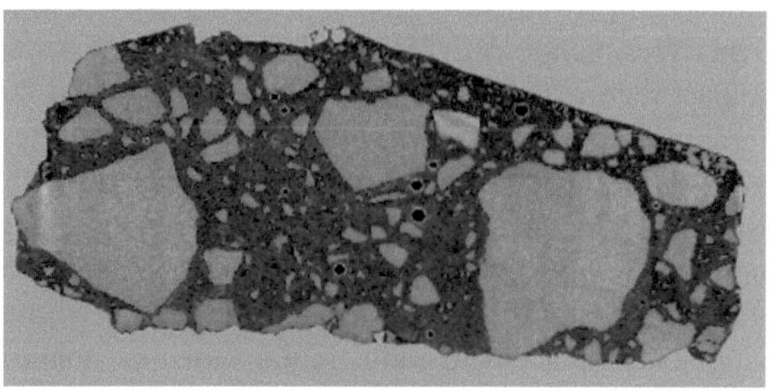

Fig. 6.1 Mesoscopic view of a recycled concrete coarse aggregate

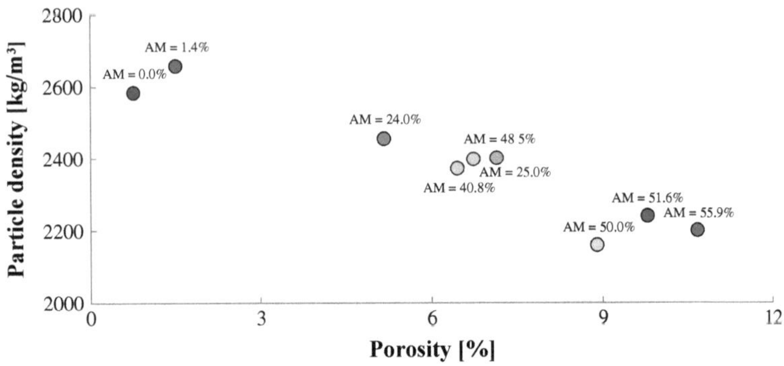

Fig. 6.2 Particle density versus porosity for RCAs with different percentages of AM

Table 6.1 Experimental campaign for RAC

Mix	w/c	RAC (%)	Moisture condition	w/c_{eff}	$R_{cm,28days}$ (MPa)
0.50_OC_Dry	0.50	0	Oven dried	0.50	49.12
0.50_OC_Sat	0.50	0	Saturated with dry surface	0.50	47.00
0.50_RAC50_Dry	0.50	50	Oven dried	0.46	47.55
0.50_RAC50_Sat	0.50	50	Saturated with dry surface	0.50	40.88
0.50_RAC100_Dry	0.50	100	Oven dried	0.39	38.57
0.50_RAC100_Sat	0.50	100	Saturated with dry surface	0.50	31.66
0.40_RAC100_Dry	0.40	100	Oven dried	0.32	42.39
0.40_RAC100_Sat	0.40	100	Saturated with dry surface	0.40	41.15
0.60_RAC100_Dry	0.60	100	Oven dried	0.47	36.60
0.60_RAC100_Sat	0.60	100	Saturated with dry surface	0.60	22.59

6.1.2 Experimental Characterisation of RACs and Hydration Process Modelling

The experimental activities carried out on RACs were aimed at understanding the impact that recycled concrete aggregates have on the physical properties of structural concrete. In particular, several experimental tests were performed to investigate the influence of most relevant parameters, such as processing procedures for RCAs, initial moisture condition of aggregates, aggregate replacement ratio and nominal water-to-cement ratio, on the resulting concrete performance. Table 6.1 describes one of the experimental campaigns performed for analysing the role of the aggregate replacement ratio, initial moisture conditions of the coarse aggregates (i.e., oven dried or saturated with dry surface, obtained by submerging the aggregates in water for 24 h) and the nominal water-to-cement ratio.

For each mixture, cube compressive strength tests ($15 \times 15 \times 15$ cm^3) were performed after 1, 3 and 28 days while cured in a water bath under isothermal conditions at a temperature 20 ± 2 °C. Moreover, the time development of temperature was measured in the centre of representative cubes, prepared for all concrete mixtures, and cured under semi-adiabatic conditions (Martinelli et al. 2013). The mean compressive strength results for this experimental campaign are also reported in Table 6.1 they clearly show that the use of RCAs in concrete, as well as their initial moisture condition, have a significant influence on the resulting concrete performance. In fact, when comparing the results of concrete mixtures characterised by the same nominal water-to-cement ratio and the same initial moisture condition, it turns out that with enhancing aggregate replacement ratio the 28 days compressive strength is decreasing. This effect becomes even more pronounced whenever employing saturated initial moisture conditions and can be explained by the higher porosities of recycled aggregates. In fact, when dry conditions are applied, the aggregates tend to absorb part of the mixing water and, because of this, reduce the actual water-to-cement ratio. Based on the above considerations and the complete analysis of the results (Pepe 2015), it becomes clear that the initial moisture condition and the aggregate porosity tend to modify the water-to-cement ratio and, for this reason, an effective w/c ratio taking into account both these parameters can be defined as follows:

$$\left(\frac{w}{c}\right)_{eff} = \frac{w}{c} + \frac{w_{add}}{c} - \delta \cdot \left(\sum_{i=1}^{n} \frac{p_i \cdot P_i}{c}\right) \tag{6.1}$$

where *w/c* is the nominal water-to-cement ratio, w_{add} is the extra water added to the mix, corresponding to a (partially or totally) soaking into the porous aggregates, p_i and P_i represent the absorption capacity and the weight in the mixture of the i-th aggregate fraction, and δ is a parameter taking into account the initial moisture condition of the aggregates and is zero in saturated conditions and 0.5 in dried ones (Pepe 2015).

Moreover, the authors recently proposed a theoretical model for predicting the time evolution of the degree of hydration from temperature measurements performed on concrete cubic samples cured under semi-adiabatic boundary conditions (Martinelli et al. 2013). Therefore, the proposed model allows to calculate of the time evolution of the degree of hydration based on semi-adiabatic temperature measurements and, based on this, may simulate the hydration reactions for different thermal boundary conditions.

Once the time evolution of the degree of hydration has been calculated, experimental results obtained from compressive tests at a certain curing age and the corresponding values of the degree of hydration were correlated. In fact, the following conceptual formula is available in the literature for correlating degree of hydration α and cubic compressive strength R_c (Lokhorst 1999):

$$R_c = R_{c,max} \frac{\alpha - \alpha_0}{1 - \alpha_0} \tag{6.2}$$

where α_0 is the minimum degree of hydration representing the moment that the strength in the mixture is starting to build up (i.e. end of setting phase), and $R_{c,max}$ the maximum cube compressive strength, ideally corresponding to a degree of hydration equal to 1. Then, the correlation between the mechanical properties and the degree of hydration may show the impact of RCAs on the resulting RAC properties. Specifically, it may express the influence of the residual anhydrous cement fractions, the porosity and the moisture conditions of the RCA attached mortar on the overall hydration process that controls the formation of the concrete microstructure and associated strength.

6.2 Physical Properties of RCAs: The Role of Attached Mortar

Based upon the experimental results achieved from RCAs, a possible linear correlation is proposed, describing the variation of the porosity p as a function of the Attached Mortar (AM) content:

$$p = [p_{NA} \cdot (1 - AM) + p_{AM}(AM)] \cdot \frac{d_0}{d} \tag{6.3}$$

where d is the diameter of the aggregate expressed millimetres and d_0 is a reference diameter of 20 mm. This formulation can be interpreted as a liner combination of the two phases NA and AM. In fact, p_{NA} represents the porosity of a natural aggregate inside RCA, while p_{AM} represents the porosity of the attached mortar residing to it. The calibration of Eq. (6.3) led to approximate values for p_{NA} and p_{AM} equal to 0.45 and 15%, respectively.

Moreover, the particle density represents another fundamental parameter to be determined for the mix design of RAC mixtures. The results reported in the previous section highlight the particle density, which complies with the general trend that its value tends to decrease with enhancing open porosity. As a first order approximation, the following correlation can be proposed for the relationship between the particle density and the open porosity:

$$\gamma = \gamma_0 \cdot (1 - \beta \cdot p) \tag{6.4}$$

where γ_0 is representing the particle density of a fictive aggregate in which the open porosity is zero, is equal to 2700 kg/m^3 and β is a constant parameter that is taken equal to 2.

Finally, as already mentioned, an autogenous cleaning process was conceived and employed using different particle fractions. Specifically, the autogenous cleaning, implemented at laboratory scale, led to interesting results in terms of enhanced properties of the crushed recycled concrete particles. Especially the attached mortar content on RCA surfaces and, consequently, their water absorption capacity was significantly reduced (Pepe et al. 2014). From this, an analytical expression is proposed for evaluating the actual open porosity, p, as a function of the "autogenous cleaning time" t, expressed in minutes:

$$p(t) = \left(1 - \frac{a \cdot t}{b + t}\right) \cdot p_{t=0} \tag{6.5}$$

where $p_{t=0}$ is the initial open porosity of the aggregates without autogenous cleaning, a is a constant equal to 0.6 for class 1 and 0.2 for the class 2 aggregates, while b = 2 in both cases.

6.3 A Rational Mix Design Approach for RAC

The first step considered in formulating a model capable to predict the mechanical strength development for RAC was intended at taking into account both the concrete mixture composition and the main parameters characterising RCAs. As a matter of the fact, a first possible correlation could be elaborated between α_0 and $R_{c,max}$ [calibrated from each produced mixture as described in Chap. 2 (Martinelli et al. 2013)] and the effective water-to-cement ratio defined by Eq. (6.1), which takes into account the "real" amount of free water available into a concrete mixture. However, the concrete strength is also affected by the aggregate replacement ratio, which is not explicitly considered in the definition of $(w/c)_{eff}$. Therefore, a further parameter is introduced that accounts for the average porosity of the coarse aggregates employed in the mixture A_{MIX}, and is defined as follows:

$$A_{MIX} = \sum_{i=1}^{n} p_i \cdot V_i \tag{6.6}$$

where p_i is the open porosity characterising the i-th fraction of the coarse aggregates and V_i is the corresponding fraction volume. This parameter is useful for describing the actual reduction in compressive strength due to the replacement of NAs by RCAs. Particularly, it can physically explain the decrease of α_0 and $R_{c,max}$ that characterises the evolution of the compressive strength of RACs with respect to $\alpha_{0,NAT}$ and $R_{c,max,NAT}$, which represents the compressive strength of a reference concrete made of only natural aggregates (and keeping the grain size distribution constant). Therefore, it is possible to define the ratios of r^* and α^* for expressing the reduction in strength and its rate development in terms of $R_{c,max}$ and α_0, respectively. Moreover, the following dimensionless analytical formulations are introduced to describe the existing relation between the above-described parameters r^* and α^* and the A_{NAT}-to-A_{MIX} ratio:

$$r^* = \frac{R_{c,max}}{R_{c,max,NAT}} = \frac{a_R \cdot \frac{A_{NAT}}{A_{MIX}}}{b_R + \frac{A_{NAT}}{A_{MIX}}} \tag{6.7}$$

$$\alpha^* = \frac{\alpha_0}{\alpha_{0,NAT}} = \frac{a_\alpha \cdot \frac{A_{NAT}}{A_{MIX}}}{b_\alpha + \frac{A_{NAT}}{A_{MIX}}} \tag{6.8}$$

where, A_{NAT} is the corresponding A_{MIX} value for an ordinary concrete mixture and a_R, b_R, a_α and b_α are constants that should, in principle, be calibrated on experimental results. Based on tested mixtures, the values of a_R and b_R can be approximated to be 1.06 and 0.06, respectively, while a_α and b_α are equal to 1.05 and 0.05.

As a result, the individual pairs of $R_{c,max}$ and α_0 can be divided by the corresponding values of r^* and α^* and, then, be represented with respect to the effective water-to-cement ratio. The analysis of these results (Pepe 2015) shows that all the points ($R_{c,max}/r^*$ vs. w/c_{eff}) can be aligned on a curve that can be analytically expressed by the well know Abram's law and, particularly, be represented by expression (6.9). Similarly, a possible linear correlation existing between α_0/α^* and w/c_{eff} can be expressed by analytical formulation (6.10).

$$R_{c,max} = \frac{A_R}{B_R^{w/c_{eff}}} \cdot r^* \tag{6.9}$$

$$\alpha_0 = A_\alpha \cdot \left(\frac{w}{c_{eff}} - B_\alpha \right) \cdot \alpha^* \tag{6.10}$$

In formula (6.9) the values of A_R and B_R can be assumed to be equal to 265 and 9.5, respectively, while, in Eq. (6.10) A_α is equal to 1.62 and B_α is equal to 0.28.

The accuracy of the proposed rational method can be estimated by analysing the results plotted in Fig. 6.3. The experimental data and the model prediction results are shown in terms of cube compressive strength for RACs at 28 days. This experimental data can be used as a way to calibrate the proposed rational model and to make it usable as a prediction assessment tool for RAC systems.

6.4 Design Charts and Model Potential

The novel methodology is proposed that is based on defining a parameter A_{MIX}, which is representing the average porosity of a concrete mixture made with recycled concrete aggregates. Whenever the composition of the concrete mixture is defined, the parameter A_{MIX} can be calculated simply, by applying the formula (6.6), and, consequently, the ratio A_{NAT}/A_{MIX} is defined as well. Figure 6.4 shows the general trend of $R_{c,max}$ and α_0 versus w/c_{eff}, for different ratios of A_{NAT}/A_{MIX}, representing

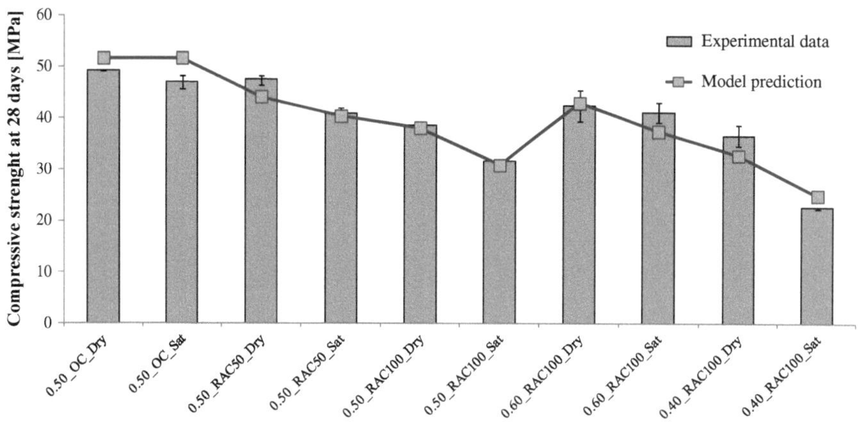

Fig. 6.3 Rational mix design approach calibration

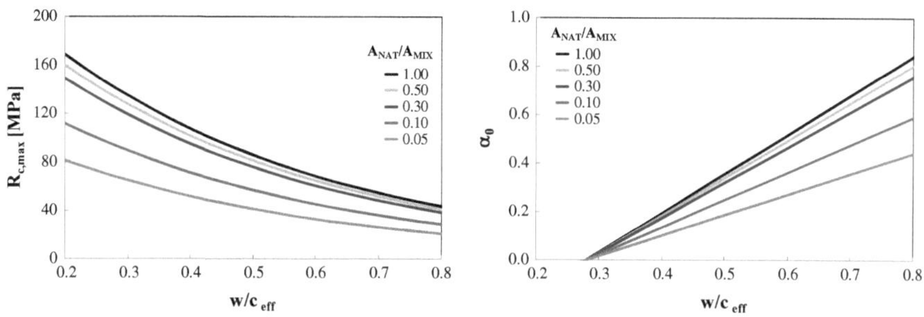

$R_{c,max}$ versus $(w/c)_{eff}$ for different A_{NAT}/A_{MIX} ratios (a)

a_0 versus $(w/c)_{eff}$ for different A_{NAT}/A_{MIX} ratios (b)

Fig. 6.4 Influence of A_{NAT}/A_{MIX} on $R_{c,max}$ and α_0

the amount of recycled concrete aggregates or, more specific, the amount of AM inside a resulting concrete mixture, and the influence of the effective water to cement ratio that takes into account the initial moisture condition RCAs.

The curves reported in Fig. 6.4, are described by the proposed formulas (6.9) and (6.10), have the meaning of a mix design abaci for RAC. In fact, the numerical model proposed by the authors allows for an easy analysis of the evolution of the compressive strength and degree of hydration. Alternatively, the degree of hydration at 28 days can be determined by means of the following relationship (Hansen 1986):

$$\alpha_{max} = \frac{1.031 \cdot \frac{W}{C}_{eff}}{0.194 + \frac{W}{C}_{eff}} \qquad (6.11)$$

Fig. 6.5 Rational mix design method potential

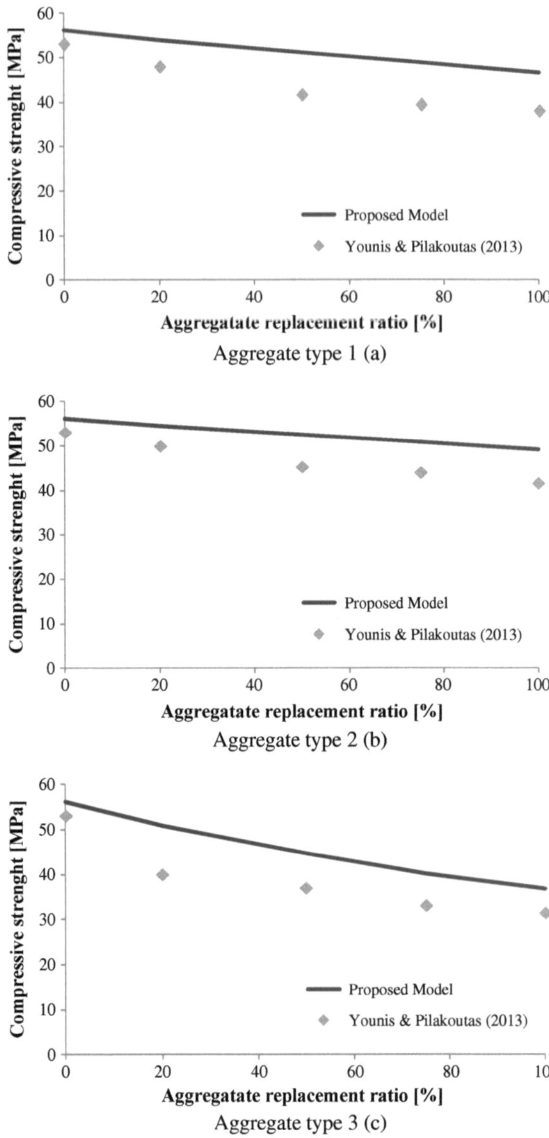

Finally, to evaluate the potential of the above described rational method, some experimental results available in literature (Younis and Pilakoutas 2013) were analyzed. The results of this analysis are reported in Fig. 6.5 and shows that even if the proposed formulation should be better calibrated with further experimental data, the introduction of the parameter A_{MIX} is able to include the changes of the mechanical performance of RACs compared with ordinary mixtures.

6.5 Conclusions

This section has proposed a rational design method to assess the compressive strength development of concretes made with recycled concrete aggregates (RCAs). The following conclusions can be drawn:

- the presence of Attached Mortar (AM) in RCAs results in a higher porosity than in NAs and influences properties, such as water absorption capacity and particle density;
- due to a higher porosity of RCAs, the resulting concrete performance is significantly affected by the following parameters: initial moisture condition of RCAs, aggregates replacement ratio and the nominal water-to-cement ratio;
- a physically based definition of effective water-to-cement ratio is proposed for taking into account the effect of the aforementioned parameters;
- time evolution of the compressive strength of RACs can be estimated by considering those parameters that are characterising recycled aggregates, i.e. attached mortar that is included in the proposed parameter A_{MIX}.

Finally, it may be worth highlighting that, in spite of the promising conceptual results emerging from the proposed model, a wider set of experimental data is needed to confirm its general applicability. This would contribute to the formulation of a quantitative method to predict the mechanical behaviour of Recycled Aggregate Concretes.

References

ASTM D4404 (2010) Standard test method for determination of pore and pore distribution of soil and rock by mercury intrusion porosimetry. Am Soc Test Mater (USA)

ASTM E1570 (2011) Standard practice for computed tomographic (CT) examination. Am Soc Test Mater (USA)

Behera M, Bhattacharyya SK, Minocha AK, Deoliya R, Maiti S (2014) Recycled aggregate from C&D waste and its use in concrete—a breakthrough towards sustainability in construction sector: a review. Constr Build Mater 68:501–516

Hansen TC (1986) Physical structure of hardened cement paste. A classical approach. Mater Struct 19(6):423–436

Lokhorst SJ (1999) Deformation behaviour of concrete influenced by hydration related changes of the microstructure. Internal report nr. 5-99-05. Delft University of Technology, Netherland, pp. 178

Martinelli E, Koenders EA, Caggiano A (2013) A numerical recipe for modelling hydration and heat flow in hardening concrete. Cem Concr Compos 40:48–58

McNeil K, Kang THK (2013) Recycled concrete aggregates: a review. Int J Concrete Struct Mater 7(1):61–69. https://link.springer.com/article/10.1007/s40069-013-0032-5. doi:10.1007/s40069-013-0032-5

Pepe M (2015) A conceptual model for designing recycled aggregate concrete for structural applications. Springer, Springer Theses Series, Berlin

Pepe M, Toledo Filho RD, Koenders EAB, Martinelli E (2014) Alternative processing procedures for recycled aggregates in structural concrete. Constr Build Mater 69:124–132

Younis KH, Pilakoutas K (2013) Strength prediction model and methods for improving recycled aggregate concrete. Constr Build Mater 49:688–701

Part II
Cementitious Materials Reinforced with Recycled or Natural Fibers—Technology, Properties, Design and Applications

Chapter 7
Introduction

Joaquim A.O. Barros and Liberato Ferrara

Fibre Reinforced Concrete (FRC) and Fibre Reinforced Cementitious Composites (FRCCs) represent nowadays a quite mature technology in the field of concrete construction industry, with applications ranging from slabs on grade and elevated slabs, to precast tunnel segments and other kinds of prefabricated structural elements. Increasing awareness of the great engineering potentials of this technology has also results into acceptance by several national and international design codes, which now already have incorporated design provisions for FRC elements (see for example *fib* Model Code 2010).

The major part of the research and applications have been done in Steel Fibre Reinforced Concrete (SFRC), thereby the available design prescriptions were, in general, supported on the results provided by these initiatives. In the last decades polymer fibres are also gaining wide popularity and proving to constitute an efficient reinforcement in several engineering applications, including structural.

Anyway, as the price for raw materials employed in the production of industrially manufactured fibres has started rising, together with the continuously increasing demand of concrete and cement based materials, mainly in developing countries, interest has likewise arisen with reference to the possibility of employing recycled fibres in concrete. Documented experiences on the use of recycled fibres from different sources can be found, including, *e.g.*, recycled plastic fibres from post-consumer goods (e.g. waste PET bottles; see Silva et al. 2005; Ochi et al. 2007; Kim et al. 2008, 2010; Pelisser et al. 2012; Borg et al. 2016), besides recycled steel fibres (RSF) from disposed tyres, and, fibres obtained as by products from other industrial processes, such as cellulose pulps and other types of

J.A.O. Barros (✉)
University of Minho, Braga, Portugal
e-mail: barros@civil.uminho.pt

L. Ferrara
Politecnico di Milano, Milan, Italy

J.A.O. Barros et al. (eds.), *Recent Advances on Green Concrete for Structural Purposes*, Research for Development, DOI 10.1007/978-3-319-56797-6_7

cellulosic/vegetable/natural fibres, generally obtained as residues of agriculture and food industry processes, which will be both dealt with in the forthcoming sections.

Not seldom, the use of "recycled fibres" my also feature added value, besides the one related to the "circular economy concepts", bringing the possibility to incorporate, through the recycled fibres, into the cementitious matrix some "non-conventional" cutting edge functionalities. This may be the case, for example, of natural fibres, which, thanks to their porous hierarchic structure, can effectively act as catalysts of self-healing.

Some kinds of these recycled fibres, as vegetable ones, are naturally available in large quantities in developing economy countries, throughout Latin America, Africa and South-East Asia, where their "secondary raw-material" characteristics may thus represent additional source of income, e.g. for rural communities. Moreover, due to the surging demand, e.g., for affordable housing in the aforementioned countries, tailored applications may be sought for this kind of fibres, thus characterizing their "signature engineering" potential. The research activities performed in the framework of the "EnCoRe" project have mainly focused on recycled steel fibres from disposed tyres, as an alternative to industrially manufactured ones, and natural fibres, also in combination with steel ones, and mainly as promoters/facilitators of self-healing functionalities in High Performance Fibre Reinforced Cementitious Composites (HPFRCCs).

In this part of the book the acronyms RSFRC and NFRC are adopted for referring the concrete reinforced with, respectively, recycled steel fibres and natural fibres.

This second part of the book reviews the main findings of the aforementioned research in the framework of the current state of art knowledge on the topic, according to the following Table of Content structure.

Chapter 8 deals with cementitious composites reinforced with recycled steel fibres from disposed tyres, and Chap. 9 presents the most significant results of the research on cementitious composites reinforced with natural fibres.

Design approaches for sustainable concrete and cementitious composites reinforced with recycled/natural fibres are presented in Chap. 10, together with inverse analysis methods and advanced numerical models employed to obtain the fundamental design properties for this kind of composites, which are respectively dealt with in Chaps. 11 and 12. Finally, examples of applications of sustainable cement based materials are presented in Chap. 13 with reference to some significant case studies.

References

Borg RP, Baldacchino O, Ferrara L (2016) Early age performance and mechanical characteristic of PET fibre reinforced concrete. Constr Build Mater 108:29–47

Kim J, Park C, Lee S, Won J (2008) Effects of the geometry of recycled PET fiber reinforcement on shrinkage cracking of cement based composites. Composites, pp 442–450

Kim S, Yi N, Kim H, Kim J, Song Y (2010) Material and structural performance evaluation of recycled PET fiber reinforced concrete. Cement Concr Compos 32:232–240

Ochi T, Okubo S, Fukui K (2007) Development of recycled PET fiber and its applications as concrete-reinforcing fiber. Cement Concr Compos 29:448–455

Pelisser F, Montedo O, Gleize P, Roman H (2012) Mechanical properties of recycled PET fibers in concrete. Mater Res, pp 679–686

Silva DA, Betioli AM, Gleize PJP, Roman HR, Gomez LA, Ribeiro JLD (2005) Degredation of recycled PET fibres in Portland cement-based materials. Cem Concr Res 35:1741–1746

Chapter 8
Cementitious Composites Reinforced with Recycled Fibres

Joaquim A.O. Barros, Cristina Frazão, Antonio Caggiano, Paula Folino, Enzo Martinelli, Hernan Xargay, Zia Zamanzadeh and Lúcio Lourenço

Abstract Pneumatic tyres are nowadays among the most widespread industrial products and, hence, handling tyres that have reached their end-of-life is indeed a critical issue. This Chapter provides readers with the key facts about tyre production and consumption. Particularly, it describes the raw materials which they are made from, and summarises the main classifications currently accepted worldwide. The typical product life-cycle is shortly outlined before analysing the possibility of recycling waste tyres for producing new products. Relevant properties of concrete reinforced with recycled fibres are presented, with special focus on the post-cracking behaviour of these composite materials. Emphasis is given to the suitability of using recycled fibres obtained from waste tyres as partial or total replacement of industrial steel fibres for obtaining fibre reinforced concrete.

8.1 Recycling Waste Tyres: Technological Issues and Applications in Concrete Production

8.1.1 Waste Tyres: Key Facts

The total production of tyres in the EU is currently around 355 million units per year and, hence, as a rule of thumb, around 0.7 tyre is produced yearly per capita, whereas it is around 1 tyre per capita in the United States (EPA 1999).

J.A.O. Barros (✉) · C. Frazão · Z. Zamanzadeh · L. Lourenço
University of Minho, Braga, Portugal
e-mail: barros@civil.uminho.pt

A. Caggiano
University of Buenos Aires, Intecin - CONICET, Buenos Aires, Argentina

E. Martinelli
University of Salerno, Salerno, Italy

P. Folino · H. Xargay
University of Buenos Aires, Buenos Aires, Argentina

J.A.O. Barros et al. (eds.), *Recent Advances on Green Concrete for Structural Purposes*, Research for Development, DOI 10.1007/978-3-319-56797-6_8

The European production corresponds to 24% of the total world production and is developed in about 90 plants. The industrial sector as a whole is worth around 45 billion Euros (2013) with more than 350,000 direct employees (ERTMA 2014). Since 1999 more than 24 million tons of "end-of life tyres" (ELTs) were recovered, and in 2010 only 4% of ELTs was disposed into landfills (ETRMA 2014).

Tyres are designed to withstand harsh weather conditions and highly demanding mechanical actions. Therefore, the durability requested during their working life makes difficult their processing and/or, in case, recycling. Moreover, due to their hollow structure, they tend to occupy a lot of space when disposed, and their lifetime in landfill is considered to be between 80 and 100 years. This casts serious doubts about the efficacy and sustainability of land filling as a solution for disposing of waste tyres.

Furthermore, tyres are made of thermosetting polymers which, as a matter of fact, makes it impossible to melt them again at their end of life and separate into their original chemical constituents. However, alternative technologies are available nowadays for processing waste tyres. In principle, the technological solution adopted for their processing can have an impact on the possible uses of the obtained materials.

As a matter of definition, a pneumatic tyre having ceased to be capable of performing its function (namely, to be installed on the wheel of a vehicle to enable mobility in a safe condition) is generally referred to as "end-of life tyre" (ELT). Conversely, "used tyre" is a more generic term, as it indicates that the tyre is not new and has been used. Used tyre should not be considered as a "waste" in the case it is reused in its actual state of wear or reconstructed; conversely, used tyre should be considered as a waste if neither reusable nor rebuildable or have been abandoned or sent for recovery or disposal (Torretta et al. 2015).

Two principal methods, based on either mechanical or thermo-chemical processes, are suitable for obtaining reusable materials from waste tyres. The former consists in cutting the tyres at ambient temperature, whereas the latter implies heating or cooling before further processing them up to the target size reduction. Beside the specific technological features of each single procedure, size reduction technologies are often classified according to the four levels described below, based on their functions (ETRA 2013):

- Level 1: destruction of the tyre structure;
- Level 2: separation of the tyre elements, which can be split into two sublevels: the former consists in obtaining shreds or chips and removing the metal parts not encased in rubber; the latter foresees a further reduction in size up to granulate or powder and removal of the metal and textile fractions as well;
- Level 3: multi-treatment technologies for modifying the separated materials by means of thermo-chemical processes, such as de-vulcanisation, reclamation, surface modification and pyrolysis;
- Level 4: material upgrading technologies for refining the output of Level 3 processes with more sophisticated technologies.

Therefore, in spite of the technical issues characterizing the end-of-life treatment, waste tyres are fully recyclable and three main materials (i.e. rubber, metal and textiles) come out of the recycling processes. Among the various alternative techniques currently available for this purpose, the following three thermochemical processes are generally carried out to convert scrap tyres into recyclable products:

- pyrolysis
- gasification
- liquefaction

Section 8.1.4 describes into details the various technological alternatives currently available for separating the constituents of tyres and turn them into reusable materials. A comprehensive model capable of assessing the actual environmental impact or benefit of various technologies and their combination in a complete recycling process has been proposed by Pehlken and Müller (2009).

As regards the size reduction of waste tyres, the following definitions are generally adopted for the materials resulting from the mainly mechanical processes mentioned above:

- Shreds, consisting in irregular particles with a maximum size ranging between 50 and 300 mm obtained via mechanical treatments;
- Chips, consisting in slightly less irregular particles of size ranging between 10 and 50 mm;
- Granulate, whose size ranges between 1 and 10 mm: they can either be obtained through mechanical processes run at ambient temperature or cryogenic treatments;
- Powders, particles smaller than 1 mm in diameters, obtained either from either ambient temperature treatments or other more complex technologies, such as pyrolysis, reclamation and de-vulcanisation;
- Very fine powders, namely particles with a diameter smaller than 500 μm, which can only be obtained by means of specialized treatments.

In EU the European Committee for Standardization (CEN) has classified products obtained from grinding of waste tyres according to their size (CEN/TS14243 2010). Particularly, the following five classes are defined:

- Cut tyres, with size bigger than 300 mm
- Shreds, with size ranging between 20 and 400 mm
- Chips, with size ranging between 10 and 50 mm
- Rubber granulate, with size ranging between 0.8 and 20 mm
- Rubber dust, smaller than 0.8 mm.

Finally, it is worth highlighting that, in principle, the EU promotes the three types of tyre management (ETRMA 2010): management model based on Extended Produced Responsibility (EPR); a tax system defining a disposal duty for producers or sellers; and the free market system that assumes the profitability of recovery and recycling of tyres. Nevertheless, there are no specific EU standards for the recovery of used tyres (Sienkiewicz et al. 2012).

8.1.2 Classification of Pneumatic Tyres

Tyres have a mixed composition and are made up of the four main parts (Torretta et al. 2015):

- the tread—designed for contact with the ground and to ensure the proper friction;
- the carcass—the structural part of the tyre on which the tread is vulcanised;
- the shoulder—which minimizes the effects of irregularities of the terrain and transfers the load due to braking and oversteering under acceleration;
- the heels—to fit the casing to the rim.

A tyre consists not only of rubber, which makes up some 70–80% of the tyre mass, but also of steel belts and textile overlays, which give the tyre its ultimate shape and fit-for-the use properties (Sienkiewicz et al. 2012). Particularly, the following composition (in percentage by weight) can be taken as a reference (ETRMA 2014):

- Rubber/Elastomers (48%);
- Carbon black (22%);
- Metal (15%);
- Additives (8%);
- Textile (5%);
- Zinc oxide (1%);
- Sulfur (1%).

This means that a wide set of natural resources is required for producing pneumatic tyres.

Pehlken and Müller (2009) confirmed the aforementioned figures: they reported that in Europe natural rubber contributes to 22% of the total weight of tyres for passenger cars and 30% for trucks; and that synthetic rubber is used for as much as 23% and 15% of weight in the two aforementioned types of tyres. Moreover, the same authors report that in Europe the amount of steel generally used in tyres sums up to about 13 and 25% of the total tyre weight for cars and trucks, respectively (Pehlken and Müller 2009).

A classification of tyres based on the purpose of the vehicle is often adopted, as it is useful in estimating both the number of tyres actually reaching the EOL and quantifying the weight and composition of recycled products (Mountjoy et al. 2015):

- Passenger tyres, including those used on passenger vehicles, motorcycles and caravans, as well as trailers for domestic use;
- Truck tyres, including those used on buses, light and heavy commercial vehicles, prime movers, trailers and semi-trailers, and fire fighting vehicles;
- Off-the-road (OTR) tyres, including those used on machinery or equipment used in industries such as agricultural, mining and construction and demolition.

Particularly, the average weight of a used passenger tyre is considered to be higher than in the past, ranging between 8.5 and 10 kg; higher and more variable weights should be considered for the other two categories.

Finally, another possible classification for tyres is based on the structure of the reinforcing "carcass" made of fibre "plies". Two main classes of tyres may be categorized according to this criterion (Michelin 2015):

- radial tyres (or, more properly, a radial-ply tyres) in which the cord plies are arranged at 90° to the direction of travel, or radially (from the centre of the tire);
- bias tyres in which the cord plies are arranged in multiple crossing layers that are generally 32° to 40° from each other from the centre line of the tread.

Figure 8.1 shows a schematic view of the two alternative arrangement of the so-called "carcass" in the two aforementioned structural solutions.

Although the radial solution is now the reference design for essentially all automotive tyres (Thomson 2001), the bias technology is still common on trailers due to their weight carrying ability and resistance to swaying when towed (Burden 2015). Moreover, bias tyres are also popular for aircrafts as they guarantee more stability at higher speeds and have stronger sidewalls (Dunlop 2013).

A more detailed scheme of a typical radial tyre is reported in Fig. 8.2, highlighting the various parts constituting the tyres structure and finishing.

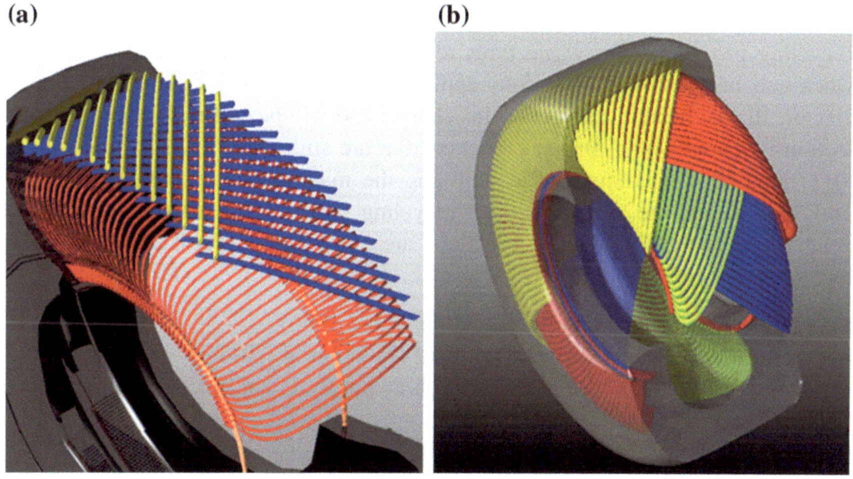

Fig. 8.1 Alternative fibre ply arrangements in tyres: **a** radial, and **b** bias (Michelin 2015)

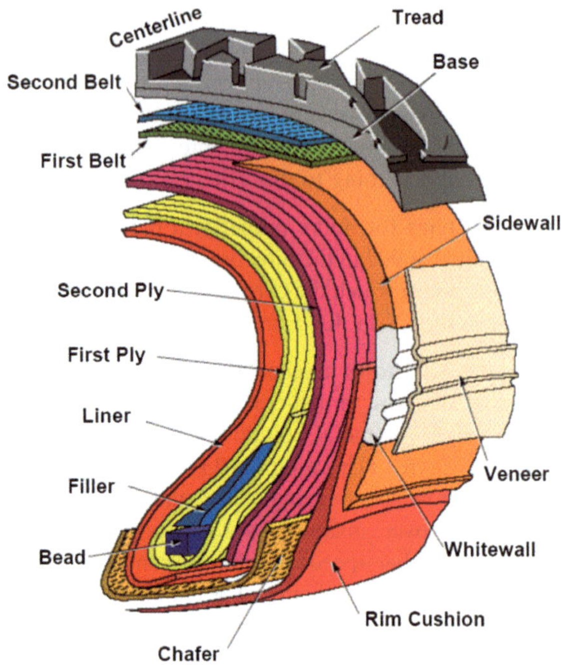

Fig. 8.2 Typical structure of a radial tyre (IWMB 2003)

8.1.3 *Alternative Recycling Processes*

Recycling procedures for waste tyres requires separating their main components, which can be employed in various applications, among which the production of concrete that is the main subject of this chapter and will be outlined in the following Sections. Beyond some further processes that are still under development (UNEP 2011), this section is intended at outlining the main technological solutions currently adopted in the aforementioned recycling procedures. Particularly, the ones summarized in the following subsections are listed below (Pehlken and Essadiqi 2005):

- Ambient or Cryogenic Grinding (Blumenthal 1996);
- De-vulcanisation and surface treatment (Eldho et al. 2011);
- Thermal conversion technologies, such as pyrolysis and microwave treatments (Juma et al. 2006);
- Energy recovery (Laboy-Nieves 2014);
- Re-treading (Sharma 2013).

However, the potential of waste tyre treatment through processes such as gasification and liquefaction is undervalued: the increase in global awareness of environmental friendly treatment methods will lead the EU countries consider these processes as future waste tyre treatment methods.

8.1.3.1 Grinding Processes

Grinding processes belong to the wide class of abrasive processes that can be performed in industrial applications for removing material layers or particles from the surface of objects. They are also used for processing waste tyres, since they are capable of removing the outer rubber layers. The two following subsections outline the main technological aspects and applicative implications of this process, by considering the two options of ambient and cryogenic conditions.

Ambient Grinding

The grinding process performed at 'ambient' temperature (namely without cooling) is referred to as "ambient grinding". In detail, tyres are preliminarily shredded and the rubber chips enter into a granulator that reduces their size to less than 10 mm in diameter: this results into separation of most of the steel and fibre particles from the rubber chips. The formers are then removed magnetically (the steel fraction) and through shaking processes (fibre fraction).

Since most applications require finer rubber particles (i.e. between 0.6 and 4.0 mm in diameter), ambient grinding plants often operate several consecutive grinding steps: secondary granulators, high-speed rotary mills and extruders, screw presses or cracker mills are the machines most commonly used to obtain the required fineness of rubber particles in ambient grinding plants.

It is worth mentioning that, ambient grinding is the reference choice for producing relatively coarse crumbs (larger than 0.6 mm) of recycled rubber. Moreover, rubber particles generated in an ambient grinding plant have a rough surface and tend to be partly devulcanised (Pehlken and Essadiqi 2005).

Cryogenic Grinding

The process referred to as "cryogenic grinding" consists in cooling down to a temperature below −80 °C either whole tyres or tyre chips by means of liquid nitrogen: this is a "glass transition temperature" below which rubber becomes almost as brittle as glass, and size reduction can be more easily performed through crushing and breaking. As a first advantage, cryogenic size reduction of rubber requires less energy and fewer pieces of machinery than ambient size reduction. Moreover, steel and fibre separation is much easier, leading to a cleaner product. Conversely, the need for liquid nitrogen (LN_2) results in more complex technological equipment and higher costs (Blumenthal 1996).

Common cryogenic processes generally start with a preliminary shredding, which is largely similar to the one carried out in ambient grinding plants. Afterwards, tyre chips are cooled down in a freezing tunnel to below −120 °C and then dropped into a quickly rotating hammer mill. In the hammer mill, the chips are shattered into a wide range of particle sizes. Since the rubber granules may be damp upon leaving the hammer mill, the material is dried before sorting into well-defined particle sizes. A secondary cryogenic grinding step is required to produce finer rubber particles (Pehlken and Essadiqi 2005).

The cryogenic grinding process is often adopted for producing rubber particles finer than the ones generally resulting from the ambient process. Due to the embrittlement, the texture of the rubber particles is different from that of the ambient particles: very sharp edges of the particles and less surface area are evident compared with the rubber particle obtained through ambient grinding (Liang and Hao 2000).

8.1.3.2 De-vulcanisation and Surface Treatment

De-vulcanization is a chemical process that consists in returning rubber from its thermoset state back into a moldable one. This result is accomplished by selectively severing the sulphur bonds in the molecular structure of rubber particles already separated from the steel and fibre fractions, which are generally present in pneumatic tyres. Therefore, both de-vulcanisation and surface treatment for rubber requires previous ambient or cryogenic processing. This processing step enables rubber manufacturers to use a much larger percentage of recycled material without compromising quality, appearance or performance characteristics of the final products. Alternative methods and technique are available for rubber de-vulcanization (Pehlken and Essadiqi 2005):

- Thermal reclamation process
- Mechanical de-vulcanisation
- De-vulcanisation with ultrasound and
- Bacterial de-vulcanisation.

In recent years, several methods have been developed to chemically treat or modify the surface of rubber crumb particles with the aim of increasing the adhesion between the rubber granules and the rubber polymer during vulcanization.

The estimated cost of producing de-vulcanized materials from waste tyres is of \$0.20–\$0.50/lb (1 lb = 0.45 kg—IWMB 2004). However, these figures are inadequately documented in the literature and, in fact, they are supposed to largely depend on local aspects, such as waste tyre disposal fees, and fees for disposal of process residues.

8.1.3.3 Thermal Conversion Technologies

Two main thermal technologies are sometimes adopted for processing waste tyres, namely pyrolysis and microwave technology, whose main aspects are outlined hereafter.

Pyrolysis

Pyrolysis consists in a thermal decomposition of organic material in the absence of air. The process is typically carried out either in a flow of inert gas or under vacuum conditions. As mentioned in the previous sections, elastomers, carbon black filler/reinforcement material and other products, such as zinc oxide and steel, are the main components of tyres. Upon heating, the rubber and other organic compounds decompose and are converted into oil and gas. The pyrolysis residue consists of the recovered carbon black filler, inorganic materials and varying proportions of carbonaceous materials formed from the rubber decomposition products. Tyre crumbs (0–7 mm) are fed into a pyrolyser and conveyed by a raking-mixing system circulating over two fixed heating plates. The pyrolysing tyre particles remain inside the reactor for approximately 12 min at 500 °C, and the pyrolysis pressure is 15–20 kPa (Pehlken and Essadiqi 2005).

The heavy oil is condensed in the first packed tower, and the lighter oil is recovered at the bottom of the second packed tower. Excess pyrolytic gas is available to generate steam, or it can be burned in a gas turbine. The amount of gas is sufficient to maintain the energy supply for the pyrolysis plant, and the excess gas can be sold.

Yields of a vacuum pyrolysis plant at 500 °C and 20 kPa with a capacity of 36,000 metric tonnes are estimated as follows (Pehlken and Essadiqi 2005):

- 34% (in weight) pyrolytic oil 9 (calorific value 43,800 kJ/kg, sulphur content 1.2%);
- 32% pyrolytic carbon black (grade similar to commercial black N300),
- 17% gas (calorific value 46,000 kJ/kg, rich in hydrogen, methane propane, butane);
- 15% steel;
- 2% fibres other then steel.

In 2010, about 3.3 million tonnes of used tyres were managed in an environmental acceptable manner in the European Union (EU), a 2% increase from 2009 (Nkosi and Muzenda 2014). About 2.7 million tons of used tyres were treated (recycled and recovered). Currently there is a great deal of research on waste tyre pyrolysis. Forty companies are estimated to be working worldwide on tyre pyrolysis (Jonathan and Kiser 2002). Despite more than 30 years of research on the topic, the pyrolysis of scrap tyres and related waste materials has not achieved commercial success in the United States, where, based on data available in the literature (IWMB 2004), unit production "costs" were typically estimated in the range of $0.20–$0.50/lb. It should be noted that these figures can be significantly influenced by "local" aspects such as waste tyre disposal fees, and fees for disposal of process residues.

Microwave Treatment

Microwave treatment of waste tyre rubber is based on reverse polymerization through a low-temperature (250–350 °C) microwave-induced sublimation process, in which reduction occurs in a nitrogen-rich environment, without producing unwanted combustion products. All tyre-reduction products are used or recovered: a typical microwave plant has a capacity of 18,000 metric tonnes per year, equivalent to about 2.2 million passenger tyres (Pehlken and Essadiqi 2005).

Tyres are unloaded or conveyed from inventory, visually inspected for dirt and stones and washed, if required, prior to loading onto a dry feed system consisting of four process lines, each one capable of reducing 1500 tyres per day. At the dry feed system, air is purged with nitrogen until the nitrogen concentration is equal to that of the reduction tunnel where the tyres are being reduced. Each reduction tunnel consists of a number of contiguous chambers (i.e. ten chambers), and each chamber is fitted with several microwave generators (i.e. fifteen per chamber) mounted on top of the tunnel. The microwaves engage the scrap tyres with energy as they advance along the stainless steel conveyor and are subsequently reduced.

The reduction tunnel is continuously purged with nitrogen and monitored for total oxygen content to ensure safe operation. The hydrocarbon vapours are removed from the reduction chamber through insulated piping exiting the top and centre walls. The remaining solid material (steel and carbon black) is removed from the reduction tunnel through a water trap. This ensures that all of the hydrocarbon vapours remain in the reduction tunnel.

The solid material remaining after the reverse polymerization process is carried on the conveyor to the water trap. The water trap has three functions:

- Prevents gases in the reduction chamber from escaping,
- Separates the majority of the carbon black from the steel, and
- Functions as a safe overpressure relief valve in an emergency

Recent studies have highlighted that microwave based processes lead to reactions quicker than in traditional heating techniques. Higher efficiency (3 kW power per 0.2 kg of tyres) can be achieved and more valuable products (such as a solid residue containing up to 92.03% of carbon, a low viscosity oil and a gas containing light hydrocarbons, hydrogen and only traces of nitrogen) characterized by a high calorific value can be also obtained (Undri et al. 2012).

8.1.3.4 Energy Recovery Solutions

Although uncontrolled fires of waste tyres is known to cause substantial air and ground pollution, the controlled high-temperature incineration of either whole tyres or rubber chips in industrial furnaces is environmentally safe.

The calorific value of tire-derived fuel (TDF) exceeds the one of coal, while the sulphur content is in the same order of magnitude or even lower. Particularly, the combustion process is spontaneous above 400 °C, highly exothermic and once turned on becomes self-supporting: waste tyres characterized by a calorific value of 7500 ± 8000 kcal/kg are used as fuel in the incinerators (Sharma et al. 2000).

The use of TDF as a fuel supplement in cement kilns is another viable means of safe disposal of large amounts of scrap tyres. It can be employed also in paper mills, but in this case tyre shred with most of the steel removed should be employed (Pehlken and Essadiqi 2005).

8.1.3.5 Re-treading

The greatest cost in the manufacture of a new tyre is the tyre body, since the tread, namely the portion of the tyre that will go into contact with the road surface, represents a relatively small portion of the total production cost (Ferrer 1997).

Therefore, "re-treading" is, in principle, a cost-effective solution for reusing end-of-life tyres and, hence, reducing the amount of waste to be disposed in landfill or handled through alternative recycling techniques (Pehlken and Essadiqi 2005).

However, not all tyres can be re-treaded: in order to meet safety requirements and only carefully inspected tyre bodies are re-treaded. Once the worn layer is removed, the new tread is bonded to the tyre body in a process very similar to the manufacture of a new tyre. Tyre re-treading is a well-established industrial activity that started in the early 1900s. Passenger cars, aircrafts, sand and gravel trucks, delivery vans, farm equipment and earthmovers can all use re-treaded tyres. Moreover, 80% of all aircraft tyres are re-treaded (ITRA 2001).

8.1.3.6 Comments About Feasibility and Convenience of the Available Technical Solutions

The feasibility of the solutions cannot be compared directly using definitive numbers for each process. Most processes are influenced by several factors including labour costs, transportation, tyre supply and product quality. A life-cycle analysis for each technical method of processing tyres is required to generate accurate data (Feraldi et al. 2013): however, this is beyond the scopes of this book.

The rubber market affects the processing of scrap tyres such that higher revenues enable improved processing technologies. Moreover, the prices of recycled tyres are also affected by crumb quality, crumb coloration, purchase quantity, competitive pricing factors, impact of subsidies, and negotiations between producers and end-users (Pehlken and Essadiqi 2005).

Definitions of quality appear to be quite diverse and are driven by customer specifications unique to different market segments. In general, a "high quality of crumb" means low fibre content (less than 0.5% of total weight), low metal content (less than 0.1%). An accepted level of moisture content is about 1% in weight (Duffey and Sunthonpagasit 2004).

Processing and shredding costs for scrap tyres include power, labour, equipment and maintenance costs are other aspects of relevance: as a general trend, the smaller the crumb size, the higher the investment and operating costs: in general, energy, labour, and other variable costs are largely a function of the product particle size, quality and quantity.

As regards thermal conversion, cryogenic processing plants have higher investment and provision costs due to the need for liquid nitrogen, but their labour and maintenance costs are lower. Plant location is important since the cost of liquid nitrogen increases when the supplier is not close by. Moreover, cryogenic plants can produce a larger quantity of fine crumb rubber and get a higher market price. The cost of transportation alone can be estimated in the order of $0.50 per tyre (Pehlken and Essadiqi 2005).

Use of scrap tyres as TDF in cement kilns or paper mills is financially positive for the processor because it replaces fossil fuel (STMC 1992).

De-vulcanisation, surface modification and thermal conversion technologies are often affected by high investment costs, but they produce a high-quality product. Their feasibility is a function of the market and the interest of consumers (Ferrer 1997).

Finally, re-treading is a highly feasible option and probably the most common because it reduces the manufacture of new tyres and replaces just part of the rubber on the tire. Unfortunately tyres cannot be re-treaded forever, and there are limitations due to quality and safety restrictions (Sharma 2013).

8.1.4 Recycling of Waste Tyres in Concrete Production

In the last years, the suitability and efficiency of using different by-products obtained from the recycling of waste tyres for improving some concrete properties has been investigated. Related research in this field is anyway so far limited, leaving wide space for further investigations (Richardson 2013; Bjegović et al. 2014).

Mainly two by-products obtained from the recycling of waste tyres have been considered to be added to concrete: steel fibres and rubber particles. In the first case, recycled fibres obtained from waste tyres are used as partial or total replacement of industrial steel fibres for the production of Fibre Reinforced Concrete (FRC) (Pilakoutas et al. 2004; Neocleous et al. 2006; Aiello et al. 2009; Nasir 2009; Centonze et al. 2012a, b; Graeff et al. 2012a, b; Jala 2012; Barros et al. 2013a; Caggiano et al. 2015; Rodrigues 2015; Martinelli et al. 2015). In the second case, rubber particles obtained from recycled tyres are used as partial or total replacement of natural aggregates for producing rubberized concrete (Siddique and Naik 2004; Papakonstantinou and Tobolski 2006; Khaloo et al. 2008; Ganjian et al. 2009; Aiello and Leuzzi 2010). In the following Sections, focus will be on the use of recycled fibres obtained from waste tyres for producing structural fibre reinforced concrete.

The characterization of these kind of fibres and the possible applications will also be reviewed. With reference to the replacement of natural aggregates by rubber shreds, it was observed that it reduces unit weight of concrete, causes a considerably decrease of compressive and flexural concrete strengths. Although further research is needed, this material appears to be suitable for different engineering applications such as pavements, and also for thermal or acoustic insulation purposes (Siddique and Naik 2004; Aiello and Leuzzi 2010).

8.1.5 Recycled Fibres from Waste Tyres for FRCC

8.1.5.1 Characterizing Recycled Fibres for Producing Steel Recycled Fibre Reinforced Concrete

So far Fibre Reinforced Concrete using recycled fibres has been produced using the same processes employed for FRC with industrial fibres. However the methodologies proposed by Barros et al. (2007) and Soltanzadeh et al. (2015) have high potentialities for being used, since both consider directly the strong perturbation that fibres introduce in the aggregates skeleton organization (Ferrara et al. 2007). In this section, the relevant characteristics of recycled fibres from disposed waste tyres are highlighted in order to better understand the possible steps in designing this relatively new material.

With reference to mix proportioning, analogously to the case of industrial steel fibres, an increasing quantity of water and superplasticizer has been generally employed with an increase in the percentage of fibres in order to improve the workability of fresh concrete (Aiello et al. 2009). On the other hand, according to the methodology of Barros et al. (2007) the paste content increases, and the maximum aggregate diameter and content of coarse aggregates decrease with the increase of the fibre content and/or fibre aspect ratio (Ferrara et al. 2007).

Contrary to the case of industrial steel fibres, recycled fibres obtained from waste tyres feature variable diameters and lengths and are characterized by irregular shapes with curls and twists, as it can be observed in Fig. 8.3. Moreover, the aspect of the recycled fibres depends not only on the type of tyre used but also on the process that was used at the recycling plant for recovering the fibres.

Therefore, before using this type of recycled fibres for FRC production, geometric properties of the fibres should be assessed. This is fundamental at the research level, due to the influence of these properties on the performance of RSFRC. In the industrial production activity, these properties should be assessed by statistical sampling in order to obtain reliable information on the batch that is being recycled.

The by-product obtained from the recycling plants, looks like the brunch in Fig. 8.3. First of all, thicker pieces of steel, which can still be attached to rubber crumbs, not suitable as fibres must be separated from the rest. Then, it is necessary to properly identify its main geometric parameters as well as the material properties.

Fig. 8.3 Typical recycled steel fibres extracted from waste tyres (Caggiano et al. 2015)

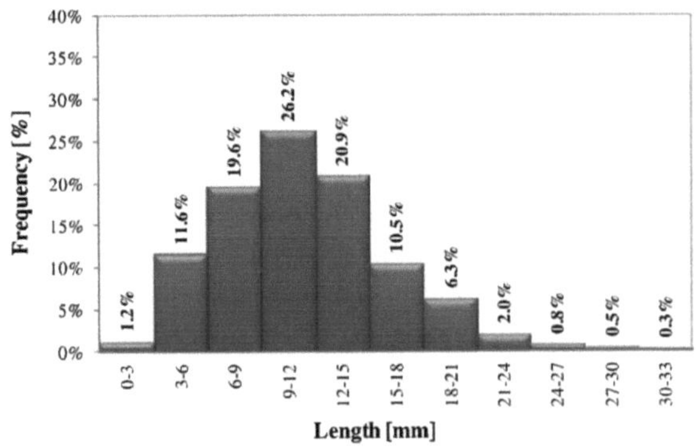

Fig. 8.4 Typical frequency of lengths measured on a group of recycled steel fibres extracted from waste tyres (Caggiano et al. 2015)

For the geometric characterization, the diameter and length of an adequate number of fibres should be measured. In the literature, the group of fibres used for this purpose is above 1000. As example, the results of a geometric characterization over 2000 samples of fibres is presented in Figs. 8.4 and 8.5 were typical frequencies of distribution of lengths and diameters can be respectively observed.

With reference to the diameter, each single fibre should be manually measured by means of a micrometer (Fig. 8.6), and it is recommended to take three measures, namely at the two ends and at the fibre mid-point, reporting an average value of diameter for each fibre.

Considering the irregular shape of these fibres and according to different recommendations in the subject (i.e. CNR-204/2006 specifications), the fibre length is considered as the distance between the outer ends of a fibre as depicted in Fig. 8.7 (see Aiello et al. 2009).

Finally, a key parameter for controlling the fibres mechanical performance in fibre reinforced concrete can be determined: the fibre aspect ratio (length-to-diameter). In Fig. 8.8, a representative distribution is shown of fibre aspect ratio measured in a group of recycled fibres.

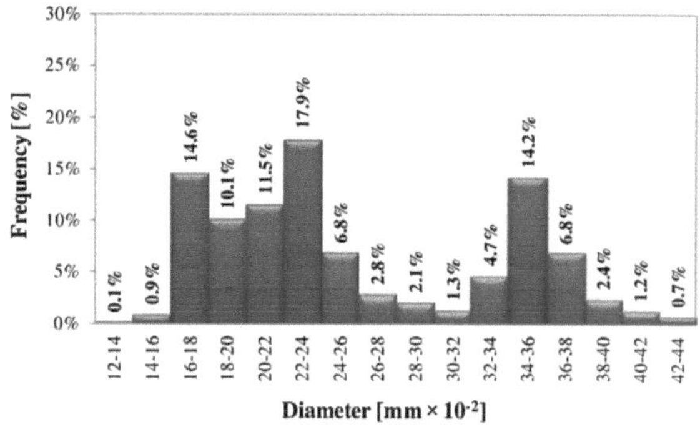

Fig. 8.5 Typical frequency of diameters measured on a group of recycled steel fibres extracted from waste tyres (Caggiano et al. 2015)

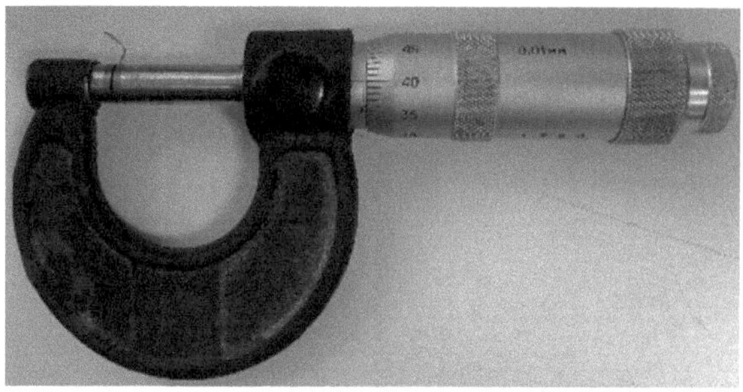

Fig. 8.6 Micrometer (Caggiano et al. 2015)

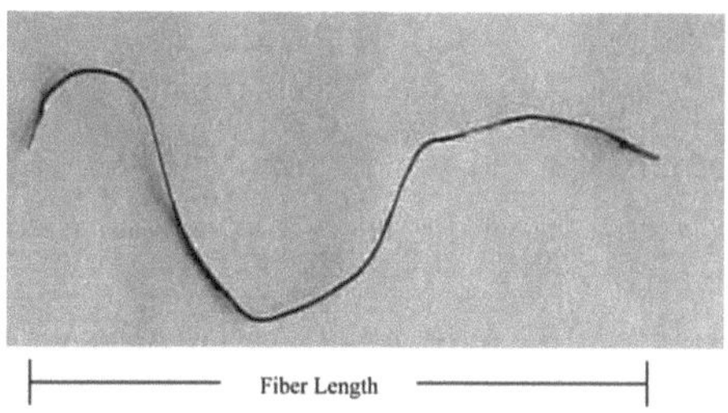

Fig. 8.7 Correct measure of fibre length (Aiello et al. 2009)

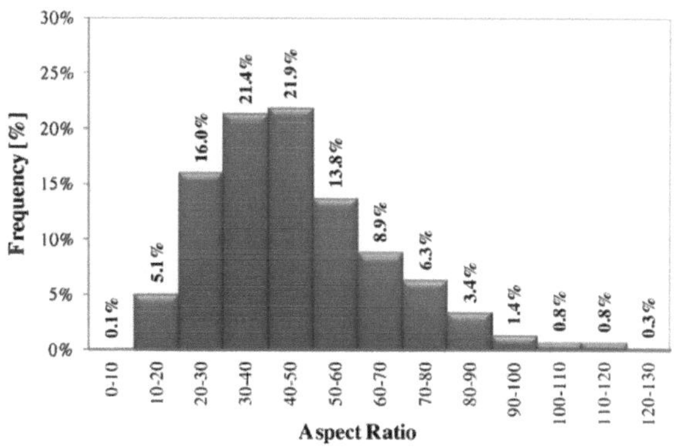

Fig. 8.8 Typical frequency of fibres aspect measured on a group of recycled steel fibres extracted from waste tyres (Caggiano et al. 2015)

The mechanical behaviour of the steel fibres from disposed tyres has to be experimentally identified through direct tensile tests. These tests can be somehow difficult, since an adequate testing machine with appropriate grips is needed for this purpose. The fibre must be kindly straighten out before being submitted to tensile stresses. In Fig. 8.9, images from a typical test are shown. The experimental results in the literature demonstrate that tensile strength of recycled steel fibres obtained from waste tyres is comparable to classical values of industrial steel fibres. In fact Aiello et al. (2009) have measured tensile strength values of 1250–2300 MPa in RSF that is an interval of values found in industrial steel fibres (ISF).

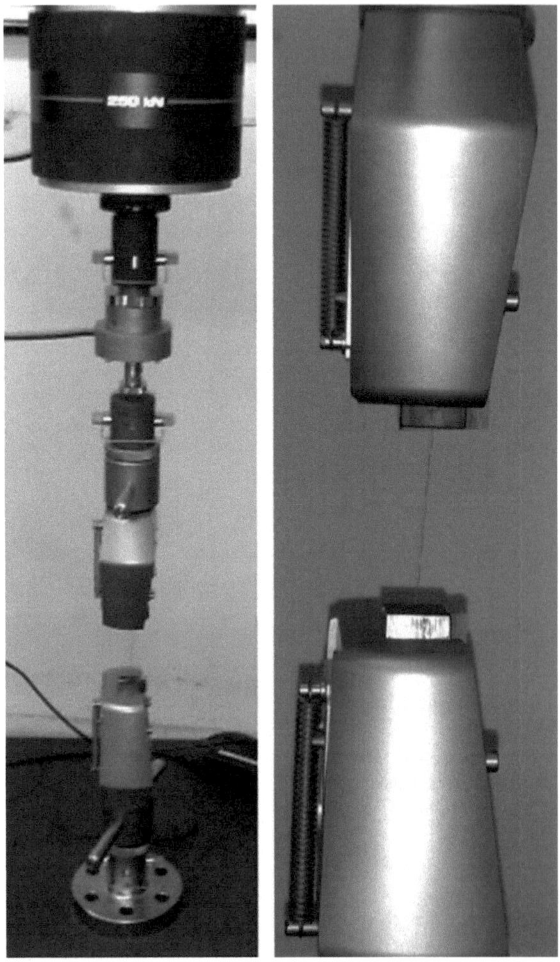

Fig. 8.9 Direct tensile test on recycled fibres. *Photo credits* Laboratory of Materials and Structures, University of Buenos Aires, 2014

8.1.5.2 Insights into the Mechanical Behaviour

The addition of recycled steel fibres obtained from waste tires improves the post-cracking behaviour of cement based materials enhancing ductility, energy dissipation and impact resistance of elements made by this type of composites (Barros et al. 2013b). Therefore, the incidence of recycled fibres on concrete properties does not substantially differ from the corresponding to industrial steel fibres. Nevertheless, experimental evidence shows that the toughness improvement as a consequence of the addition of recycled steel fibres (RSF) is smaller than that produced by industrial steel fibres (ISF) (Martinelli et al. 2015). These aspects will be deeper treated in Sect. 8.2.

For assessing the reinforcing mechanism provided by RSF two types of tests can be adopted: (i) Pull-out tests and (ii) Bending tests. Pull-out tests are used for characterizing the bond behavior fibres-cement paste, but no standard is still available. These tests can provide the "critical length", which is the one above that the tensile rupture of the fibre is attained. A local bond slip law can also be obtained, if the purpose is using this information for design and modelling. In Fig. 8.10 it can be observed the typical layout of a pull-out test on RSF, while in Fig. 8.11 some results for different embedment lengths are shown. As expected, by increasing the fibre length promotes the tendency for the tensile rupture of the fibre, a failure mode that should avoided since this affects detrimentally the fibre reinforcement benefits in terms of energy absorption capacity and ductility. Similar effect is reported in the bibliography when the fiber's inclination increases towards the crack plane. The fibre pull-out mechanism shown in Fig. 8.11 is the one that should be promoted, and the fibre reinforcement performance is as higher as larger is the area under the force-slip curve.

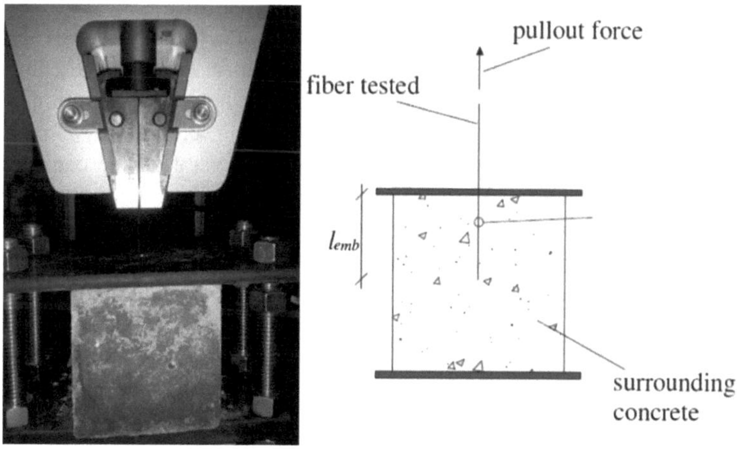

Fig. 8.10 Layout of pull-out test on recycled fibres (Caggiano et al. 2015)

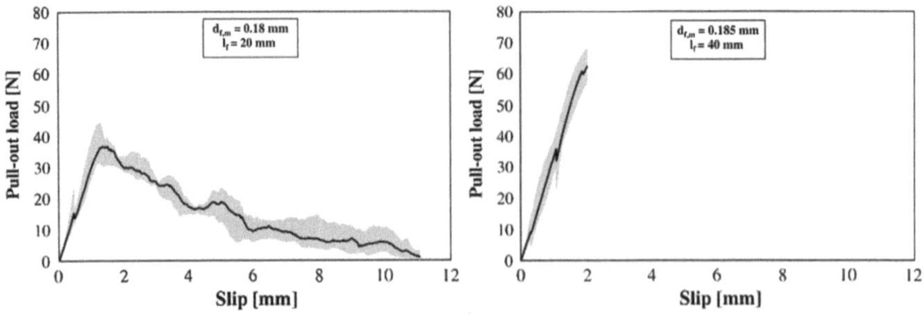

Fig. 8.11 Pull-out test results on recycled fibres for 20 and 40 mm of embedment lengths, respectively (Caggiano et al. 2015)

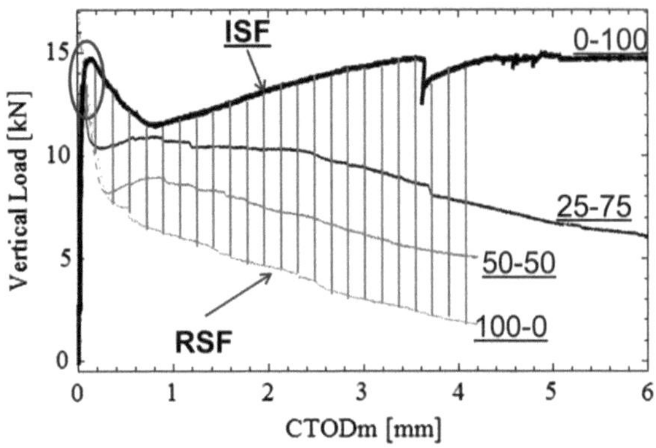

Fig. 8.12 Direct tensile test on recycled fibres (Martinelli et al. 2015)

It is worth to remark that even though RSFs are characterized by smaller diameters than ISFs, both fibres lead to bond strengths of the same magnitude.

Since RSF are not conceived with anchorage mechanisms it is expected to have larger slip than ISF designed with slipping resisting mechanisms. This can have a detrimental effect in terms of limiting the crack with at serviceability limit state conditions. Therefore, when selecting RSF for an application aiming to take advantage of these fibres for limiting the crack opening, preliminary experimental tests should be executed for comparing the residual flexural tensile capacity up to the crack width limit intended to be used in the design of the RSF structure.

In Fig. 8.12 it can be observed the results of Four Point Beam Tests corresponding to a total fibre content of 0.50% in volume. Each curve corresponds to a different replacement percentage of ISF by RSF. It can be noted that the most abrupt post peak corresponds to the case when only RSF are used, and on the contrary, the most ductile behaviour corresponds to 100% of ISF.

In the next section, the mechanical properties of recycled steel fibre reinforced concrete are explained in detail.

8.2 Mechanical Properties

8.2.1 Introduction

Mechanical behaviour of fiber reinforced concrete produced with recycled fibers (RSFRC—Recycled Steel Fibre Reinforced Concrete) has been investigated by means of the same test methods currently employed for the characterization of concrete reinforced with commercially available (steel) fibers. This will be henceforth denoted as Industrial Steel Fiber Reinforced Concrete (ISFRC). This section will review the aforementioned test methods and representative available results on the mechanical behaviour of RSFRC are provided and commented.

8.2.2 Compression

According to Pereira (2006), the four distinct consecutive stages of crack localization and propagation can be identified in concrete under uniaxial compressive load (Shah et al. 1995), as sketched in Fig. 8.13:

- Stage I—below about 30% of the peak stress. The initiation of internal cracks is insignificant, and the stress-strain relationship may be considered as linear;
- Stage II—from about 30 to 80% of the peak stress. Micro-cracks initiate in the weaker spots of the matrix, which is the interfacial transition zone (ITZ), and, upon increase of the stress, they starts to propagate and new micro-cracks develop. At approximately 60% of the peak stress, micro-cracks start forming also throughout the cement matrix. However, all these cracks are isolated and randomly distributed over the material volume;

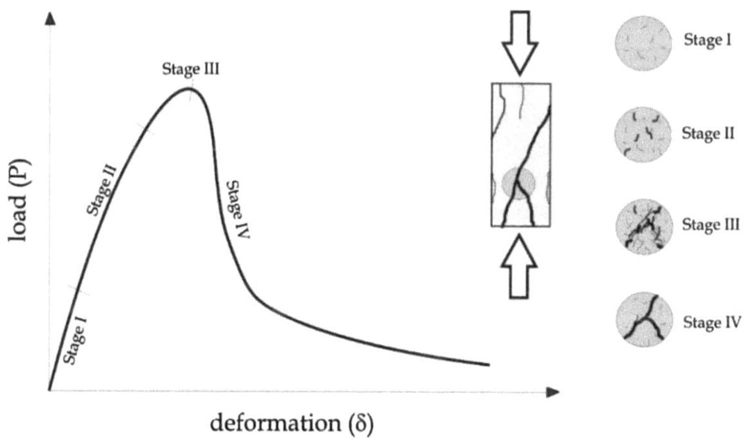

Fig. 8.13 Strain localization in concrete behaviour under compression (Pereira 2006)

- Stage III—between about 80% and up to 100% of the peak stress. At this stage, internal micro-cracks become unstable and start to localize into major cracks. This phenomenon is referred to as damage localization or strain localisation. Anyway, in a short term monotonic tests, the crack growth is stable until the peak is reached, which means that cracks continue to propagate only if the load is increased.
- Stage IV—for deformation levels beyond the one corresponding to the peak stress. The major cracks continuously propagate, even though the load is decreasing (softening). As pointed out by van Mier (1997), this stage is highly influenced by the geometry of the specimen, as well as by testing conditions and measurement set-up.

The values of the compressive stress ratios indicated above as boundaries of the different crack propagation stages depend on the concrete strength class (van Mier 1997). In FRC, the amplitude of stages II and III is generally larger than for the parent plain concrete matrix. This amplitude depends, mainly, on the fibre type and content. However, the main influence of the fibre reinforcement mechanisms is especially visible in stage IV (compressive strain softening), where, depending on the fibre characteristics and concrete properties, a significant increase on the energy absorption capacity can be obtained (area under the stress-strain relationship in this stage) and the specimen failure mode shifts from a localized shear-band type, as the one shown in Fig. 8.13, to a failure mode characterized by many finer parallel cracks (Bencardino et al. 2008). In general, the higher the fibre factor $V_f\ l_f/d_f$ (with V_f fibre volume fraction, l_f fibre length and d_f fibre diameter), the more pronounced this change in failure mode (Ou et al. 2012) and the higher is the energy absorption capacity in the post-peak stage.

On the other hand, the influence of the fibres, no matter their type and origin, on the compressive strength is scant, in some cases even detrimental because they may act as defects into the concrete matrix, mainly in the case of an improper spatial distribution. This can be also captured in the fresh state of FRC due to the negative effects that fibres may have on the workability of the matrix, especially if no adjustment is made to the mixture composition (Bayasi and Soroushian 1992), such as the rearrangement of the (coarse) aggregate grading to take into account the effects of fibres on the void ratio of the solid particle skeleton (Ferrara et al. 2007).

This may be the case of recycled steel fibres (RSF), which also due to their highly irregular shape and scattered geometric characteristics, can even severely affect the workability, upon progressively increasing the fibre dosage, (Micelli et al. 2014; Centonze et al. 2012a, b; Aiello et al. 2009; Mohammadi et al. 2008; Yazici et al. 2007; Tlemat et al. 2006) which also has a direct detrimental consequence on the compressive strength of RSFRC. In this respect Meddah and Bencheikh (2009) studied the effects of the incorporation of various types of waste metallic fibres (WMF) and polypropylene fibres (WPF) on the mechanical properties of fibre reinforced concrete. The waste metallic fibres were a by-product collected from local metal lathe workshop, while the waste polypropylene fibres were obtained from polypropylene storage bags. The results have shown that the addition of WMF up to 1.5% by volume did not affect the compressive strength. However, adding a higher volume fraction of fibres (more than 2% by volume) has resulted in a slight decrease of the compressive strength. Pilakoutas et al. (2004) suggested that the maximum content (by weight) of recycled steel fibres, which could be effectively used in concrete, is 6 and 2% for polypropylene and steel fibres, respectively. Rossli and Ibrahim (2012) suggested a value of 0.4% (by volume) for recycled steel fibres.

Anyway, by using a suitable mix design strategy (Ferrara et al. 2007; Barros et al. 2007), and through adequate type and dosage of (super)plasticizer, the same workability classes as for plain concrete (PC) can be also achieved. This can also have significant impact on the compressive strength of the RSFRC, since values higher than those of the parent plain matrix as well as of similar FRC reinforced with industrial fibres were also reported.

8.2.3 Direct Tension

Direct tensile tests are the most appropriate to experimentally identify the tensile behaviour of any type of cement based material, but their appropriate execution requires very stiff testing rigs with sophisticated testing control protocols, like the one shown in Fig. 8.14 (Barros et al. 1994).

To assess the effect of fibre distribution and orientation on the tensile behaviour, prismatic specimens can be extracted from a beam, like the one shown in Fig. 8.15,

Fig. 8.14 Photo of a testing rig used in tensile tests (Barros et al. 1994)

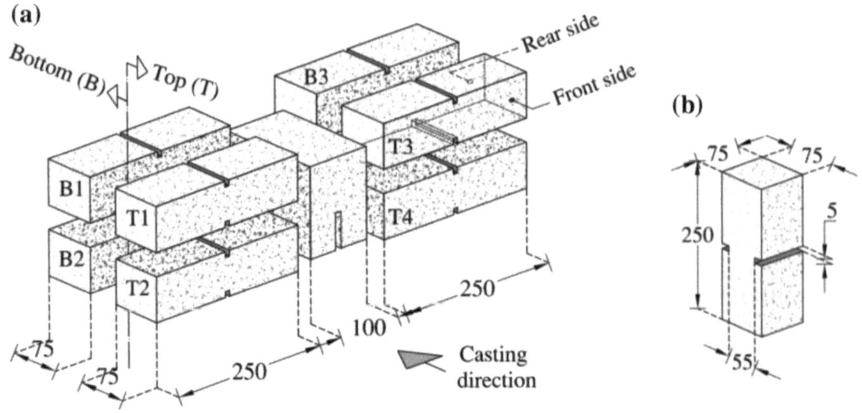

Fig. 8.15 **a** Extracting process of direct tensile specimens from tested 3PNBBT; **b** Typical tensile specimen (dimensions in mm) (Salehian and Barros 2015)

whose geometry corresponds to the three point notched beam bending test (3PNBBT) proposed by CEB-FIP Model Code 2010 (2011), hereafter abbreviated by MC2010 (2011). This strategy has been used by extracting specimens from notched beams after have been tested in bending (Salehian and Barros 2015).

Due to the softening behaviour of most RSFRCs, a notch is executed in the middle length of the specimen in order to localize the crack initiation and propagation through the notched plane. Direct tensile tests allow the tensile strength, f_{ct}, to be determined, together with the Young's modulus, E_c, the shape of the softening branch of the stress-crack opening diagram, $\sigma - w$, and the mode I fracture energy, G_{fI}, which is defined as the area under the stress-crack opening diagram up to the complete exhaustion of the specimen's load carrying capacity. The stress-crack opening relationship can be obtained from a direct tension test, as schematized in Fig. 8.16, where δ is the average displacement measured by displacement transducers placed bridging the notched plane, Fig. 8.17 (Barros et al. 1994).

For the characterization of the tensile behaviour of steel fibre reinforced concrete (SFRC), RILEM TC 162-TDF (2003) recommends the tests setup represented in Fig. 8.18, which employs a cylinder specimen of 150 mm diameter and 150 mm height, width a circumferential notch of 15 mm depth at middle height. The details about the monitoring system and testing conditions can be found in Vandewalle et al. (2001).

Due to the relative low content of RSF used for the reinforcement of cement based materials, the tensile strength of the RSFRC is not, in general, affected by the tensile properties of the RSF.

Recently Zamanzadeh (2017) have performed direct tensile tests for the characterization of the tensile behaviour of RSFRC developed with the main purpose of preparing thin panels for the shear strengthening of RC beams (Zamanzadeh et al. 2015b). For the preparation of these panels and specimens, a SIFCON type technology was adopted, where 3.8% by volume RSFs were introduced, as dispersed as

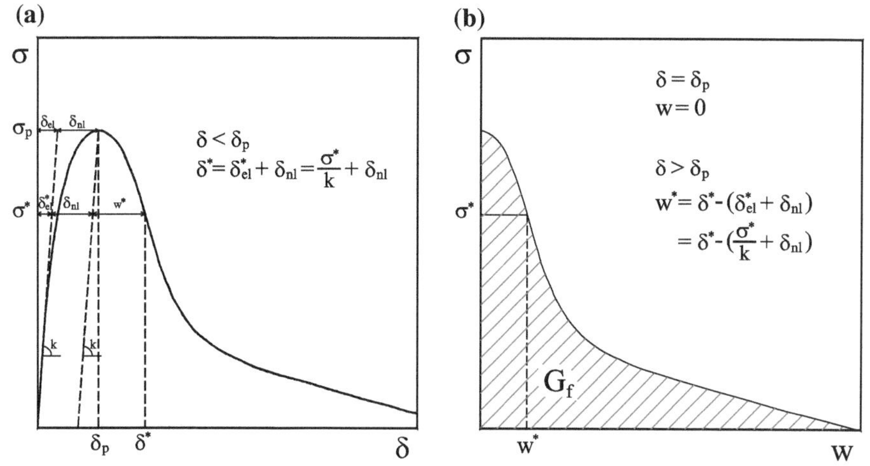

Fig. 8.16 Typical relationships obtained from direct tensile tests: **a** stress versus axial deformation; **b** stress versus crack opening

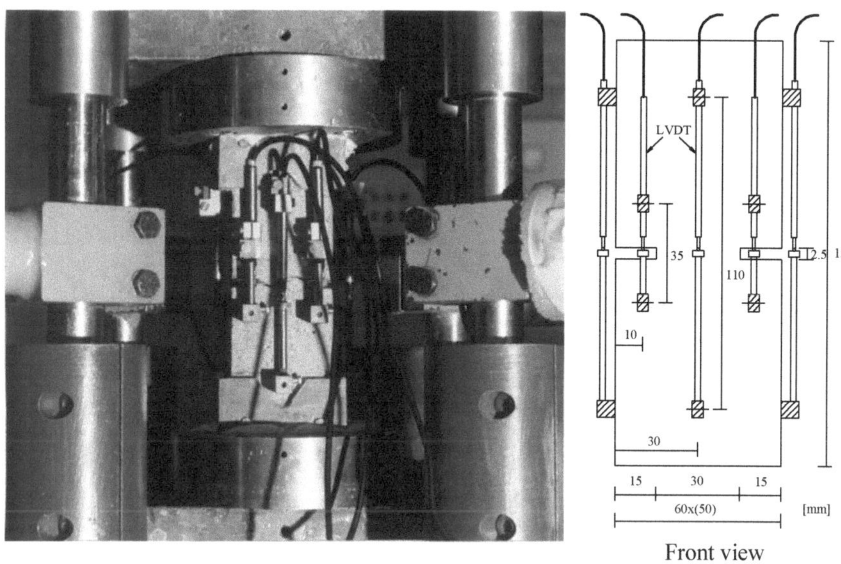

Fig. 8.17 Arrangement of the measuring devices to assure tensile stable tests

possible, in a mould, and then a high fluid mortar was poured on the fibre mesh up to complete mould filling, while being vibrated for ensuring a proper enveloping of the fibres (Fig. 8.19; Zamanzadeh 2017). The composition of this mortar reinforced with RSF is indicated in Table 8.1. At 28 days RSFRC developed an average compressive strength of 44.37 MPa and a Young's modulus of 19.64 GPa (Zamanzadeh 2017).

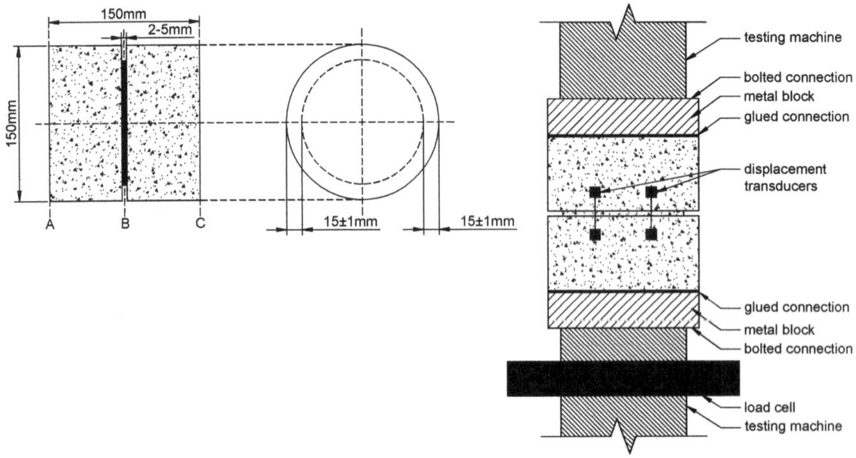

Fig. 8.18 Test setup recommended by RILEM TC 162 TDF for the characterization of the uniaxial tensile behaviour of FRC materials (Vandewalle et al. 2001)

Fig. 8.19 Process of preparing RSFRC (Zamanzadeh 2017)

Table 8.1 Mix proportions (kg) per mortar m^3 (Zamanzadeh 2017)

C	FA	LF	FS	VMA	SP	W	W/B
546	669	–	437	1.710	11	318	0.26

C Cement; *FA* Fly Ash; *LF* Limestone Filler; *FS* Fine Sand; *VMA* Viscosity modifying admixture; *SP* Superplasticizer; *W* Water; *W/B* Water/Binder ratio (B = C + FA)

From the panels cast as above, rectangular specimens (RDTT) were extracted as shown in Fig. 8.20a; dog-bone type specimens (DBDTT) were also cast in appropriate moulds, using the same RSFRC manufacturing process adopted for the panel. The geometry of the specimens is shown in Fig. 8.20b, while the test setup for the RDTT and DBDTT specimens is represented, respectively, in Fig. 8.20c, d.

The average tensile stress versus tensile strain curves obtained from three DBDTT specimens, as well as from five RDTT specimens extracted at 0° (RDTT0), and four RDTT specimens extracted at 45° (RDTT45) are shown in Fig. 8.21.

The average stress at crack initiation was about 2.5 MPa. After this stage, all the specimens developed a tensile strain hardening phase corresponding to the formation of several cracks up to an average tensile strain of about 1%, at which an average tensile strength of 3.34 MPa was measured. The specimens RDTT0 developed a stiffer behaviour and attained a higher tensile strength than the other type specimens due to the better alignment of the fibres with respect to the applied tensile loading, while the dog-bone type specimens featured the smallest tensile strength. Above the strain corresponding to the tensile strength, all the specimens showed a smooth tensile softening phase. At a tensile strain equal to about 2.7%, at which all the specimens still carried out a post-cracking tensile strength of about 2.31 MPa (69% of the tensile strength) the curve of dob-bone specimens started exceeding the ones of rectangular specimens. For tensile strain hardening FRCs, un-notched specimens are recommended, like the ones used in the experimental investigation herein reported, and the tensile behaviour of this kind of FRC materials can be modelled by the stress-strain relationship proposed by MC2010 (2011).

Due to the inherent difficulties in performing direct tensile tests, most of the available results on the characterization of the tensile behaviour of RSFRC have been obtained by performing indirect tensile and flexural tests.

8.2.4 Indirect Tensile Tests

8.2.4.1 Brazilian Type Test

Indirect tensile tests, also designated splitting tensile tests, have been used to assess the tensile behaviour of RSFRC. The original test setup, currently known as Brazilian indirect tensile test, described in the ASTM C496/C496M-11 (2011), has been widely employed. Anyway, as demonstrated by Abrishambaf et al. (2016), this test setup is not the most appropriate for deriving the stress-crack width relationship, even applying a relatively sophisticated monitoring system like the one represented in Fig. 8.22. The main drawback relies in the fact that most of the specimen volume is subjected to a biaxial stress field, with compressive stresses acting cross-wise to tensile ones all along the critical region of the specimen. Due to this, no pure tension is applied to the material, even executing a notch along the

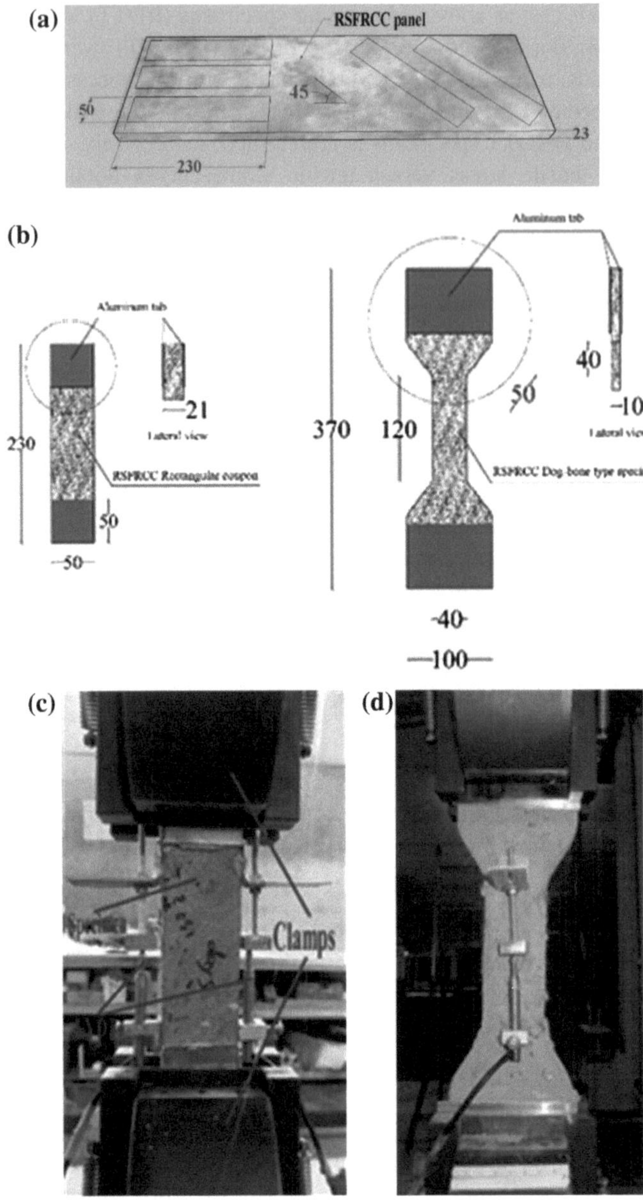

Fig. 8.20 Direct tensile tests with RSFRC: **a** configuration for the extraction of the rectangular-type specimens from the panel; **b** geometry of the tensile specimen's type; **c** test setup of the rectangular-type specimen; **d** test setup of the dog-bone type specimen (dimensions in mm)

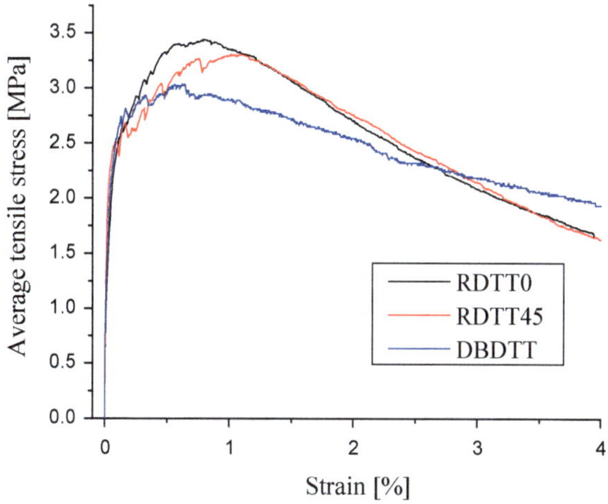

Fig. 8.21 Direct tensile tests in RSFRC (RDTT0 = rectangular-type specimens extracted at 0° from the panel; RDTT45 = rectangular-type specimens extracted at 45° from the panel; DBDTT = Dog-bone type specimens) (Zamanzadeh 2017)

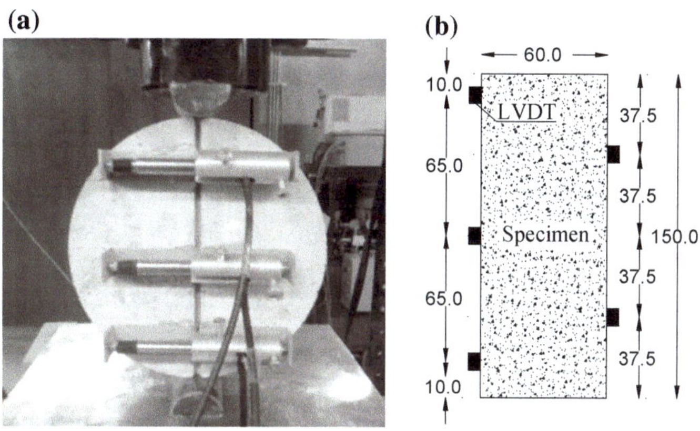

Fig. 8.22 Geometry of the specimen and setup of the splitting tensile test (dimensions are in mm): **a** specimen front view (*top of the panel*), **b** specimen lateral view (Abrishambaf et al. 2016)

loading plane, as shown in Fig. 8.22. Therefore, only performing inverse analysis on the results obtained from these tests, and using advanced numerical models capable of simulating both the nonlinear behaviour in tension and compression, is possible to derive with acceptable accuracy the uniaxial tension stress-crack width relationship, which is needed for modelling the post-cracking behaviour of fibre reinforced cement based materials (Abrishambaf et al. 2016).

Table 8.2 Splitting tensile strength of RSFRC

RSF type	V_f (%)	Cylinder dimensions(mm)	Splitting tensile strength (MPa)
Bdour and AlKhalayleh (2010)			
Steel cord recovered from waste tyre	0	⃠ 150*300	2.74
	2	⃠ 150*300	2.60
	4	⃠ 150*300	2.54
	6	⃠ 150*300	2.43
	8	⃠ 150*300	2.31
	10	⃠ 150*300	2.01
Papakonstantinou and Tobolski 2006			
Steel beads recovered from waste tyre	0	⃠ 100*200	2.82
	2	⃠ 100*200	2.75
	4	⃠ 100*200	2.64
	6	⃠ 100*200	2.55
	8	⃠ 100*200	2.47
Rossli and Ibrahim 2012			
Obtained from the tyre shredding process	0	⃠ 150*300	3.88
	0.2	⃠ 150*300	3.39
	0.4	⃠ 150*300	3.98
	0.6	⃠ 150*300	3.90
	0.8	⃠ 150*300	3.50
	1.0	⃠ 150*300	4.43
Abdul Awal et al. 2013			
Obtained from the tyre shredding process	0	⃠ 100*200	3.80
	0.5	⃠ 100*200	6.65
	1	⃠ 100*200	7.10
	1.5	⃠ 100*200	8.15
	2	⃠ 100*200	9.45

In spite of these restrictions, this test setup has been used by several authors, but mainly to assess the splitting tensile strength $(f_{ct,sp})$ of RSFRC (Lion 2012; Rossli and Ibrahim 2012; Abdul Awal et al. 2013; Bdour and AlKhalayleh 2010; Papakonstantinou and Tobolski 2006). Representative results of splitting tensile strength of RSFRC are included in Table 8.2. A decrease of the splitting tensile strength with the increase of the content of fibres is reported (Bdour and AlKhalayleh 2010, Papakonstantinou and Tobolski 2006), which would have been caused by the decrease of the mechanical performance of the matrix due to an increase of air voids and rubber particles existing in some recycled fibres. An almost null influence on the splitting tensile strength with the content of recycled fibres (RF) was reported by Rossli and Ibrahim (2012), while an abnormal high increase with the fibre content of RF was reported by Abdul Awal et al. 2013. Papakonstantinou and Tobolski (2006) have proposed the equation $f_{ct,sp} = 0.44\sqrt{f_c'}$

for determining the splitting tensile strength from the corresponding compressive strength, but since this equation was calibrated only on the results obtained by the same authors, its general applicability to RSFRC is questionable.

8.2.4.2 Double Edge Wedge Splitting Test (DEWST)

This indirect tensile test was proposed by di Prisco et al. (2013) for straightforward evaluation of the uniaxial tension stress *versus* crack opening relationship and mode I fracture properties of fibre reinforced concrete. The test setup, schematically represented in Fig. 8.23, was conceived in order to introduce in the central part of the specimen a stress field mainly governed by tensile stresses almost orthogonal to the plane defined by the two notch tips.

The normal component of the force (F_n), as deviated from the loading axis because of two V-grooves cut at specimen edges with an angle of 45°, results in a splitting tensile force (F_{sp}) acting on the fracture plane that is function of the total applied compressive load (F):

$$F_{sp} = \eta F \tag{2.1}$$

where η is a reduction factor that depends on the angle of the V-notch and the concrete-to-steel roller friction coefficient (di Prisco et al. 2013). The tensile stress normal to the fracture surface is obtained from:

$$\sigma_t = \frac{\eta F}{h_{fs} \cdot l_{fs}} \tag{2.2}$$

where h_{fs} and l_{fs} are the height and length of the fracture surface. For the η parameter a value close to 0.9 has been proposed (Salehian 2015).

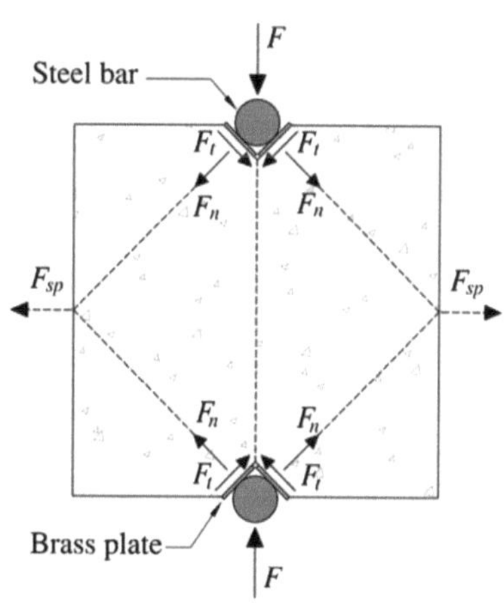

Fig. 8.23 Rigid body force diagram in double edge wedge splitting test (Salehian 2015; adapted from di Prisco et al. 2013)

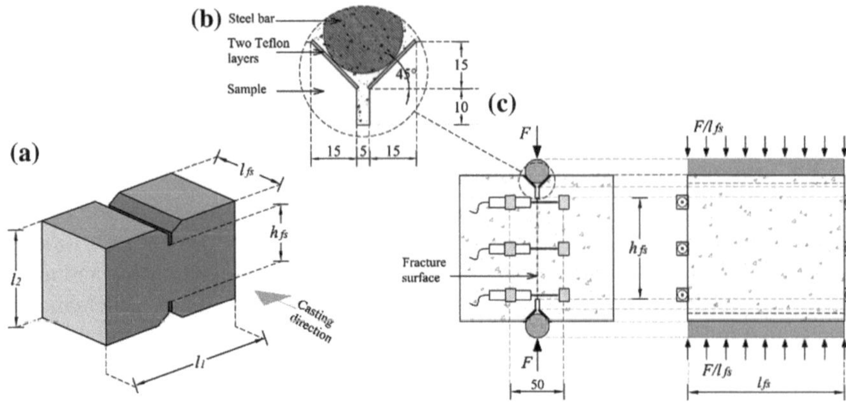

Fig. 8.24 Details of the double edge wedge splitting test; **a** sample shape, **b** geometry of the "V" notch, **c** test setup (dimensions in mm) (Salehian et al. 2014)

Ferrara et al. (2012), di Prisco et al. (2013) and Salehian et al. (2014) have proposed the introduction of a notch in the tip of the V notch in order to better localize the fracture plane in the alignment of the notches, Fig. 8.24.

The DEWST is also very appropriate to evaluate the influence of the fibre distribution and orientation on the "in-structure" behaviour of FRCs, since cylindrical type specimens can be extracted from these members, and fracture planes with tailored orientations can be easily cut, as shown in Fig. 8.25 (Ferrara et al. 2012a, b; Salehian 2015; Abrishambaf et al. 2013; Lameiras 2016).

In spite of the advantages of DEWST setup for the characterization of the tensile behaviour of FRC, no experimental data is still available for RSFRC.

8.2.5 Bending Tests

Since direct tensile tests are relatively time consuming and expensive, and the indirect tensile tests for the direct derivation of the mode I fracture parameters still need to be better assessed and calibrated, as discussed with reference to the DEWST in the previous section, three point notched beam bending tests (3PNBBT) are being used to obtain information that is required for defining the constitutive laws of FRC.

The 3PNBBT was initially proposed by RILEM TC 162 TDF (2003) (Vandewalle et al. 2002) for steel fibre reinforced concrete, and later extended by MC2010 (2011) for any FRC. Figure 8.26 represents schematically the 3PNBBT setup, as well as the measuring system used to determine the deflection (LVDT1 in Fig. 8.26a), and/or the crack mouth opening displacement, CMOD (LVDT4 in Fig. 8.26c), and/or the crack tip opening displacement, CTOD (LVDT3 in Fig. 8.26b). Sometimes compressive deformation is also measured (LVDT2 in Fig. 8.26b) in order to obtain, together with

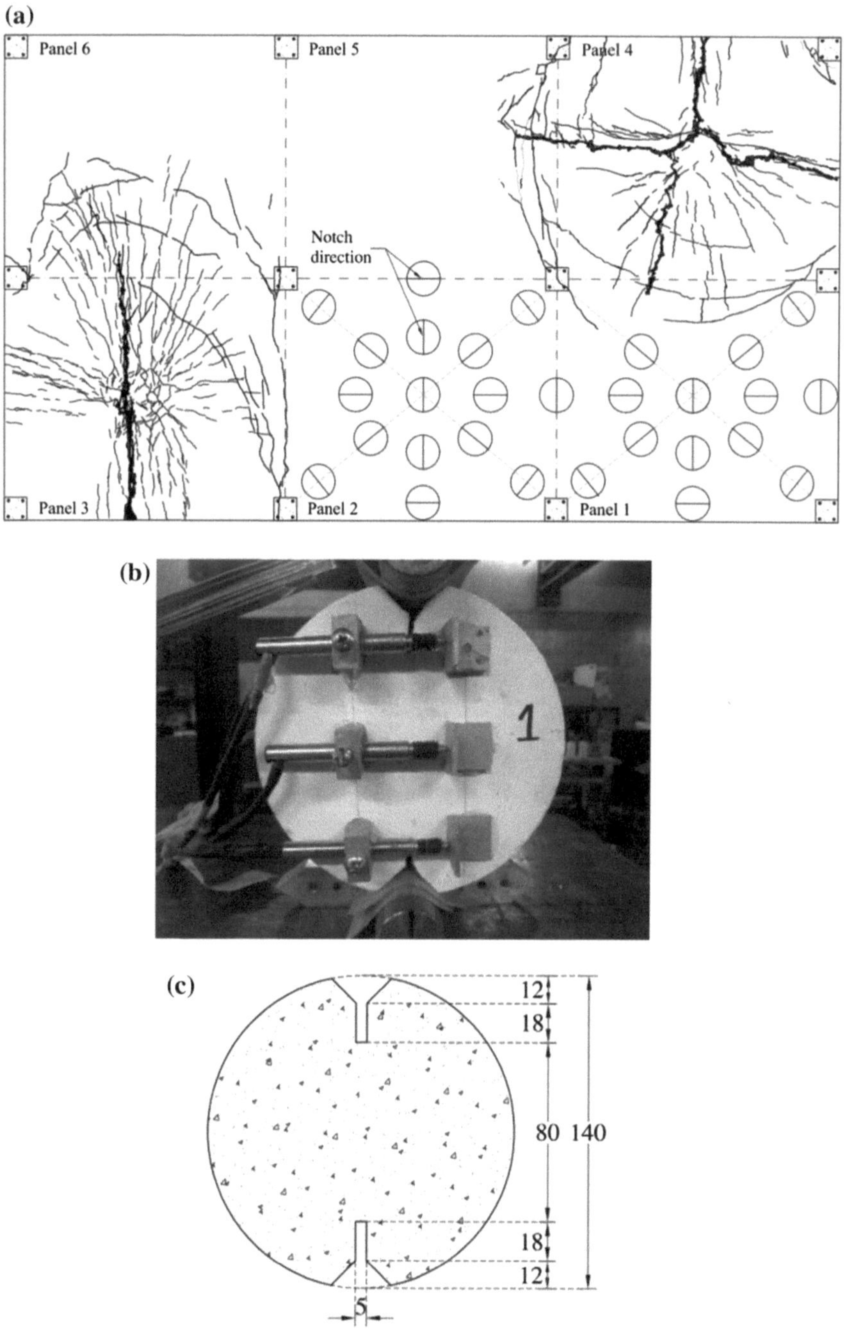

Fig. 8.25 **a** Plan of extracted core from elevated slab of steel fibre reinforced self-compacting concrete (E-SFRSCC); **b** DEWST specimens and **c** (dimensions in mm) (Salehian 2015)

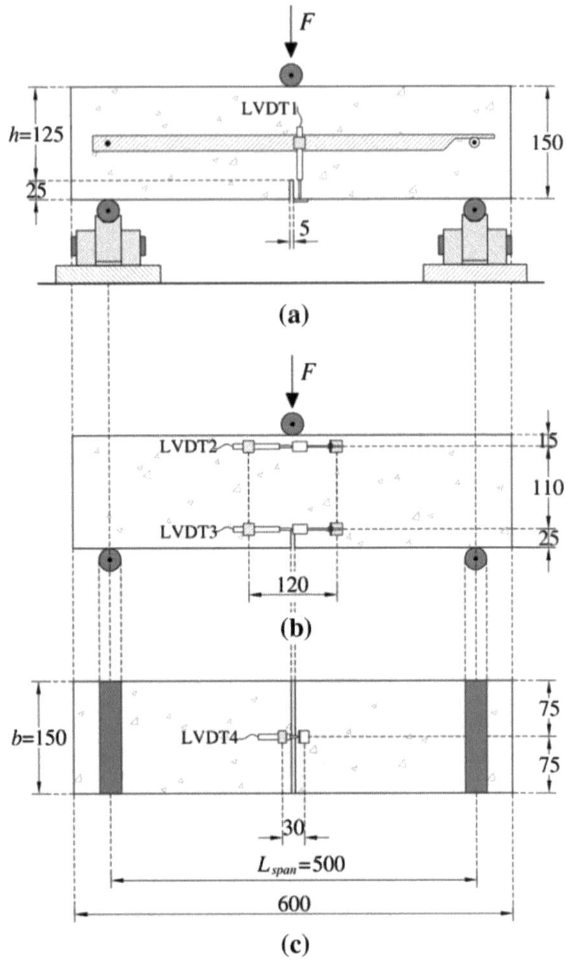

Fig. 8.26 Three-point notched beam bending test setup: **a** front side; **b** rear side; and **c** bottom side (dimensions in mm) (Salehian 2015)

the displacement in LVDT3, the rotation and the curvature in a zone where the nonlinear behaviour due to cracking and inelastic deformation in compression is likely to occur.

Both the RILEM TC 162 TDF (Vandewalle et al. 2002) and MC2010 (2011) propose the concept of residual flexural strength parameters (f_{Rj}) to characterize the post-cracking behaviour of FRC. These parameters are determined from the force-CMOD curves obtained from 3PNBBTs, whose typical relationship is depicted in Fig. 8.27. Based on the force values, F_j, corresponding to the $CMOD_j$ (j = 1 to 4), the f_{Rj} are determined from the following equation:

$$f_{Rj} = \frac{3\,F_{\mathrm{j}}\,L}{2\,b\,h_{\mathrm{sp}}^2} \tag{2.3}$$

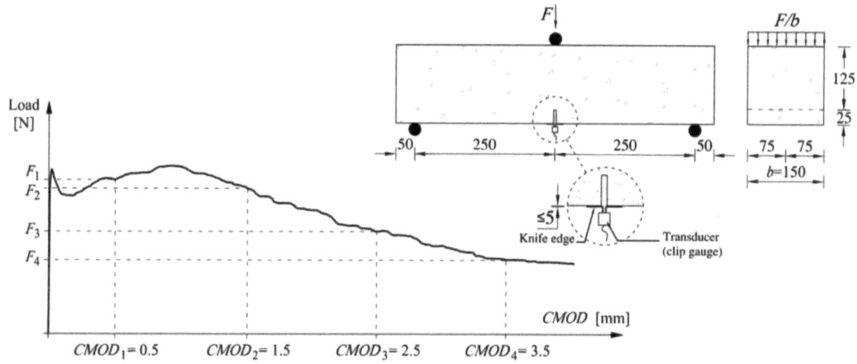

Fig. 8.27 Typical load versus CMOD curve of FRC (MC2010)

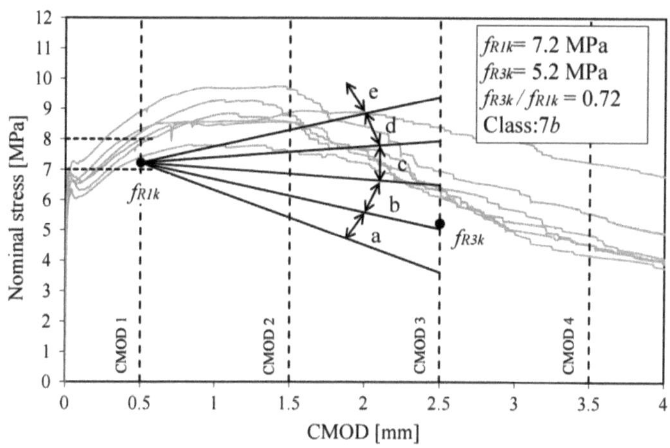

Fig. 8.28 Concept of toughness class for FRC based on the relationship between the nominal flexural stress and CMOD (MC2010)

where b(= 150 mm) and L(= 500 mm) is the width and the span of the specimen, respectively, and h_{sp}(= 125 mm) is the distance between the tip of the notch and the top of the cross section.

For structural applications with normal and high-strength concrete, the FRC classification proposed by MC2010 (2011) is based on the post-cracking residual strength. For this purpose, a value representing the characteristic residual flexural strength f_{R1k} (referred to the strength interval) and a letter *a*, *b*, *c*, *d* or *e* (representing the f_{R3k}/f_{R1k} ratio—*k* means characteristic value) are considered. For instance, a material denoted as "7b" has a strength f_{R1k} ranging between 7 and 8 MPa, and a f_{R3k}/f_{R1k} ratio ranging between 0.7 and 0.9 (Fig. 8.28).

Zamanzadeh et al. (2015a) have assessed the potentialities of RSFRC for the shear reinforcement of RC beams. For this purpose these authors have started

Table 8.3 Designation of the series of tests of the experimental program (Zamanzadeh et al. 2015a)

Mix	Type of fibres	Series name	Number of specimens	Content of steel fibres (kg/m^3)
MRSF_45	RSF	RSFRC45	10	45
MRSF_60	RSF	RSFRC60	10	60
MRSF_90	RSF	RSFRC90	4	90
MISF_45	ISF	ISFRC45	4	45
MISF_60	ISF	ISFRC60	4	60
MISF_90	ISF	ISFRC90	4	90

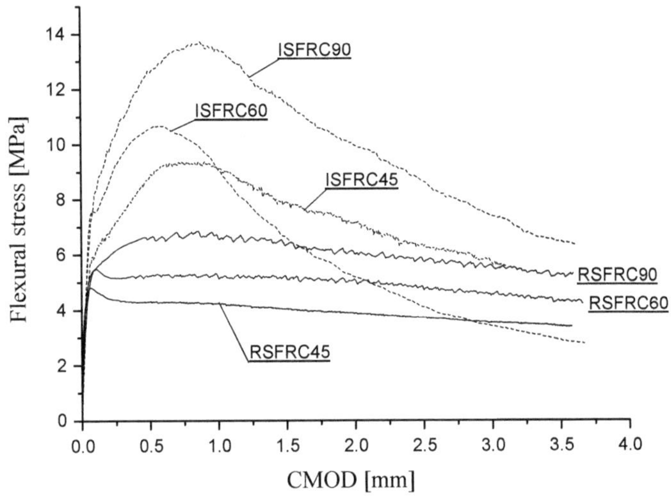

Fig. 8.29 Comparison of the flexural behaviour of ISFRC and RSFRC (Zamanzadeh et al. 2015a)

comparing the flexural performance of RSFRC with the one of an analogous FRC reinforced with industrial steel fibres (ISF). Table 8.3 includes the designation of the mix compositions of the experimental programs, and their principal characteristics. The comparison between the flexural behaviour of RSFRC and ISFRC are shown in Fig. 8.29 and the corresponding f_{eq} (obtained according to the recommendations of RILEM TC 162 TDF (2003) (Vandewalle et al. 2002) and $f_{R,i}$ are represented in Fig. 8.30. This comparison is also provided in Tables 8.4 and 8.5. From the obtained results it is evident that the deflection hardening phase registered in the ISFRC specimens (from crack initiation up to the flexural tensile strength) was not developed in the RSFRC specimens. This indicates that fibre bridging mechanisms across the crack surfaces for relatively small crack widths were not as effective in the RSF as in ISF due to the geometry and surface characteristics of the formers. However, in the post-peak stage the RSFRC specimens have almost retained the maximum flexural tensile strength up to the ultimate crack width recorded in the tests (3.5 mm).

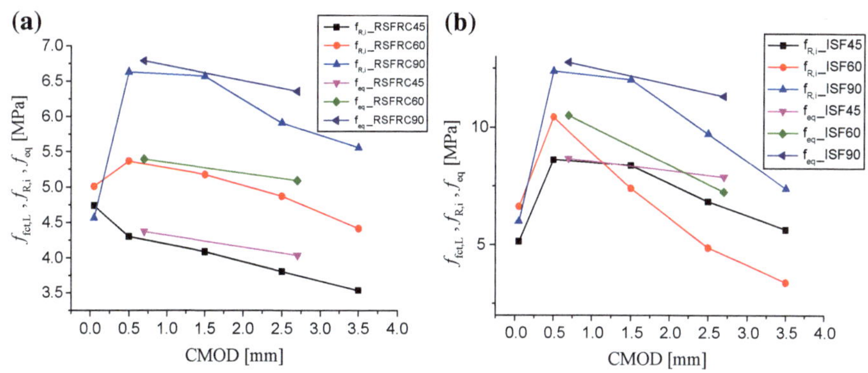

Fig. 8.30 Representation of the feq and fR,i parameters for the series: **a** RSFRC; **b** ISFRC (Zamanzadeh et al. 2015a)

Table 8.4 Equivalent and residual flexural tensile strength parameters for RSFRC (MPa)

Series		$f_{fct,L}$	$f_{eq,2}$	$f_{eq,3}$	$f_{R,1}$	$f_{R,2}$	$f_{R,3}$	$f_{R,4}$
RSFRC45	Average	4.73	4.28	3.90	4.16	3.94	3.69	3.43
	COV (%)	18.6	28.9	30.8	24.6	32.4	33.4	33.5
RSFRC60	Average	5.00	5.39	5.08	5.36	5.17	4.86	4.41
	COV (%)	11.4	15.2	16.8	13.6	17.2	18.6	20.7
RSFRC90	Average	4.56	6.78	6.35	6.62	6.56	5.90	5.55
	COV (%)	9.5	8.3	9.3	7.6	9.1	11.9	12.2

Table 8.5 Equivalent and residual flexural tensile strength parameters for ISFRC (MPa)

Series		$f_{fct,L}$	$f_{eq,2}$	$f_{eq,3}$	$f_{R,1}$	$f_{R,2}$	$f_{R,3}$	$f_{R,4}$
ISFRC45	Average	5.14	8.66	7.87	8.61	8.36	6.83	5.64
	COV (%)	4.9	25.5	24.3	25.0	23.0	24.3	21.9
ISFRC60	Average	6.62	10.49	7.24	10.43	7.39	4.86	3.40
	COV (%)	6.7	12.8	13.0	13.3	19.1	21.7	20.4
ISFRC90	Average	5.99	12.75	11.31	12.37	12.00	9.71	7.38
	COV (%)	10.5	12.1	22.7	11.9	25.3	34.2	37.2

Figure 8.31 shows the relationship between $f_{eq,2}$ and $f_{eq,3}$ obtained in RSFRC specimens. A clear linear relationship emerges between these two parameters, which is in agreement with previous research on ISFRC specimens (Barros et al. 2005).

The relationships between $f_{eq,2}$ and $f_{R,1}$, and between $f_{eq,3}$ and $f_{R,4}$ are represented in Fig. 8.32. Also a linear trend is shown between these parameters.

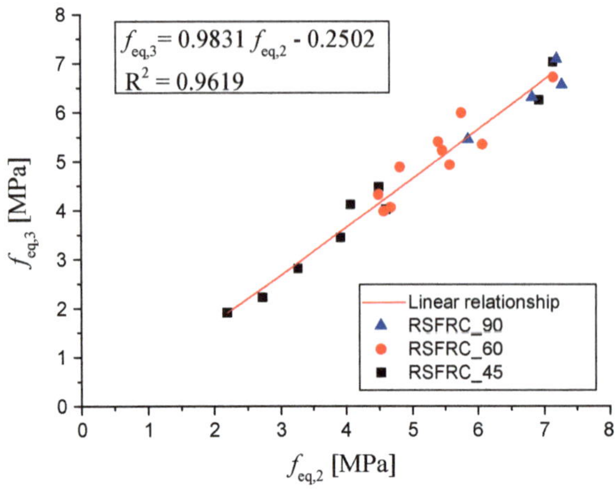

Fig. 8.31 Relationship between $f_{eq,2}$ and $f_{eq,3}$ (Zamanzadeh et al. 2015a)

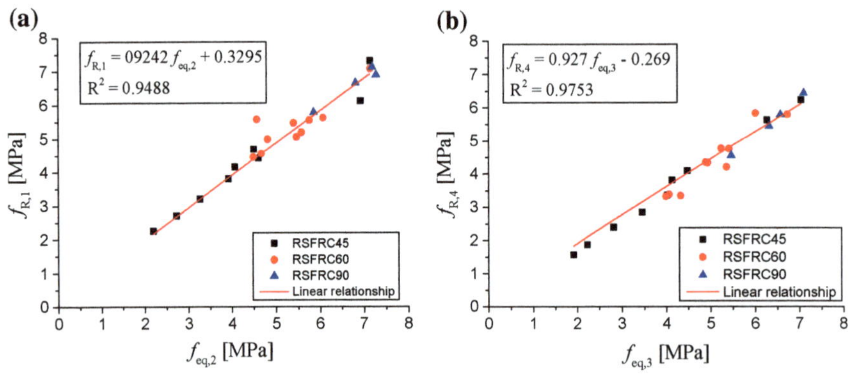

Fig. 8.32 Relationship between: **a** $f_{eq,2}$ and $f_{R,1}$; **b** $f_{eq,3}$ and $f_{R,4}$ for RSFRC (Zamanzadeh et al. 2015a)

For a wider comparison between RSFRC and SFRC, the database (DB) collected by Neto et al. (2013) in terms of f_{Ri} values was used. This DB includes f_{Ri} values of ISFRC, reinforced with hooked end ISFs and featuring tensile strain softening behaviour, which is also the type of behaviour of both RSFRC and ISFRC used by Zamanzadeh et al. (2015a, b). Figure 8.33 compares $f_{R1} - f_{R3}$ and $f_{R1} - f_{R4}$ from the experimental results of DB with those obtained from RSFRC specimens. A similar trend is evidenced between the RSFRC and the DB results in terms of $f_{R1} - f_{R3}$ and $f_{R1} - f_{R4}$.

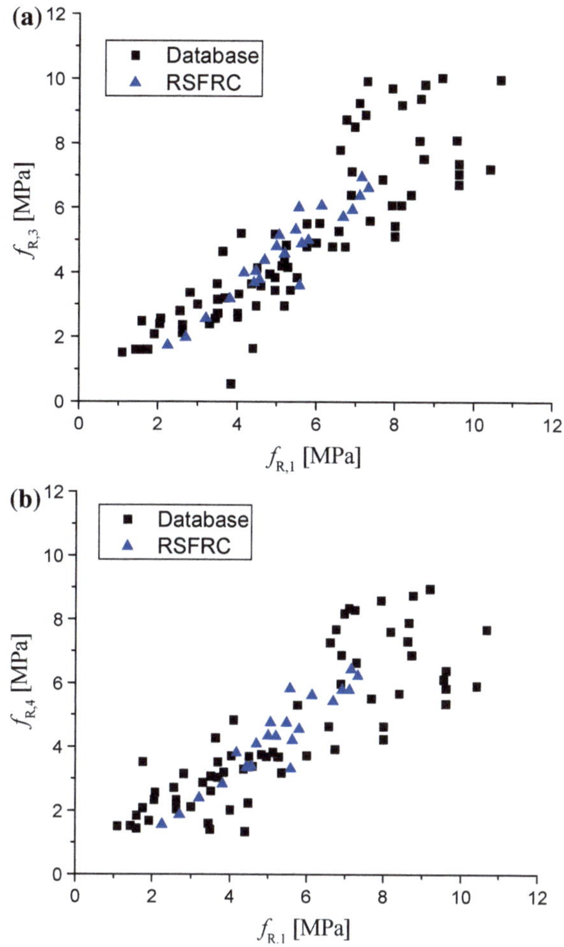

Fig. 8.33 Relationship between: **a** $f_{R,1}$ and $f_{R,3}$ and **b** $f_{R,1}$ and $f_{R,4}$ (Zamanzadeh et al. 2015a)

Figure 8.34 compares $f_{R,i}$ and V_f from the experimental results of the DB and from those obtained from the RSFRC specimens tested by Zamanzadeh et al. (2015a). It is confirmed that the increase of the fibre volume percentage in SFRC specimens of the DB leads to a higher increase of the residual flexural tensile strength parameters than for the tested RSFRC specimens. This means that, for the investigated SFRC compositions, the increase of the fibre reinforcement effectiveness with the fibre content is higher when using ISF than when adopting RSF. However, the development of mix design strategies for the RSFRC able to guarantee proper fibre dispersion up to fibre contents used in structural applications might mitigate this different fibre reinforcement effectiveness. In any case, the $f_{R,i}$ values obtained for the developed RSFRC are sufficiently high to create good perspectives for the use of these composites in some applications as shown in Sect. 8.2.7.

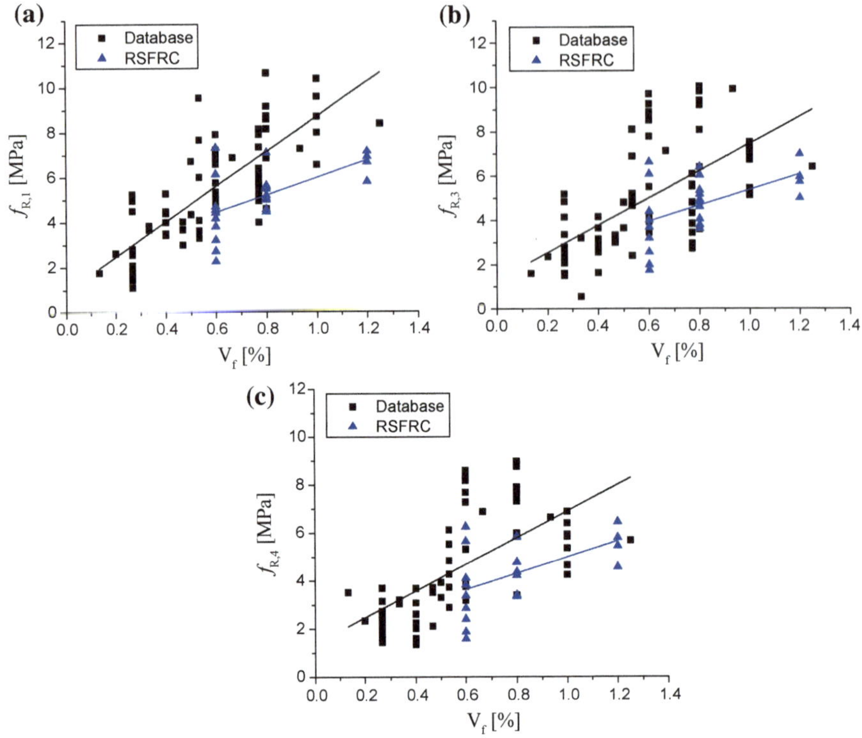

Fig. 8.34 Influence of V_f on: **a** f_{R1}; **b** f_{R3}; and **c** $f_{R,4}$ (Zamanzadeh et al. 2015a)

The characterization of the flexural behaviour of RSFRC has been assessed by several researchers (Centonze et al. 2012a, b; Lion 2012; Buratti et al. 2011; Olivito and Zuccarello 2010; Neocleous et al. 2006; Tlemat et al. 2005; Pilakoutas and Strube 2001). In general considerable benefits provided by RSFs have been highlighted in terms of post-cracking tensile capacity, energy absorption capacity and deflection performance. By executing four point beam bending tests with RSFRC, Telmat et al. (2005) verified that these composites had a post-cracking flexural behaviour similar to the corresponding ones reinforced with ISF (Fig. 8.35), which is in the same trend of the results obtained by Pilakoutas and Strube (2001), who furthermore reported fibre pull-out failure mechanisms in all the tested specimens.

The flexural performance of RSFRC and ISFRC was also compared by Aiello et al. (2009). The average load-CTOD relationships obtained by these authors from four point beam bending tests are shown in Fig. 8.36. It clearly appears that ISFRC developed a pseudo-plastic response up to a CTOD of about 1.5 mm, while for similar content of fibres the RSFRC developed a deflection softening behaviour, but with appreciable flexural capacity up to a CTOD equal 1.5 mm (about 50% of the flexural capacity of ISFRC). The apparent contradictory relative level of reinforcement efficiency of RSF versus ISF from different studies may be attributed to different contents and type of fibres used and to mix design strategies adopted.

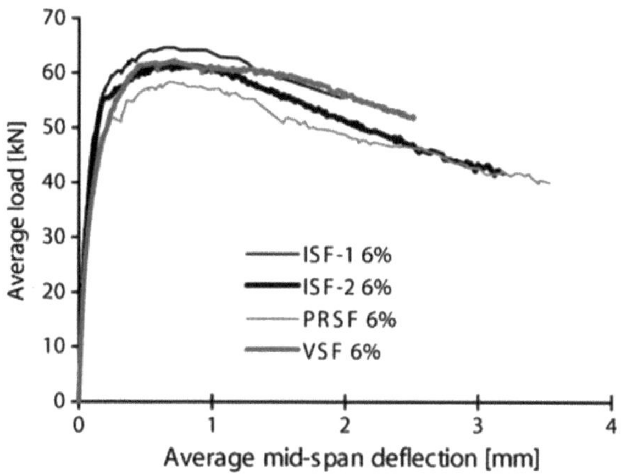

Fig. 8.35 Four point beam bending tests with concrete reinforced with 6% by weight of ISF, PRSF (pyrolysed recycled steel fibre), and VSF (chopped tyre-cord), (Tlemat et al. 2005)

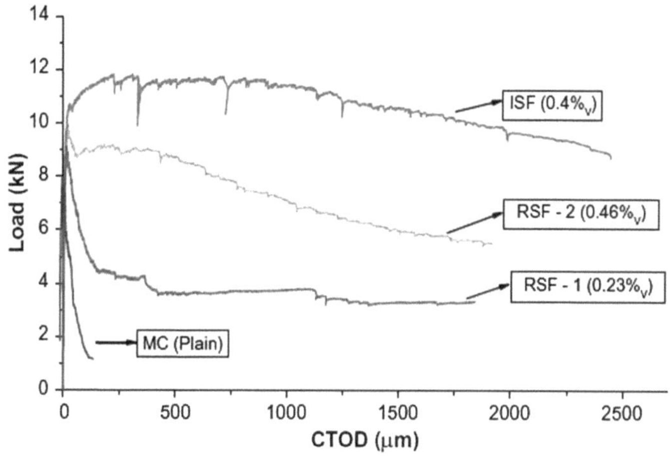

Fig. 8.36 Comparison of the flexural load versus crack tip opening displacement (CTOD) of plain, RSFRC and ISFRC (Aiello et al. 2009)

Meddah and Bencheikh (2009) have demonstrated that a hybrid recycled fibre reinforcement (combining recycled steel fibres and recycled polypropylene fibres—HRFC) is more effective in terms of post-cracking behaviour and load-carrying capacity than adopting exclusively RSF. Generally, the results have shown that ductility, toughness, and especially the post-cracking behaviour of the RFRC are significantly improved when using a mix of short and long recycled fibres compared to composites reinforced with a single type of fibers. Due to the difficulties in guaranteeing a homogeneous fibre distribution of RSF of relatively large length, with the consequent detrimental effect on post-cracking mechanical performance, these authors have recommended to limit the length of these fibres to 10 mm.

8.2.6 Direct Shear

The Iosipescu test is recommended by the ASTM D-5379 (1993) for the characterization of the shear behaviour of polymer based materials. The specimen for the Iosipescu test has a double V-edge notched region with an angle (α) and minimum height (h_0) at the section of null bending moment (Fig. 8.37).

Recently Baghi and Barros (2016) have extended the use of this test to determine the fracture mode II parameters of strain hardening cement composites (SHCC).

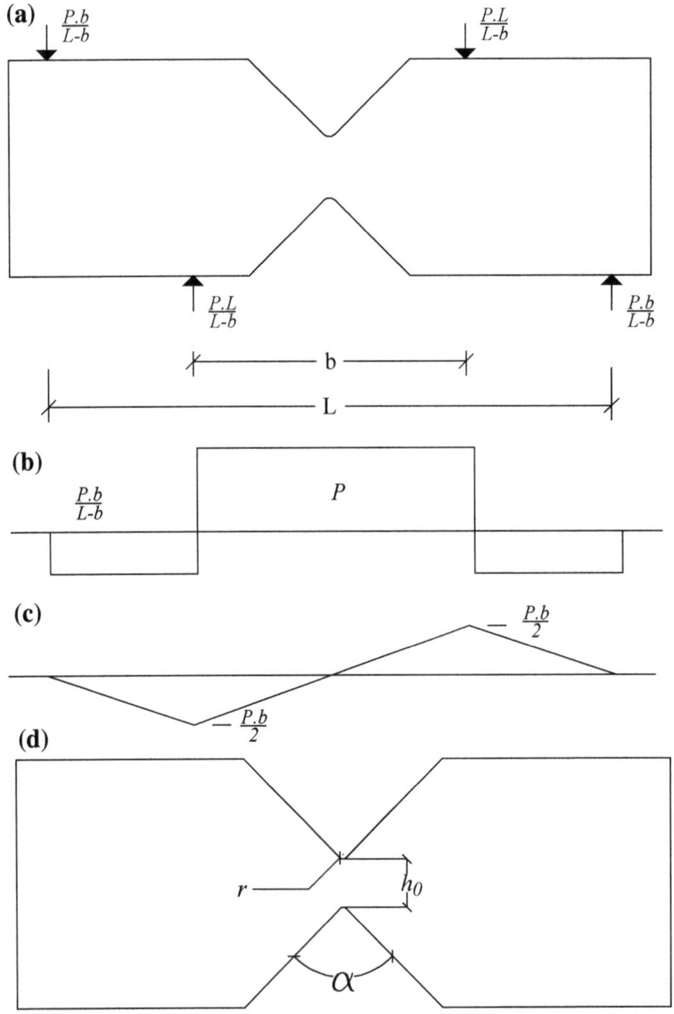

Fig. 8.37 Iosipescu test: **a** specimen's applied forces; **b** shear force diagram; **c** bending moment diagram; **d** specimen geometric parameters

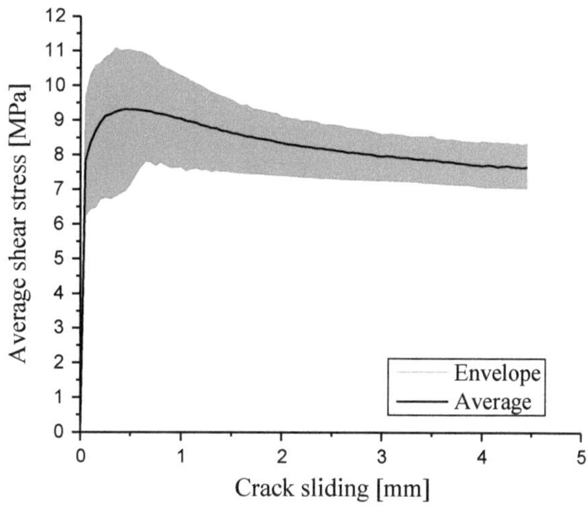

Fig. 8.38 Envelope and average shear stress versus crack sliding curves from Iosipescu tests on RSFRC specimens (Zamanzadeh 2017)

The dimensions of the specimens were 380 × 140 × 14.5 mm^3 with a depth of the critical cross section (h_0) equal to 25 mm, angle of notch root of (α) 90°, and tip radius at notches (r) equal to 2.5 mm. By using the same test apparatus and specimen geometry adopted by Baghi et al. (2016) has characterized the shear behaviour of RSFRC. Twelve RSFRC specimens were prepared according to the process described in Sect. 8.2.3. The tensile and compressive properties of this RSFRC are those indicated in this Sect. 8.2.

The envelope and average shear stress ($\tau = P/A$, where P is the applied load and A is the cross sectional area of the specimen's mid-span section) versus crack sliding curves obtained are represented in Fig. 8.38 (Zamanzadeh 2017).

By comparing these results with those obtained by Hadi and Barros (2016) where the same test setup and geometry of specimens were adopted, but a SHCC reinforced with PVA fibres was investigated (Fig. 8.39), it can be concluded that the developed RSFRC has much higher shear strength and shear sliding energy absorption capacity. The investigated SHCC featured an eminent shear softening behaviour, while RSFRC presented an almost plastic response above the sliding corresponding to shear strength. In the RSFRC, at a shear sliding of about 10 times higher the sliding at the average shear strength, the shear capacity was still almost 80% of the average shear strength

8.2.7 *Round and Square Panel Tests*

Tests with square panel continuously supported on its contour (8.39a), and tests with round panels, continuously, or simply supported (Fig. 8.39b), are especially appropriate for slab and shell type FRC structures, since cracks of different

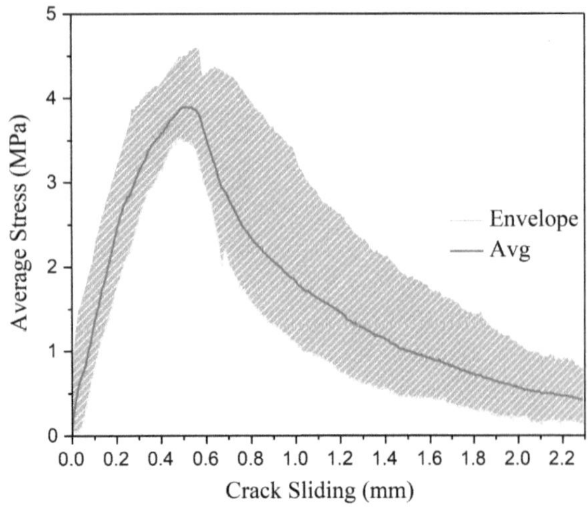

Fig. 8.39 Average shear stress versus crack sliding in SHCC (Baghi and Barros 2016)

orientation are generally formed due to the predominant 2D stress field. These tests therefore mobilize the fibre reinforcement mechanisms in a way that more closely replicate the one occurring in this type of structures, where several cracks of different orientation are generally reported (Salehian and Barros 2015; Mobasher et al. 2015). Figure 8.41a represents the geometry of the square panel proposed by EFNARC (2002), while the geometry of the round panel recommended by ASTM C-1550-03a (2003) is shown in Fig. 8.41b (Salehian et al. 2014).

From these tests force versus centre-point deflection diagrams are obtained, from which the energy absorption (area under the force-deflection curve) can be calculated. The energy absorption of the panel is the parameter that featured the highest increase with the fibre reinforcement efficiency.

Recently Salehian et al. (2014) have proposed an analytical model that can predict the stress-crack width relationship of FRC from the panel test results, being also quite appropriate for deriving the fracture mode I parameters to be used in the material nonlinear analysis of slab and shell type structures.

Square panel tests on RSFRC and ISFRC have been reported by Micelli et al. (2014) and showed similar failure mechanisms: 4–6 flexural cracks that propagate from the centre loaded region towards the edges and/or the corner regions. The experimental results in terms of load versus deflection for both RSFRC and ISFRC slabs were comparable, confirming the effectiveness of the RSF. By comparing the load carrying capacity of panels reinforced with conventional steel mesh, RSFRC and ISFRC panels featured an increase of 20 and 40%, respectively.

The energy absorption capacity of RSFRC was evaluated by Centonze et al. (2012a, b) by performing flexural tests on square panels of the type shown in Figs. 8.40a and 8.41a, with spans of 500 mm and reinforced with 0.46% by volume of RSF. For comparison purposes, three concrete panels reinforced with ISF were

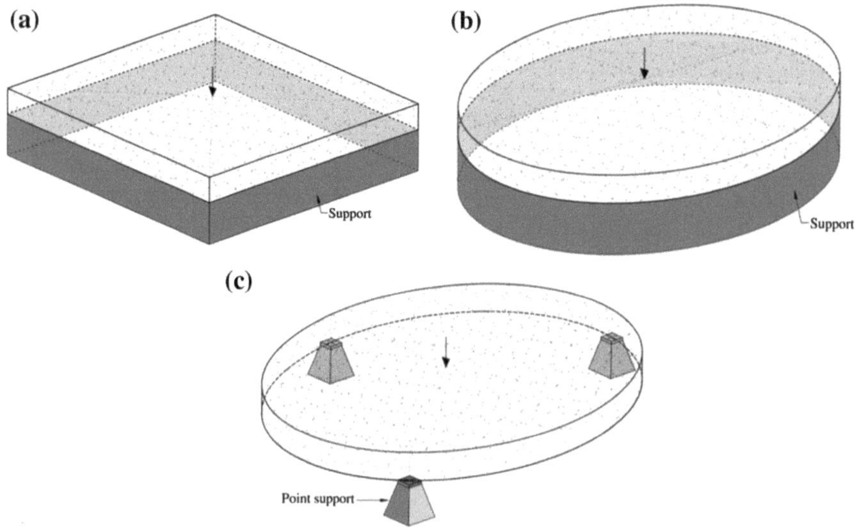

Fig. 8.40 Panel tests recommended in the literature: **a** square panel test simply supported continuously in its contour, **b** round panel test with simply supported continuously in its contour, and **c** round panel test with three point supports (Salehian et al. 2014)

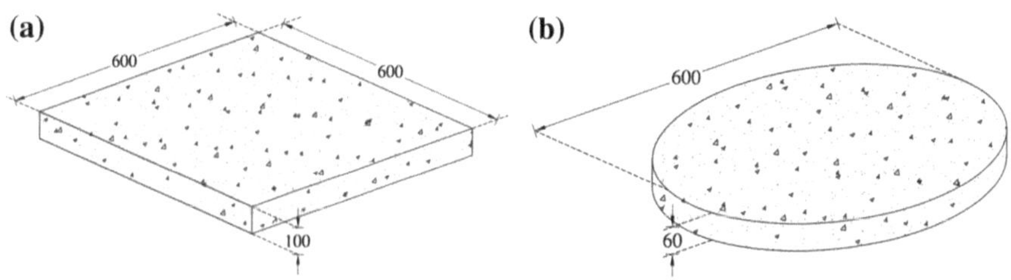

Fig. 8.41 Geometric properties of **a** square panel (SP), and **b** round panel (RP), dimensions in mm, (Salehian et al. 2014)

also tested. The main results are presented in Fig. 8.42, where it is possible to verify a large scatter on the results, regardless the type of fibre (RSF or ISF). This has been justified by the different fibre orientation and dispersion amongst the tested panels, observed in the fracture planes at the pseudo-yield lines. The results have also demonstrated that ISF provided a slightly higher peak load and post-peak load carrying capacity. In fact, after peak load the resisting capacity of ISFRC panels has an almost linear decreasing trend with the increase of the deflection, while in the RSFRC panels a more abrupt load decay with the increase of the deflection is observed, which also results in a smaller energy absorption capacity.

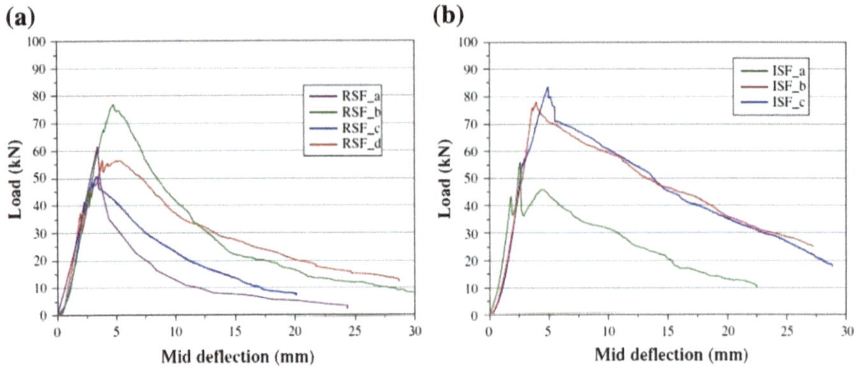

Fig. 8.42 Load versus mid-span deflection in SP of: **a** RSFRC; **b** ISFRC (Centonze et al. 2012a, b)

8.2.8 *Fatigue Behaviour*

Graeff et al. (2012a, b) carried out an experimental research with the aim of examining the potentialities of RSF for the reinforcement of concrete pavements under fatigue loading. Concrete prisms were subjected to cyclic three point bending tests. Two types of mixes, conventional and roller compacted concrete, and two recycled fibre contents, 2 and 6% by mass of concrete, were used. For comparison purposes, plain concrete and fibre reinforced concrete with industrial fibres were also tested. Cold drawn wire industrial fibres of 1 mm diameter and 54 mm length, with a cone forged at each end and 1100 MPa tensile strength were used. The RSF had a diameter of 0.2 mm and a length that varied between 3 and 22 mm, and a tensile capacity up to 2000 MPa. The results demonstrated that the recycled fibres have improved the fatigue behaviour of concrete by restraining the degeneration of micro-cracks into meso- and macro-cracks, whilst industrially produced fibres were more efficient in arresting the propagation of macro-cracks. Therefore, these authors have recommended to combine recycled and industrial steel fibres for a more efficient performance of these composites under fatigue loading. Based on predictive models developed using a probabilistic approach, the results showed that the use of recycled steel fibres may contribute to reduce up to 26% the pavement thickness when considering the influence of fatigue alone.

8.2.9 *Bond Performance of RFRC-Steel Bar Reinforcement*

Tests on the bond behaviour between concrete matrix and reinforcing steel bars showed that the bond failure for all tested specimens (plain and fibre reinforced concrete), with a cover/nominal bar diameter ratio (*c/d*) ranging in 0.94–2.19, occurred by concrete splitting (Micelli et al. 2014; Centonze et al. 2013). The bond mechanisms did not change when recycled steel fibres are added to the concrete mix. For all tested specimens three different stages could be distinguished, related to chemical adhesion, mechanical interlocking and friction between bar and concrete. The presence of steel fibres in the concrete mix did not affect the maximum bond strength, but was able to improve the bond performance after the peak bond stress has been attained. It was also found that the crack width of the RSFRC samples was lower than the corresponding plain concrete values.

8.2.10 Durability Properties of RSFRC

Steel fibres have been successfully used in concrete to improve its mechanical properties, such as post-cracking load bearing capacity and energy absorption performance. Fibres are also used to limit the crack width, with beneficial consequences in terms of concrete durability.

In an environment containing chlorides, an important aspect of FRC durability is the corrosion resistance (ACI 544.5R-10 2010; Bentur and Mindness 2007; Granju and Balouch 2005; Balouch et al. 2010). In fact, by limiting the crack width to 0.3 mm the corrosion is limited to the fibres located at the surface, which has only an aesthetic drawback (Balouch et al. 2010; Frazão et al. 2015, 2016). Therefore, FRC seems to be an appropriate composite material for structures in corrosive environments such as concrete breakwaters (Vondran 1991).

Corrosive agents, liquid or gaseous, may penetrate the concrete through one of the three transport mechanisms: diffusion, capillary transport and permeation. The permeation is highly dependent on the concrete cracking process. An increase in the crack width not only produces a highly permeable concrete, but also enhances the possibility of fibre corrosion. However, the concrete permeability decreases significantly by increasing the fibre content due to the reduction of shrinkage cracks (Singh and Singhal 2011).

The insufficient knowledge on the deterioration mechanisms caused by fibre corrosion contributes to a conservative design philosophy, which limits the mobilization of the full potential of SFRC (Solgaard et al. 2010). In fact, some design guidelines recommend not to take into account the contribution of fibre reinforcement in the outer layer of a given thickness for the evaluation of the flexural resistance of SFRC members (RILEM TC 162-TDF 2003).

It is widely reported that in case of SFRC, steel fibre corrosion is much less severe as compared with steel rebar reinforcement of concrete structures (Balouch et al. 2010; Berrocal et al. 2013, 2015a, b; Sadeghi-Pouya et al. 2013). Published experimental observations indicate that steel fibres improved corrosion resistance of rebars moderately, which was mainly attributed to the reduced ingress of chlorides due to arrested crack growth provided by fibre reinforcement mechanisms (Berrocal et al. 2015a, b). Due to their large surface area to volume ratio, steel fibres are more effectively protected by the lime rich layer than the large diameter bars used in conventional reinforced concrete. However, the corrosion of fibres can produce micro-spalling of concrete, as well as the reduction of the sectional area of the fibres that may justify some concerns on the long term material and structural performances of SFRC structures (Granju and Balouch 2005).

The chloride diffusivity depends of the concrete pore structure and all the factors that determine it, such as, mix design parameters (W/C ratio, type and proportion of mineral admixtures and cement, compaction, curing, etc.) and presence of cracks (Shi et al. 2012). Literature is mainly focused on corrosion arising from cracking process (Granju and Balouch 2005; Yoon 2012; Nordstrom 2005). There are three main consequences of corrosion in SFRC associated to the cracking process: (1) a decrease of the carrying capacity and energy absorption performance of the SFRC element (more brittle behaviour) (Berrocal et al. 2015a, b); (2) due to the rust formation from fibre corrosion process, fibre-paste friction can increase, enhancing the fibre pull-out response, with beneficial effects in terms of load carrying capacity and energy absorption of the SFRC element (Granju and Balouch 2005); (3) if crack width is small enough to be sealed by the SFRC self-healing capacity, fibre corrosion has negligible influence in terms of structural and durability performance (Granju and Balouch 2005). Granju and Balouch (2005) verified that fibre corrosion leads to a decrease of its cross section, which has resulted into a decrease of the load carrying capacity of SFRC elements. They also observed that the rate of degradation of SFRC due to fibre corrosion is dependent on several parameters, such as: crack width, exposure environmental, and type of fibres.

The carbonation penetration rate is determined by the permeability of the concrete and starts at the concrete surface and continues inwards, as long as there is enough carbon dioxide available. In practice it has been found that the carbonation front is stopped when it reaches the steel fibre by the large supply of the lime around the fibre. The single fibre may lose its protective passivating layer in the longer term, but the fibres at deeper position remain free from corrosion (Corinaldesi and Moriconi 2004).

Corrosion is significantly affected by the type of cement used. The corrosion in concretes with blast furnace slag cement and aluminous cement was less severe than in concrete with Portland cement. As a consequence, in concretes made with Portland cement, such as ordinary, high-early-strength and moderate heat types, there were many recognizable fine cracks and fine spalling on the surface at the age of 20 years. On the other hand, cracks and spalling found on the surface of the concretes using blast furnace slag cement B-class and aluminous cement were lower in number (Fukute and Hamada 1993). According to the study performed by Fukute and Hamada (1993) on the durability of concrete exposed in marine environment for 20 years, concrete using blast furnace slag cement and aluminous cement show excellent performance against sea-water attack and corrosion of reinforcement. Furthermore, sea water as mixing water of concrete has little effect on the durability of concrete and corrosion of reinforcement, when concrete has been exposed in tidal zone for long time (Fukute and Hamada 1993).

Surface corrosion affects negatively the appearance of SFRC but does not affect its mechanical properties (Graeff et al. 2009). This can be prevented by having both W/C $\leq$ 0.5 and a minimum cover of the fibres as large as 0.2 mm (Balouch et al. 2010). SFRC possesses extremely low resistivity due to high electrical conductivity of steel fibres, which increases the possibility of fibre corrosion (Tsai et al. 2009). Concrete resistivity is also affected negatively by the moisture and fibre content (Michel et al. 2009). The exposure of FRC to high temperatures increases the risk of reinforcement corrosion (Khalil 2006; Lourenço 2012).

Several researchers have investigated the effect of corrosion in SFRC (Granju and Balouch 2005; Baliuch et al. 2010; Frazão et al. 2015, 2016; Solgaard et al. 2010; Berrocal et al. 2013, 2015a, b; Sadeghi-Pouya et al. 2013; Shi et al. 2012; Yoon 2012; Nordstrom 2005; Corinaldesi and Moriconi 2004; Graeff et al. 2009; Tsai et al. 2009; Michel et al. 2009; Chen et al. 2015; Nemegeer et al. 2003; Waweru 2011). However, the conclusions presented in these works are not sufficient to attest the corrosion resistance of RSFRC.

Graeff et al. (2009) investigated the corrosion durability of ISFRC and RSFRC for pavement applications. Since corrosion is a long term process, corrosion accelerating methods by wet-dry cycles in salt solution were used to evaluate the losses in terms of flexural and compressive strength. During the curing phase of the specimens reinforced with RSF it was visually observed that some fibres at the surface of the specimens already presented signs of corrosion. After 5 months of accelerated corrosion procedure, it was noticed that the specimens reinforced with RSF showed more superficial corrosion effects. From the visual analysis, a scale of deterioration was proposed by Graeff et al. (2011), as shown in Table 8.4. The ISFRC usually present the lowest amount of rust (in the range of 0–3) on the surface, whilst RSFRC fall in the highest levels of deterioration (in the range of 2–5) (Graeff et al. 2011) Table 8.6.

By observing the fracture surface of the samples, it was assessed that signs of corrosion were limited to a concrete cover of a thickness do not exceeding 10 mm in case of specimens reinforced with RSF, while corrosion signs are limited to the fibres at the surface of the samples in case of ISF.

After 5 months of accelerated corrosion procedures, in spite of the occurrence of concrete micro-spalling, the flexural and compressive strength capacity had not decreased because of corrosion effects when compared to the control samples (wet cured and tested at 28 days), as can be seen in Fig. 8.43. The increased capacity of

Table 8.6 Scale of superficial deterioration of SFRC due to rust (Graeff et al. 2011)

Scale of deterioration	0	1	2	3	4	5
External appearance		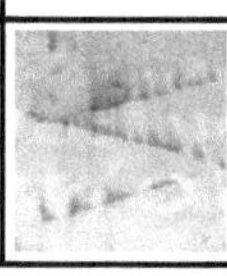	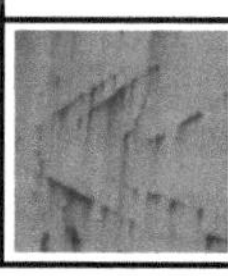	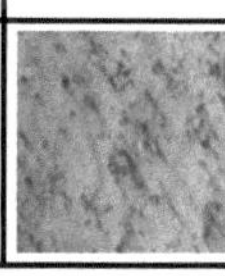	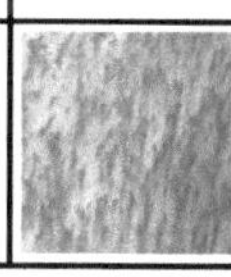	

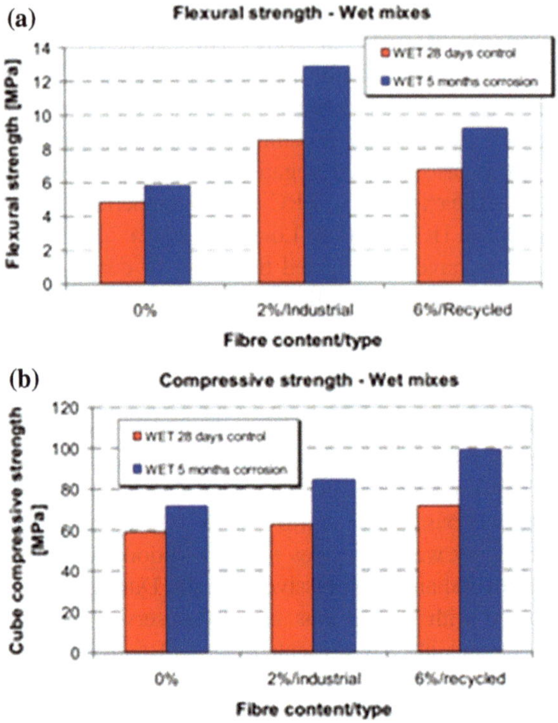

Fig. 8.43 **a** Flexural strength and **b** compressive strength results (Graeff et al. 2009)

corroded wet mixes when compared to those at 28 days is probably due to the age of the specimens (Fig. 8.43). The ISFRC presented a better performance than RSFRC in flexural tests. However, specimens reinforced with RSF showed higher compressive strength than specimens reinforced with ISF (Graeff et al. 2009).

The corrosion characteristics and the effects of corrosion degree on the mechanical properties of steel fibres are directly related to the performances of SFRC. Waweru (2011) performed mechanical tests on concrete specimens with pre-corroded steel fibres to 0, 12.5 and 50% reduction in the minimum fibre diameter, induced by drying–wetting cycles. Increasing the degree of corrosion, the failure type of steel fibres changed from typical pull-out to their rupture. The experimental results demonstrated that fibre corrosion corresponding to a diameter reduction lower than 12.5% had no effect on the shear capacity, while 50% reduction in the minimum fibre diameter led to a decrease of 24% on the beam shear strength. Furthermore, the toughness and ductility of SFRC were affected by corrosion to a much higher extent than the shear strength (Waweru 2011).

Based on the experimental results carried out by Chen et al. (2015), the steel fibres corroded by wetting and drying cycles to 0.47–33.38% in weight loss rate (measured by a scale after cleaning the fibres to remove corrosion) exhibiting pitting corrosion, while the steel fibres corroded to 1.81–87.48% by the galvanostatic method showed a relatively uniform corrosion. As the corrosion degree increased, the weight loss increased, and both of the nominal tensile strength and rupture elongation of corroded steel fibres decreased. The results of tensile tests showed that the actual tensile strength of steel fibres was reduced in the case of pitting, while it was marginally affected by uniform corrosion. The local section loss at the pits is the main reason for further degradation of the mechanical properties of steel fibres, which led to a brittle failure (Chen et al. 2015).

References

Abdul Awal ASM, Yee LL, Hossain MZ (2013) Fresh and hardened properties of concrete containing steel fibre from recycled tire. Malays Civ Eng J 25

Abrishambaf A, Barros JAO, Cunha VMCF (2013) Relation between fibre distribution and post-cracking behaviour in steel fibre reinforced self-compacting concrete panels. Cem Concr Res 51, 57–66

Abrishambaf A, Cunha V, Barros JAO (2016) The influence of fibre orientation on the post-cracking tensile behaviour of steel fibre reinforced self-compacting concrete. Fract Struct Integrity J 31(1):38–53

ACI 544.5R-10 (2010) Report on the physical properties and durability of fibre-reinforced concrete. ACI Committee 544

Aiello MA, Leuzzi F (2010) Waste tyre rubberized concrete: properties at fresh and hardened state. Waste Manag 30(8):1696–1704

Aiello MA, Leuzzi F, Centonze G, Maffezzoli A (2009) Use of steel fibres recovered from waste tyres as reinforcement in concrete: pull-out behaviour, compressive and flexural strength. Waste Manag 29(6):1960–1970

ASTM C 1500-03a (2003) Standard test for flexural toughness of fibre reinforced concrete (using centrally loaded rounded panle). ASTM Int, 10

ASTM C496/C496M-11 (2011) Standard test method for splitting tensile strength of cylindrical concrete specimens. ASTM Int, 5

ASTM ASTM D 5379 (1993) Standard test method for shear properties of composite materials by the V-notched beam method. Am Soc Test Mater

Bayasi MZ, Soroushian P (1992) Effect of Steel fiber reinforcement on fresh mix properties of concrete. ACI Mater J 89(4):369–374

Baghi H, Barros JAO (2016) Shear properties of the strain hardening cementitious composite materials. ASCE J Mater Civ Eng. doi:10.1061/(ASCE)MT.1943-5533.0001603

Balouch S, Forth J, Granju J (2010) Surface corrosion of steel fibre reinforced concrete. Cem Concr Res 17:410–414

Barros JAO, Figueiras JA (1999) Flexural behavior of steel fibre reinforced concrete: testing and modelling. ASCE Mater Civ Eng J 11(4):331–339

Barros JAO, Figueiras JA, Veen CVD (1994) Tensile behaviour of glass fibre reinforced concrete. In: Silva Gomes JF et al (ed) Recent advances in experimental mechanics, Lisbon, Portugal, vol 2, pp 1073–1080

Barros JAO, Cunha VMCF, Ribeiro AF, Antunes JAB (2005) Post-cracking behaviour of steel fibre reinforced concrete. RILEM Mater Struct J 38(275):47–56

Barros JAO, Pereira EB, Santos SPF (2007) Lightweight panels of steel fiber reinforced self-compacting concrete. ASCE Mater Civ Eng J 19(4):295–304

Barros JA, Zamanzadeh Z, Mendes PJ, Lourenço L (2013a) Assessment of the potentialities of recycled steel fibres for the reinforcement of cement based materials. In 3rd Workshop: The New Boundaries of Structural Concrete, Session C-New Scenarios for Concrete, ACI Italy Chapter. Università degli Studi di Bargamo, pp 1–11

Barros JAO, Baghi H, Dias SJE, Ventura-Gouveia A (2013b) A FEM-based model to predict the behaviour of RC beams shear strengthened according to the NSM technique. Eng Struct J 56:1192–1206

Bdour N, AlKhalayleh A (2010) Innovative application of scrap-tyre steel cords in concrete mixes. Jordan J Civ Eng 4(1)

Bencardino F, Rizzuti L, Spadea G, Swamy RN (2008) Stress-strain behavior of steel fiber-reinforced concrete in compression. ASCE J Mater Civ Eng 20(3):255–263

Bentur A, Mindess S (2007) Fibre reinforced cementitious composites, 2 edn. Taylor & Francis, London

Berrocal C, Lundgren K, LÖfgren I (2013) Influence of steel fibres on corrosion of reinforcement in concrete in chloride environments: a review. In: International conference: FC2013-fibre concrete 2013, Prague, Czech Republic, 12–13 Sept 2013

Berrocal CG, Lundgren K, Löfgren I (2015a) Corrosion of steel bars embedded in fibre reinforced concrete under chloride attack: state of the art. Cem Concr Res 80:69–85

Berrocal CG (2015b) Chloride induced corrosion of steel bars in fibre reinforced concrete. Thesis for the degree of licentiate of engineering, Department of Civil and Environmental Engineering, Chalmers University of Technology, Goteborg, Sweden

Bjegović D, Pečur IB, Štirmer N, Serdar M, Milovanović B, Rukavina MJ, Baričević A (2014). Construction materials roadmap for R&D. In Future Trends in Civil Engineering

Blumenthal MH (1996) Producing ground scrap tire rubber: a comparison between ambient and cryogenic technologies. In: Proceedings of the 17th Biennial Waste Processing Conference ASME 1996. Available online at www.rma.org/download/scrap-tires/processing/PRS-005%20%281%29.pdf. Accessed 30 Oct 2015

Buratti N, Mazzotti C, Savoia M (2011) Post-cracking behaviour of steel and macro-synthetic fibre-reinforced concretes. Constr Build Mater 25(5):2713–2722

Burden T (2015) Trailer tire basics. Available online at www.westmarine.com/WestAdvisor/Trailer-Tire-Basics. Accessed on 30 Oct 2015)

Caggiano A, Xargay H, Folino P, Martinelli E (2015) Experimental and numerical characterization of the bond behavior of steel fibres recovered from waste tires embedded in cementitious matrices. Cem Concr Compos 62:146–155

CEN/TS14243 (2010) Materials produced from end of life tyres. Specification of categories based on their dimension(s) and impurities and methods for determining their dimension(s) and impurities. European Technical Specification. European Committee for Standardization, Brussels April 2010

Centonze G, Leone M, Aiello MA (2012a) Steel fibres from waste tires as reinforcement in concrete: a mechanical characterization. Constr Build Mater 36:46–57

Centonze G, Leonard M, Aiello MA (2012b) Steel fibres from waste tyres as reinforcement in concrete: a mechanical characterisation. Constr Build Mater J 36(12), 2012

Centonze G, Leone M, Vasanelli E, Aiello MA (2013) Interface analysis between steel bars and recycled steel fibre reinforced concrete. In: VIII international conference on fracture mechanics of concrete and concrete structures, FraMCoS-8

Chen G, Hadi MNS, Gao D, Zhao L (2015) Experimental study on the properties of corroded steel fibres. Constr Build Mater 79:165–172

CNR-204/2006. CNR Consiglio Nazionale delle Ricerche: Istruzioni per la progettazione, l'esecuzione ed il controllo di strutture di calcestruzzo fibrorinforzato; 2006 [in Italian]

Corinaldesi V, Moriconi G (2004) Durable fibre reinforced self-compacting concrete. Cem Concr Res 34:249–254

di Prisco M, Ferrara L, Lamperti ML (2013) Double edge wedge splitting (DEWS): an indirect tension test to identify post-cracking behaviour of fibre reinforced cementitious composites. Mater Struct 46(11):1893–1918

Duffey MR, Sunthonpagasit N (2004) Scrap tires to crumb rubber: feasibility analysis for processing facilities. Resour Conserv Recycl 40:281–299

Dunlop (2013) Aircraft tyre technology. Available online at www.dunlopaircrafttyres.com/tech_support/aircraft-tyre-technology.aspx. Accessed on 30 Oct 2015

EFNARC (2002) Specification and guidelines for self-compacting concrete, p 32
Eldho A, Bibin MC, Elbi PA, Laly AP, Sabu T (2011) Recent developments in polymer recycling. In: Fainleib A, Grigoryeva (eds) Recent advances in the recycling of rubber waste transworld research network, Kerala, India, pp 47–100
EPA (1999) State scrap tire programs. A Quick Reference Guide: 1999 Update, United States Environmental Protection Emergency. Available online at www.epa.gov. Accessed on 30 Oct 2015
ETRA (2013) Introduction to tyre recycling: 2013. European Tyre Recycling Industry. Available online at http://www.etra-eu.org/joomla/publications. Accessed on Oct 2015
ETRMA (2010) End of life tyres – a valuable resource with growing potential. European Tyre Rubber Manufactures Assosation. Available online at http://www.etrma.org/uploads/Modules/Documentsmanager/2010_etrma_elt_management_brochure_final.pdf. Accessed on Oct 2015
ETRMA (2014) European tyre and rubber industry—statistics. European Tyre Rubber Manufactures Association. Available online at http://www.etrma.org/statistics-2, Accessed on Oct 2015
Feraldi R, Cashman S, Huff M, Raahauge L (2013) Comparative LCA of treatment options for US scrap tires: material recycling and tire-derived fuel combustion. Int J Life Cycle Assess 18 (3):613–625
Ferrara L, Park YD, Shah SP (2007) A method for mix-design of fiber reinforced self compacting concrete. Cem Concr Res 37:957–971
Ferrara L, Faifer M, Toscani S (2012a) A magnetic method for non destructive monitoring of fiber dispersion and orientation in steel fiber reinforced cementitious composites—part 1: method calibration. Mater Struct 45:575–589
Ferrara L, Faifer M, Muhaxheri M, Toscani S (2012b) A magnetic method for non destructive monitoring of fiber dispersion and orientation in steel fiber reinforced cementitious composites—part 2: correlation to tensile fracture toughness. Mater Struct 45:591–598
Ferrer G (1997) The economics of tire remanufacturing. Resour Conserv Recycl 19:221–255
Frazão C, Camões A, Barros J, Gonçalves D (2015) Durability of steel fibre reinforced self-compacting concrete. Constr Build Mater 80:155–166
Frazão CMV, Barros JAO, Camões AFLL, Alves A, Rocha L (2016) Corrosion effects on pullout behavior of hooked steel fibres in self-compacting concrete. Cem Concr Res 79:112–122
Fukute T, Hamada H (1993) A study on the durability of concrete exposed marine environment for 20 years. Report of the Port and Harbour Research Institute, vol 31, issue 5, Nagase, Yokosuka, Japan, 1993
Ganjian E, Khorami M, Maghsoudi AA (2009) Scrap-tyre-rubber replacement for aggregate and filler in concrete. Constr Build Mater 23(5):1828–1836
Graeff A, Pilakoutas K, Lynsdale C, Neocleous K (2009) Corrosion durability of recycled steel fibre reinforced concrete. Article No. 7, Intersections/Intersec_ii 6(4)
Graeff AG, Pilakoutas K, Neocleous K, Lynsdale C (2011) Behaviour of concrete reinforced with recycled steel fibres exposed to chloride contaminated environment. In: 6th international conference fibre concrete 2011, Prague, Czech Republic, 8–9 Sept 2011
Graeff AG, Pilakoutas K, Neocleous K, Peres MVN (2012a) Fatigue resistance and cracking mechanism of concrete pavements reinforced with recycled steel fibres recovered from post-consumer tyres. Eng Struct 45:385–395
Graeff A, Pilakoutas K, Lynsdale C, Neocleous K, Peres M (2012b) Fatigue resistance and cracking mechanism of concrete pavements reinforced with recycled steel fibres recovered from post-consumer tyres. Eng Struct 45:385–395
Granju J, Balouch S (2005) Corrosion of steel fibre reinforced concrete from the cracks. Cem Concr Res 35(3), 572–577
ITRA (2001) Understanding retreading—Guideline. International Tire and Rubber Association Foundation Inc.
IWMB (2003) Assessment of markets for fibre and steel produced from recycling waste tires. Integrated Waste Market Board, Sacramento, CA, US

IWMB (2004) Evaluation of waste tire devulcanization technologies. Integrated Waste Market Board, Sacramento, CA, US

Jala M (2012) Compressive strength enhancement of concrete reinforced by waste steel fibres utilizing nano SiO_2. Middle East J Sci Res 12(3):382–391

Jonathan VL, Kiser JVL (2002) Scrap-tire pyrolysis—the impossible dream? Magazine Archive. September/October 2002

Juma M, Korenova Z, Markoš J, Annus J, Jelemenský L (2006) Pyrolysis and combustion of scrap tire. Petrol Coal 48(1), 15–26. Available online at www.vurup.sk/pc. Accessed on 30 Oct 2015

Khalil W (2006) Influence of high temperature on steel fibre reinforced concrete. J Eng Dev 10 (2):139–150

Khaloo AR, Dehestani M, Rahmatabadi P (2008) Mechanical properties of concrete containing a high volume of tire–rubber particles. Waste Manag 28(12):2472–2482

Laboy-Nieves EN (2014) Energy recovery from scrap tires: a sustainable option for small islands like Puerto Rico. Sustainability 6:3105–3121. doi:10.3390/su6053105

Lameiras RM (2016) Sandwich structural panels comprising thin-walled SFRSCC and GFRP connectors: from material features to structural behaviour. PhD Thesis, University of Minho, 2016

Liang SB, Hao YC (2000) A novel cryogenic grinding system for recycling scrap tire peels. Adv Powder Technol 11(2):187–197

Lion Y (2012) Mechanical properties of recycle steel fibre reinforced concrete. MSC thesis, Faculty of Civil Engineering of University Technology, Malaysia, 2012

Lourenço L (2012) Fibre reinforced concrete: applications and inspection techniques and reinforcement of structural elements affected by the action of a fire. PhD Thesis, University of Minho, Guimarães, Portugal, 2012 [in Portuguese]

Martinelli E, Caggiano A, Xargay H (2015) An experimental study on the post-cracking behaviour of Hybrid Industrial/Recycled Steel Fibre-Reinforced Concrete. Constr Build Mater 94: 290–298

MC2010, CEB fib Model Code 2010 - Final Draft, 2011

Meddah M, Bencheikh M (2009) Properties of concrete reinforced with different kinds of industrial waste fibre materials. Constr Build Mater 23:3196–3205

Micelli F, Leone M, Centonze G, Aiello M (2014) Go green: using waste and recycling materials. In: Y Lu (ed) Infrastructure corrosion and durability sustainability study, OMICS Group eBooks, USA, 2014

Michel A, Geiker M, Stang H, Olesen J, Solgaard A (2009) Numerical modelling of reinforcement corrosion, influence of steel fibres and moisture content on resistivity and corrosion current density. In: 3th international RILEM PhD student workshop on modelling the durability of reinforced concrete, Guimarães, Portugal, 22–24 Oct 2009

Michelin (2015) Radial versus bias technologies. Available online at http://www.michelinag.com/Innovating/Radial-vs.-Bias-technology. Accessed on 30 Oct 2015

Mobasher B, Barros JAO, Naaman AE, Destree X et al (2015) Report on design and construction of steel fibre-reinforced concrete elevated slabs. ACI 544.6R-15, Reported by ACI Committee 544, 2015

Mohammadi Y, Singh SP, Aushik SK (2008) Properties of steel fibrous concrete containing mixed fibres in fresh and hardened state. Constr Build Mater 22(5):956–965

Moraes-Neto BN, Barros JAO, Melo GSSA (2013) A model for the prediction of the punching resistance of steel fibre reinforced concrete slabs centrically loaded. Constr Build Mater J 46:211–223

Mountjoy E, Hasthanayake D, Freeman T (2015) Stocks & fate of end of life tyres—2013–14 Study—Final Report, Hyder Consulting Pty Ltd. Available online at http://www.nepc.gov.au/system/files/resources/8f17c03e-1fe7-4c93-8c6d-fb4cdc1b40bd/files/stocks-and-fate-end-life-tyres-2013-14-study.pdf. Accessed on 30 Oct 2015

Nasir B (2009) Steel fibre reinforced concrete made with fibres extracted from used tyres. Master of Science in Civil Engineering dissertation, AAU

Nemegeer D, Vanbrabant J, Stang H (2003) Brite-euram program on steel fibre concrete, durability corrosion resistance of cracked fibre-reinforced concrete. Test and design methods for steel fibre reinforced concrete-background and experiences. In: Schnütgen B, Vandewalle L (eds) RILEM technical committee 162, TDF Workshop, Proceedings Pro, vol 31, pp 47–66, 2003

Neocleous K, Tlemat H, Pilakoutas K (2006) Design issues for concrete reinforced with steel fibres, including fibres recovered from used tyres. Mater Civ Eng 18(5):677–685

Nkosi N, Muzenda E (2014) A review and discussion of waste tyre pyrolysis and derived products. In: Proceedings of the world congress on engineering, vol II, WCE 2014, July 2–4, 2014, London, UK. Available online at http://www.iaeng.org/publication/WCE2014/WCE2014_pp979-985.pdf. Accessed on 30 Oct 2015

Nordstrom E (2005) Durability of sprayed concrete – steel fibre corrosion in cracks. Doctoral Thesis, Department of Civil and Environmental Engineering, Division of Structural Engineering, Luleå University of Technology, Sweden, 2005

Olivito RS, Zuccarello F (2010) An experimental study on the tensile strength of steel fibre reinforced concrete. Compos B Eng 41(3):246–255

Ou YC, Tsai MS, Liu KY, Chang KC (2012) Compressive behavior of steel fiber reinforced concrete with a high reinforcing index. ASCE J Mater Civ Eng 24(2):207–215

Papakonstantinou CG, Matthew JT (2006) Use of waste tire steel beads in Portland cement concrete. Cem Concr Res J 36:1686–1691

Papakonstantinou CG, Tobolski MJ (2006) Use of waste tire steel beads in Portland cement concrete. Cem Concr Res 36(9):1686–1691

Pehlken A, Essadiqi E (2005) Scrap tyre recycling in Canada. CANMET Materials Technology Laboratory, MTL 2005–08(CF). Available online at https://www.nrcan.gc.ca/ sites/www.nrcan.gc.ca/files/mineralsmetals/pdf/mms-smm/busi-indu/rad-rad/pdf/scr-tir-rec-peh-eng.pdf. Accessed on 30 Oct 2015

Pehlken A, Müller DH (2009) Using information of the separation process of recycling scrap tires for process modeling. Resour Conserv Recycl 54:140–148

Pereira ENB (2006) Steel fibre reinforced self-compacting concrete: from material to mechanical behaviour. MSC Thesis, University of Minho, 2006

Pereira EB, Barros JAO, Camões AFFL (2008) Steel fibre reinforced self-compacting concrete—experimental research and numerical simulation. ASCE Struct Eng J 134(8):1310–1321

Pereira ENB, Fischer G, Barros JAO (2012) Direct assessment of tensile stress-crack opening behavior of strain hardening cementitious composites (SHCC). Cem Concr Res 42:834–846

Pilakoutas K, Strube R (2001) Reuse of tyre fibres in concrete, in recycling and reuse of used tyres. In: Proceedings of the international symposium, Dundee, 2001

Pilakoutas K, Neocleous K, Tlemat H (2004). Reuse of tyre steel fibres as concrete reinforcement. In: Proceedings of the ICE: engineering sustainablity, London, pp 131–138

Richardson A (2013) Recycled materials in concrete. In: Reuse of materials and byproducts in construction, Springer, London, pp 67–109

RILEM TC 162-TDF (2003) Test and design methods for steel fibre reinforced concrete. σ-e-design method. Final Recommendation. Mater Struct 36, 560–567

Rodrigues JPC (2015) Betão com Adição de Fibras de Aço e Têxteis Recicladas de Pneu Sujeito a Altas Temperaturas (Master thesis dissertation, Universidade de Coimbra)

Rossli SA, Ibrahim IS (2012) Mechanical properties of recycled steel tire fibres in concrete. Technical Report, Faculty of Civil Engineering, University Technology Malaysia

Sadeghi-Pouya H, Ganjian E, Claisse P, Muthuramalingam K (2013) Corrosion durability of high performance steel fibre reinforced concrete. In: 3rd international conference on sustainable construction materials and technologies, Japan Concrete Institute, Kyoto, Japan, August 2013

Salehian H (2015) Evaluation of the performance of steel fibre reinforced self-compacting concrete in elevated slab systems; from the material to the structure. PhD in Civil Engineering, University of Minho, p 308

Salehian H, Barros JAO (2015) Assessment of the performance of steel fibre reinforced self-compacting concrete in elevated slabs. Cem Concr Compos 55:268–280

Salehian H, Barros JAO, Taheri M (2014) Evaluation of the influence of post-cracking response of steel fibre reinforced concrete (SFRC) on load carrying capacity of SFRC panels. Constr Build Mater J 73:289–304

Schellekens JCJ (1990) Interface elements in finite element analysis. TU-Delft report 25.2-90-5-17/TNO-IBBC report BI-90-165, October 1990

Sena-Cruz JM, Barros JAO, Ribeiro AF, Azevedo AFM, Camões AFFL (2004) Stress-crack opening relationship of enhanced performance concrete. In: 9th Portuguese conference on fracture, ESTSetúbal, Portugal, pp 395–403, 18–20 Feb 2004

Shah SP, Swartz SE, Ouyang C (1995) Fracture mechanics of concrete: applications of fracture mechanics to concrete, rock and other quasi-brittle materials. John Wiley & Sons, Inc.

Sharma A (2013) Retreading of tyres. Int J Eng Adv Tech (IJEAT) 2(6):143–145

Sharma VK, Fortuna F, Mincarini M, Berillo M, Cornacchia G (2000) Disposal of waste tyres for energy recovery and safe environment. Appl Energy 65(1–4):381–394

Shi X, Xie N, Fortune K, Gong J (2012) Durability of steel reinforced concrete in chloride environments: an overview. Constr Build Mater 30:125–138

Siddique R, Naik TR (2004) Properties of concrete containing scrap-tire rubber–an overview. Waste Manag 24(6):563–569

Sienkiewicz M, Kucinska-Lipka J, Janik H, Balas H (2012) Progress in used tyres management in the European Union: a review. Waste Manag 32:1742–1751

Singh A, Singhal D (2011) Permeability of steel fibre reinforced concrete influence of fibre parameters. Proc Eng 14:2823–2829

Solgaard A, Kuter A, Edvardsen C, Stang H, Geiker M (2010) Durability aspects of steel fibre reinforced concrete in civil infrastructure. In: 2nd international symposium on service life design for infrastructure, Delft, The Netherlands, 4–6 Oct 2010

Soltanzadeh F, Barros JAO, Santos RFC (2015) High performance fiber reinforced concrete for the shear reinforcement: experimental and numerical research. Constr Build Mater J 77:94–109

STMC (1992) The use of scrap tires in cement rotary kilns, Scrap Tire Management Council, Washington, DC, USA. Available online at http://www.cmtirerecyclingequipment.com/Public/15663/cement_kiln_report.pdf. Accessed on 30 Oct 2015

Thomson J (2001) What we should know about tires: a historical background. Available online at www.jags.org/TechInfo/2001/05May01/tires/historyoftires.htm. Accessed on 30 Oct 2015

Tlemat H, Pilakoutas K, Neocleous K (2005) Stress-strain characteristic of SFRC using recycled fibres. Mater Struct RILEM J 39(3):365–377

Tlemat H, Pilakoutas K, Neocleous K (2006) Stress-strain characteristic of SFRC using recycled fibres. Mater Struct 39(3):365–377

Torretta V, Rada EC, Ragazzi M, Trulli E, Istrate IA, Cioca LI (2015) Treatment and disposal of tyres: two EU approaches. A Rev Waste Manag 45:152–160

Tsai C, Li L, Chang C, Hwang C (2015) Durability design and application of steel fibre reinforced concrete in Taiwan. Arab J Sci Eng 34(1B), 2009

Undri A, Meini S, Rosi L, Frediani M, Frediani P (2012) Microwave pyrolysis of polymeric materials: Waste tires treatment and characterization of the value-added products. J Anal Appl Pyrol 103:149–158

UNEP (2011) Revised technical guidelines for the environmentally sound management of used and waste pneumatic tyres, United Nations Environment Programme. Available online at http://www.moew.government.bg/files/file/Waste/Nasoki_rakovodstva/TG_waste_tyres.pdf. Accessed on 30 Oct 2015

van Mier J (1997) Fracture processes of concrete: assessment of material parameters for fracture models. CRC Press, New Directions in Civil Engineering

Vandewalle L, et al (2001) Uni-axial tension test for steel fibre reinforced concrete. Mater Struct J 34, 3–6 Jan–Feb 2001

Vandewalle L et al (2002) Test and design methods for steel fibre reinforced concrete—final recommendation. Mater Struct RILEM J 35:579–582

Ventura-Gouveia A (2011) Constitutive models for the material nonlinear analysis of concrete structures including time dependent effects. PhD Thesis, Department of Civil Engineering, University of Minho, 2011

Vondran GL (1991) Applications of steel fibre reinforced concrete. Concr Int 13(11):44–49

Waweru R (2011) The effect of fibre corrosion on shear capacity of steel fibre reinforced concrete beams and an initial investigation on alkali-silica reaction in steel fibre reinforced concrete. Master Thesis in Civil Engineering, University of Texas at Arlington, 2011

Yazici S, Inan G, Tabak V (2007) Effect of aspect ratio and volume fraction of steel fibre on the mechanical properties of SFRC. Constr Build Mater 21(6):1250–1253

Yoon I (2012) Chloride penetration through cracks in high- performance concrete and surface treatment system for crack healing. Hindawi Publishing Corporation, Advances in Materials Science and Engineering, Article ID 294571, 8, 2012

Zamanzadeh Z (2017) Cement based materials reinforced with recycled steel fibres for the shear reinforcement and strengthening of RC beams. PhD Thesis, University of Minho

Zamanzadeh Z, Lourenço LAP, Barros JAO (2015a) Recycled steel fibre reinforced concrete failing in bending and in shear. J Constr Build Mater 85:195–207

Zamanzadeh Z, Lourenço LAP, Barros JAO (2015b) Thin panels of cement composites reinforced with recycled fibres for the shear strengthening of reinforced concrete elements. Fibre Concrete 2015, FC2015, Prague, Czech Republic, 10–11 Sept 2015b

Chapter 9
Cementitious Composites Reinforced with Natural Fibres

Liberato Ferrara, Saulo Rocha Ferreira, Visar Krelani, Paulo Lima, Flavio Silva and Romildo Dias Toledo Filho

Abstract Natural fibres as dispersed reinforcement in cement-based materials may represent an interesting alternative to industrial fibres, because of their good mechanical properties and inborn sustainability signature. In this section the mechanical properties of Natural Fibre Reinforced Cementitious Composites (NFRCCs) will be reviewed. Cutting-edge topics will be finally addressed, focusing, on the one hand, on the use of nano-sized cellulose-based constituents in cementitious composites, and, on the other, on the self-healing capacity that natural fibres may bring to cementitious composites. An appendix summarizing the main properties of the natural fibres so far most commonly employed in cementitious composites will complement this information.

9.1 Introduction

The use of natural fibres as a discontinuous and randomly dispersed reinforcement in cement based construction materials may represent an interesting alternative to the use of other types of industrially manufactured fibres, either new or recycled,

L. Ferrara (✉) · V. Krelani
Politecnico Di Milano, Milan, Italy
e-mail: liberato.ferrara@polimi.it

S.R. Ferreira · R.D.T. Filho
Federal University of Rio de Janeiro, Rio de Janeiro, Brazil

P. Lima
State University of Feira de Santana, Feira de Santana, Brazil

F. Silva
Pontifical University of Rio de Janeiro, Rio de Janeiro, Brazil

S.R. Ferreira
Department of Civil Engineering, Universidade Federale de Lavras (UFLA), Lavras, MG, Brazil

V. Krelani
University for Business and Technology (UBT), Kalabria, Prishtine, Kosova

J.A.O. Barros et al. (eds.), *Recent Advances on Green Concrete for Structural Purposes*, Research for Development, DOI 10.1007/978-3-319-56797-6_9

such as metal and/or polymeric/synthetic ones. Such an opportunity has been already keenly investigated and quite successfully exploited in other fields of material and industrial engineering, with reference, e.g., to polymer matrix composites (Faruk and Sain 2015). Natural fibres are in fact lightweight, biodegradable and renewable, have good mechanical properties and are relatively abundant, not seldom available as a by-product of food industry and agriculture related activities, mainly in rural areas and/or developing countries, for whose communities they may represent an interesting source of income. Moreover, they have neutral CO_2 emissions, since the cultivation of plants, from which they are obtained, depends primarily on solar energy and only small amounts of energy, from fossil fuels but even from any other renewable source, are needed for the production and extraction processes. Interestingly, it is also worth remarking that natural fibres are likely to be not abrasive to the concrete mixing and casting equipment, which may resulted into an increased service life of the latter.

It has anyway to be remarked, that the physical and mechanical properties of natural fibres are highly affected by the natural conditions in which the mother plants are grown as well as by the production and extraction processes and by the further processes needed to have the same fibres suitably ready to be incorporated into a cementitious matrix. This is likely to result into a dispersion of the aforementioned properties quite higher than for industrially manufactured fibres, and into a likewise highly scattered fibre to matrix bond, all of which may affect the reliability of the mechanical performance of the NFRCCs.

In addition, due to their organic composition, the compatibility of natural fibres with the alkaline environment of the surrounding cementitious matrix has to be carefully evaluated, mainly with reference to the long-term stability of the performance of the fibre reinforced composite. In this respect, techniques have been conceived and validated to improve the aforementioned compatibility and the resulting durability of the fibre reinforced cementitious composites, with main reference to the degradation over time of its mechanical performance.

Besides the properties and relative proportions of fibres and matrix and their bond interaction, the mechanical and structural performance of NFRCCs is likely to be affected by several other factors, including the casting procedure and, in case, the related dispersion and flow-induced orientation of the fibres, as well as by the adopted curing procedure. In this respect, the hierarchically porous structures of natural fibres, and their hydrophilic behaviour, result into an absorption of water from the matrix. This affects the workability of the material and the effectiveness of the designed casting process; early age behaviour, in terms of both autogenous and drying shrinkage, is also likely to be affected, thus risking to jeopardize the effectiveness of the whole scheduled curing procedure.

The content of this section is structured as follows:

First of all, the knowledge developed over the past years on this topic by the authors has formed a solid background on which the project research activities have been developed. This with main reference to the experimental characterization of the mechanical performance of NFRCCs, also considering their durability and stability over time because of the implemented technologies aimed at improving the compatibility between the same fibres and the alkaline cementitious matrix. This *review* is of the utmost importance also with the goal of filling a critical gap of current design codes on Fibre Reinforced Concrete (FRC) structures, which, as for example *fib* Model Code 2010, explicitly do not cover those fibre materials with a Young modulus which is significantly affected by time and/or thermos-hygrometric variations, which both apply to natural fibres. Such an appraisal is critical to pave the way for a broader and solidly founded use of NFRCCs for high end engineering applications, but also to identify new trends of development, *e.g.* with the use of *nano-sized* natural constituents, such as nano-cellulose powder and/or crystals.

Secondly, the potentials of natural fibres of transporting water throughout the cementitious matrix and of thus activating crack-sealing and-healing processes has been investigated in deep details. This with the clear purpose of enriching the characterization of the sustainability and multi-functional durability signature of the category of advanced cement based materials at issue.

9.2 Physical and Mechanical Properties of Natural Fibres

In principle, all vascular plants that can be found in nature can be used as sources of natural (also called cellulosic) fibres. The use of a particular plant as a source of fibre for a given application will depend on its availability and on the costs of extraction. According to their origin and composition, natural fibres are classified as non-wood and wood fibres (Faruk and Sain 2015).

The non-wood fibres can be further classified into four main groups, depending on the part of the plant used to extract the fibres:

- bast fibres: jute, hemp, flax, ramie, kenaf, piassava;
- leaf fibres: sisal, henequen, pineapple leaf and curaua, banana and abaca, oil palm leaf fibres;
- seed fibres: coir/coconut, cotton and oil palm biomass.
- stalk fibres: straws, as rice, also including rice husk, wheat, maize, oat, barley and rye; cane, such as sugarcane bagasse; reeds, as bamboo; and grass, as esparto and elephant grass;

Wood fibres, also known as lignin-cellulosic fibres because they have a higher lignin content than non-wood fibres, are grouped, depending on their origin, into softwood fibres, e.g. from pines, firs, etc., and hardwood fibres, as e.g. from birch tree, eucalyptus, beech etc.

Apart from their origin, natural fibres employed as reinforcement in cement-based materials, can be also classified by the function of their form (Fig. 9.1). Thus, natural fibres can be found as strands (long fibres with lengths between 200 mm and 1 m), staple fibres (shorter length fibres, which can be spun into yarns) or pulps (very short fibres with length ranging from 1 to 10 mm and which should be dispersed into water to separate them). The strands or staple fibres are obtained from crop or wild plants directly or after a water retting process.

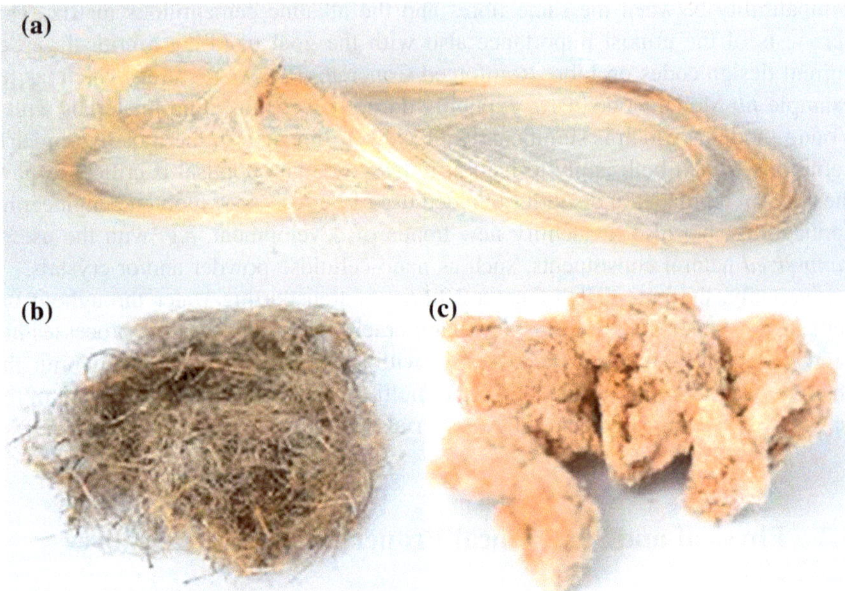

Fig. 9.1 Images of natural fibres in different forms: **a** strands, **b** staple, **c** pulp

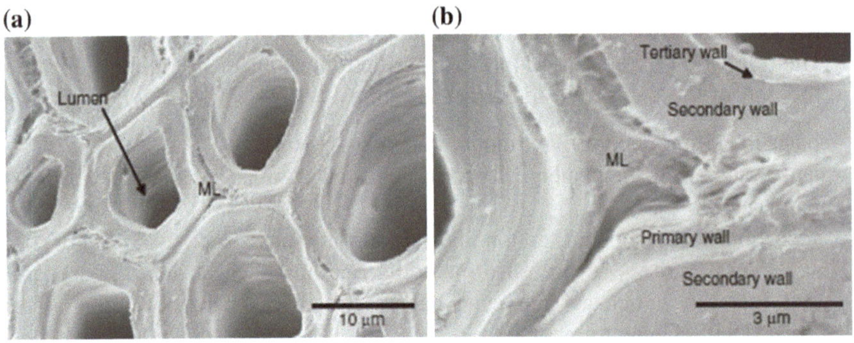

Fig. 9.2 Sisal fibre microstructure showing **a** fibre cells with lumen and middle lamellae (*ML*) and **b** detail of ML and cell-walls (de Andrade Silva et al. 2010a, b, c, d, e)

In this group also are included cellulose fibres traditionally employed by the textile industry, which are characterized by high aspect ratio and low linear mass. The pulps are generally obtained from wood sources by a pulping process, which, depending on the treatment used to destroy or weaken the inter-fibre bonds, can be mechanical, thermal, chemical or a combination of them.

All natural fibres generally share similar morphology, arranged according to a hierarchical structure. Each fibre consists of several fibre cells (micro-tubes), the diameter of which ranges from a few (5–6) to a few tens (up to 30) µm.

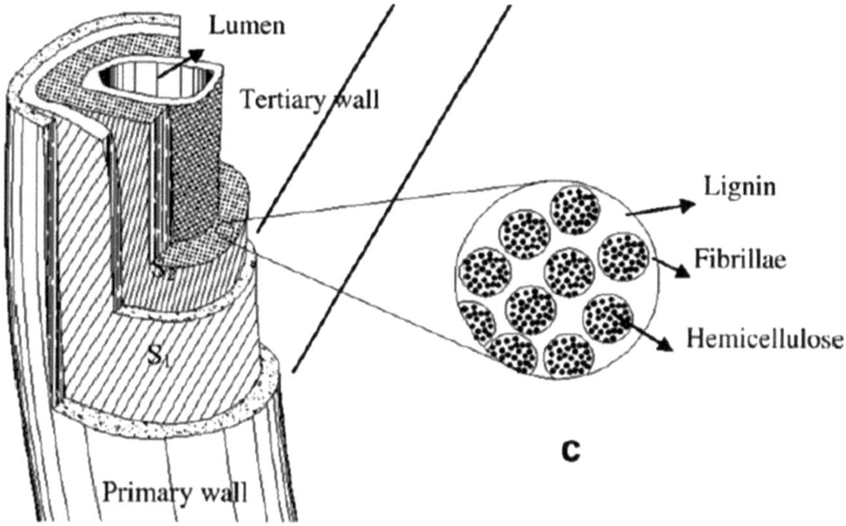

Fig. 9.3 Schematic of the microstructure of an individual fibre cell (de Andrade Silva et al. 2009)

Each fibre cell is made up of four main parts, namely the primary wall, the thick secondary wall, the tertiary wall and the lumen (Fig. 9.2). The fibre cells are linked together by means of the middle lamellae, which consist of hemicellulose and lignin. The lumen varies in size but is usually well defined.

The walls are arranged in several fibrillated layers, consisting of fibrillae which are linked together by lignin (Fig. 9.3). In the primary wall the fibrillae have a reticulated structure while in the secondary wall the fibrillae are arranged in spirals. In details: in the outer secondary wall (S1 in Fig. 9.3) the spiral angle is of about 40° in relation to the longitudinal axis of the individual fibre; in the inner thicker secondary wall (S2) the spiral slope is sharper, equal to about 18–25°. The thinner innermost tertiary wall has a parallel fibrillae structure and encloses the lumen. The fibrillae are, on their hand, built of micro-fibrillae, with a thickness of a few tens of nm (about 20 nm). The micro-fibrillae are composed of cellulose chains with a thickness of around 0.8 nm, and a length of a few μm, and are linked together by means of hemicellulose.

A wide variety of natural fibres, from different forms and origins, has been so far used to reinforce cementitious matrices, as can be seen from a (non-exhaustive) summary reported in Appendix.

Pulp has been so far the most common type of fibre reinforcement form, followed by staples (Ardanuy et al. 2015). This is because cellulose pulps are cheap raw materials, used generally by the paper industry, which can be easily dispersed in water, a basic constituent for the preparation of cement based materials. The majority of cellulosic pulps used to reinforced cement based materials are provided from wood resources and are obtained chemically (kraft pulp). Pinus pulp and eucalyptus pulp have been successfully used in several studies (see references reported by Ardanuyu et al. 2015), with percentages in the range of 4–10 wt% to reinforced Portland cement matrices. Pinus pulps from chemical and thermo-mechanical treatments with percentages around 4 wt% have also been used.

The physical properties of pulps, as reported by Ardanuy et al. (2015) depend not only on the fibre source, but also on the pulping process used to obtain the fibres. In general terms, the mechanical behaviour of the composite will depend on the fibre type, length, diameter, aspect ratio and texture of the fibres. The use of pulped fibres facilitates two-dimensional and homogeneous distribution of the fibres in the cementitious matrix, which, in comparison to the use of short of flock fibres form, may also result into a better fibre-matrix bond and a greater reinforcing efficiency.

Apart from short fibres in the form of pulp, other studies reported the preparation and characterization of cementitious composites with sisal fibres, curaua, piassava, coir, malva, jute, or hemp in staple or flock form.

Finally, only a few works have used continuous reinforcement in the form of fibre strands. In this case it is possible to perform a structural reinforcement and the mechanical properties of the composite are considerable improved.

Cementitious composites reinforced with short fibres, staple or pulps, randomly dispersed into the matrix are usually classified as Fibre Reinforced Cementitious Composites (FRCCs), which will be dealt with in this chapter, whereas those with fibres in the form of aligned strands or textile structures, are classified as Textile Reinforced Composites (TRC).

The reinforcement capacity of the Fibre Reinforced (Cementitious) Composites depends on the geometric and mechanical properties and on the type of the used fibre, as well as on its adhesion to the matrix and on the dispersion of the reinforcement into the matrix and its alignment relative to the applied stress.

Different fibres differ by the number of fibre cells, the cell walls size and the fibre cross section area. Therefore, each fibre presents different characteristics and mechanical properties. This fact results into large discrepancies among values reported for tensile strength and Young modulus: since natural fibres generally feature variable and irregular cross sections, their measurement can lead to huge errors in the computation of stress. Moreover the gage length, strain rate, gripping, resolution of load cell and actuator precision can also play an important role in the final results. Finally, the methodology for measuring the modulus of elasticity is of great importance as the compliance of the machine should be taken into account.

Mechanical behaviour of natural fibres is generally characterized in terms of the tensile strength, Young modulus and strain to failure of the fibres as received.

In accomplishing the aforementioned task, not only, as remarked above, care should be taken in assessing the effects of the compliance of the gripping device and of the testing machine on the measured tensile behaviour. Moreover, a preliminary characterization of the statistical variability of the fibre tensile strength should be performed, as also correlated to the microstructure of the fibres.

Tailored metal grips clamping the fibre ends (Fig. 9.4) have been employed (de Andrade Silva et al. 2008). As an alternative, Ferreira et al. (2014) have

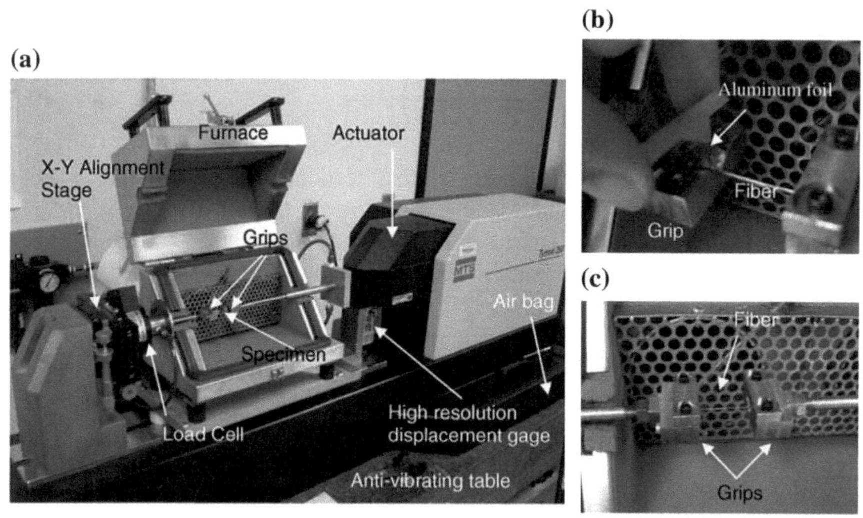

Fig. 9.4 Tytron® micro-force testing system employed by de Andrade Silva et al. (2008) for the tensile testing: **a** general view, **b** and **c** detail of the grip and fibre

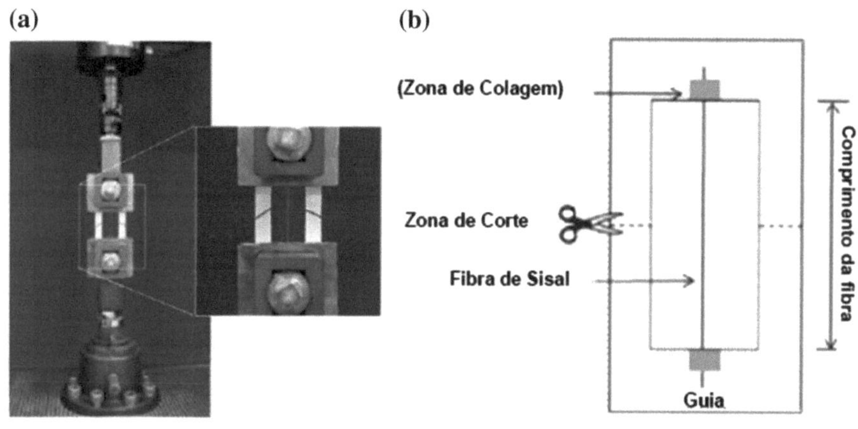

Fig. 9.5 Detail of the paper framework test set-up for testing natural fibres in direct tension (**a**) and schematic of the paper framework (**b**) (Ferreira et al. 2014)

proposed to attach the fibres with aluminium tapes to a paper frame, as shown in Fig. 9.5, which can be then inserted into the grips of the testing machine.

With reference to the influence of the compliance of the loading and gripping system, the methodology proposed by Chawla et al. (2005) can be employed, which consists in performing tensile tests on fibres with different gauge lengths, measuring the cross-head displacement δ_t and the applied force F.

The total cross-head displacement can be expressed by

$$\frac{\delta_t}{F} = \left(\frac{1}{EA}\right)\ell + c \tag{9.1}$$

with E Young modulus of the fibres, A cross sectional area of the fibre, ℓ gauge length and *c* compliance of the loading apparatus. A plot of δ_t/F versus the gage length l will yield a straight end of slope 1/EA and intercept c.

Natural fibres exhibit variability in tensile strength because of the distribution of defects within the fibre and/or on the fibre surface. Weibull statistics can be used to rank the relative fibre strength versus probability of failure of the fibres to obtain a measure of the variability in fibre strength, as hereafter detailed.

According to the Weibull analysis, the probability of survival of a fibre at a stress σ is given by:

$$P(\sigma) = exp\left[-\left(\frac{\sigma}{\sigma_0}\right)^m\right] \tag{9.2}$$

where σ is the fibre strength for a given probability of survival and m is the so called Weibull modulus. σ_0 is defined as the characteristic strength, which corresponds to $P(\sigma)$ = 1/e = 0.37. The higher the value of m, the lower the variability in strength. Ranking of the fibre strength is performed through an estimator given by:

$$P(\sigma_i) = 1 - \frac{i}{N+1} \tag{9.3}$$

where $P(\sigma_i)$ is the probability of survival corresponding to the *i*th strength value and N is the total number of fibres tested. Substituting Eq. (9.3) into Eq. (9.2) yields:

$$\ln\,\ln\left[\frac{N+1}{N+1-i}\right] = m\,\ln\left(\frac{\sigma}{\sigma_0}\right) \tag{9.4}$$

Thus the plot of $\ln\,\ln\left[\frac{N+1}{N+1-i}\right]$ versus $\ln\left(\frac{\sigma}{\sigma_0}\right)$ yields a straight line with slope of m. Generally, increasing gage lengths resulted in a decrease in the Weibull modulus. Tables 9.1 and 9.2 report values of Weibull modulus for different types of fibres.

An example of plotted experimental results can be seen in Fig. 9.6. Remarkably, even though the curves have different slopes, i.e. different values of m, they all intercept at the same point, i.e. they have the same mean strength. This behaviour of

Table 9.1 Weibull modulus of different types of natural fibres (Alves Fidelis et al. 2013)

Type of fibre	Sisal	Curaua	Piassava	Coir	Jute
Weibull modulus (40 mm gauge length)	3.70	2.22	3.68	2.74	2.74

Table 9.2 Effect of gage length on Weibull modulus of natural fibres

Gauge length (mm)	5	10	15	20	25	30	35	40	50	60
Sisal fibres[a]		4.6		3.7		3.6		3.0		
Jute fibres[b]		1.59		2.09		3.78		2.74	2.27	2.33
Jute fibres[c]	2.7	4.3	3.0		3.9		4.6			
White coir fibres[c]	6.0		5.8	6.0		6.0	5.8			
Brown coir fibres[c]	8.0		9.3		5.5		3.7			
Bamboo fibres[c]	4.2		9.3		3.8		3.5			

From [a]de Andrade Silva et al. (2008); [b]Alves Fidelis et al. (2013); [c]Defoirdt et al. (2010)

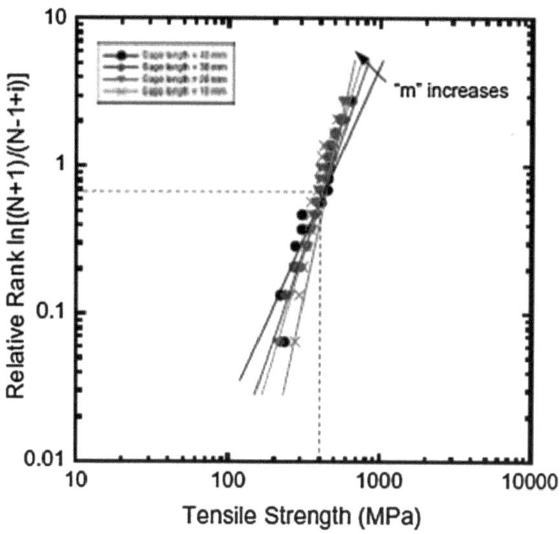

Fig. 9.6 Weibull distribution of the sisal fibre tensile strength for different gage lengths (de Andrade Silva et al. 2008)

different materials having the same mean strength but different Weibull modulus can be explained as follows, according to Chawla (1993).

The mean defect size for all materials controls the mean strength, but the number of defects controls the Weibull modulus. Thus, while the fibres with a lower gage length have a smaller number of defects, the average defect size is the same, irrespective of the gage lengths. The question of why strength is not affected by gage length, while the ductility is, can be explained as follows. This mean flaw size does not change with gage length. Once a crack is formed at the largest flaw, how quickly the linkage between flaws occurs will determine the ductility. Thus, of a larger number of flaws exists, the linkage between flaws will be quicker, and the ductility lower.

Mechanical behaviour of fibres is affected by the fibre microstructure and its response to the loads. Correlation of tensile properties with micromechanical structure of the fibres shows (Fig. 9.7) that fibre fracture generally occurs by three different processes: (i) fracture of the fibre cells; (ii) delamination within the fibre cells,

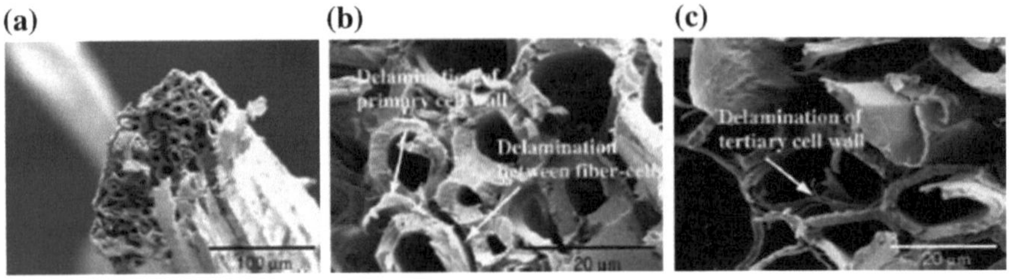

Fig. 9.7 Fractography of a sisal fibre: **a** overall view, **b** and **c** details of fibre-cell fracture, fibre-cell and fibre-cell/wall delamination (from de Andrade Silva et al. 2008)

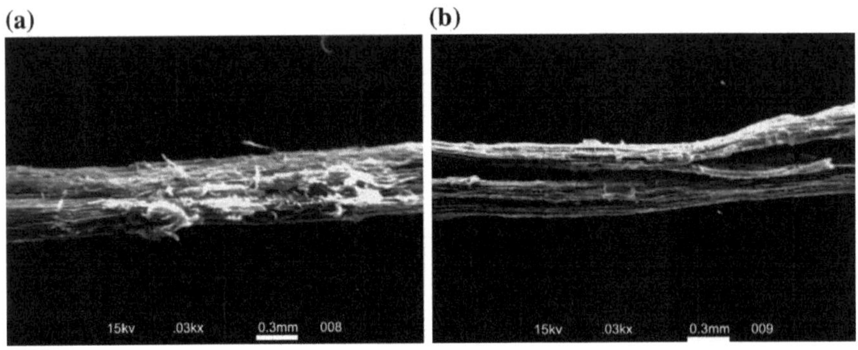

Fig. 9.8 SEM micrographs of a bagasse fibre: **a** untreated and **b** treated in a 5% NaOH solution (Cao et al. 2006)

and (iii) delamination between fibre cells. Fibre cells may fracture in different planes, possibly due to the variability in strength of the individual fibre cells. Delamination of one or more fibre cells may result in a non-linear portion of the stress-strain diagram prior to fracture. Delamination of the tertiary and primary wall of fibre cells, also observed in several cases, which may occur before or after the fracture of the various fibre cells, may also contribute to the change in the slope of the stress-strain curve.

Natural fibres have a relatively low adhesion with matrices based on Portland cement, the bond strength, according to the literature and the type of fibres, varying in the range of a few tenths of MPa. Associated with low chemical adhesion, the high water absorption capacity of natural fibres also causes variations in their volume, with the change in moisture content of the composite resulting in loss of physical contact with the matrix.

Chemical and physical treatments, both in the fibres and the matrix, can be employed both to improve the fibre-matrix adhesion and to reduce the water absorption capacity of the natural fibres.

Alkaline solutions can be employed to clean the surface of vegetable fibres, removing impurities (Figs. 9.8 and 9.9) and increasing the area of contact between the fibres and the matrix (Cao et al. 2006; Rahman et al. 2007).

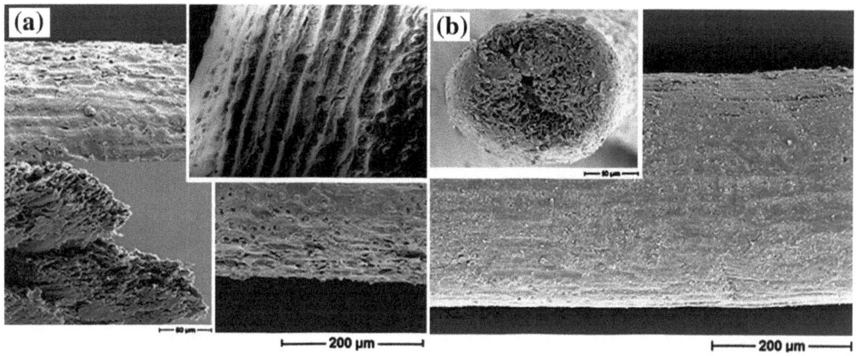

Fig. 9.9 SEM micrographs of a coir fibre: **a** untreated and **b** treated in a 20% alkali solution (Rahman et al. 2007)

Simpler treatments such as soaking the fibres in distilled water and drying them after this process also results in improvements in fibre-matrix adhesion and hence in enhanced mechanical performance of the fibre reinforced composite. Washing in warm water has been shown to promote a modification of the surface layer of the fibres, increasing the contact area by exposing the fibrils and protrusions.

Wetting and drying cycles in water or in acetic acid is able to promote lower volumetric expansion as well as lower changes in the mechanical properties of the fibres and of the fibre reinforced composite. This is due to a stiffening of the polymeric structure present in the fibre-cells (a process called hornification). The polysaccharides cellulose chains are arranged in a tighter way with the removal of water during the drying process and thereby the micro-fibrils join to each other in the dry state as a result of the tighter packing. The fibre capillary voids and the lumens are gradually closed with the cycles (Fig. 9.10) and can no longer be completely reopened with re-humidification. The whole process does not only results in a reduction of the dimensional change of the fibre between the dry and the saturated state but also in a significant reduction in the variation of the cross sectional area of the fibre, in an improvement of its tensile strength and strain capacity and in an increased fibre to matrix bond strength and stiffness (Table 9.2).

Natural fibres can suffer degradation when exposed to an alkaline environment. The extent of the attack, as determined by strength loss of the fibres, depends on the kind of alkaline solution. The possible causes of deterioration of natural fibres in the alkaline solutions are associated with two different mechanisms (Fig. 9.11) (de Melo Filho et al. 2013) (Table 9.3).

The main mechanism is the crystallization of calcium in the lumen, in the walls of the individual fibres as well as in middle lamellae, leading to a decrease in the fibre flexibility and strength. The other one is the chemical decomposition of the lignine and hemicellulose in the middle lamellae, which breaks the link between the individual fibres and then their reinforcing capacity. This effect mainly occurs above a temperature of 75 °C, because of the high degree of polymerization in cellulose.

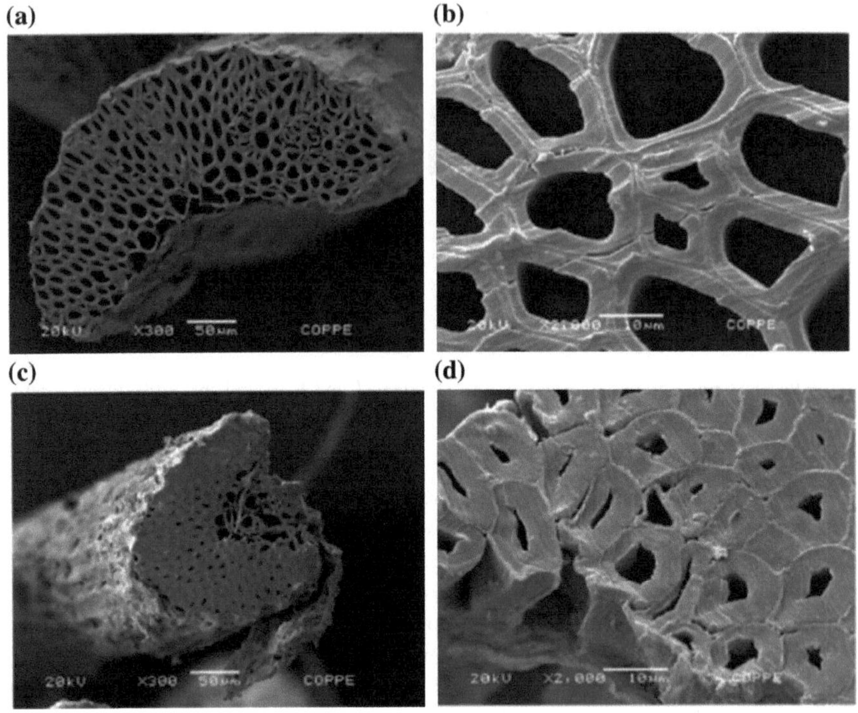

Fig. 9.10 Sisal fibre cross section showing details of the fibre-cells before (**a**, **b**) and after (**c**, **d**) hornification (Ferreira et al. 2014)

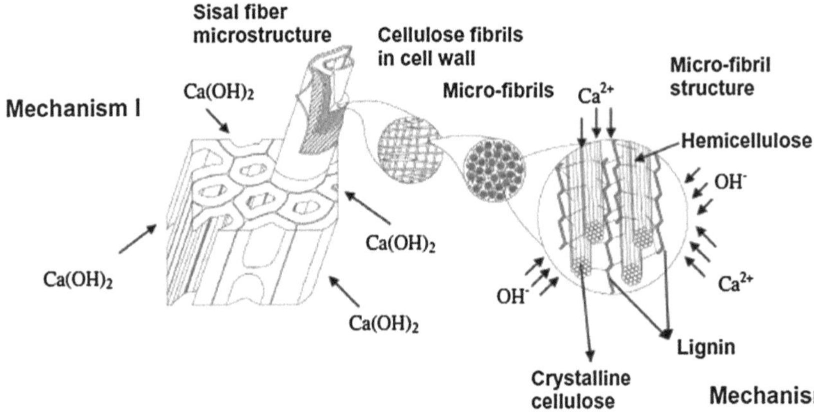

Fig. 9.11 Mechanisms of natural fibre degradation in a Portland cement matrix (de Melo Filho et al. 2013)

Table 9.3 Mechanical properties of untreated and treated sisal fibres (Ferreira et al. 2014)

Treatment	Tensile strength (MPa)	Tensile strain (mm/mm)	Young modulus (GPa)	Adhesion strength (MPa)	Frictional strength (MPa)	Bond stiffness[a] (N/mm)
Untreated	447.20	0.030	19.3	0.30–0.32	0.18	9.3–33.8
Treated	470.25	0.043	17.6	0.42–0.43	0.27	20.6–47.3

[a]Depending on fibre length. Higher values for 50 mm long fibres, lower for 25 mm ones

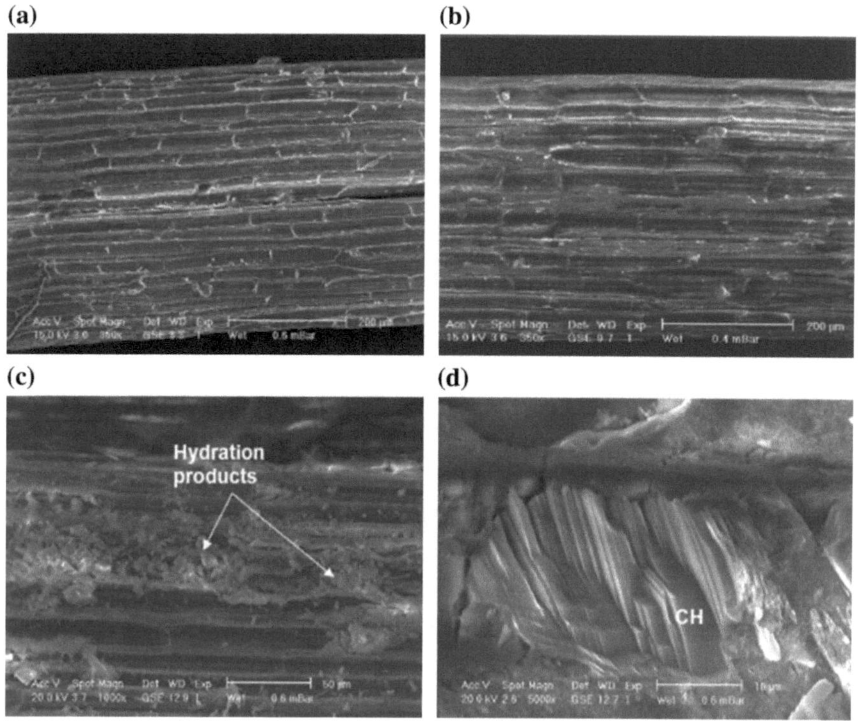

Fig. 9.12 Magnification of surface of sisal fibres in natural stage (**a**), after extraction from a matrix consisting of Portland cement and pozzolanic material (**b**) and after extraction from a pure Portland cement matrix (**c**, **d**) (de Melo Filho et al. 2013)

Special care shall be taken in mitigating the effect of this degradation in order to guarantee the expected performance of the fibre reinforced cementitious composite along time. Reduction of calcium hydroxide in the matrix, e.g. through the replacement/substitution of cement with pozzolanic binders, is a viable solution to reduce and/or temper the effects of the aforementioned fibre degradation (Figs. 9.12 and 9.13), which also results into degradation of the mechanical performance of the fibre reinforced composite as a whole.

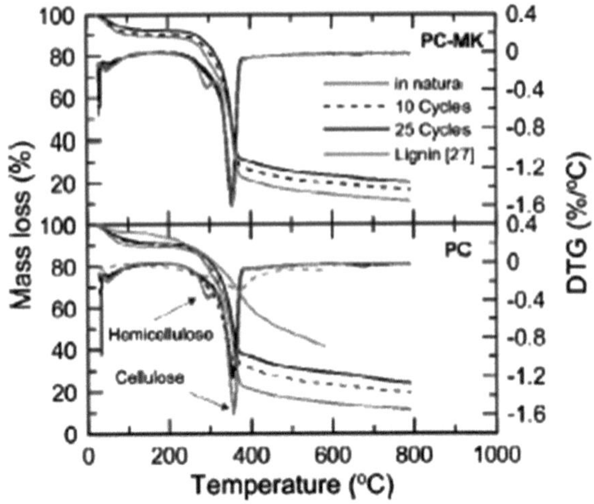

Fig. 9.13 TG and DTG analysis on fibres extracted from a pure Portland cement matrix (*PC*) and from a matrix made of Portland cement and metakaolin (PC_MK) (de Melo Filho et al. 2013)

As a side effect, it has also been shown that the partial replacement of cement by slag or microsilica can also result in an increase of the pull-out adhesion strength between the fibre and the matrix, most likely due to the fineness of the employed cement substitute which reduces the porosity of the transition zone and improves the fibre-matrix adhesion.

9.3 Mechanical Properties of NFRCCs

The main goals to achieve in order to develop natural/cellulose fibre reinforced cementitious composites with well-balanced mechanical properties can be summarized as follows:

- a low porosity of the matrix;
- an optimized percentage of the fibres, enough to reinforce the material while allowing the continuity of the matrix;
- a homogeneous dispersion of the fibres in the matrix;
- a well-balanced fibre-matrix interaction to allow fibre pull-out.

All the aforementioned requisites depend, aside, as remarked above, from the individual properties and relative dosage of the fibres and of the matrix, also on the manufacturing process.

The majority of the fabrication methods for cement composites reinforced with cellulose fibres in the pulp form are based on the Hatschek process, patented by L. Hatschek in 1900. It is a semi-continuous process comprised of three steps: sheet formation, board formation, and curing. In the first step, a conveyor belt is soaked in a mixture of fresh fibre cement supplied by a roller from a tank under continuous

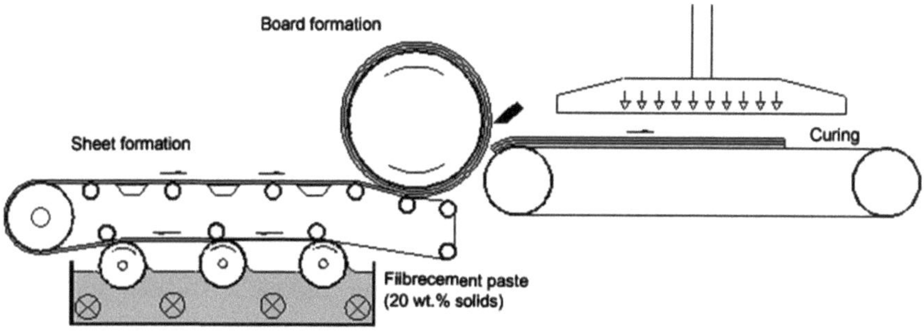

Fig. 9.14 Schematic of the Hatschek process (from Claramunt et al. 2011a, b)

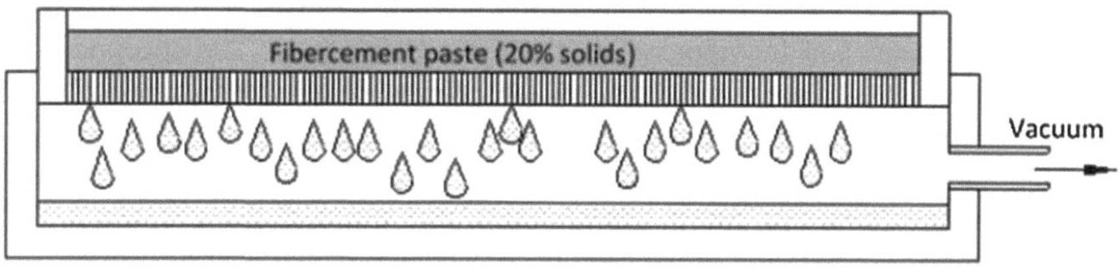

Fig. 9.15 Schematic of the process employed by Savastano and co-workers

agitation. Using a vacuum system, a significant portion of the mixing water is removed from the slurry, forming a very thin sheet (about 1 mm). The board formation is made in a large cylinder which receives the sheet from the previous step and rolls up in successive layers until the required thickness is achieved. Following this, a guillotine cuts the boards and deposits them on a press to compress and mould the board to the desired shape. Finally, the boards are cured under air or steam conditions—autoclave—(Fig. 9.14). These processes produce composites with an adequate percentage of fibres well dispersed into the matrix.

The procedures for preparing cellulose cement composites reported in literature can be divided into two main groups, depending on the fibre form: fibres randomly dispersed in the matrix, and aligned fibres or fibrous structures.

With reference to composites reinforced by fibres randomly dispersed in the matrix, Warden and Coutts (2001), Savastano et al. (2003, 2005, 2006, 2009), Tonoli et al. (2009a, b, 2010a, b, 2011), D'Almeida et al. (2013) have successfully used a variation of the Hatschek process (the slurry vacuum de-watering technique) to prepare cement composites with pulp fibre mass fractions of around 8 wt% (approximately 10% by volume content). In this technique the matrix materials are stirred, with the appropriate amount of fibres dispersed in water, to form a slurry with approximately 20% solid materials. Then the slurry is transferred to a drilled mould and a vacuum is applied as can be seen in Fig. 9.15. Afterward, the board is pressed until it has a thickness up to 15 mm.

A similar procedure was also used by Soroushian et al. (2012).

Fig. 9.16 Process employed by Claramunt et al. (2011a, b): mold and press system used to prepare the specimens (from *left* to *right* molds, vibrating table, pressing system and demolding)

Another possibility based on de-watering with pressure was used to successfully prepare cellulose composites incorporating until 4 wt% of cotton and pine pulp (Claramunt et al. 2011a, b). The cement paste consists of a mixture of cement, sand, water and fluidizer with a water/cement ratio around 1:1. The cellulose fibres are dispersed in water and incorporated in the paste beforehand. The specimens are prepared in a micro-grilled mould to allow the evacuation of water with a minimum loss of cement and sand. The mould is placed on a plate, then compacted on a vibrating table, and pressed for 24 h up to 4 MPa (Fig. 9.16).

Mohr et al. (2005a, b, 2006a, 2007, 2009) also successfully prepared cement mortar composites without pressuring or a vacuum using a cast-in-place mix methodology with a 4% fibre volume fraction. In this methodology the pastes are prepared by firstly mixing the pulp fibre with water (approximately 50% water) and superplasticizer. Subsequently, cement is added, and mixing continues to allow uniform fibre dispersion. In some cases, to improve the dispersion of the fibres these were treated by a process with cationic starch and fly ash.

Soroushian et al. (2006) used the extrusion procedure to prepare cellulose pulp cement composites with up to 8 wt% fibres. These authors used a laboratory-scale de-airing ceramic extruder which allowed the production of flat specimens with 60 mm width and 8 mm thickness. This process allows the alignment of the fibres in the extruder direction generating higher reinforcement capacity.

Tolêdo Filho et al. (2000, 2003, 2005) also used the casting methodology to prepare cement reinforced composites with short sisal and coconut fibres. In this case, after casting, a conventional vibration table is used.

Recently, Ferrara et al. (2015b, 2017), exploiting the self-compactability properties of the fresh NFRCC mixtures (obtained through a balanced mix-design) have successfully been able, also thanks to a tailored casting process (Fig. 9.17) to govern the alignment of the fibres along the casting flow. This results into an improved mechanical performance of the composite if stressed parallel to the direction of the preferential fibre alignment, and in a poorer performance otherwise.

Other authors, taking into account the limitations of the reinforcement capacity of the pulp or short fibres randomly dispersed in the matrix, have been developing other manufacturing processes for semi-finished products, such as aligned fibres, sheets or nonwovens, which can allow a higher level of reinforcement under flexural or tensile loads. In this sense, Toledo Filho et al. (2009), de Andrade Silva et al. (2009, 2010b, c, d) and de Melo Filho et al. (2013) have successfully prepared

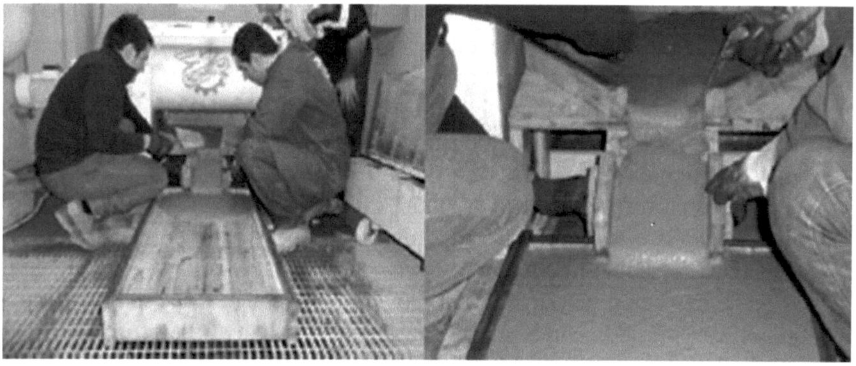

Fig. 9.17 Casting process employed by Ferrara et al. (2015a, b) for self-compacting NFRCC

high-performance cement composites reinforced with aligned sisal strands with the following methodology: firstly the long fibres are cut to the size of the moulds, weighed and separated into different layers, resulting in a total volume fraction of 10%. The fibres are stitched to make homogeneous spacing between the fibres and to facilitate the moulding process. Then, laminates are produced by placing the mortar mix into the mould one layer at a time, followed by single layers of long unidirectional aligned fibres (up to 5 layers). The samples are consolidated using a vibrating table and, after casting, compressed at 3 MPa for 5 min. Cement composites reinforced with pulp fibres in the form of aligned perforated sheets were prepared by Mohr et al. (2006b). To prepare the composites, firstly a mixture of cement, sand and water is added to the mould and vibrated to level the mortar surface. Afterwards, the fibre sheet is placed in the mould and lightly tamped to remove trapped air voids beneath the fibre sheet. Finally, the remaining mortar is slowly added to the mould, taking care to keep the fibre sheet plane. For the samples with multiple fibre sheet addition, a thin layer of mortar is spread between the fibre sheets to ensure bonding between the sheets.

The vast majority of work dedicated to the study of the mechanical behaviour of cement composites with pulp or short natural fibres randomly dispersed in the matrix has analysed their tensile and/or flexural properties, respectively by means of direct tensile tests and either three or four point bending tests.

The importance of the tensile properties of fibre reinforced cement composites is self-evident and many experimental tensile tests (direct or indirect) have been used for their identification: however, no tensile test standard has been established to date. Several technical organizations (ASTM, JCI, RILEM) have developed flexural tests meant to simulate the characteristics of FRC composites in tension. However, such tests are insufficient to characterize the true tensile response, particularly when strain-hardening behaviour and multiple cracking in tension are present. Similarly, the indirect split cylinder tensile strength test (ASTM C496/C496M 2011 or EN12390-6 2000) for FRC specimens are difficult to interpret after the first matrix cracking because of the unknown stress distribution (Schrader 1978). The precise identification of the first crack in the split cylinder test can also be difficult to determine without strain gauges or other sophisticated means of crack detection, such as acoustic emission or laser holography (Maji and Shah 1988; Miller et al. 1988) or techniques employing Digital Image Correlation (DIC) (Pereira et al. 2012; Baril et al. 2016). The relationship between splitting tensile strength and direct tensile strength or modulus of rupture has not been determined, though consistent research on this topic is on going (Salehian et al. 2014).

While no standard test exists to determine the direct tensile properties of FRC composites, it is essential to ensure that a standard test allows the measurement of the key properties being investigated Moreover, it is recommended to record crack distribution, spacing, and width at maximum post-cracking stress or ultimate.

de Andrade Silva et al. (2009) tested thin prismatic specimens, 400 mm long, 50 mm wide and 12 mm thick, using a gage length of 300 mm with fixed-ends boundary conditions. Aluminium thin sheets were glued on both ends of the specimen and the pressure of the hydraulic grips was adjusted to 1.37 MPa in order to minimize stress concentration and damage. The tensile load, cross-head displacement and strain were recorded. Tensile strains were also measured by a strain gage glued on the centre of the specimen (Fig. 9.18).

By evaluating cracking patterns at regular time intervals, crack development throughout the loading cycle of tensile test was recorded. A digital Pulnix camera with a 10× macro zoom lens and frame grabber captured images of 480 × 640 in resolution at 60 s intervals. Images were used to measure the crack formation during bending and tension tests.

Image processing was done by the digital processing toolbox of MATLAB. In order to measure the crack spacing as a function of the applied strain, a two-step approach was used. During the first step, newly formed cracks of each image were traced and added to data from previous loading increment. The second step measured the crack spacing from the traced cracks using the following procedure. An image consisting of a series of parallel lines was generated. The crack spacing was measured in pixels, and the image was calibrated using conventional techniques to convert the size of a pixel to length measures. By conducting a binary "AND" operation the points of intersection of the series of parallel lines with the cracks were identified. A second binary operation of "OR" between the intersection points and the parallel spacing lines, broke up the straight lines into segments representing crack spacing distribution. The distribution of the length segments and statistical parameters of crack spacing were computed next.

Figure 9.19 shows a typical tensile stress strain response of the sisal fibre reinforced composite system. Two measures of tensile strain are used including the localized strain measured from the electrical resistance gage and the nominal strain defined by dividing the cross-head stroke displacement by the specimen length. Comparison of the two strain measurements allows one to evaluate the early age cracking phase and make a distinction among all stages of cracking.

These responses differ at various stages of loading as they correspond to crack initiation, propagation, distribution, opening, and localization.

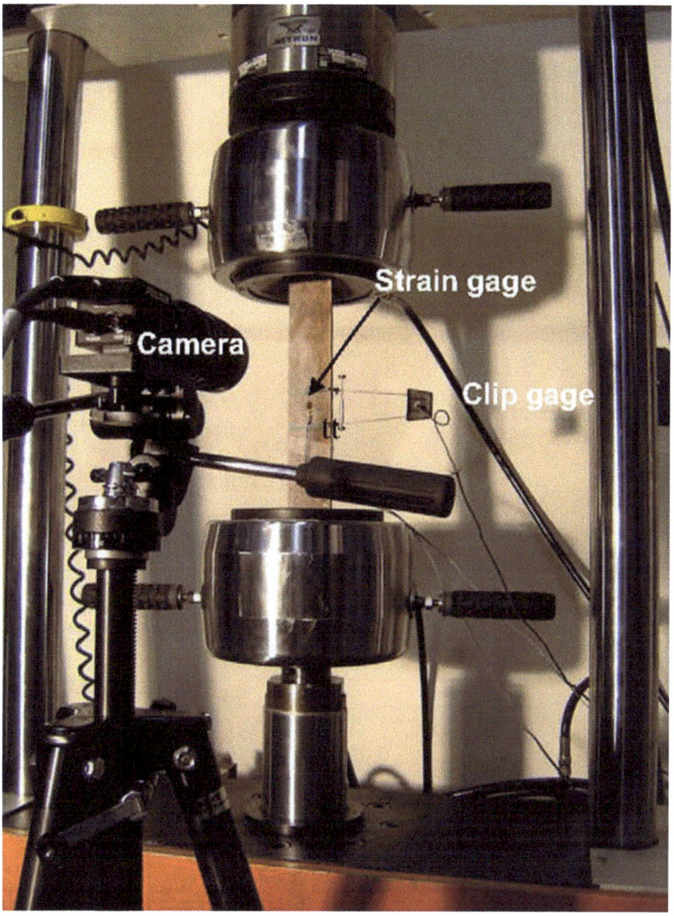

Fig. 9.18 Direct tension test setup employed by de Andrade Silva et al. (2009) for sisal fibre reinforced concrete

From a macroscopic perspective, the bend over point (BOP) corresponds to the formation of matrix cracking. Five distinct zones are identified using roman numerals with two zones prior to and three zones after the bend over point (BOP).

Figure 9.20 shows the relationship between the strain gage reading and the nominal strain obtained from the measured cross-head stroke displacement divided by the specimen length. Zone I corresponds to the elastic-linear range where both matrix and the fibre behave linearly. Due to low volume fraction of fibres ($\leq 10\%$) the stiffness of the composite is dominated by matrix properties and this zone is limited to strain measures of up to 150–175 $\times$ 10^{-6}. The initial stress–strain response is marked by a limited range of linear-elastic portion as the two strain measures are almost the same, and the specimen exhibits the highest stiffness. The deviation from linearity occur at around 150 $\times$ 10^{-6} due to initiation and

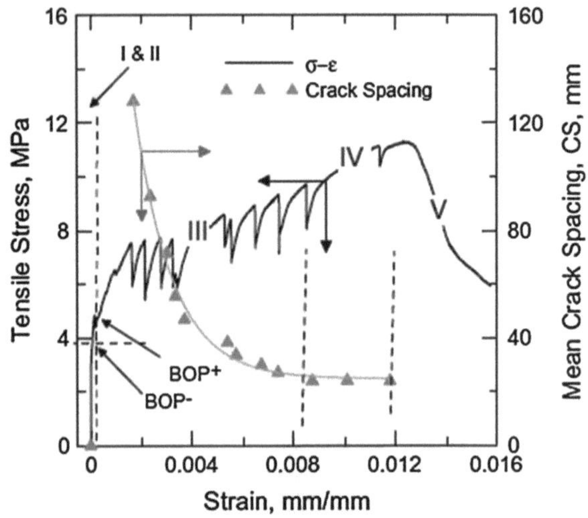

Fig. 9.19 Tensile response of the sisal fibre reinforced composite system: tensile stress and crack spacing versus strain (from de Andrade Silva et al. 2009)

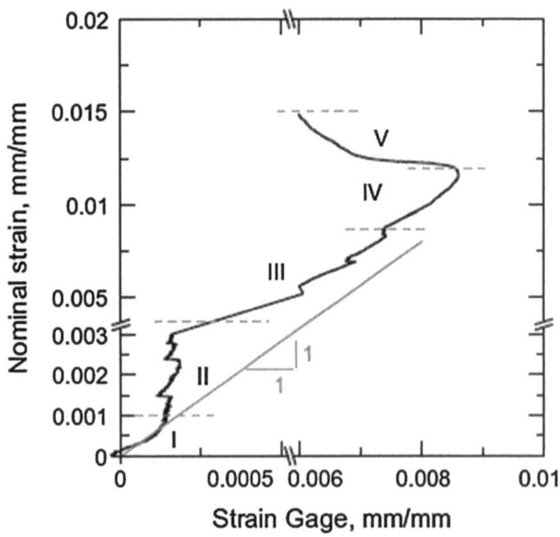

Fig. 9.20 Relationship between the strain gage and strain measured by the stroke (from de Andrade Silva et al. 2009)

propagation of first cracks. Note that the sensitivity of the stroke displacement in this range is within the instrumentation error, therefore, the strain gage response is far more reliable than the stroke in this range. The linear zone is terminated by initial crack formation in the matrix phase (denoted as Bend Over Point, BOP—see Fig. 9.19).

After the initiation of cracks in the matrix, its load carrying capacity does not vanish as the cracks are bridged by the longitudinal fibres. Immediately after the initiation of the first matrix crack, other matrix cracks also initiate throughout the specimen at approximately regular intervals and begin to propagate across the width

(Mobasher et al. 1990). The strain recorded by the strain gage remains relatively constant in this range, which indicates a steady state condition of several cracks that initiate and propagate across the width of the specimen. The strain range within Zone II is associated with formation of matrix cracks, however, no single crack has traversed the entire width. The term defined as BOP+ (Fig. 9.19) corresponds to the stress level at which the first matrix crack completely propagates across the width. Zone II is therefore defined as the stable cracking range between the two stress levels σ_{BOP-} and σ_{BOP+}, respectively corresponding to points denoted as BOP− and BOP+ in Fig. 9.19

The post BOP stage is characterized by formation of distributed cracking in Zone III. In this homogenization phase, as the applied strain increases, more cracks form and the spacing decreases in an exponential manner as presented empirically by Eq. (9.5). The strain measured by the strain gage remains constant while several cracks form throughout the section. The decrease in crack spacing can be empirically represented as a function of three parameters and its initiation is represented by parameters S0, S1, a, and ε_{mu}:

$$S(\varepsilon_i) = S_1 + S_0 e^{-\alpha(\varepsilon i - \varepsilon mu)} \quad \varepsilon_i > \varepsilon_{mu} \tag{9.5}$$

where $S(\varepsilon_i)$ = crack spacing as a function of strain, ε_{mu} = average strain at the BOP level; ε_i = independent parameter representing strain in the specimen, S_0 and α = constants representing the initial length of the specimen and rate of crack formation as a function of strain, and S_1 = saturation crack spacing. The stiffness of the sisal fibre reinforced cement composite system is sufficiently high and keeps the newly formed cracks from widening; thus promoting multiple cracking behaviour. This stiffness affects the rate of reduction of crack spacing. Significant variations in the value of the α parameter were observed in the tensile tests.

The crack spacing measurements as shown in Fig. 9.19 show a general reduction in spacing during loading until a steady state condition is reached. This zone covers a large range at the end of Zone II, and III and remains constant throughout Zone IV. This constant level of crack spacing is defined as saturation crack spacing. Beyond this point, reduction in crack spacing is not observed since no new cracks form, while, as verified by the pictures of the specimen under the load, additional imposed strain results in widening of the existing cracks. It can be seen from Fig. 9.19 that in the early stages of Zone III the cracking spacing drastically decreases from an initial value of 130 to 45 mm. Then, during the multiple crack formation the crack spacing further decreases until a point (beginning of Zone IV) where it becomes constant at 23 mm.

Zone IV corresponds to the completion of cracking phase and initiation of de-bonding. As the cracking saturates in the specimen, Zone IV is dominated by progressive damage and characterized by a crack widening stage ultimately leading to failure by fibre pull-out. This zone is asymptotically terminated at the saturation crack spacing represented by parameter S_1. The dominant mechanism of failure during stage IV is crack widening which is associated with fibre de-bonding and pull-out.

The post peak response occurs in Zone V, with unstable localization of the major crack, where still residual load bearing capacity was observed.

With reference to the flexural behaviour, it has first of all to be considered that in the case of NFRCCs, as for High Performance Fibre Reinforced Cementitious Composites in general, potential intended applications are characterized by thin sections, exploiting the one of a kind strain/deflection hardening signature behaviour of the material with substantial material, dead-load and cost savings.

In this respect, though for the experimental identification of the flexural behaviour of FRCs standard tests do exist (EN 14651 2005, ASTM C1399/1399M 2015, ASTM C1609/1609M 2012), it has to be remarked that the behaviour observed in the standard tests can deviate substantially from the behaviour of the corresponding FRC in the structural element or structure. Thus the manufacturing method and the concrete consistency should be taken into account by the designer. Tailored tests should be designed to determine the effect of fibre orientation due to casting and compaction in real structural elements, employing structural specimens better reproducing the behaviour of the material in the structural elements to be (Ferrara et al. 2011).

In this framework de Andrade Silva et al. (2009) employed prism specimens, measuring 400 mm (length) × 70 mm (width) × 12 mm (thickness/depth) in order to characterize the bending performance of a cementitious composite reinforced with sisal fibres to be employed in compression moulded laminates. The matrix was produced using a bench-mounted mechanical mixer of 20 L capacity. The cementitious materials were homogenized by dry mixing for 30 s prior to addition of sand and 5% by volume of wollastonite. The powder material was mixed for additional 30 s prior to addition of superplasticizer and water. The mixture was blended for 3 min.

Laminates were produced by placing the mortar mix in a steel mould, one layer at a time, followed by single layers of long unidirectional aligned fibres (up to 5 layers). The samples were consolidated using a vibrating table operated at a frequency of 65 Hz, and resulted in a sisal fibre volume fraction of 10%. After casting, composites were compressed at 3 MPa for 5 min. The compression load was applied in the face that the matrix was placed. The specimens were covered in their moulds for 24 h prior to moist curing for 28 days in a cure chamber with 100% RH and 23 ± 1 °C.

Displacement controlled four-point bending tests were performed at a 0.5 mm/min rate and on 300 mm span specimens. Strain-gages were glued at the centre bottom and top surface of the specimen in order to compute the extreme fibre strain and the neutral axis (Fig. 9.21). The bending load, cross-head displacement and strains were recorded and crack formation and propagation along the loading path was monitored by taking photos employing a mirror positioned at 45° with respect to the specimen. Image processing was done by the digital processing toolbox of MATLAB, as already said above, with reference to direct tension tests employed to characterize the mechanical performance of the same natural fibre reinforced cementitious composite.

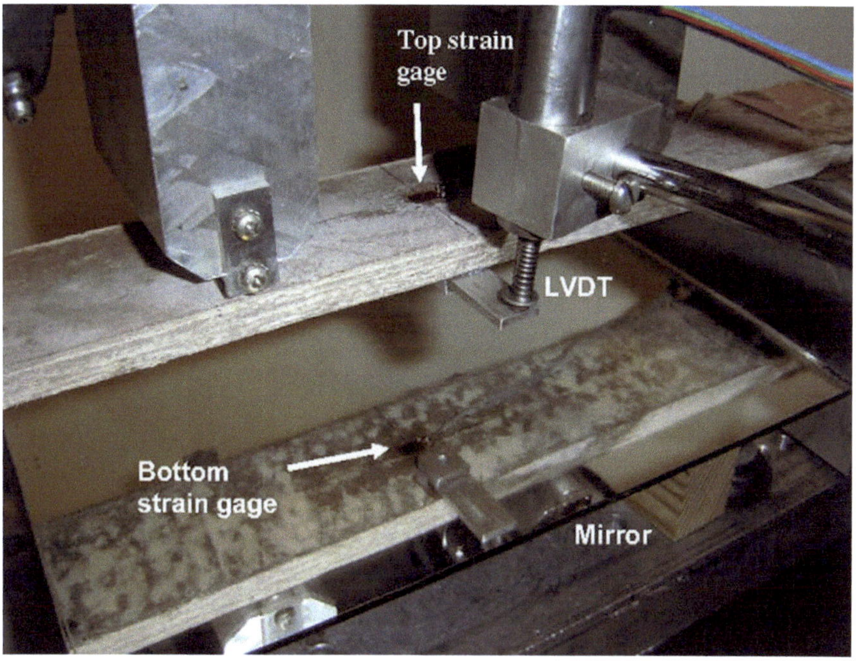

Fig. 9.21 Four point bending test set-up employed by de Andrade Silva et al. (2009)

Example discussion of results and their interpretation is given hereafter, with reference to the typical bending response of the sisal fibre-reinforced composite and its crack spacing measurements (Fig. 9.22a).

Using the same methodology as for the direct tension, the bending curve was divided into 5 regions identified by roman numerals.

Zone I corresponds to the elastic-linear range where both matrix and the fibre behave linearly. The lower and upper bounds of the limit of proportionality (LOP) delimit the Zone II. The post LOP range (Zone III) is characterized by multiple cracking formation and can be represented by an exponential decay function.

Figure 9.22a shows that the crack spacing initially drops abruptly from 180 to 60 mm up to the deflection of 10 mm. The crack spacing saturates at the end of Zone IV (40 mm) which happens at a deflection of approximately 23 mm. Zone V is characterized by the strain softening response due to the localization and widening of a major crack. No new cracks appear and crack spacing remains constant.

The bending and tensile responses with their respective crack distribution are compared in Fig. 9.22b. The inset plot shows the relationship between LOP versus BOP and MOR (modulus of rupture) versus UTS (ultimate tensile stress).

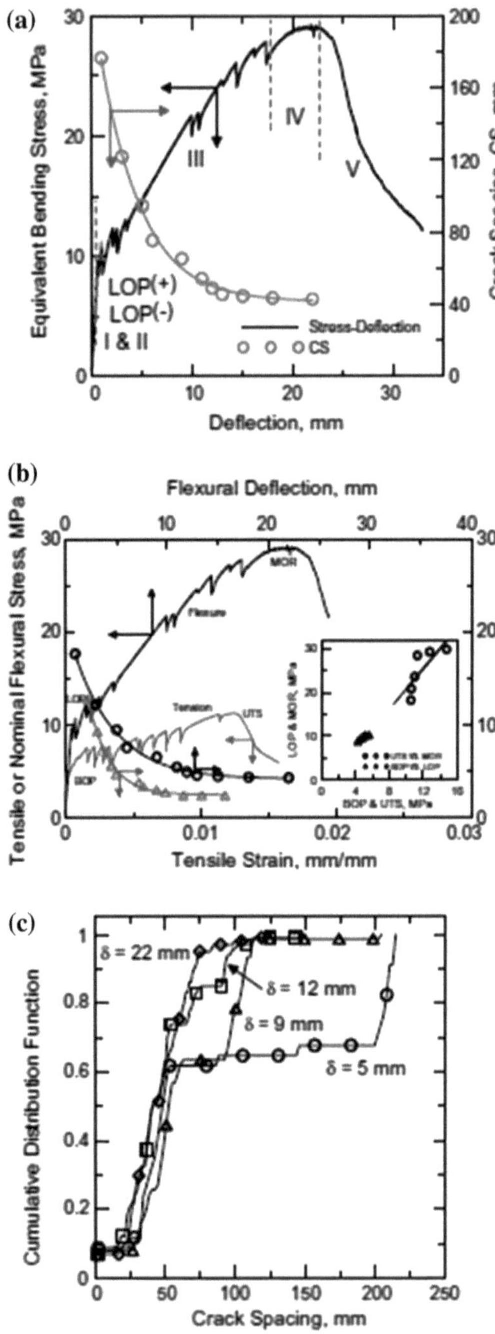

Fig. 9.22 Bending response of the sisal fibre reinforced composite: **a** bending stress and crack spacing versus displacement, **b** comparison of flexural versus tensile response and **c** cumulative distribution function for crack spacing (from de Andrade Silva et al. 2009)

It can be seen that under bending, loads associated with the formation of the first crack occur at stress levels twice as those observed for the direct tension tests. The correlation of tensile and bending test results has been documented through theoretical modelling by Soranankom and Mobasher (2008). Under flexural loads, the saturation crack spacing is twice as large as that of tensile loads.

Figure 9.22c shows the cumulative distribution function for crack spacing under bending at four different levels of displacement beyond the LOP range. As the displacement level increases from 5 to 12 mm, the variability in measured crack widths decreases. As the displacement level approaches the crack saturation level, a more homogenized set of data with less scatter is observed. At this level 99% of the measurements are below 75 mm.

The compressive and tension strains during a four-point bending test are shown in Fig. 9.23a. A distinct behaviour is observed after the first crack formation.

Three different zones are identified for the compressive strain. Zone I is the linear-elastic region which yields the same stiffness (29.55 GPa) as the tension strain. Zone II begins after the first crack formation and is characterized by an increase in stiffness during the multiple cracking. Although not shown, the unloading process is defined by Zone III. At this range a modulus of 8.0 GPa is reported.

The tensile region of the flexural sample experiences five ranges of behaviour. Zone I is characterized by a linear-elastic response. After the complete formation of the first crack, Zone II begins. Stiffness degradation is caused by the crack formation, and results in a significant drop in the modulus from 29.55 to 0.50 GPa. A decrease in tensile strain occurs in zone III and IV as a result of crack formations in the vicinity of the strain gage and shear lag mechanisms.

At Zone IV the strain gage response of the matrix between two cracks results in a stiffness of 10.7 GPa. This stage characterizes the degraded matrix's contribution to the composite overall response. Finally, Zone V is the unloading region that presents a stiffness of 22.94 GPa which is similar to Zone I.

The data acquisition of strains from the compression and tension faces during the bending test allowed the computation of the neutral axis (normalized with specimen depth, d) as seen in Fig. 9.23a, b. In the elastic stage, the normalized neutral axis depth (NA) remains constant at centroid location (0.5d) until the onset of the first cracking, which happens at 0.00023 mm/mm. A sudden drop of the NA to 0.18d is observed as tensile cracking results in load redistribution across the thickness. Further on, the crack spacing starts to decrease with an increase in the NA until a strain level of 0.004 mm/mm (Fig. 9.23b).

9.4 Durability of NFRCCs

No standards exist about how to measure the durability of NFRCCs, and in principle the same standards and procedures as for conventional plain concrete and conventional FRC are applicable. In the following some experiences related to pilot

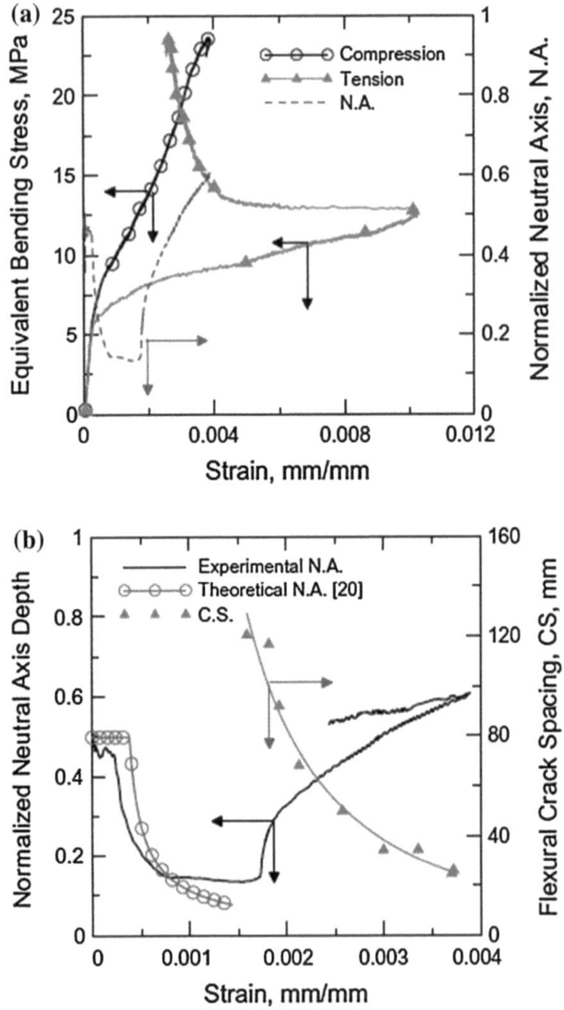

Fig. 9.23 Neutral axis computation: **a** compression and tension strains under four-point bending loading and **b** crack spacing versus neutral axis (from de Andrade Silva et al. 2009)

projects and applications will be referred so to form a useful set of informative guidelines for all future applications.

Vegetable fibre–cement composites produced with ordinary Portland cement matrices undergo an aging process in humid environments during which they may suffer a reduction in post-cracking strength and toughness. The aging process is due to fibre mineralization and results in reducing the tensile strength of fibres and decreasing the fibre pull-out ligament after fracture. This process is a result of migration of hydration products (mainly $Ca(OH)_2$) to the fibre structure.

To guarantee the durability of the composites de Andrade Silva et al. (2010a, b, c, d, e) and Toledo Filho et al. (2009) developed a calcium hydroxide free matrix using high contents of pozzolanic materials (up to 50% of cement replacement by mass), therefore also reducing the CO_2 emissions associated to cement production. Thus the low environmental impact of this material is achieved by both the use of a renewable reinforcement and a green cementitious matrix. In order to characterize the durability of the newly developed composite, accelerated aging test through hot-water immersion (T = 60 °C) have been performed during 6 months. Scanning Electron Microscopy (SEM) investigating the micro-structure of the composites before and after aging should complement mechanical tests.

The matrix was designed using the Portland cement CPII F-32 defined according to Brazilian standards as composed with filler (in mass: 85% < clinker < 91%; 3% < gypsum < 5%; 6% < limestone filler < 10%) with 32 MPa of compressive strength at 28 days. Following the recommendations of previous studies, in order to increase the durability of the composites, the Portland cement was replaced by 30% of metakaolin (MK) and 20% of calcined waste crushed clay brick (CWCCB) (Toledo Filho et al. 2009). By replacing 50% of cement by the aforementioned materials it was possible to develop a matrix that was free of calcium hydroxide (CH) at 28 days of age. The employed mortar featured a mix composition of 1:1:0.4 (cementitious material: sand: water by weight). For comparative evaluation two matrices were used: one free of calcium hydroxide (CH free composite) and the other made of Portland cement with no additions (PC composite). Mineral micro-fibre of wollastonite JG obtained from Energyarc was used as micro-reinforcement (V_f = 5%). The wollastonite fibre featured a density of 2900 kg/m^3.

The Flow Table Spread (FTS) test was used to determine the content of superplasticizer to be added to matrices in order to guarantee a fluidity index (FI) higher than 70%. This index is defined by FI = (Sf − Si)/Sf; where Sf is the final spread measured after the table drops and Si is the initial spread measured before the table drops. The matrix free of CH presented a fluidity index of 73% whereas the reference matrix (M0) presented a FI of 76%. Both matrices presented characteristics of self compacting mixes. The optimum fibre volume fraction (V_f) of the sisal fibre was determined as 10%.

Typical four-point bending curves of CH free and PC based composites after 6 months of hot-water immersion (at 60 °C) are shown in Fig. 9.34. In opposition to the PC, the CH free composite did not suffer any mechanical degradation with the accelerated aging as can be seen in Figs. 9.24 and 9.25. A multiple cracking behaviour still can be observed for the aged CH free composite. Average toughness and MOR values of 28 kJ/m^2 and 23.25 MPa were reported for the aged CH free composite, respectively. These values are in the same magnitude as the ones observed in the non-aged CH free composite. The high deformation capacity of the aged CH free composite can be seen in the insight of Fig. 9.24. The picture shows the deflected specimen at the ultimate load. The aged PC composite presented a single cracking formation with a fragile post-peak behaviour. The average toughness value of this composite decreased from 18.12 to 0.66 kJ/m^2 with the aging process whereas the MOR decreased from 21.08 to 6.06 MPa. A correlation between degraded MOR versus UTS for aged PC and CH free composites is shown in the inset of Figs. 9.5 and 9.6, respectively. It can be seen that after aging the

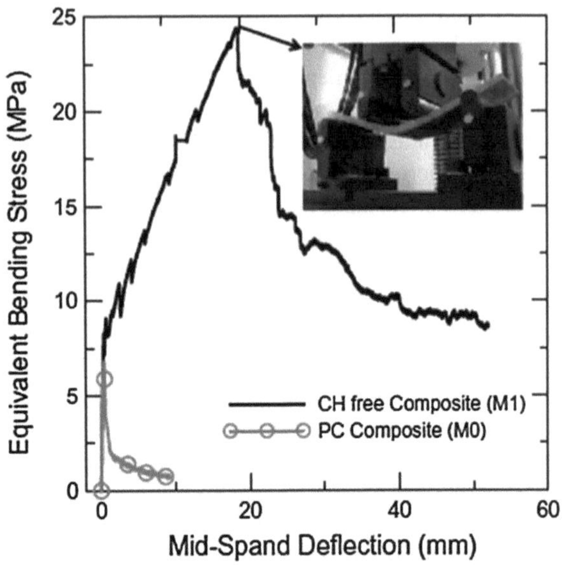

Fig. 9.24 Effect of hot-water immersion at 60 °C in the four-point bending response of CH free and PC composites (from de Andrade Silva et al. 2010a, b, c, d, e)

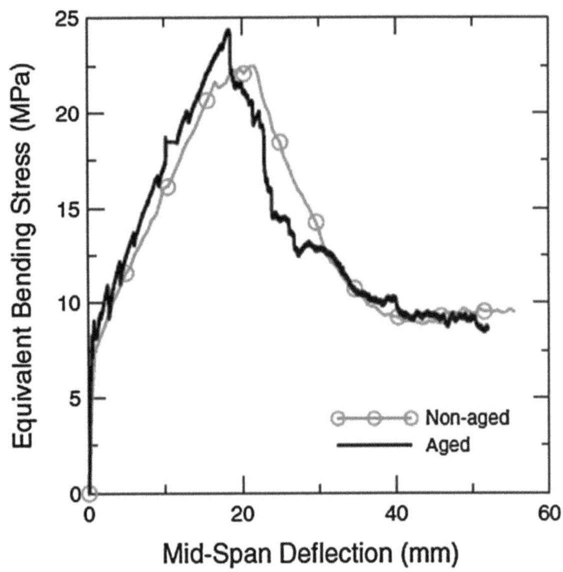

Fig. 9.25 Effect of hot-water immersion on aged CH free composite. The plot shows a non-aged CH free composite tested at 28 days and a CH free composite tested after being subjected to 6 months under hot-water immersion at 60 °C (de Andrade Silva et al. 2010a, b, c, d, e)

relationship for PC composites drops from 2 to about 0.7, whereas the correlation for the CH free composite remains the same.

A microstructural investigation was performed in the CH free and PC aged composites to assess the fibre degradation process. Figure 9.26 shows SEM images of the CH free aged composite. The fibre structure remains intact with no signs of mineralization (Fig. 9.26a–c). A calcium (Ca) Energy Dispersive Spectroscopy

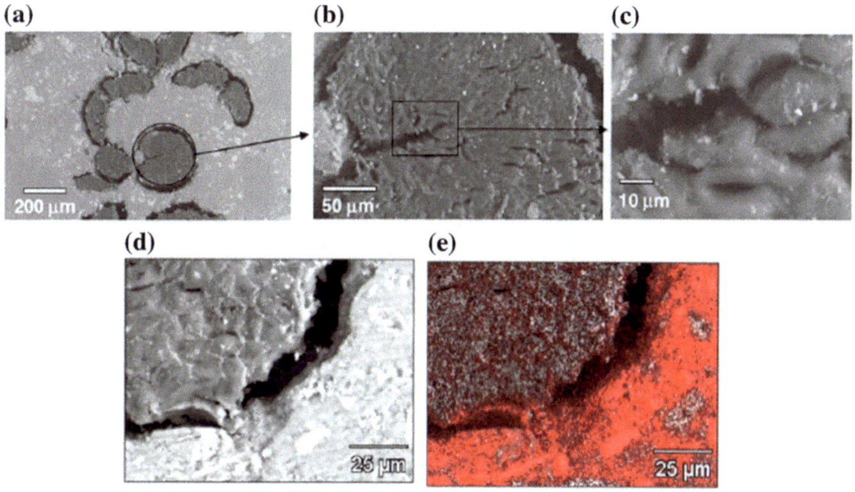

Fig. 9.26 SEM images of the aged CH free composite: **a** overall view of the composite **b** sisal fibre **c** detail of fibre-cells, **d** EDS mapped region and **e** EDS mapping for Ca in the fibre–matrix interface (de Andrade Silva et al. 2010a, b, c, d, e)

(EDS) mapping was performed in the fibre–matrix interface to verify if leached calcium, from the CH, could have migrate to the interior of the sisal (Fig. 9.26d, e). It was observed a light red color that indicates low amount of Ca.

Figure 9.27 shows SEM images of the PC aged composite. The sisal fibre micro-structure was found to be mineralized as can be seen in Fig. 9.27a, b and d. An EDS analysis was performed inside the sisal fibre that showed high Ca concentration (see Fig. 9.27c, d). A Ca EDS mapping was performed in the fibre–matrix interface and its results have shown a high Ca concentration as can be seen in Fig. 9.27e, f. This micro-structure analysis confirms that in the PC composite the fibre-cells are mineralized possible due to the high Ca concentration. In the CH free composite it was observed no signs of fibre deterioration and much lower Ca concentration inside the fibre-cells.

The addition of MK and CWCCB have successfully sustained the energy absorption capacity of the CH free composite, increased its first crack strength and maintained its ultimate strength through accelerate aging, proving to be a good solution for the durability issues of natural fibre as reinforcement in cement composites. The obtained results using the accelerated aging technique are in accordance with those obtained with wetting and drying accelerated aging.

Controlled wetting and drying cycles were performed by Toledo Filho et al. (2009) to evaluate the durability of the same composite as above. For the drying cycles a forced air flow chamber (FAFC) was used. The FAFC was designed in order to allow the control of the wind velocity and air temperature enabling a simulation of the environmental conditions to which the material can be subjected in real life (see Fig. 9.28). In the present study the FAFC was set to a temperature of

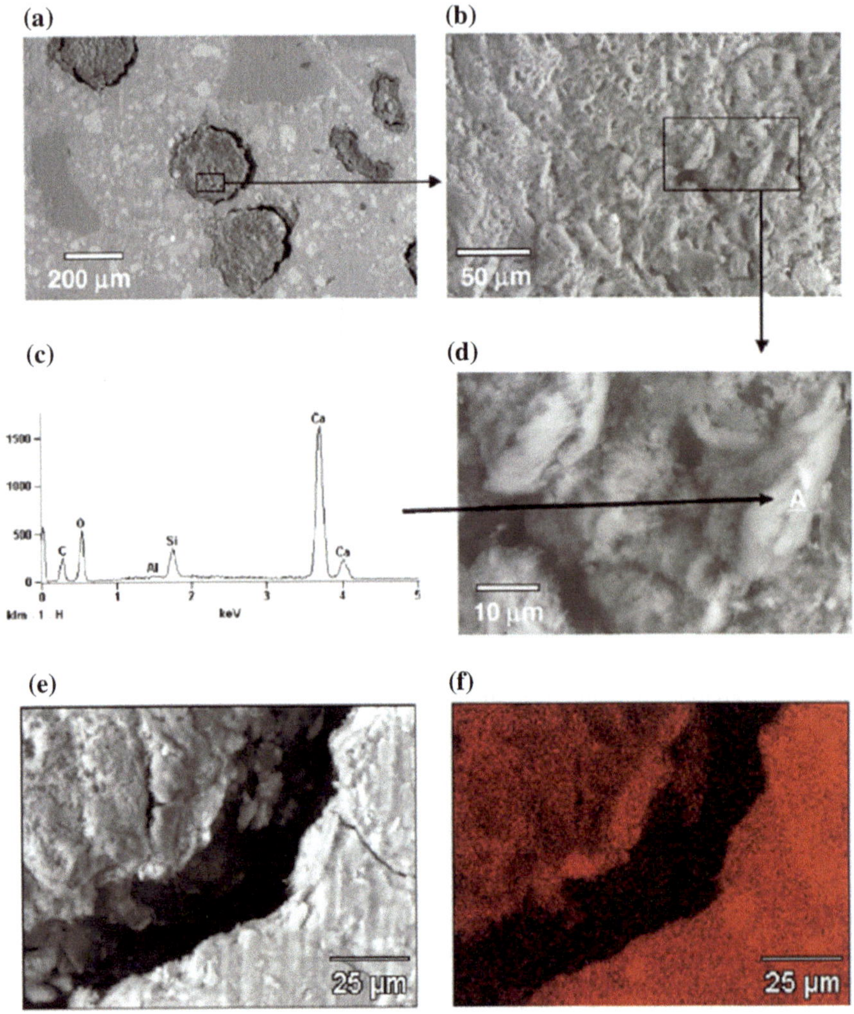

Fig. 9.27 SEM images of PC aged composite: **a** overall view of the composite **b** Sisal fibre **c** EDS analysis of point A **d** region analyzed by EDS **e** EDS mapped region and **f** EDS mapping for Ca in the fibre–matrix interface (de Andrade Silva et al. 2010a, b, c, d, e)

36 ± 1 °C and wind velocity of 0.5 m/s. In order to define the length of the wetting and drying cycles a sample was completely saturated in water at 30 °C and left to dry in the FAFC. Considering that after 24 h of water immersion the sample absorbed about 90% of its total saturation capacity and that after 48 h it lost about 70% of the gained mass, a three day cycle was chosen (1 day under water followed by 2 days of drying under the FAFC condition).

The performance of conditioned specimens was evaluated by means of four point bending tests carried out at a crosshead rate of 0.5 mm/min. Three specimens

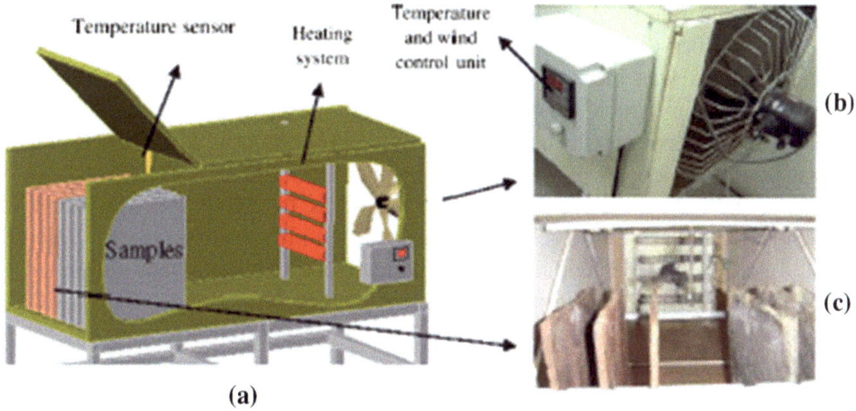

Fig. 9.28 Forced air flow chamber (FAFC): **a** schematic view; **b** fan, temperature and wind velocity control unit; **c** specimen holder and temperature sensor (Toledo Filho et al. 2009)

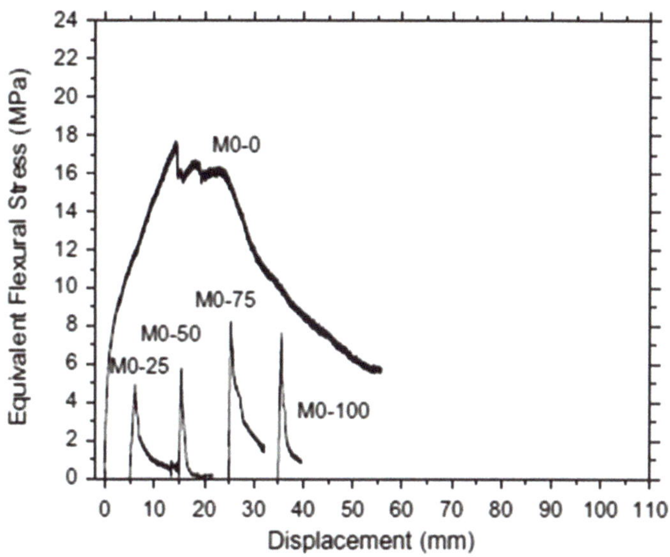

Fig. 9.29 Typical four point bending curves of aged and non-aged PC composites (Toledo Filho et al. 2009)

with geometry of 400 mm × 100 mm × 12 mm (length × width × thickness) were tested under bending (300 mm span). Displacements at mid span were measured using electrical transducers (LVDT) and were continuously recorded, together with the corresponding loads, using a 32-bit data acquisition system taking four readings per second.

Figure 9.29 presents typical equivalent flexural stress–displacement curves obtained from the reference composite specimens (M0-0) as well as for specimens made with the reference materials after 25, 50, 75 and 100 cycles respectively (M0-25, M0-50, M0-75 and M0-100). The first crack strength (FCS) of the reference specimens was increased up to 45%. The reference specimen presented a ductile behaviour which was drastically decreased by the aging process. It is important to observe that the deterioration in the PC composite toughness is already complete after 25 cycles of wetting and drying.

Figure 9.30 shows the typical cracking patterns for M0-0 specimens and for M0-25 specimens. It was observed that whereas the non-aged specimens presented a multiple-cracking behaviour under bending loads, all specimens submitted to cycles of wetting and drying presented a single crack formation similar to M0-25.

The failure surface presented in Fig. 9.31 shows that the fibre pull-out length was severely affected by the accelerated aging processing.

Figure 9.32 presents typical examples of equivalent flexural stress–displacement curves obtained from M1-0 and M1-25, M1-50, M1-75 and M1-100 specimens. A ductile behaviour is observed not only for the reference specimens but also for the aged ones. For the free CH composites, it was also observed an increase in the First Cracking Strength, up to 65% after 100 cycles. The obtained results indicate that the increase in FCS was higher for the specimens containing pozzolans. This behaviour may be related to a twofold action that combines the degradation of the sisal fibres in the OPC mixtures with the aging process and the late pozzolanic reaction in the CH-free mixtures. It is well known that sisal fibres mineralization leads to a complete reduction in its strength and strain capacity (Berhane 1999; Canovas et al. 1992; Tolêdo Filho et al. 2000, 2003) and, therefore, it can be

(a)

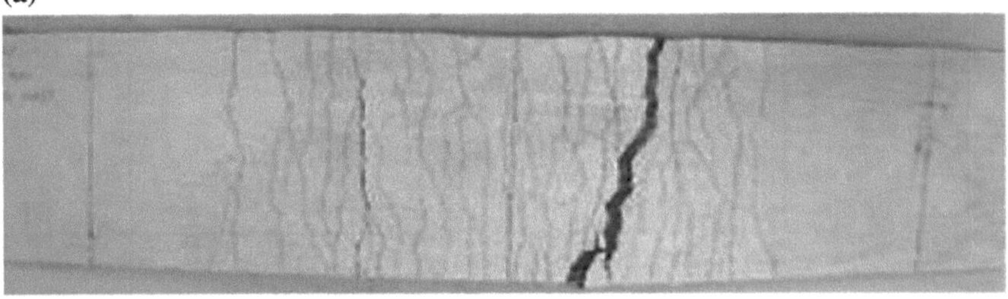

(b)

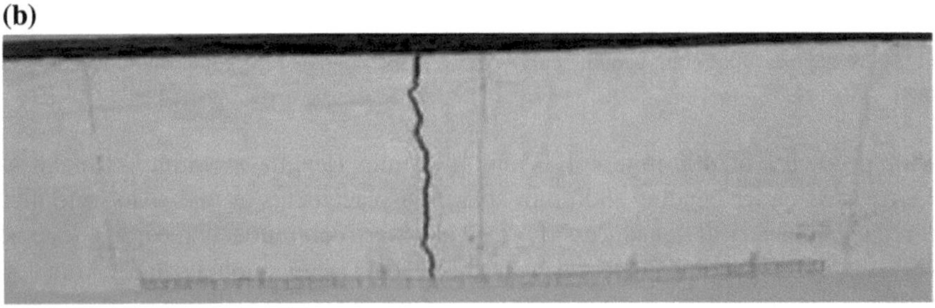

Fig. 9.30 Typical cracking patterns: (*top*) non-aged PC specimens and (*bottom*) aged PC specimens submitted to 25 cycles of wetting and drying (Toledo Filho et al. 2009)

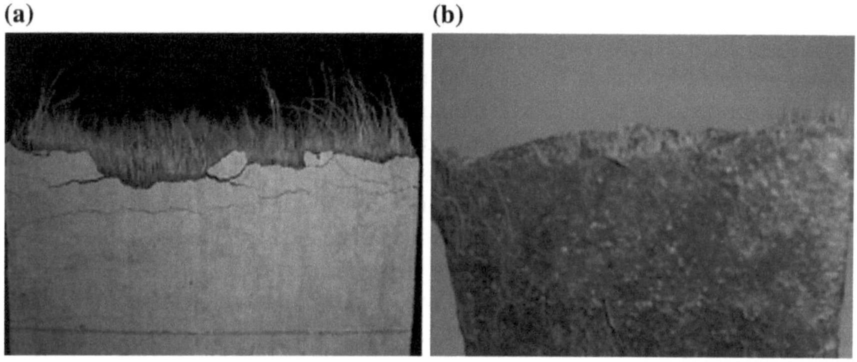

Fig. 9.31 Failure surface of the composites before and after aging. Pull-out length of the fibres in **a** non-aged PC specimens and in **b** aged PC specimens submitted to 25 cycles of wetting and drying (Toledo Filho et al. 2009)

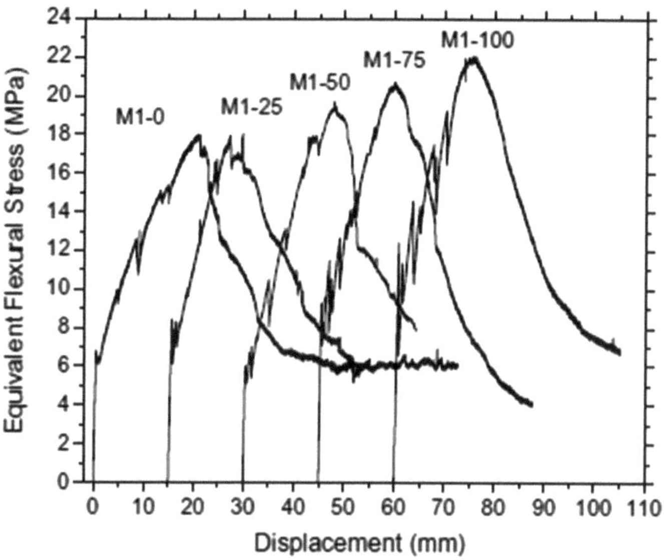

Fig. 9.32 Typical four point bending curves of non-aged and aged free CH composites (de Andrade Silva et al. 2009)

considered that a reduction in the homogenized session of the OPC composites occurs after submitting the specimens to the wetting and drying cycles. Pozzolanic reactions can also contribute in some extent to the increase of the FCS values once the hydration reactions of the OPC still progress at late ages and the CH produced is consumed by the pozzolan along with the evolution of these reactions. The use of 50% of calcined clay as cement replacement is enough to consume all CH formed at late ages. Considering that all specimens have been cured for 180 days before starting the accelerated aging process, it is the authors' opinion that the elevated increase in FCS experienced by both composites cannot be only related to the thermo-activated hydration process and that the understanding of this behaviour demands further and deeper investigations. The Post Cracking Strength of the reference specimens have increased up to 29% with the accelerated aging regime.

Regarding to the bending toughness parameter a slightly tendency in its reduction was observed which lies in the same range of its standard deviation. Observing the stress-displacement curves it can be noted an increase in the slope of the curves between the FCS and PCS as well as in its descending branch with the wet/dry cycling process. This behaviour may indicate an increase in the fibre-matrix adhesion with the wetting and drying cycles. The fibres, however, were not deteriorated during the used accelerated aging process.

The multiple-cracking patterns of the aged specimens shown in Fig. 9.33 and the long fibre pull-out lengths (P30 mm) shown in Fig. 9.34 for specimens M1-25 and M1-100 indicates that the CH-free matrix was capable of keeping the sisal fibres integrity with the aging process. Visual observation of the fibres extracted from aged specimens indicates that they kept their original flexibility.

Figure 9.35 shows micrographs of PC composites after 25 cycles of wetting and drying. The cavities presented in Fig. 9.35 parts (a) and (b) appeared as a result of the specimens sawing during sample preparation for SEM analysis. This behaviour

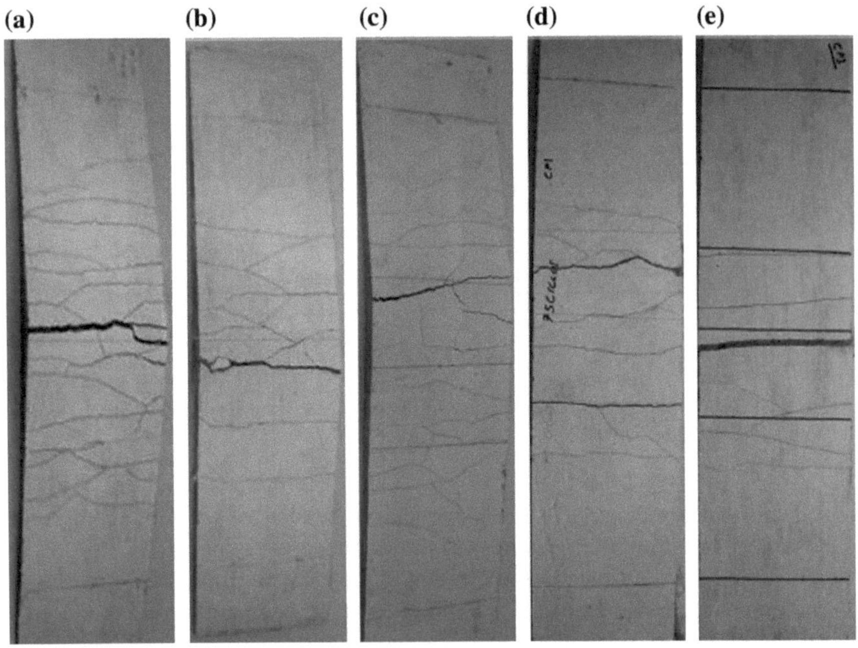

Fig. 9.33 Typical cracking patterns of CH-free composites: **a** non-aged, **b** aged for 25 cycles, **c** aged for 50 cycles, **d** aged for 75 cycles and **e** aged for 100 cycles (Toledo Filho et al. 2009)

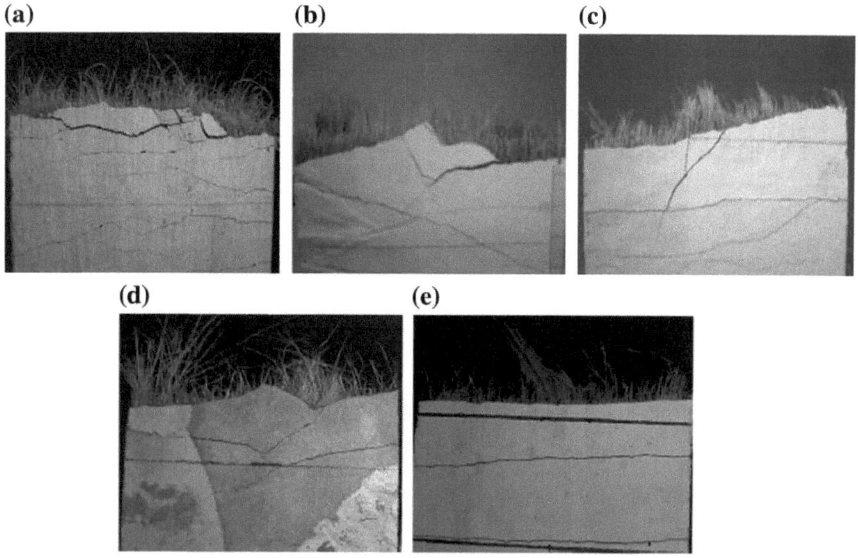

Fig. 9.34 Failure surface of the CH-free specimens before and after aging. Pull-out length of the fibres for the reference specimens (**a**); for specimens submitted to 25 (**b**); 50 (**c**); 75 (**d**) and 100 wet/dry cycles (**e**) (Toledo Filho et al. 2009)

indicates that the fibres have lost their strength and flexibility due to the attack by calcium hydroxide. Signs of fibre mineralization are evident in Fig. 9.35c. The same behaviour was observed in the specimens subjected to 50, 75 and 100 cycles of wetting and drying. There was no indication of fibre mineralization in all the micrographs taken from the aged free CH composites.

No signs of fibre degradation can be seen in Fig. 9.36a, b. A close up view of the sisal fibre-cells in specimens subjected to 50 cycles of wetting and drying are shown in Fig. 9.36c, d and clearly indicates that the fibres did not suffer any process of mineralization. Similar results were obtained also by de Melo Filho et al. (2013) who tested a composite similar to the one tested by Toledo Filho et al. (2009) as well as by de Andrade Silva et al. (2010a, b, c, d, e).

The only difference consisted in the matrix composition, in which 50% of cement was replaced all by amorphous metakaolin. The durability of the two composites (one with a pure Portland Cement—PC—matrix and another one with a matrix consisting of Portland Cement and Metakaolin PC–MK) was assessed by submitting specimens to 5, 10, 14, 20 and 25 aging cycles and then testing them under a four point bending configuration. The wetting and drying cycles were performed as already explained and employing the same equipment and protocols as in Toledo Filho et al. (2009).

Figure 9.37 shows one representative flexural stress–deflection curve for the PC–MK and PC composites tested at 28 days of age. In general it can be seen that both PC–MK and PC composites present a deflection hardening response with a

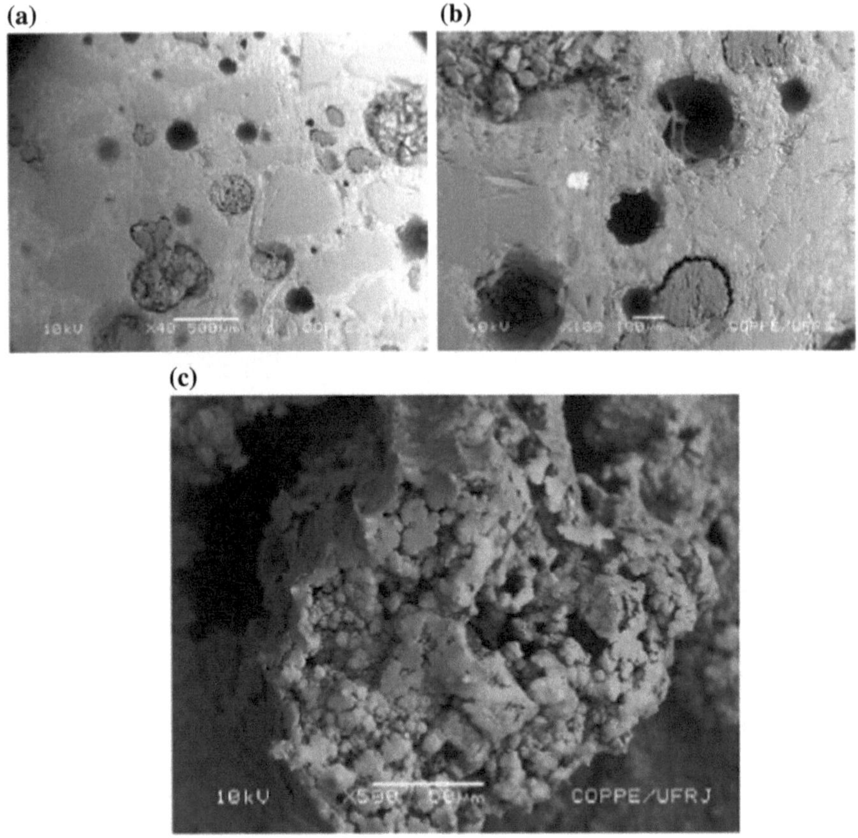

Fig. 9.35 Micrographs of aged PC specimens subjected to 25 cycles of wetting and drying: **a** and **b** cavities in the matrix due to fibres breaking and pull-out; **c** mineralized fibre inside the specimen (Toledo Filho et al. 2009)

multiple cracking formation. The flexural behaviour begins with an elastic-linear range where both matrix and the fibre behave linearly up to a point where the matrix cracks (zone I). The PC–MK composite presented a limit of proportionality (LOP) of 6.67 MPa while for the PC composite a value of 4.95 MPa was reported. Standard deviations varied from 0.45 to 0.74 MPa. The post LOP range was characterized by a multiple cracking formation (zone II) until no more cracks can form. The widening of the existing cracks takes place in region III where the crack spacing becomes constant. The computed crack spacing was 19.25 mm for the PC–MK and 5.44 mm for the PC composite. The failure of the composites occurred after a mid-span deflection of around 27 mm (PC composite) and 31 mm (PC–MK composite). The failure was followed by a strain softening response due to the localization and widening of one of the existing cracks (zone IV). The ultimate flexural strength values observed for the PC and PC–MK composites were 18.73 and 26.86 MPa, respectively.

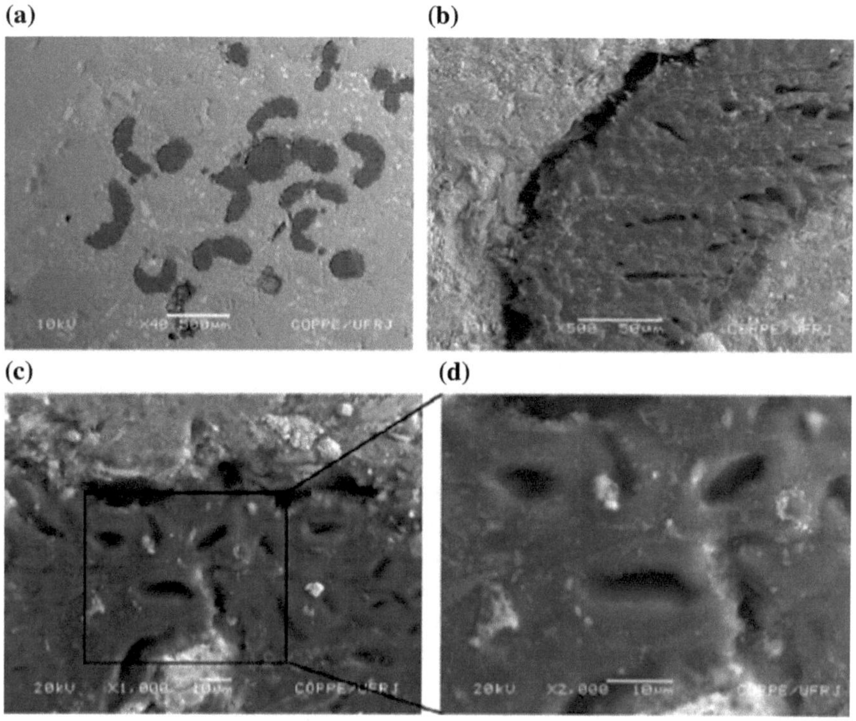

Fig. 9.36 Micrographs of CH-free specimens: **a** specimen subjected to 25 cycles of wetting and drying **b** zoom in a sisal fibre subjected to 25 cycles **c** and **d** close up view of the sisal fibre-cells in specimens subjected to 50 cycles of wetting and drying (Toledo Filho et al. 2009)

Fig. 9.37 Flexural and fracture behaviour of PC–MK and PC composites performed at 28 days of age (de Melo Filho et al. 2013)

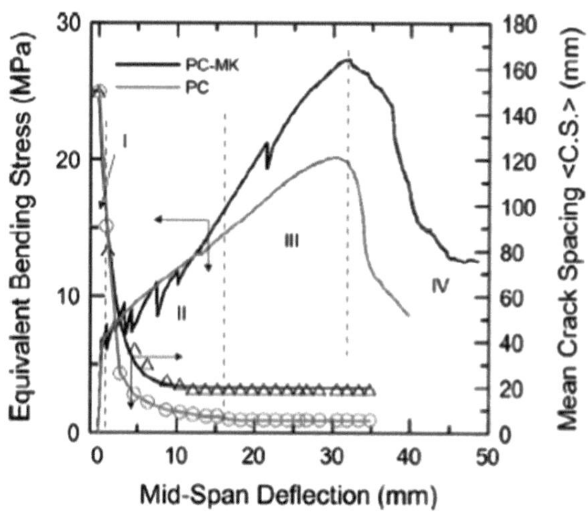

The toughness, measured as the area under the load versus deflection curve up to the ultimate bending strength, obtained by the PC–MK composite was 22% higher than that of the PC. Moreover a superior number of cracks with lower width was observed for the PC composites (34 cracks) in comparison to the PC–MK (15 cracks). This behaviour may indicate a higher fibre–matrix bond adhesion for the PC composite system.

Figure 9.38 shows one representative flexural stress–deflection curve for the PC–MK and PC composites tested under bending load after being exposed to 5, 10, 15, 20 and 25 cycles of wetting and drying. For the PC–MK system a ductile behaviour with a multiple cracking formation is observed also for the aged specimens. On the other hand the PC composites presented a progressive degradation process with the cycles. It can be seen that both the ductility and bending strength (Fig. 9.39) reduced to the same level of an unreinforced matrix after 25 wet-dry cycles.

It can be also observed that the accelerated aging process is more severe in the first 10 cycles for the PC composite. The PC system toughness lowered from 13.94 to 3.67 kJ/m^2 at 10 cycles and further decreased to 0.07 kJ/m^2 after 25 cycles. The reduction in the bending strength reached the same value (7.4 MPa) of the unreinforced matrix after 25 cycles. From these results it can be first concluded that 10 cycles is a threshold level where most of the degradation in the fibre and fibre–matrix interface takes place and that the sisal fibre completely degrades when the PC composite is aged at 25 cycles. For the PC–MK composite the toughness values decreased from 17.09 to 10.31 kJ/m^2. The bending strength presented a reduction in the first 10 cycles from 28.86 MPa to 21.64 MPa. After 10 cycles of aging the variation in the mean bending strength values were not statistically representative. As no degradation was observed in the sisal fibres extracted from the composite,

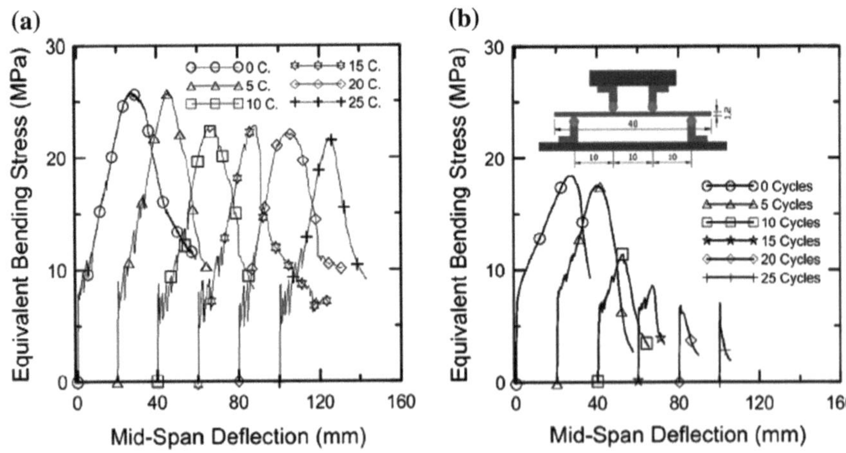

Fig. 9.38 Effect of the wetting and drying cycles on the mechanical behaviour under bending load of **a** PC–MK composite and **b** PC composite (de Melo Filho et al. 2013)

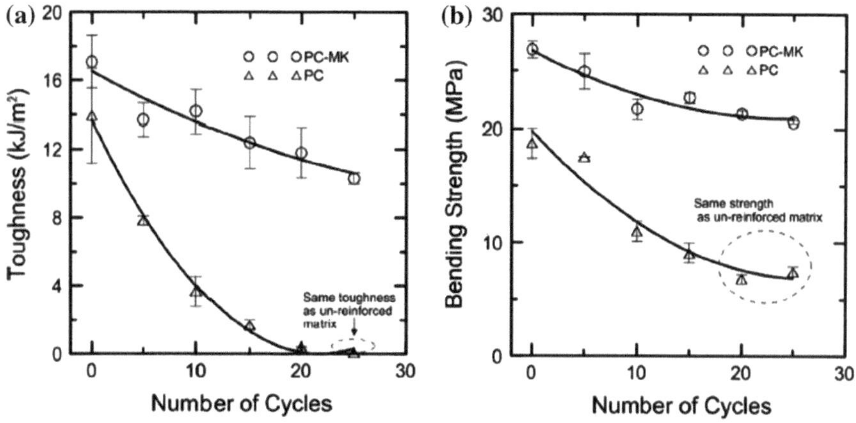

Fig. 9.39 Effect of the wetting and drying cycles on the **a** toughness and **b** bending strength of PC–MK and PC composites (de Melo Filho et al. 2013)

no reduction in the stiffness of the post-cracking region was observed and no changes in the cracking pattern occurred, the authors explain the decrease in bending strength and ductility as follow. In the first 10 cycles there was a change in the fibre–matrix interfacial shear mechanisms. The frictional stress may have lowered after the first 10 cycles as a result of an interface fatigue mechanism. This is due to the volumetric variation of the sisal fibres as a result of the wetting and drying cycles. After 10 cycles the fibres undergo a hornification process (Ferreira et al. 2014) which is characterized by the stiffening of the polymeric structure presented in the fibre cell wall as a result of the water removal. This process reduces the fibre volumetric variation mitigating the damage in the fibre–matrix interface.

The effect of the aging cycles in the crack spacing is presented in Fig. 9.40. The aging process seems not to affect the fracture behaviour of the PC–MK composite. Its crack spacing is around 19 mm with a standard deviation that ranges from 3 to 7 mm. In a different manner the cycles are affecting the fracture process of the PC composite and changing its behaviour from a multiple to a single cracking formation. Again it seems that there is a threshold level around 10 cycles for the PC composite. After this point the number of cracks drastically reduces from 34 to 15 and then to one (after 20 cycles).

Figure 9.41 presents the typical curves of four-point bending tests performed in PC–MK composites from 28 days to 5 years: the LOP increases from 9.66 to 11.33 MPa and the bending strength did not present any representative alteration. From Fig. 9.50 it is noticed an increase in the post-cracking region (zone II) as a function of age which results in a reduction of the composite ductility. The stiffness increased from 1.71 to 2.37 GPa for ages of 28 days and 5 years. A reduction in the number of cracks from 17 (at 28 days) to 5 (at 5 years) was reported. This behaviour can be traced back to a change in the elastic bond component of the system. During debonding, the shear stress intensity in the elastically bonded region may

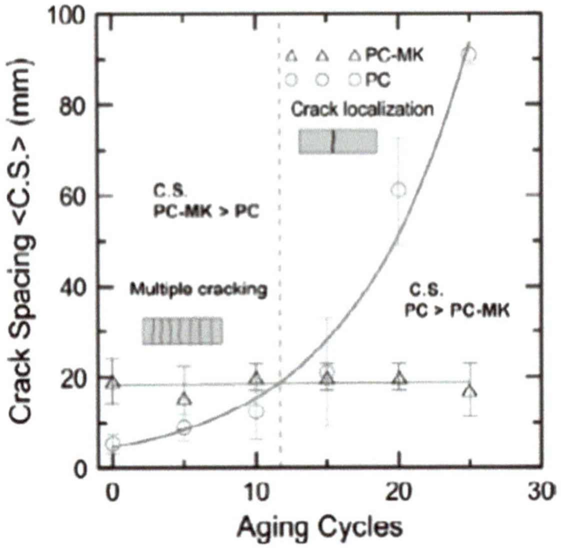

Fig. 9.40 Effect of wetting and drying cycles on the crack spacing of PC–MK and PC composites (de Melo Filho et al. 2013)

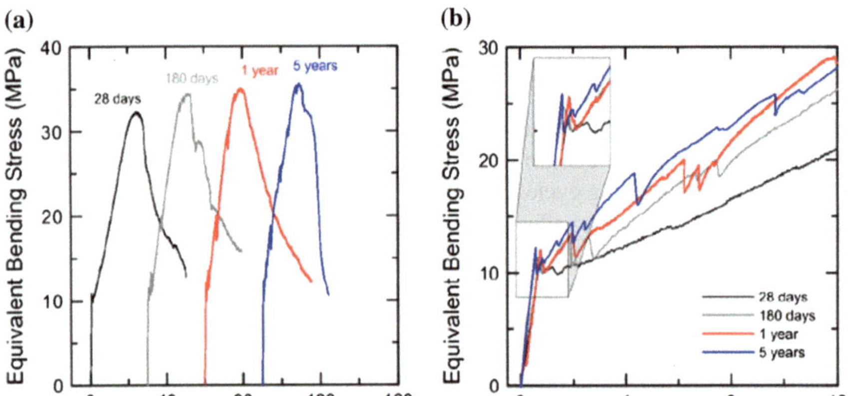

Fig. 9.41 Effect of natural aging on **a** the ultimate flexural strength and **b** on the postcracking stiffness of PC–MK composites subjected to flexural loading (de Melo Filho et al. 2013)

increase with age as a result of fibre–matrix densification. The frictional bond strength is not altered with age as the bending strength is remaining constant.

The effects of fibre hornification on the bending behaviour of cementitious composites reinforced with natural fibres has been confirmed by dedicated study performed by Ferreira et al. (2014). Figure 9.42 shows the typical bending

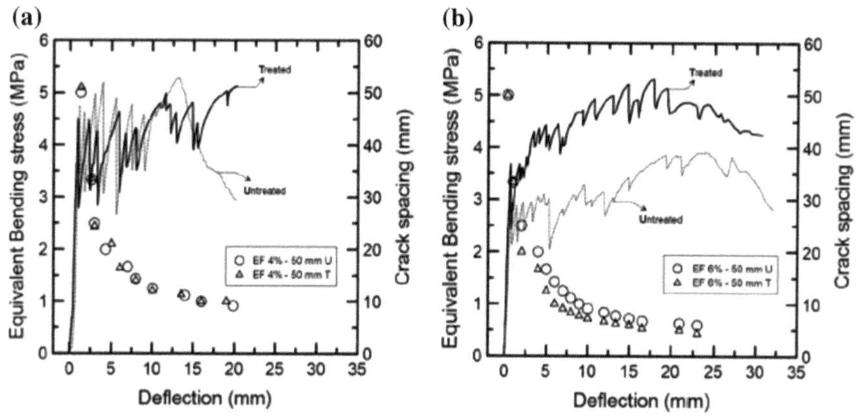

Fig. 9.42 Effect of the wetting and drying treatment and fibre volume fraction on the bending behaviour of sisal fibre cement composite and on its cracking spacing: **a** composites reinforced with 4% and **b** with 6% of sisal fibres (Ferreira et al. 2014)

behaviour of the composites reinforced with treated and untreated sisal fibres in volume fractions of 4 and 6%. All studied composites presented a multiple cracking behaviour with deflection hardening. The results indicate that the hornification improved the mechanical behaviour of the composite under bending loads. Increments in the values of fist crack and maximum bending strength are observed for both volume fractions used but the benefits of the treatment are best visualized in the tests with higher volumetric fraction, 6% (see Fig. 9.42b). Observing the stress-displacement curves it can be noticed an increase in the slope of the curves between the composites with treated and untreated sisal fibres.

This behaviour is related with the increase in the fibre-matrix adhesion due to the wetting and drying cycles carried out on the fibres. As observed in the pullout test, the treated fibre has better adhesion to the matrix, thereby promoting better mechanical properties to the composite. This improved adhesion is responsible for the largest number of cracks, bending resistance and less crack spacing.

9.5 Natural Fibres as Promoters of Self-healing in Cementitious Composites

Self-healing cement based materials are a valuable asset for the XXI century civil engineering, able to provide a technically effective and economically competitive solution to the increasingly urging problem of deteriorating and hence repair-needing building structures and infrastructure facilities (Mihashi and Nishiwaki 2012; Van Tittelboom and De Belie 2013).

Evidence of the ability of different self-healing techniques to seal the cracks, depending on their width, exposure conditions etc. has been widely proved in the laboratory. In very recent years interest and concern has also arisen in the engineering community about the efficacy of these techniques not only to seal the cracks and restore material imperviousness to the ingress of water and other aggressive harmful substances but also to "heal" the material. This refers to the ability of providing a (partial) recovery of the pristine level of engineering and mechanical performance, in terms of, e.g., load and deformation capacity, stiffness etc., which are evidently impaired by cracks and damages (Yang et al. 2009a, b).

As a matter of fact such a capacity of healing the material properties will depend not merely on the amount of crack healing but also on the nature of the healing products, and hence on the healing mechanisms, whether driven by delayed hydration producing stronger CSH crystals or by carbonation resulting into weaker $CaCO_3$ crystals, by the age of the healing products and of the substrate which also affects the degree of reciprocal compatibility and hence the effectiveness of the "through crack" material continuity restoration.

The interest versus the recovery of material mechanical performance has increased with the advent and spread in the construction market of Fibre Reinforced Concrete (FRC) and Cementitious Composites (FRCCs). This is not merely due to the positive interaction between fibres and self-healing, due to the capacity of the formers to restrict crack width and enhance the formation of multiple tiny cracks, which will be easier to heal. As a matter of fact, these materials exhibit a not negligible post-cracking residual strength, in some cases even featuring a strain hardening tensile behaviour. These tensile strength and toughness parameters do explicitly enter into design approaches for FRC structures (di Prisco et al. 2009). In view of the aforementioned concept, the capacity of the material to regain a higher post-cracking residual strength or retaining its target design value for a longer time rightly due to crack healing phenomena may represent an added value not only in terms of better and longer durability but also in terms of a longer lasting structural performance, which results into delayed first repair time, and into an overall lower life cycle cost.

Ferrara et al. (2014) have tried to correlate the amount of recovery of load bearing capacity and stiffness of ordinary concrete to the amount of crack sealing. They showed that the strength recovery is lower than stiffness one and that anyway a crack-sealing threshold (e.g. 50–60% of its initial amount) has to be overcome to start appreciating some recovery of the mechanical performance. Qualitatively and quantitatively similar results have been obtained by Roig Flores et al. (2015, 2016) with reference to the correlation between recovery of water flow tightness and crack sealing.

In this framework, High Performance Fibre Reinforced Cementitious Composites (HPFRCCs—Naaman and Reinhardt 2006) represent a broad category of ideal candidate materials both for building new and repairing old and/or deteriorated structures and infrastructures. Their peculiar material composition, designed through micromechanics concepts based either on crack-tip toughness or on the balance between matrix cracking strength and fibre bridging strength (Li et al. 1993), makes it possible to have, under tensile actions, an extended phase of stable multiple cracking, with each single crack narrower than 100 μm and even more, before the unstable crack localization. On the other hand, that same composition, characterized by high contents of cement and/or cement substitutes and low water/binder ratios, makes the material highly conducive to self-healing, as it has been demonstrated in a number of quite recent experimental studies (Mihashi and Nishiwaki 2012; Van Tittelboom and De Belie 2013; Yang et al. 2009a, b; Li et al. 1998; Sahmaran and Li 2008; Garas et al. 2009; Qian et al. 2009, 2010; Kan et al. 2010; Schlangen et al. 2010; Li and Li 2011; Kan and Shi 2012; Wu et al. 2012; Ozbay et al. 2013; Schlangen and Sangadji 2013; Kim et al. 2014; Lv and Chen 2014; Sahmaran et al. 2015; Snoek and de Belie 2015, 2016; Tang et al. 2015; Yildrim et al. 2015a, b; Ferrara et al. 2016a, b).

In this respect, the possibility of engineering the aforementioned autogenous healing capacity would also be of the utmost interest, enhancing the durability and sustainability signature of this category of advanced cement based materials.

Besides the use of cement substitutes, which, as above said, is quite common in HPFRCCs, few recent studies have addressed this issue with respect to a few different self-healing engineering techniques. Combined use of Super Absorbent Polymers (SAPs) and fibres has been reported (Snoeck and de Belie 2016). Synergy effects between fibres and crystalline admixtures, already effectively employed as self-healing promoters in normal strength concrete (Ferrara et al. 2014; Roig-Flores et al. 2015a, b, 2016), have been likewise studied, resulting in a likely activation of a sort of "chemical micro-prestressing" (Ferrara et al. 2016c) from which the overall mechanical performance of the material can greatly benefit.

In this framework also natural fibres, if employed as hybrid reinforcement together with other kinds of industrial fibres, such as steel or polypropylene ones, can act as facilitators and promoters of crack-sealing and healing reactions. As a matter of fact, thanks to their porous hierarchical micro-structure, natural fibres can absorb water, either in a pre-saturation stage or in the mixing stage or even when into the hardened mix from the environment, and release it to the surrounding matrix "on demand". When used as a dispersed reinforcement into a cement based matrix, they can thus form a well-structured, even if discontinuous, porous network, through which water can be effectively and uniformly distributed to the entirety of the material inside a specimen or a structural element. The effectiveness of this mechanism in counteracting, e.g., the autogenous shrinkage of HPFRCCs has been addressed (Ferrara et al. 2015a).

In the framework of the research project EnCORe, this peculiar role of natural fibres, used in a hybrid HPFRCC mix together with steel ones, has been investigated. The healing capacity of the fibre reinforced cement based composite has been assessed with reference to its effects on the recovery of mechanical properties, as measured by means of bending tests, tailored to the identification of mechanical parameters relevant to the structural applications of the material itself (e.g. thin slab/roof/façade panels). The methodology herein employed is based on the one proposed by Ferrara et al. (2014) for Normal Strength Concrete, with adaptions tailored to the peculiar behaviour of the investigated materials.

Healing capacity will be quantified by means of "healing indices". These account for the recovery of load bearing and deformation capacity (toughness/ductility) and stiffness. Healing indices will be correlated to the amount of crack closure, evaluated through image processing of crack patterns acquired by means of an optical microscope. This will be also instrumental at paving the way to define new significance for classical design parameters such as crack width, in an attempt to incorporate self-healing concepts and outcomes into a durability based design framework.

Reference has been made to a HPFRCC reinforced with a hybrid mix of steel and sisal fibres (Table 9.4), elaborated from a steel-only HPFRCC (Ferrara et al. 2011, 2012a, b; Ferrara 2015a, b) by replacing half of the steel fibre dosage with an equal volume percentage of sisal fibres.

Straight high carbon steel fibres, 13 mm (0.5 in.) long and 0.16 mm (0.006 in.) in diameter (lf/df = 80) were employed, featuring a wire tensile strength equal to 2000 MPa. The main characteristics of the employed sisal fibres are summarized in Table 9.5.

The composition of the tested materials is highly conducive to a superior performance in the fresh state, as witnessed by the self-compacting consistency achieved for both the steel only and the steel + sisal hybrid HPFRCC mixes (Fig. 9.17). Thanks to the achieved self-compactability, a tailored alignment of the fibres along the casting flow direction was obtained.

In view of this, slabs 1 m long × 0.5 m wide × about 30 mm thick were cast by pouring the mix in the center of a short edge and allowing it to freely flow, until complete mold filling, along the long edge. From the hardened slabs, beam specimens 500 mm long × 100 mm wide were cut, to be further tested in 4-point bending. In order to discriminate the effect of flow induced fibre alignment on the performance of the FRCC, the beam specimens were cut with their axis either parallel or orthogonal to the casting flow direction (Fig. 9.43). The reference

Table 9.4 Mix proportions of investigated HPFRCC mixes

Constituent	Dosage (kg/m^3)	
	Steel HPFRCC	Hybrid HPFRCC
Cement type I 52.5	600	600
Slag	500	500
Water	200	200
Superplasticizer	33	33
Sand 0–2 mm	982	982
Steel fibres	100	50
Sisal fibres	–	7

Table 9.5 Characterization of sisal fibres

Fibre type	Density (kg/dm^3)	Cellulose (%)	Hemicellulose (%)	Lignine (%)	Length (mm)	Sect. area (mm^2)
Sisal	1.01	60.5	25.7	12.1	13	0.023

performance of the material was investigated by means of 4-point bending tests performed according to the schematic also shown in Fig. 9.43 The adopted testing configuration is in accordance with Italian guidelines CNR-DT 204 (2006), which recommend the characterization of the material through test on un-notched specimens in the case of thin elements (with a cross section depth lower than 150 mm), and/or with a hardening type behaviour, which both apply to the present study. Tests were displacement controlled, at a rate of 5 μm/s and the crack opening (COD —Crack opening Displacement), was measured by means of two LVDT transducers (Fig. 9.43) over a gauge length equal to 200 mm.

The measured mechanical performance, for reference specimens monotonically tested about 30 days after casting, is shown in Fig. 9.44: it clearly appears that a

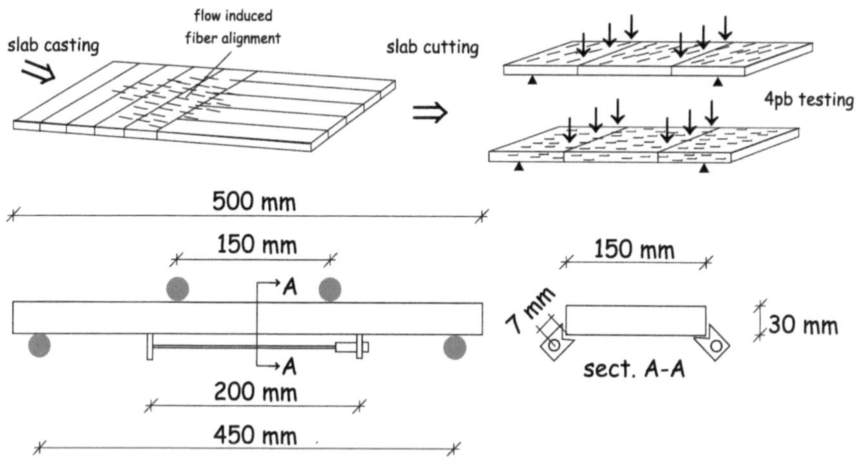

Fig. 9.43 Slab casting scheme and beam specimen cutting and test set-up

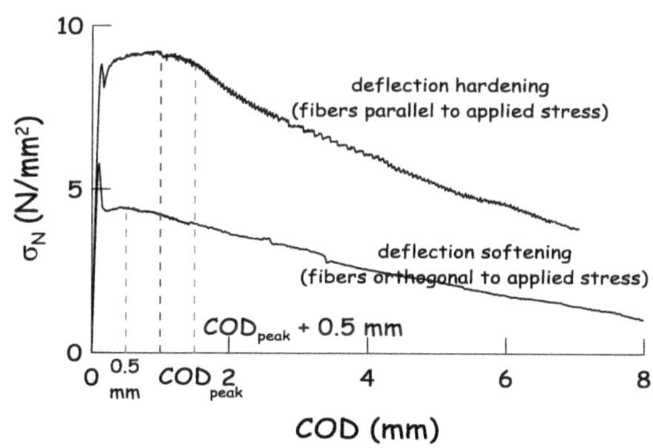

Fig. 9.44 Steel + sisal HPFRCCs: monotonic 4pb test results (COD pre-cracking thresholds highlighted)

deflection hardening response is obtained when the fibres are aligned parallel to the applied bending stress, with the formation of multiple cracks in the central region of the specimen subjected to constant bending moment (Fig. 9.45a). On the other hand, a deflection softening response is obtained otherwise, with the formation of a single crack (Fig. 9.45b).

The replacement of a percentage of steel fibres with sisal ones obviously resulted in a decreased performance, in terms of load bearing capacity and ductility, mostly for deflection hardening specimens (Figs. 9.44, 9.45, 9.46 and 9.47), which anyway featured a stable multi-cracking process, even if with a number of fibres lower than in the case of a steel fibre only HPFRCC. It is also worth remarking that with a

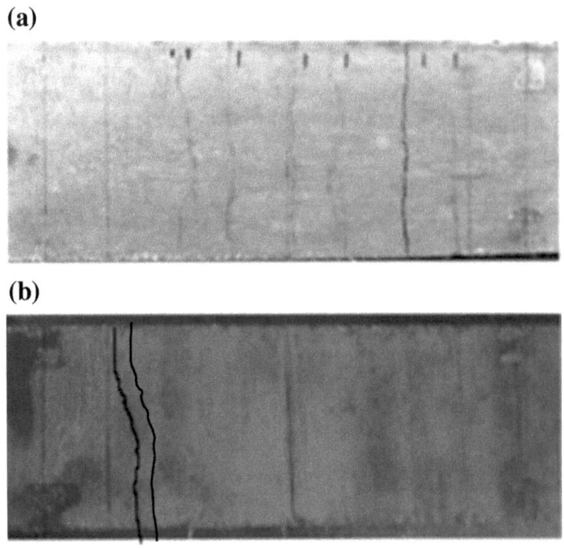

Fig. 9.45 Crack pattern for deflection hardening (**a**) and softening (**b**) hybrid HPFRCC specimens

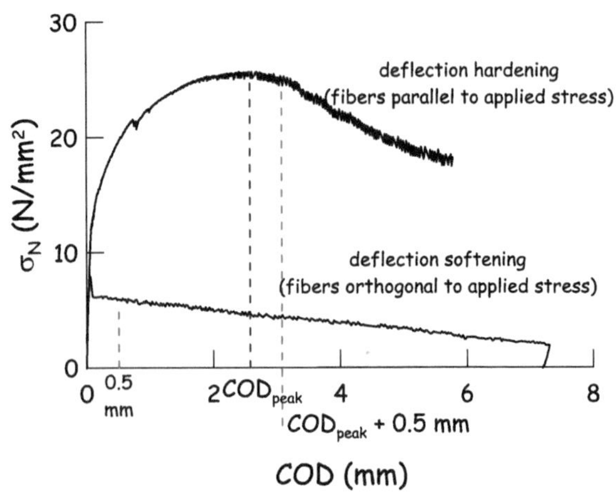

Fig. 9.46 Steel HPFRCCs: monotonic 4pb test results (COD pre-cracking thresholds highlighted)

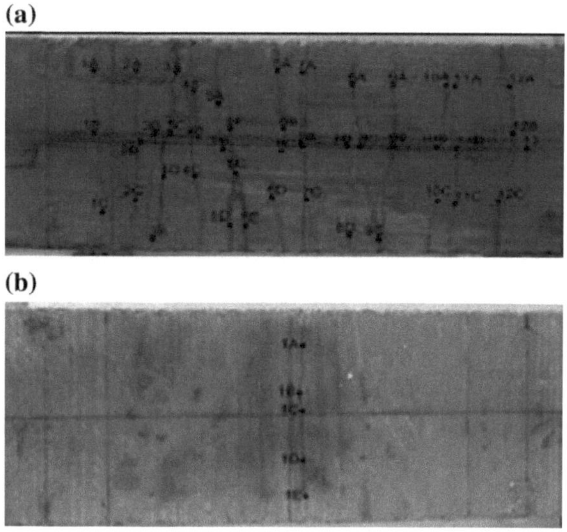

Fig. 9.47 Crack pattern for deflection hardening (**a**) and softening (**b**) steel HPFRCC specimens

0.64% volume fraction of steel fibres only, deflection hardening hardly could be obtained (Ferrara et al. 2012a, b).

For the self-healing experimental campaign, specimens were pre-cracked, employing the same 4pb testing scheme and set-up as shown in Fig. 9.43. Specimens featuring a deflection softening response (i.e. with fibres perpendicular to the beam axis) were pre-cracked up to a Crack Opening Displacement (COD) value equal to 0.5 mm. On the other hand specimens with fibres parallel to the axis, featuring a deflection hardening response, were pre-cracked in the post-peak regime equal to (COD_{peak} + 0.5 mm), where COD_{peak} denotes the measured value of the COD in correspondence of the peak stress.

With reference to this issue it is worth here remarking that for one reference deflection hardening specimen, the multi-cracking process was monitored during the test by placing a cell-phone camera under the specimen intrados and then processing the recorded photograms. In Fig. 9.48 the processed values of openings of each single crack have been plotted on the nominal bending stress versus COD graph measured for the specimen. It can be observed that all the cracks undergo a stable growth until the peak load bearing capacity of the specimen is attained; henceforth, only one single crack localizes in an unstable way and undergoes abrupt further opening, being moreover responsible of almost all the further crack opening which is measured after the peak. This, together with the processed COD values, also confirms the reliability of the "equivalent localized crack opening assumption" herein adopted, coherently with Ferrara et al. (2011), underlying the choice of the pre-cracking values targeted for deflection softening and hardening specimens.

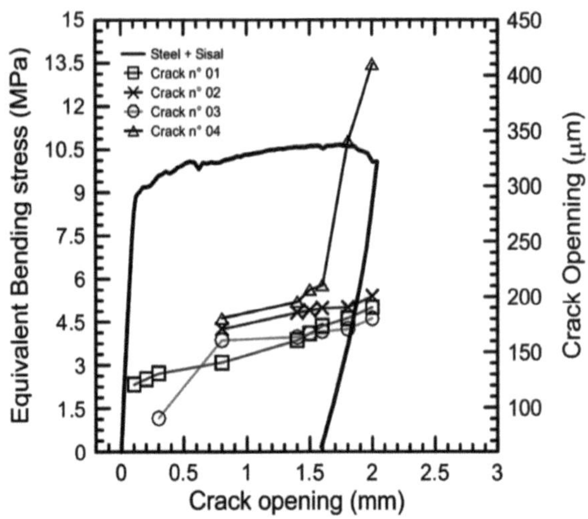

Fig. 9.48 Evolution of the opening of each single crack in a deflection hardening specimen

After pre-cracking specimens were "conditioned" in four different "environments":

- immersion in water (wet);
- dry (in a chamber with 20 °C temperature and 50% relative humidity);
- moist (in a chamber with 20 °C temperature and 90% relative humidity);
- wet and dry cycles: the cycles were made by alternating one day of immersion of specimens in water, at T = 20 °C, and one day of exposure in a dry environment at T = 20 °C and 50% relative humidity.

At the end of the scheduled conditioning period, specimens were tested up to failure according to the same 4-point bending set-up shown in Fig. 9.43. "Superposition" between pre-cracking and post-conditioning nominal flexural strength σ_N versus COD curves allowed the self-healing capacity and its effects on mechanical performance of the material to be evaluated. Three tests per each deflection behaviour, exposure condition and duration (3, 6 or 24 months) were performed.

In Figs. 9.49 and 9.50 different levels of crack healing can be appreciated, depending on exposure conditions and durations. Interestingly, in one case, for a deflection hardening specimen immersed in water for six months, in the post-conditioning test the sealed crack didn't reopen and a new crack was formed elsewhere (Fig. 9.51). This clearly highlights that the delayed hydration reactions which occurred in the specimen, were responsible of the crack closure and of a reconstructed through-crack material continuity, to such an extent that the strength of the healed crack was even higher than the one of the bulk material, somewhere else, because of the randomness of the material strength. Moreover, it can be

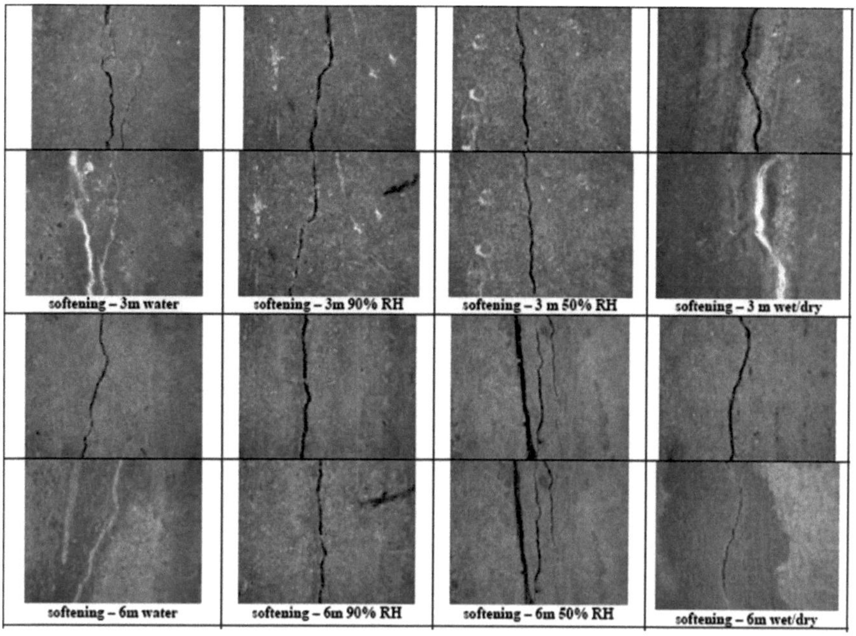

Fig. 9.49 Images of cracks before and after healing for deflection softening (50×) specimens and for different exposure conditions and durations

confirmed that the observed recovery of the mechanical performance, in terms of load bearing capacity, is rightly attributable to crack healing and not to a generalized bulk delayed hydration. This in fact mostly occurred in the larger crack, where water could more easily penetrate and not in correspondence of smaller cracks or in the un-cracked zone, where water could hardly reach the inner core of the specimen.

This one of a kind performance was never observed in steel only HPFRCC specimens (Ferrara et al. 2016b, c). It can be reliably attributed to the structure of the natural fibres, which absorbed water and transported it through diffusion to the surrounding zones, promoting better and more widespread delayed hydration, which may have toughened the healed crack also through a better fibre-matrix bond. This is also witnessed by microscopy images of a natural fibre pulled out from a healed crack (Fig. 9.52a) which was completely covered by the delayed hydration products, being almost undistinguishable from the surrounding matrix. On the other hand a steel fibre, in a steel HPFRCC specimen which underwent the same history as the hybrid one, the amount of delayed hydration products covering the fibre was far more limited (Fig. 9.52c). Anyway a toughening effect of the healing products on the fibre-matrix bond could also be appreciated, comparing the neat fibre

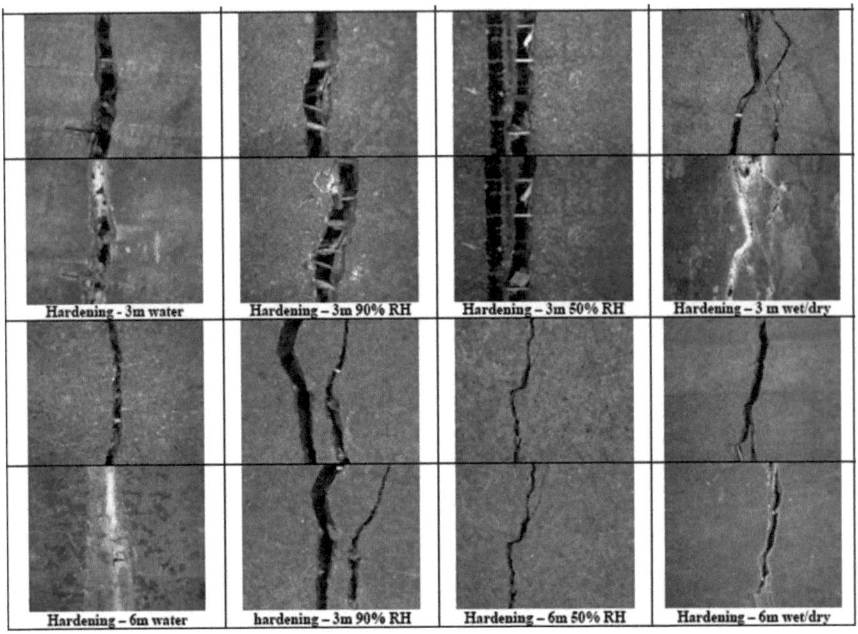

Fig. 9.50 Images of cracks before and after healing for deflection hardening specimens (200×) and for different exposure conditions and durations

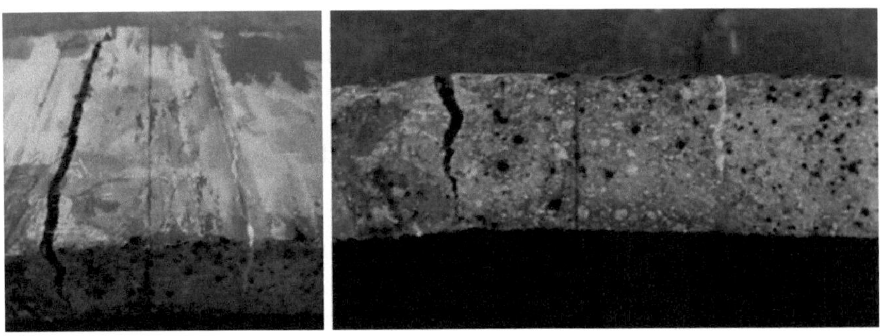

Fig. 9.51 Healed crack and new crack in post-healing tests in sisal + steel HPFRCC

pull-out in the pre-cracking test (Fig. 9.52b) with the highly fractured matrix in the specimen after healing (Fig. 9.52c).

By comparing for each and all the specimens, the σ_N-COD curves in the pre-cracking regime with that in the post-conditioning one, suitable indices have been defined to quantify the effects of healing on the recovery of the load bearing capacity and flexural stiffness, as hereafter detailed. More information can be found in Ferrara et al. (2017).

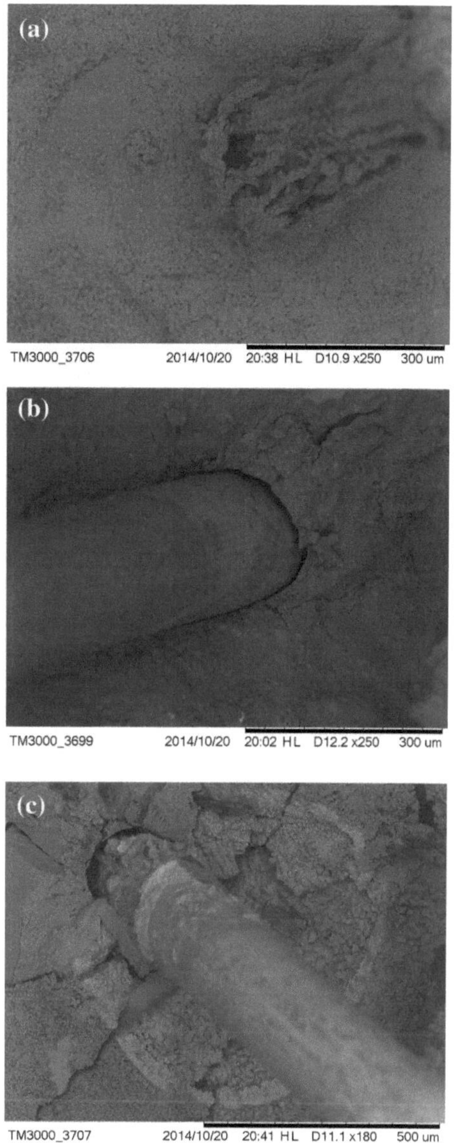

Fig. 9.52 Fibre-matrix interface: **a** healed sisal + steel, **b** non healed steel and **c** healed steel HPFRCC

9.5.1 Index of Strength Recovery

9.5.1.1 Deflection-Softening Specimens

In the case of deflection-softening specimens, the effectiveness of healing in promoting the recovery of the "through crack" residual load-bearing capacity can be evaluated by calculating the amount of strength gained after the conditioning period, with respect to the residual strength featured at the maximum pre-crack opening, and comparing it to the stress loss exhibited by the same specimen when pre-cracked up to the aforementioned crack opening threshold (Fig. 9.53). With reference to the notation in the same Figure, the Index of Strength Recovery (ISR) is defined as follows:

$$\text{Index of Strength Recovery ISR} = \frac{f_{peak,\,post\ conditioning} - \sigma_{N\ unloading,\,pre-cracking}}{f_{peak,\,pre-cracking} - \sigma_{N\ unloading,\,pre-cracking}} \quad (9.6)$$

From the plots shown in Fig. 9.54 it holds that all specimens featured a recovery of the load bearing capacity continuing over time, even up to quite longer exposure times, and to different extents, depending on the conditions of exposure. Higher availability of water resulted in higher levels of performance recovery, with value of the Index even higher than one: this would correspond to a performance of the healed specimen even better than in the virgin state. For specimens subjected to wet and dry cycles, a very high recovery in the initial stages of exposure was observed, followed, for longer exposure times, by a performance similar to those of specimens exposed to a humid conditioning environment. This could be justified by considering that during the dry stages of each cycle, as it will be later confirmed, calcium carbonate is likely to form in the cracks, together with delayed hydration products formed during the wet stages. Upon prolonged cycling, the calcium carbonate, responsible of a fast initial strengthening, may dissolve, thus bringing down the performance recovery.

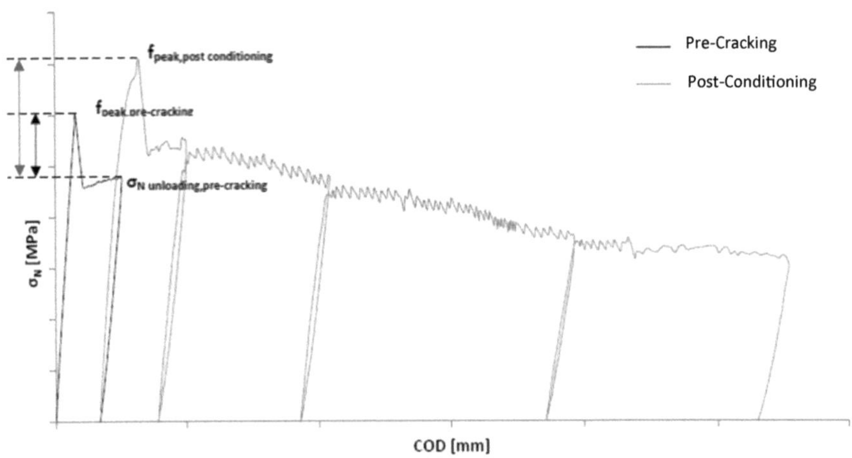

Fig. 9.53 Notation and definition of the parameters for the index of strength recovery for deflection-softening specimens

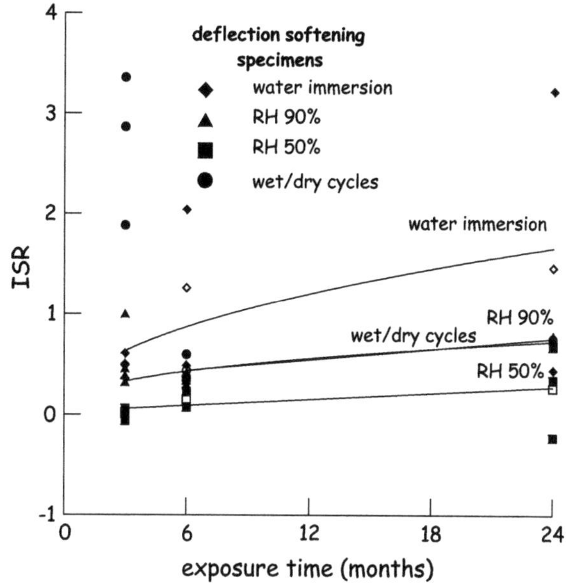

Fig. 9.54 Index of strength recovery versus conditioning time for deflection softening specimens (*hollow markers* single tests, *solid markers* average of nominally identical tests)

9.5.1.2 Deflection-Hardening Specimens

In the case of deflection-hardening specimens pre-cracked after the peak, the value of the ISR computed as per Eq. 9.6 does incorporate the major effects of healing of the single localized crack and also those, likewise significant, all the other multiple cracks which have been formed up to the peak.

The trends of ISR for deflection-hardening specimens for the different pre-crack opening cases herein investigated, are shown in Fig. 9.56. The influence of different exposure times and conditions, as already observed with reference to deflection softening specimens, is confirmed. Furthermore, it can be observed, as already with reference to deflection-softening specimens, that the on-going healing of cracks was also instrumental to overcome the damage that, in some cases, was caused by some non-favorable exposure conditions (exposure to dry environment may have exacerbated shrinkage resulting in an early loss of performance) (Fig. 9.55).

Quite interestingly, the distinctive feature of natural fibres as facilitators and "triggerers" of healing phenomena, once again clearly appears with reference to specimens subjected to wet and dry cycles, which actually featured the highest and continuing increasing recovery of the performance. It is believed that this exposure condition does really challenge the role of natural fibres as promoters of self-healing: water is absorbed during the wet phases and continues to be gradually released throughout the specimen during the drying phase. The multiple cracking enhances this effect and the tight opening of each single crack makes it possible for the high and continuously increasing levels of the measured recovery of mechanical performance. This may also contribute to neutralize the negative effects of some kind of calcium carbonate dissolution which have been hypothesized above for

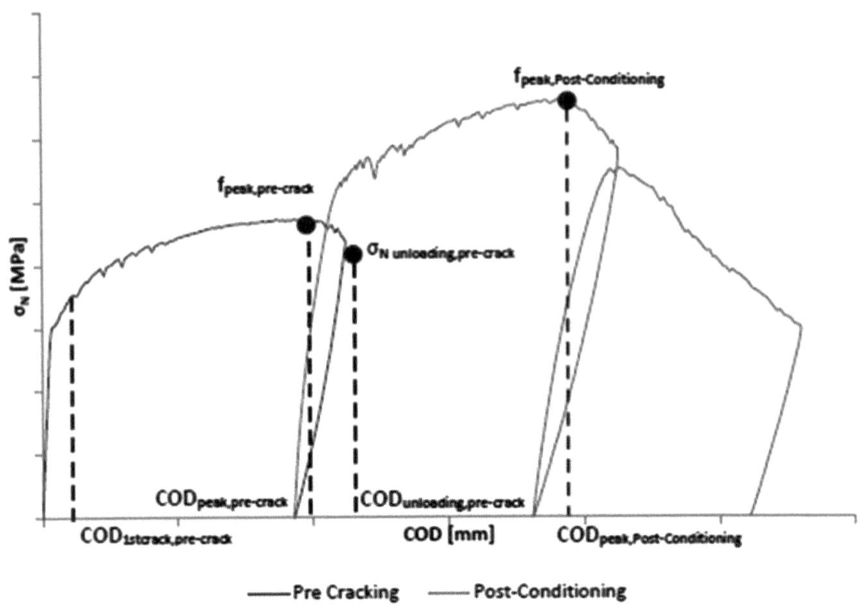

Fig. 9.55 Definition of parameters to calculate the index of strength recovery and index of ductility recovery for deflection-hardening specimens pre-cracked in post-peak regime

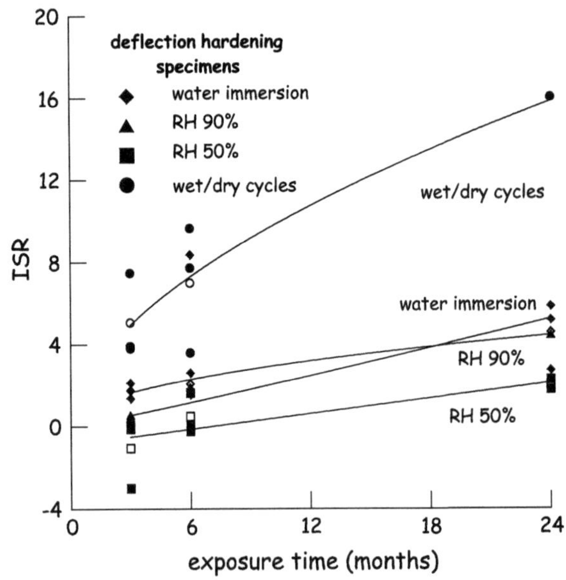

Fig. 9.56 Index of strength recovery versus conditioning time for deflection-hardening specimens pre-cracked up to 0.5 mm after the peak (*hollow markers* refer to values of single tests, *solid markers* represent average values of nominally identical tests)

deflection softening specimens. Comparison with the performance of similarly conditioned and tested steel only HPFRCCs, which will be shown further in this section, will provide further evidence to support the aforementioned educated guess.

9.5.2 Index of Stiffness/Damage Recovery

Thanks to unloading-reloading cycles performed both during pre-cracking and post-conditioning tests, the values of secant unloading and tangent reloading stiffness were evaluated (Fig. 9.57). It is worth remarking that the unloading process always took place in a matter of few tenths of seconds, which is quite short for any significant relaxation to have occurred. From them, an Index of Damage Recovery was calculated as:

$$\text{Index of Damage Recovery IDaR} = \frac{K_{reloading, postconditioning} - K_{\text{unloading, pre-cracking}}}{K_{loading, pre-cracking} - K_{\text{unloading, pre-cracking}}} \tag{9.7}$$

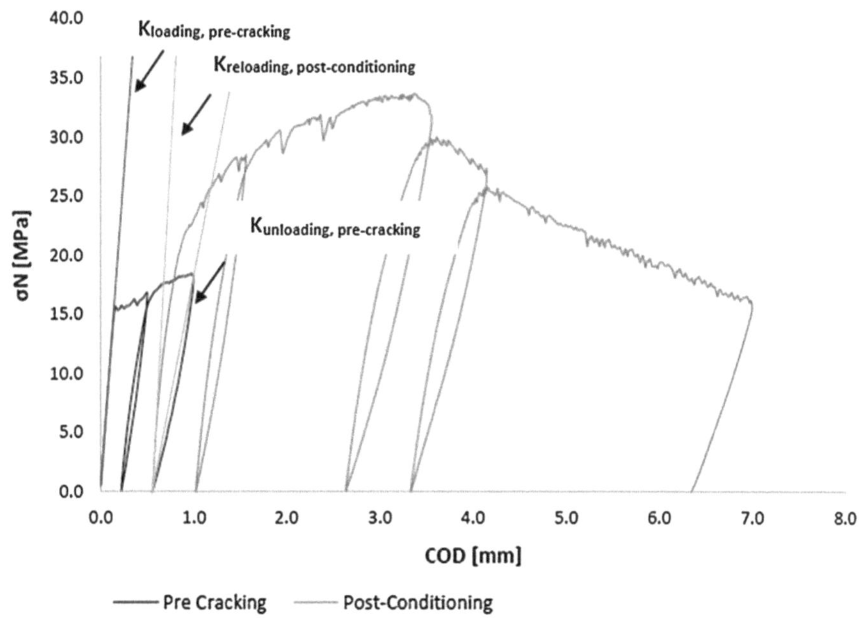

Fig. 9.57 Notation for the calculation of index of damage recovery

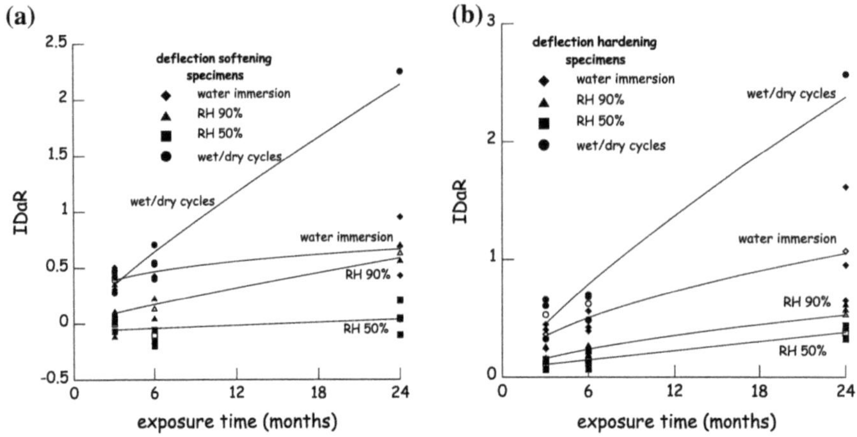

Fig. 9.58 Index of damage recovery versus conditioning time for deflection softening (**a**) and hardening (**b**) specimens (*hollow markers* refer to values of single tests, *solid markers* represent average values of nominally identical tests)

whose plots are shown in Fig. 9.58a, b, respectively for deflection-softening and hardening specimens, and appear to be coherent with the previously discussed trends with reference to other indices.

Specimens exhibited a stiffness recovery ranging from scant to significant, as a function of exposure conditions, wet and dry cycles and exposure to open air providing the best results and exposure to dry environment the worst. When correlated to the Index of Strength Recovery (Fig. 9.59), interesting the Index of Damage Recovery features a square root trend.

9.5.3 Crack Opening Measurements and Crack Sealing Evaluation

9.5.3.1 Image Analysis Processing of Cracks Observed by Optical Microscopy

The measurement of crack closure parameters is based on the processing, made through the Adobe Photoshop CS6 software®, of pictures showing the cracks all along their length, as obtained by means of digital optical microscope before and after the healing exposure at different magnifications (×50, ×200).

An example of the crack pattern surveying procedure for both deflection softening and hardening specimens is shown in Fig. 9.60. It is worth remarking that for both deflection-softening and deflection-hardening specimens pre-cracked in the

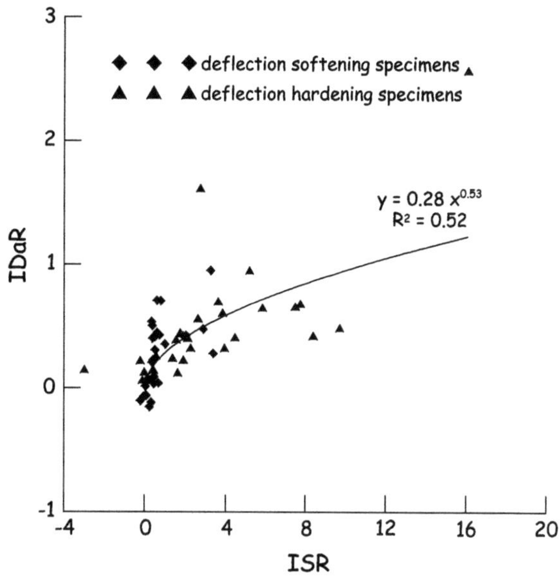

Fig. 9.59 Correlation between index of damage versus index of strength recovery

post-peak regime only one crack had to be measured, which is the single crack or the crack unstably localizing after the peak respectively.

The employed method is more precise and reliable than the measurements which can be obtained by using the transparent calibration ruler since through the use of the graphical software is possible to choose the real width by-passing the error related to the human eye, which in small cracks, as those studied in this research, becomes more relevant. Two different measures were taken:

- Measure of crack width:
  w_{avg}, average crack width: by means of graphic software, crack width (in millimeters) was determined at sixty positions. For each crack, all over its length, 20 photograms were acquired and for each photogram the crack was measured at three different positions (approximately located 10 mm from the edges of the pictograph and at mid-length of the surveyed portion of the crack) and averaged (Fig. 9.61a);
- Measure of black pixel area:
  Apx, measuring black pixels: by using graphic software, black pixels in the image are counted (Fig. 9.61b). The calculation of the crack pixel area involved the use of the graphical software to quantify the number of black pixels inside the pictures. The basic concept is that the cracks are shown as a black area, thus a higher number of black pixels correspond to a greater pixel area of the crack. In order to avoid that pores or other dark parts of the pictures that may affect the quantification of black pixels, all the panoramas pictures have were been cleaned by a specific tool of the software.

(a)

(b)

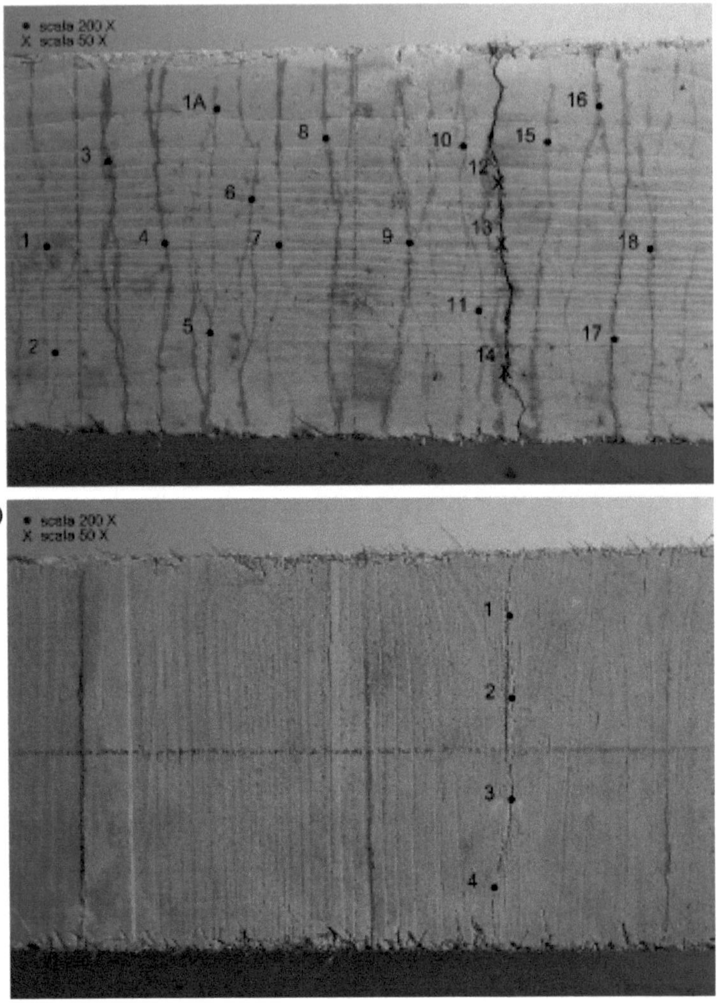

Fig. 9.60 Photographic mapping of cracks and location of magnified images locations

9.5.3.2 Index of Crack Healing

From the values of crack-opening, measured and/or estimated both in the pre-cracking and in the post-conditioning regime an Index of Crack Healing has been defined as follows:

$$\text{Index of Crack Healing ICH} = 1 - \frac{COD_{post\text{-}conditioning}}{\text{COD}_{\text{pre-cracking}}} \quad (9.8)$$

The Indices of Crack Healing calculated from visual crack processing are shown in Figs. 9.62a, b and 9.63a, b, for the two different crack width measurement methodologies and respectively for deflection softening and hardening specimens.

The whole corpus of the results provides a further confirmation to the statements exposed above with reference to the indices of recovery of mechanical properties. Visual proofs of crack healing complemented the aforementioned analysis (Figs. 9.49 and 9.50) but also provided material for a quantitative analysis of crack closure, which can be interestingly correlated to the quantified recovery of the mechanical properties to further corroborate the reliability of the proposed healing assessing approach.

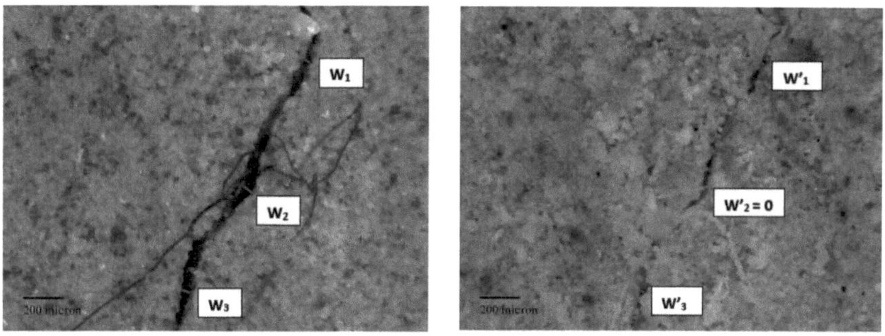

(**a**) Measure points for calculation of average crack width before (left) and after conditioning (right)

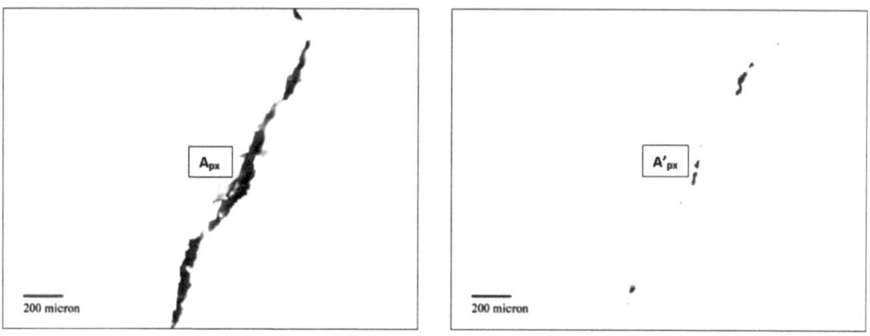

(**b**) Filtered crack images for pixel area evaluation before (left) and after conditioning (right)

Fig. 9.61 Example of visual image processing for evaluation of crack closure due to self-healing phenomena: average crack width (**a**) and pixel area (**b**)

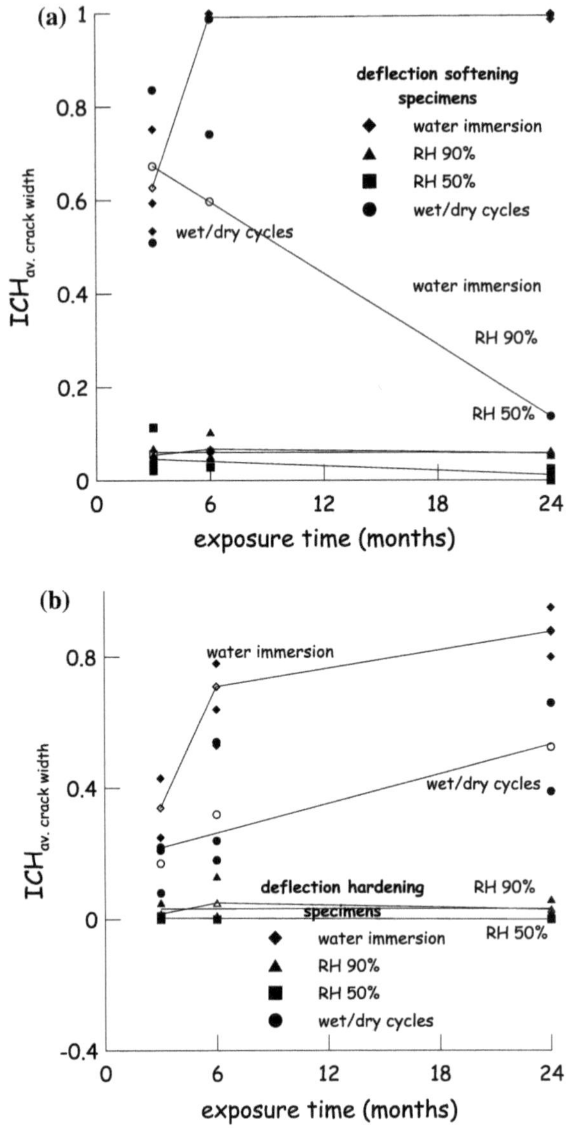

Fig. 9.62 Index of crack healing, evaluated through the average crack width measurement, versus conditioning time for deflection softening (**a**) and hardening (**b**) specimens (*hollow markers* refer to values of single tests, *solid markers* represent average values of nominally identical tests)

9.5.4 Comparison Indices of Mechanical Properties Recovery and Index of Crack Healing

The different indices of recovery of material mechanical properties have been finally compared with the Index of Crack Healing, evaluated as in Sect. 9.5.1. The results are shown in Figs. 9.64 and 9.65.

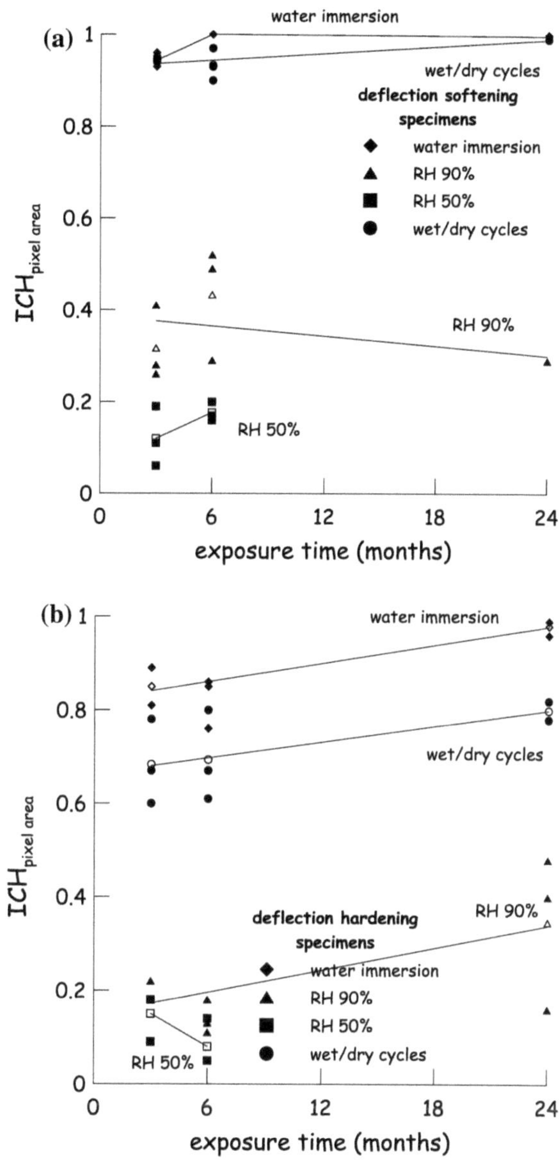

Fig. 9.63 Index of crack healing, evaluated through the pixel area measurement, versus conditioning time for deflection softening (**a**) and hardening (**b**) specimens (*hollow markers* refer to values of single tests, *solid markers* represent average values of nominally identical tests)

The sparsity of the results, processed as a whole, should not be deceptive about the reliability and significance of this analysis, since it is evidently due to the wide set of experimental variables which have been considered in this investigation (deflection hardening/softening behaviour, crack-opening, exposure conditions and durations). The analysis yields clearer outputs if the five herein considered exposure conditions are separately dealt with.

Fig. 9.64 Correlation between index of strength recovery and index of crack healing evaluated through average crack width (**a**) and pixel area (**b**) measurements

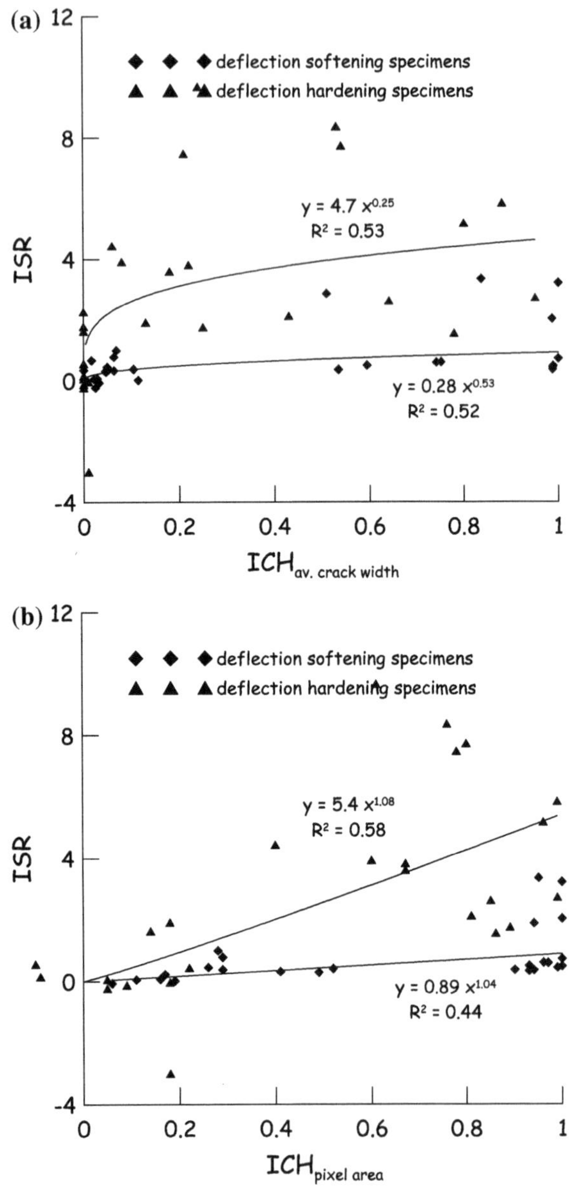

With reference to both the evolution of the strength and stiffness recovery with crack closure, it can be clearly observed that the trend is coherent. Moreover, the influence of exposure conditions is as expected and as analyzed above, with reference to both the amount of recovery and to the scattering of the results, and hence to the neatness of the inferred trends.

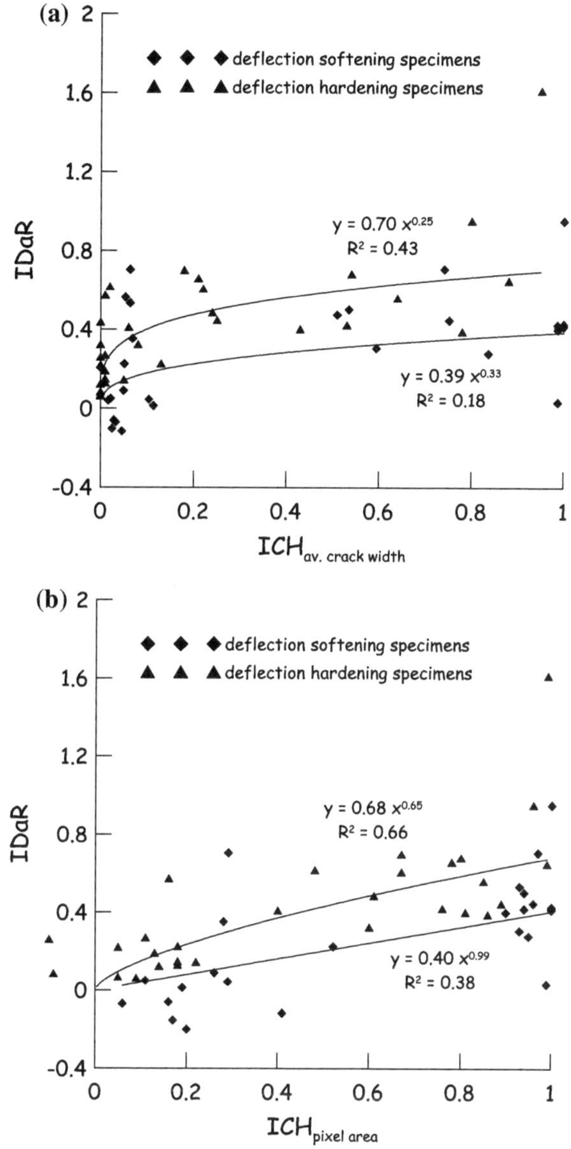

Fig. 9.65 Correlation between index of damage recovery and index of crack healing evaluated through average crack width (**a**) and pixel area (**b**) measurements

9.5.5 Microscopical Characterization

In order to further clarify the different processes and mechanisms characterizing and governing the healing process, Thermo-Gravimetric Analyses (TGA) have been performed on four samples extracted from different parts of the cracked/healed surface of a specimen, after the post-conditioning failure tests (Fig. 9.66) on a sample subjected to wet and dry cycles for 3 months. The first part was extracted

Fig. 9.66 Location of two of the four samples collected on the cracked/healed surface of a specimen subjected to three months wet/dry cycles and tested up to failure and employed for TGA analyses

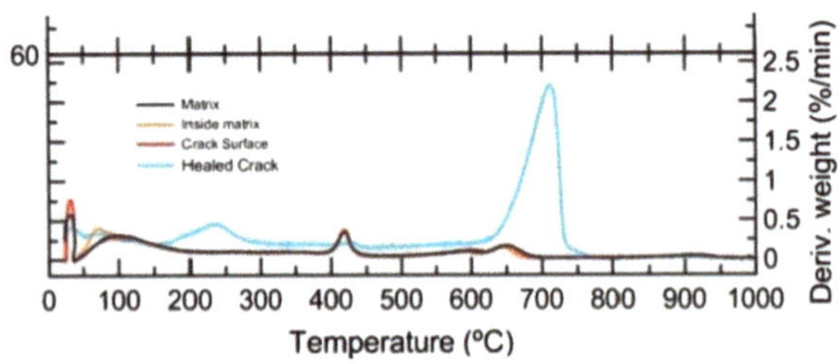

Fig. 9.67 Differential thermogravimetric curves for four examined samples collected from bulk material specimens) (matrix), from the intrados outside of the crack (crack surface) as well as at different locations on the cracked/healed surface

from inside the matrix in what was believed to be the compressed part of the cracked cross section of the specimen. The second part was from the same specimen but outside of the cracked surface, whereas the third part was obtained from a cracked/healed portion of the examined cross section. The last sample was obtained from a separated material, which underwent the same curing and conditioning.

Thermogravimetric experiments were performed in a thermos-balance TGA/DTG/DSC, model SDT Q600 manufactured by TA Instruments ©. In each experiment, 10 mg of undried material was spread uniformly on the aluminium bottom of the thermal analyser. The pyrolysis process were performed at heating rates of 5 °C/min in a nitrogen flow of 100 ml/min and the temperature of the furnace was programmed to rise from room temperature to 1000 °C. In the analysis the derivative thermogravimetric curves (DTG) have been used to determine the weight loss and to identify the decomposition of the material at given temperatures. The results obtained (Fig. 9.67) highlight that all the samples, in a similar way, featured mass loss peaks related to typical hydration products, such as ettringite (80 °C), CSH (104–140 °C) and CH (460 °C). On the other hand, the sample collected from the healed part of the crack surface, shows peaks in correspondence of the de-hydroxylation of CSH (230–240 °C) and of the de-carbonation (620–645 °C). The presence of the aforementioned substances confirms that the products appearing in the cracks are products of the delayed hydration and carbonation reactions, which are the cause of the healing produced during the wet and dry stages respectively of the wet and dry cycles. On the other hand, the similarity of mass losses for ettringite, CSH and CH, confirms that the recovery of mechanical performance observed when testing specimens in the post-conditioning stage, is not attributable to a generalized prolonged hydration of the binder. As a matter of fact, because of the high compactness of the hardened matrix, the water, which is necessary to promote the aforementioned delayed hydration, can hardly penetrate into the matrix but through the cracks, even in completely immersed specimens.

9.5.6 Comparison Between Healing Capacities of Hybrid Steel + Sisal and Steel Only HPFRCCs

Evaluation on the effectiveness of natural sisal fibres as facilitators and enhancers of the healing capacity of hybrid steel + sisal HPFRCCs has been hereafter completed through a one of a kind comparison with the healing performance of steel only HPFRCCs. Reference has been made for this purpose to the results recently obtained by the first authors and his co-workers (Ferrara et al. 2016a, b, c). Trends of the Indices of Recovery of the different mechanical properties have been compared for the two investigated HPFRCCs as shown in Figs. 9.68a, b and 9.69.

With reference to the trend of the Indices of Stress Recovery, it can be observed that, in the cases of exposure to wet and dry cycles for both deflection softening and hardening specimens, the performance of hybrid steel + sisal HPFRCCs always significantly overpasses that of steel only HPFRCC. This fact is always confirmed with reference to the recovery of all the investigated mechanical properties, ranging from pre- and post-peak ductility to stiffness and toughness. For other exposure conditions, with few exceptions, the calculated trends of Indices of Recovery are absolutely coherent for both HPFRCCs.

As already remarked in Sect. 9.5.1, exposure to wet and dry cycles actually represents the most challenging condition in order to test the efficacy of natural fibres as vehicles of water throughout the material and hence facilitate and trigger the healing reactions. The afore-discussed comparison with the performance of steel only HPFRCC is likely to confirm that natural fibres at a cracked site absorb the water in the wet stages of the cycles and transport it throughout the bulk material in the vicinity of the same crack. This contributes to an enhancement of the delayed hydration and carbonation reactions which are responsible of self-healing, thanks to the higher and longer availability of water, as retained by the fibres all along the drying stage of each cycle. This may also like to contribute to "shadow" the negative effect of calcium carbonate dissolution, precipitating during the same drying stages of each cycle, which is also responsible of self-healing and which may dissolve upon prolonged exposure, mainly in wider cracks, as observed.

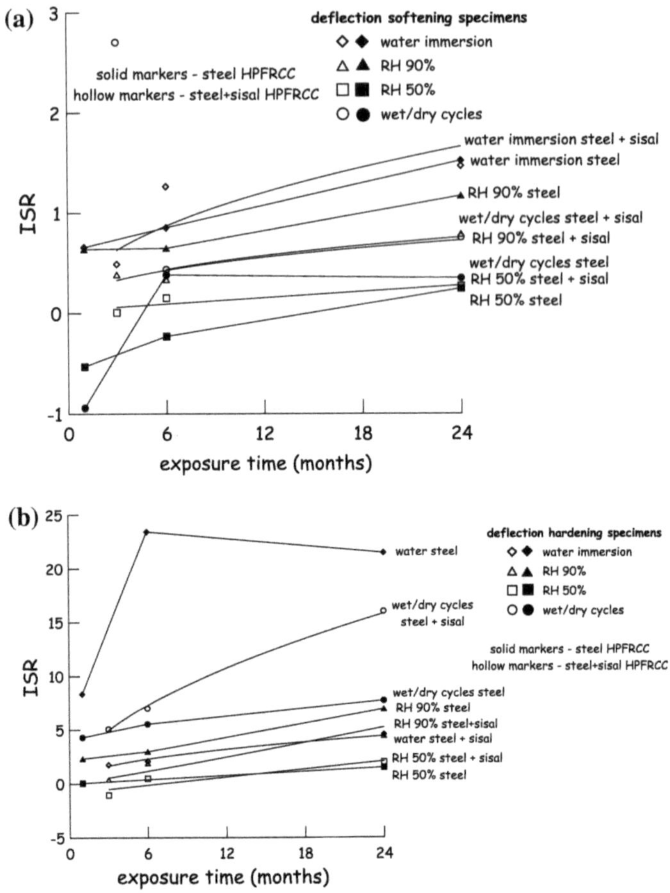

Fig. 9.68 Comparison of index of strength recovery for deflection softening (**a**) and hardening (**b**) specimens made with steel only and hybrid steel + sisal HPFRCCs

Tailored microstructure and microscopy investigation has also provided a preliminary hint about the effect of healing reaction of the fibre-matrix bond, which, together with the restored cross-crack material continuity, also contributes to the recovery, if any, of the mechanical performance of the material, as observed in terms of load-bearing, deformation and energy absorption capacity and stiffness.

Correlation between the recovery of mechanical properties and the contemporary occurring crack closure has revealed that the former initially proceeded quite faster than the latter, further stabilizing at a lower pace. This also seems to be a distinctive feature due to the use of natural fibres, which, thanks to their role, may have promoted, not only at a crack but also in the vicinity of it, more prompt delayed hydration reactions, if any. As a matter of fact, for steel only HPFRCCs, an appreciable recovery of mechanical properties could started being observed only after about 50–60% crack closure (Ferrara et al. 2016b, c).

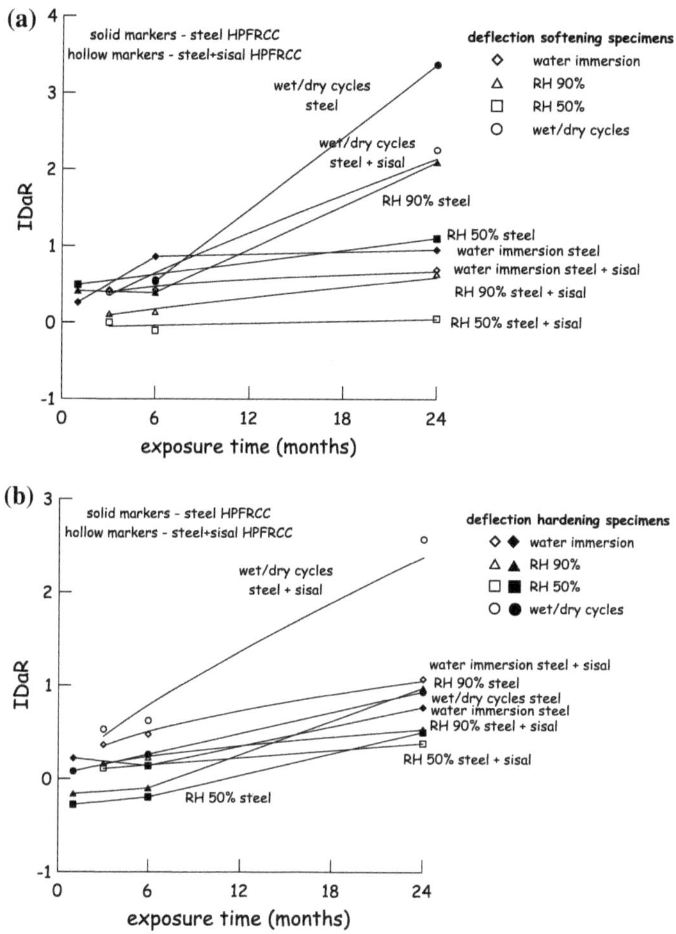

Fig. 9.69 Comparison of index of damage recovery for deflection softening (**a**) and hardening (**b**) specimens made with steel only and hybrid steel + sisal HPFRCCs

9.6 Cellulose-Based Micro- and Nano-reinforcement in Cementitious Composites

9.6.1 Cellulose

Natural fibres are constituted by several cellulose microfibrils, where cellulose chains are arranged in two types of conformations: an amorphous conformation, characterized by its flexibility; and a crystalline conformation in which the cellulose chains are disposed in organized and repetitive arrays to form crystals (Fengel and Wegener 1984; Bledzki and Gassan 1999; Abe and Ryo 2005; Yousif and El-Taayeb 2007; Azubuike and Esiaba 2012; Azwa et al. 2013).

Cellulose is considered to be one of the most abundant organic compound derived from biomass. The annual worldwide production of cellulose is estimated to be between 1010 Mtons (Azizi Samir et al. 2005), only a small part of which is processed by industries such as paper, textile, material and chemical industries (Simon et al. 1998). Although cellulose is the main building material out of which woods are made, there are other major sources such as plant fibres, marine animals (tunicate), or algae, fungi, invertebrates, and bacteria. Regardless of the plant sources, cellulose can be present in the leaf (e.g., sisal), in the fruit (e.g., cotton) or in the stalk or the rigid structure of plants (e.g., flax). Irrespective of its source, cellulose is a white odorless fibre-like structure with a density of 1.5 Mg/m^3.

Since its discovery by Payen (1838), the physical and chemical aspects of cellulose have been intensively studied. Today, its unique hierarchical structure no longer holds any secrets: cellulose is a linear homopolysaccharide of β-1.4-linked anhydro-D-glucose units (Habibi et al. 2010; Siqueira et al. 2010) with a degree of polymerization of approximately 10,000 for cellulose chains in nature and 15,000 for native cellulose cotton (Simon et al. 1998). The basic chemical structure of cellulose (Fig. 9.70), shows a dimer called cellobiose that appears as a repeated segment. The monomer, named anhydroglucose unit, bears three hydroxyl groups. These groups and their ability to form strong hydrogen bonds confer upon cellulose its most important properties, in particular its (i) multi-scale microfibrillated structure, (ii) hierarchical organization (crystalline vs. amorphous regions), and (iii) highly cohesive nature (with a glass transition temperature higher than its

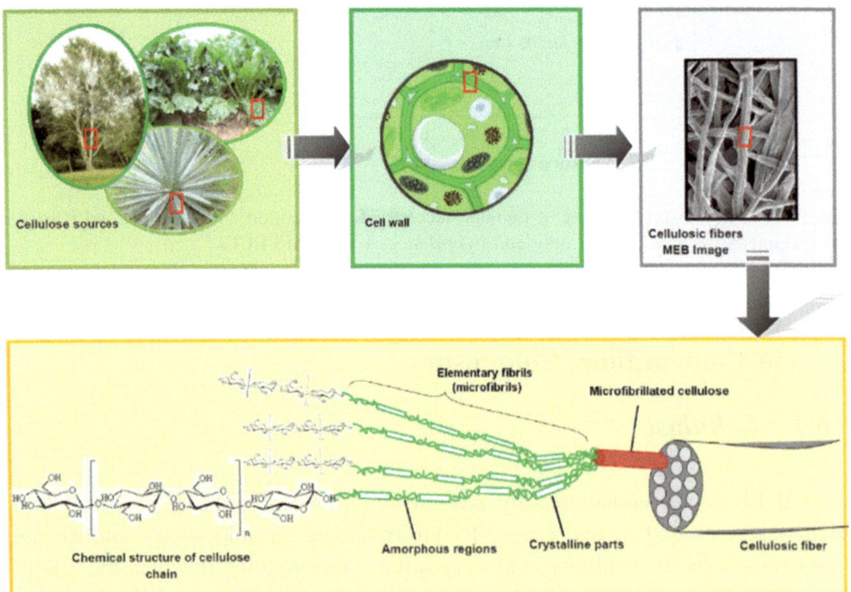

Fig. 9.70 From the cellulose sources to the cellulose molecules: details of the cellulosic fibre structure with emphasis on the cellulose microfibrils (Lavouine et al. 2012)

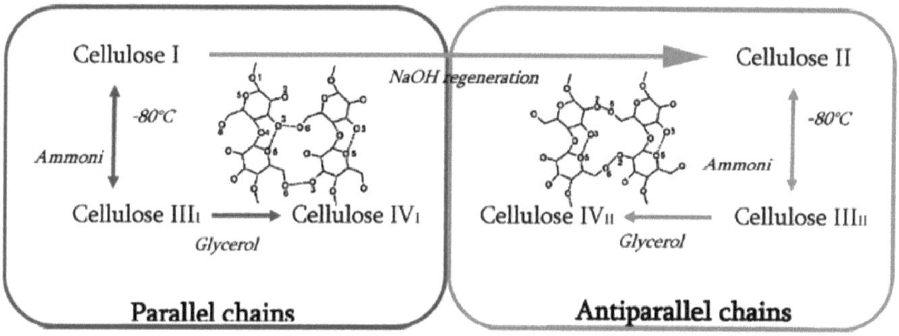

Fig. 9.71 Polymorphs of cellulose and the main steps to obtain them (Lavouine et al. 2012)

degradation temperature). There are four different polymorphs of cellulose: cellulose I, II, III, and IV (Fig. 9.80). Cellulose I, native cellulose, is the form found in nature, and it occurs in two allomorphs, Iα and Iβ. Cellulose II, or regenerated cellulose, emerges after re-crystallization or mercerization with aqueous sodium hydroxide. It is the most stable crystalline form (Simon et al. 1998). The major distinction between these two forms of cellulose lies in the layout of their atoms: cellulose II has antiparallel packing, whereas the chains in cellulose I run in a parallel direction (Payen 1838). Cellulose III is obtained by ammonia treatment of cellulose I and II, and with the modification of cellulose III, cellulose IV is finally produced (Fig. 9.71). In this review, only native cellulose I is considered. This semicrystalline fibrillar structure is the main source of nanocellulose.

9.6.2 *Nano-cellulose*

According to Habibi et al. (2010), about 36 individual cellulose molecules are brought together by biomass into larger units known as elementary fibrils or microfibrils, which are packed into larger units called microfibrillated cellulose. The latter are in turn assembled into familiar cellulose fibres, which are presented schematically in Fig. 2.3.80. The diameter of elementary fibrils is about 5 nm whereas the microfibrillated cellulose (also called nanofibrillated cellulose) has diameters ranging from 20 to 50 nm. The microfibrils are formed during the biosynthesis of cellulose and are several micrometers in length. Each microfibril can be considered as a flexible hair strand with cellulose crystals linked along the microfibril axis by disordered amorphous domains (Azizi Samir et al. 2005). The ordered regions are cellulose chain packages that are stabilized by a strong and complex network of hydrogen bonds (Habibi et al. 2010) that resembles nanocrystalline rods. This is why the two main types of nanocellulose are: (i) cellulose nanocrystals and (ii) cellulose microfibrils.

9.6.3 Cellulose Nanocrystals

Concerning cellulose nanocrystals, the first reference to the existence of definite crystalline zones in the amorphous structure of cellulose materials originates with Nageli and Schwendener, in 1870. They confirmed the optical anisotropy of plants both in cell walls and in fibres. Many decades later, Rånby and Ribi (1950), were the first to produce a stable suspension of colloidal cellulose crystals via the sulfuric acid hydrolysis of wood and cotton cellulose. The nanocrystals obtained were 50–60 nm in length and had a diameter of about 5–10 nm. Since then, an exponential number of researchers have focused their work on such materials. Pioneering work in cellulose nanocrystals mainly comes from Canada (Marchessault, Gray) and France (Dufresne, Cavaillé), as shown by the date and the number of scientific papers of each author. Many terms are used to refer to cellulose nanocrystals: rod-like colloidal particles, nanocrystalline cellulose, cellulose whiskers, cellulose microcrystallites, microcrystals, microfibrils, etc. The recent review by Habibi et al. (2010) gives an overview of these nanocelluloses. In this review, the term "cellulose nanocrystals" (CNC) will be used. As described by Azizi Samir et al. (2005), cellulose nanocrystals are cellulose structures grown under controlled conditions, leading to the formation of high-purity single crystals. The main process in the preparation of CNC is based on strong acid hydrolysis under strictly controlled conditions of temperature, agitation, and time. The amorphous regions, which are considered as structural defects, are attacked under acid hydrolysis, leaving the crystalline regions, the more resistant domain, intact. The resulting suspension is washed by successive centrifugations and dialysis is performed using distilled water to remove any free acid molecules. Different sources of cellulose were studied and used for producing CNC: tunicin (Sturcova et al. 2005), valonia (Sugiyama et al. 1994), cotton (Revol et al. 1994), wood pulp (Boluk et al. 2011), microcrystalline cellulose (Bondeson et al. 2006), sugar-beet pulp (Azizi Samir et al. 2004), etc. A recent study (Julien et al. 2011) shows the influence of sources on the dimensions and structure of cellulose nanocrystals. Several researchers (Boluk et al. 2011; Beck-Candanedo et al. 2005; Dong et al. 1998) have already underlined the impact of the hydrolysis conditions on the size and stability of CNC. To strike a balance between all the parameters involved, Boluk et al. (2011) studied the optimal process conditions. They determined that with a sulfuric acid concentration of 63.5% (w/w) and a reaction time of approximately 130 min, CNC can be produced from microcrystalline cellulose with a length ranging between 200 and 400 nm, a width narrower than 10 nm, and a yield that is 30% of the initial weight. As first reported in 1959 by Marchessau et al. cellulose nanocrystals also exhibit chiral nematic liquid crystalline alignments, which are seen as a flow of birefringence between two crossed polarizing films. The precise morphological characteristics of CNC are typically studied using microscopic techniques (TEM, AFM, SEM) or light scattering techniques (SANS, polarized and depolarized dynamic light scattering, DLS, DDLS). Recently, a study showed that intrinsic viscosity measurements could also be used to calculate the aspect ratio of CNC rods

(Revol et al. 1994). Concerning mechanical properties, the Young's modulus of CNC was estimated to lie between 130 GPa (Sakurata et al. 1962) and 250 GPa (Zimmermann et al. 2004), a value that is close to the modulus of a perfect crystal of native cellulose (which has a modulus of 167.5 GPa, according to (Tashiro and Kobayashi 1991; Habibi et al. 2010). Modeling strategies have also been developed, as detailed in a very recent review (Moon et al. 2011).

9.6.4 Cellulose Nanofibrils

Cellulose nanofibrils are produced through a fibrillation of cellulose fibres. Nanofibrils have diameters of roughly less than 100 nm, lengths of several micrometers and have thus a large aspect ratio. Nanofibrils are highly crystalline, which provides extremely high mechanical properties. Having a large aspect ratio, a large specific surface area with reactive OH-groups proposes the nanofibrils as an interesting material for several applications. Since its introduction, a series of application areas have been suggested for the nanofibrillated material, e.g. for rheology adjustment and emulsion stabilizer of food, paint and cosmetics. Recently, applications as strength enhancer in paper and composite materials have been explored (Eriksen et al. 2008; Henriksson and Berglund 2007; Morseburg and Chinga-Carrasco 2009). Cellulose nanofibrils seem to be most adequate for barrier applications in novel packaging concepts (Aulin et al. 2010; Hult et al. 2010; Minelli et al. 2010). Novel applications have been foreseen within medicine, where cellulose nanofibrils may e.g. be applied as scaffolds for tissue or bone (Klemm et al. 2006).

9.6.5 Micro-cellulose

Micro-fibrillated cellulose (MFC), also called cellulose micro-fibril, micro-fibrillar cellulose, or more currently, nano-fibrillated cellulose (NFC), has been reviewed quite recently, particularly in terms of nanocomposite applications (Siqueira et al. 2010; Siro and Plackett 2010; Klemm et al. 2011). MFC is a cellulosic material, composed of expanded high-volume cellulose, moderately degraded and greatly expanded in surface area, and obtained by a homogenization process (Herrick et al. 1983; Turbak et al. 1985; Lu et al. 2008).

Herrick et al. (1983) were the first to patent a new process to produce a new kind of cellulose, which they named microfibrillated cellulose (MFC). In that process, a wood fibre suspension was passed several times through a narrow gap under high pressure, thus forming a viscous gel. This low-cost and totally new kind of cellulose already showed attractive applications such as a binder for paper or as a viscosity modifier. This patent marked the emergence of many significant research studies focusing on either optimizing such a mechanical process, or analyzing newly found properties and applications. Very recently, the industrial production of MFC has been announced by fibre companies such as Booregaard, UPM, and Innventia. Moreover, very important European projects such as SUNPAP (2009) and FLEXPAKRENEW (2007) are focusing on the scaling up of MFC productions and new possible applications. The first cellulosic material that was used to produce MFC was wood, as reported by Herrick et al. (1983) and Klemm et al. (2011). Wood pulp was disintegrated many times in a high-pressure homogenizer in order to obtain a viscous and shear thinning aqueous gel at a very low concentration (between 2 and 7%, w/w). This is one of the two main characteristics of such a nanomaterial, and the other is its ability to form a transparent film once it is dried. Both these key properties are linked to its high specific area (at least ten times larger than that of cellulose fibres) and its extensive hydrogen-bonding ability. Increasingly, various mechanical treatments have been used to manufacture MFC, depending on the type of MFC that is desired. In the 1980s, research groups used a Gaulin homogenizer with a pressure of 8000 psi (Fig. 9.72) (Herrick et al. 1983, 1984a, b).

In this approach, the wood pulp is passed through a small-diameter orifice under considerable pressure. It is exposed to a high shearing action followed by a high-velocity decelerating impact. Several passes of the suspension are necessary in order to obtain a substantially stable gel: 8–10. In order to maintain a product temperature in the range of 70–80 °C during the homogenization treatment, cooling water is used.

Other equipment was subsequently produced that offered alternative approaches, such as micro-fluidizers, in which, the wood pulp is forced through thin z-shaped chambers under pressures as high as 30,000 psi (Siqueira et al. 2010), thus enabling to produce more uniform nano-cellulose. In this approach, though, the repetition of the mechanical treatment with different sizes of chambers remains a necessary step in order to increase the degree of fibrillation (Klemm et al. 2011).

MFC actually consists of aggregates of cellulose microfibrils. Its diameter is in the range 20–60 nm and it has a length of several micrometers. To a microfibrils with 2–10 nm thick fibrous cellulose structure and a length of several tens of

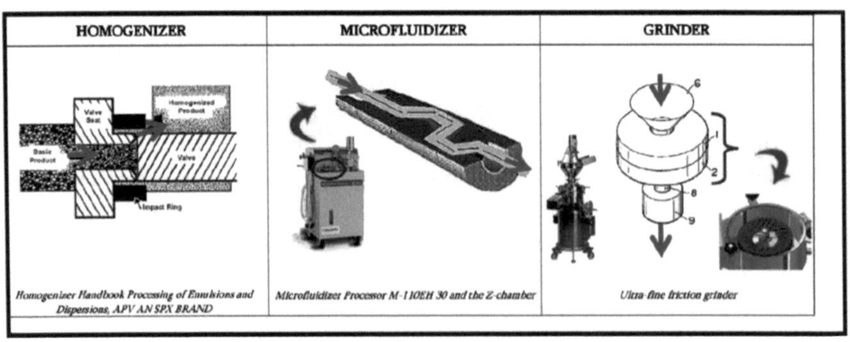

Fig. 9.72 The most applied mechanical treatment processes used in the fabrication of microfibrillated cellulose: homogenizer, microfluidizer and grinder (Lavouine et al. 2012)

microns (Klemm et al. 2006), then MFC is composed of 10–50 microfibrils. Unlike CNC, MFC exhibits both amorphous and crystalline parts and presents a web-like structure (Lu et al. 2008). Furthermore, the ratio L/d of MFC is very high, which endows it with a very low percolation threshold (Fig. 9.73). Microcrystals of cellulose microfibrils are highly hydrophilic, have high water retention capability and excellent mechanical properties: an elastic modulus of about 150 GPa, superior to glass fibres (between 70 and 76 GPa) and aramide fibres (65 GPa) modulus (Azubuike and Esiaba 2012; Azizi Samir et al. 2005). Microfibrillated cellulose thus has a very good ability to form a rigid network. Hydrolyzing the cellulose chains, MFC can be extracted from vegetable matter by means of a diluted mineral acid at its boiling temperature. The hydrolysis process removes most of the amorphous fraction and destroys the fibrillar morphology of cellulose as follows: first vegetable mater is selected and cut, then is hydrolyzed converting insoluble hydroxides, oxides and sulfates of vegetable mater in soluble compound that are removed by a filtration process. Finally the filter cake is suspended in water and subjected to a spray drying process, yielding the MFC with a size distribution and moisture content according to the conditions of the last two stages (Reier and Shangraw 1966). The MFC have been widely utilized in the food, cosmetics and medicine industries; as stabilizers for aqueous suspensions, flow controllers and as reinforcement of final product (Azubuike and Esiaba 2012; Lu et al. 2008; Majeed et al. 2013). Emerging applications for this kind of materials in cement based matrices have been investigated, as reviewed hereafter.

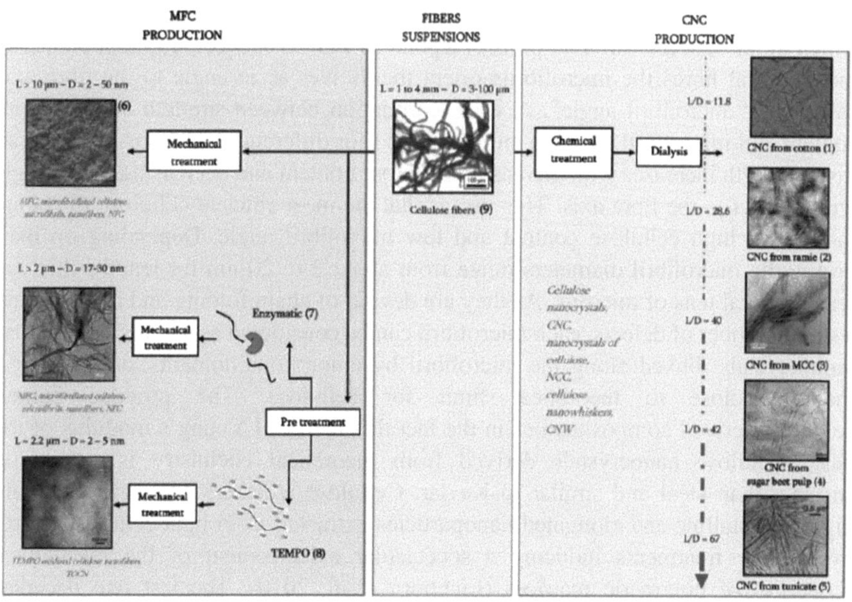

Fig. 9.73 From fibres suspensions to nanocelluloses with their various terminologies (Lavouine et al. 2012)

9.6.6 Use of Micro and Nano-cellulose in Pastes for Cementing Oil Wells

The incorporation of micro and nano cellulose fibres is gaining attention due to the benefits in mechanical behaviour of cement based materials. In literature is possible find the incorporation of microfibres in self-compacting concrete (Mohamed et al. 2010) and ultra-high performance concrete (Peters 2009), both showing an increase in mechanical behaviour. The application in cement mortars also showed good results (Claramunt et al. 2011a, b).

Nano and micro fibres was also utilized to elaboration in oil well cement.

The exploitation of cellulose fibres and/or nano-whiskers constitutes a rapidly growing issue, because it fits very well with the marked tendency of using renewable biodegradable raw materials (Belgacem and Gandini 2008a; Dufresne 2010). Unfortunately, natural organic fibres display two major limitations when used as reinforcing elements in composite materials, which are: (i) high sensitivity to water, moisture and alkaline environment; and (ii) poor compatibility with the hydrophobic matrices generally used in this field (Polymer and cement based matrix). The consequences of such features are dramatic in a composite materials context, because the moisture adsorption induces a loss of their mechanical properties and the poor compatibility yields weak interfacial adhesion and dispersibility, and consequently low performance composites. Lignocellulosic fibres consist of cellulose microfibrils embedded in a cementing matrix of other, mostly hemicelluloses and lignin, polymers. The properties of natural fibres are strongly influenced by many factors, particularly chemical composition and internal fibre structure, which differ between different parts of a plant as well as between different plants. In most natural fibres the microfibrils orient themselves at an angle to the fibre axis called the "microfibril angle". A weak correlation between strength and cellulose content and microfibril or spiral angle is found for different plant-fibres. In general, fibre strength increases with increasing cellulose content and decreasing spiral angle with respect to the fibre axis. This means that the most efficient cellulose fibres are those with high cellulose content and low microfibril angle. Depending on their origin, the microfibril diameters range from about 2 to 20 nm for lengths that can reach several tens of microns. As they are devoid of chain folding and contain only a small number of defects, each microfibril can be considered as a string of cellulose monocrystals, linked along the microfibril by amorphous domains, and having a modulus close to theoretical limit for cellulose. The promise behind cellulose-derived composites lies in the fact that the axial Young's modulus of the basic cellulose nanocrystals derived from theoretical chemistry is potentially stronger than steel and similar to Kevlar. Cellulose whiskers or nanocrystals are highly crystalline and elongated nanoparticles extracted from lignocellulosic fibres by specific treatments inducing a spectacular enhancement of the mechanical properties of polymeric matrices (Eichhorn et al. 2010). The last two decades, researchers have developed wide research activities based on the use of cellulose fibres and whiskers from different origins and morphologies, including the investigation of several modification approaches, the improvement of thermoplastic and thermosetting polymeric matrices, the preparation of a variety of nanocomposites, involving the use of cellulose and chitin-based nano-whiskers, starch-based nano-platelets. The main aims of the cellulose fibres surface modification strategies are: (i) to provide them an efficient hydrophobic barrier; (ii) to minimize their interfacial energy with the non-polar matrix, thus generating optimum adhesion and dispersion; and (iii) to graft coupling agents bearing the matrix in mind, in order to yield the best interface between the two phases or even to create fibre-matrix covalent bonds. Different modification strategies have been developed using experimental condition preserving the fibre integrity in order to take advantage from their reinforcing character (Belgacem and Gandini 2005, 2008b). The main objective of using nanoscale reinforcing elements is to achieve further improved fibres and composites by eliminating the macroscopic flaws by disintegrating the natural grown fibres, and separating the almost defect-free highly crystalline fibrils. Indeed, a big variation of properties is observed for lignocellulosic fibres, which is inherent to natural products. When changing the dimensions of the reinforcing phase, some important changes occur. The specific area is considerably increased from few $m^2\ g^{-1}$ to few 100 $m^2\ g^{-1}$. The average inter-particles distance decreases as their size decreases, allowing enhanced particle-particle interactions. Improved properties can be reached for low filler content without detrimental effect on properties such as impact resistance and plastic deformation. Finally, a reduction of gas diffusion can be observed (barrier effect). Experience from practical applications has shown that significant improvements in the mechanical properties of cementitious materials can be readily achieved by incorporation of fibres. For example fibre-reinforced concrete is widely used when specific property criteria beyond that of conventional concrete are required. Fibre reinforcement increase tensile strength and strain capacity, flexural and shear strength, ductility, toughness, and resistance to cracking induced by thermal effects, shrinkage or other causes. Furthermore, fibres act to arrest crack growth and transfer stresses across cracks. In general, the properties of the fibres, fibre length and volume fraction, interfacial bond strength, orientation of fibres and aspect ratio. In the case of geothermal cements, fibres added for reinforcement are also required to demonstrate durability and thermal compatibility in the well environment (Berndt and Philippacopoulos 2002). It is essential that well cements maintain structural performance and sealing capability throughout the lifetime of the well. Experience from practical applications has shown that significant improvements in the mechanical properties of cementitious materials can be readily achieved by incorporation of fibres. In many industrial processes, the pipeline systems are lined with a protective layer of cement mortar. In petroleum wells, cement slurry is placed in a wellbore to be hardened into an impermeable mass that seals the annulus from fluid flow and protects the casing from corrosion for the life of the well. When uniform linings of neat cement fail in tension, one or more large cracks are formed and the pressurizing fluid or mud easily flows through the cracks. The necessity to check the damaging effect of plastic shrinkage in cement mortar, and thus the formation of cracks, has called for further studies in this topic. In the past, the most common research topic has been in the areas of polymer fibres that are expensive and environmentally unacceptable. In the quest of pursuing technologies that are environmentally friendly, inexpensive, and innovative. Hair waste has been used as a new natural fibre to reinforce mortar and cement and improve their impermeability. The results show that human hair fibres are effective in reducing the plastic shrinkage cracks area of mortar by a remarkable percentage up to 92% (Al Darbi et al. 2006). Fibres can also be utilized in geothermal well cements. Their ability to improve withstand higher tensile stresses than conventional cements show a great potential value. Investigations about additions of fibre was made. Steel, stainless steel, carbon, basalt and glass fibres where used to improve mechanical behaviour in oil well cement. According to Belgacem and Gandini (2005), fibre reinforcement can be used to increase the tensile strength of standard, lightweight and latex-modified geothermal well cements. The magnitude of increase depends on the type of fibres used and the fibre volume fraction.

9.6.7 Dispersion of Nano-and Micro-cellulose Fibres in Cement Pastes

The capability of processing composite materials is crucial for gaining acceptance in various applications. Strain, magnitude of tensile and others behaviour capacities are closely associated with fibre dispersion uniformity. Notably, most fibre reinforced composites utilize short, randomly distributed fibres. During mixing, fibres are added after the other ingredients, i.e. cementitious and pozzolanic ingredients, sand, water, and admixtures, are mixed and have achieved a consistent mortar state. The processing details such as mixer type, mixing speed, time and sequence, and mixing personnel's experience level can influence the rheological properties of mortar, which may strongly affect fibre dispersion uniformity, size distribution of entrapped air pores in the cementitious matrix, and bonding properties at the fibre/matrix interface in the hardened state. Published literature on the correlation between cementitious composite rheological properties and fibre dispersion, has typically focused on rigid steel fibres and PVA fibres. It has been shown that increasing fiber volume fraction and aspect ratio decreases the workability of cementitious composite material (Hughes and Fattuhi 1976; Rossi 1992; Grunewald 2004; Dhonde et al. 2007; Martinie et al. 2010; Banfill et al. 2006; Markovic 2006; Bui et al. 2003; Kuder et al. 2007; Ozyurt et al. 2007; Boulekbache et al. 2010; Ferrara et al. 2007, 2008; Chung 2005; Li and Li 2013).

Understanding the rheology of cementitious suspensions is important for the design, execution and evaluation of a primary cementing job (Rahimirad 2012). About MFC, the dispersion follow these steps: according (Hoyos et al. 2013), the cellulose microcrystalline particles provided by Sigma Aldrich were saturated for 2 days in water. The following methodology was applied to estimate the amount of water absorbed by MFC and replace it to elaborate the fresh cement mix:

a proportion of 0.5 g of MFC were mixed with 3 ml of water and left standing for 3 days at 25 °C, to promote saturation. After 3 days the saturated solution, was separated into two phases by centrifugation at 3000 rpm during 25 min. Following several tests, it was found that 0.5 g of MCC absorbed about 1.05 ml of water, which corresponds to 100% of its own volume and 230% of its own mass. The MFC bulk density was calculated by a volumetric method. It occurs that 0.5 g of MFC occupied a volume of 1.09 ml, therefore MFC have a bulk density of 0.459 g/ml. Another dispersion process is to make hydro-pulper. Cellulose fibres are mixed in a blender with potable water and processed for a time to promote best pulp dispersion. According (Carvalho et al. 2008) the time to promote a best pulp dispersion is around 30 min and the concentration of pulp is 6 g of fibre per litre of water. The excess is drained off through a sieve and a fine strainer (causing the pulp to re-agglomerate). Lastly, the pulp, with a moisture content of about 75%, was again mixed for 5 min in a low speed laboratory electric mixer and stored in plastic bags in a refrigerator. Similar process was used by Savastano et al. (2003). The fibres were pre-dispersed in water, to form a slurry of approximately 20% solids by mass. Using short fibres is possible to have higher number of fibres per volume or weight in relation to long fibres and, therefore, to reduce the fibre-free areas, i.e. the distance between the fibres. Additionally, the smaller the fibre length (which generally relates to lower aspect ratio), the easier is the fibre dispersion Chung (2005). It is noticeable that the length of the fibres can modify the dispersion process. According to Jarabo et al. (2012),the number the fibres per gram on hemp pulps are higher than that on pine pulp. This could improve their dispersion in the matrix.

9.6.8 Effect of Cellulose Nanopulp and Other Micro- and Macro-cellulose Based Fibres on Autogenous and Drying Shrinkage of Cement Based Composites

Ferrara et al. (2015a) investigated the effect of cellulose nano-pulps, and, comparatively, of other different scale natural fibre reinforcement on the autogenous and drying shrinkage of cement pastes formulated from typical HPFRCC composition. Together with nano-pulp, micro fibre of eucalyptus and natural sisal fibres were investigated, besides a "conventional" high carbon steel fibre reinforcement. The main characteristics of the employed fibres are summarized in Table 9.6.

Table 9.6 Characteristics of natural fibres employed by Ferrara et al. (2015a)

Fibre type	Density (kg/dm^3)	Cellulose (%)	Hemicellulose (%)	Lignin (%)
Nanopulp	1.4	96.5	–	–
Eucalyptus	1.68	77.1	18.9	3.8
Sisal	1.04	60.5	25.7	12.1

The cellulose nano-pulp and the eucalyptus micro fibres used in this study were manufactured by SUZANO-Brazil. Accordingly, the nano-pulp fibres have a nominal diameter between 20 and 50 nm and a density of 1.4 kg/dm^3. The concentration of solids is 1.5% per liter, (15 g of nano fibres per litre of suspension—Fig. 9.74). Regardless of the source, in an aqueous environment, the rheological properties of the nano-pulp suspension can be described in terms of pseudo-plasticity and shear thinning behaviour.

The eucalyptus micro fibres were extracted by kraft pulping method and bleached treatment. The length of the fibres, according to the company, was 0.8 mm. In Fig. 9.75b, the cross section of the eucalyptus fibre can be observed. The eucalyptus fibre has tubular shape with a hollow internal cavity, i.e. the lumen, a typical feature of plant fibres, oval cross section and thick wall. The same sisal fibres as for the self-healing investigation were employed (Fig. 9.75a).

High carbon straight brass coated steel fibres, 13 mm long and 0.16 mm in diameter, and with a tensile strength equal to 2000 MPa were employed.

The cellulose nano-pulps and different types of natural and steel fibres were employed to cast fibre reinforced cement mortars, formulated from the composition of a typical HPFRCC (Table 9.7). The reference mortar consisted Portland Cement

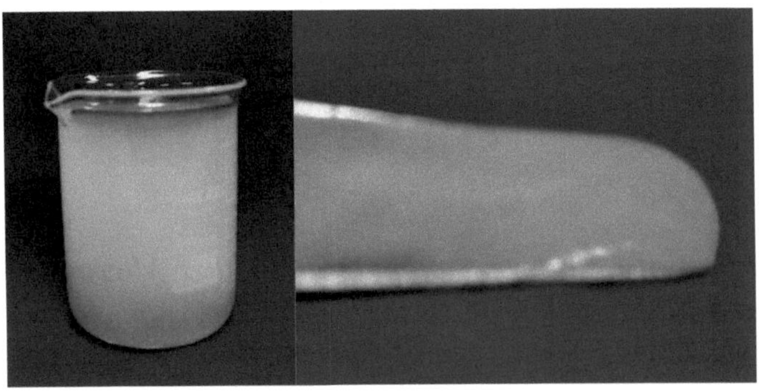

Fig. 9.74 Nano cellulose pulp suspensions: concentration of 1.5% (Ferrara et al. 2015a)

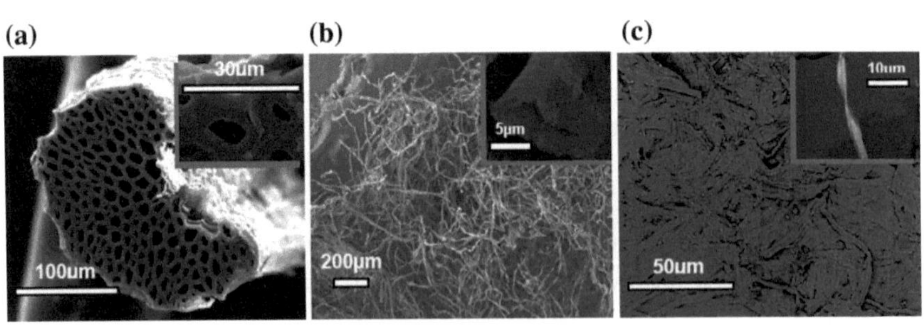

Fig. 9.75 SEM image of sisal fibre (**a**); eucalyptus micro fibres (**b**); dried nano cellulose pulp (**c**) (Ferrara et al. 2015a)

Table 9.7 Mix composition of cement composites investigated by Ferrara et al. (2015a)

Constituent	Dosage (g/dm^3)				
Mix type	REF	Nanopulp	Eucalyptus	Sisal	Steel
Cement type I 52.5	600	600	600	600	600
Slag	500	500	500	500	500
Sand 0–2 mm	983	983	983	983	983
Water	200	187	200	200	200
Superplasticizer	33	33	33	33	33
Nanopulp	–	14	–	–	–
Eucalyptus	–	–	0.8	–	–
Sisal	–	–	–	14	–
Steel	–	–	–	–	100

52.5 typeR, blast furnace slag, water and super-plasticiser (Glenium ACE 300®) with a solid content of 31%. The mixtures were produced in a room with controlled temperature (21 ± 1 °C) using a 2 l mixer and following the mixing protocol detailed in Ferrara et al. (2011). Starting from a reference steel fibre dosage equal to 100 kg/m^3, dosage of sisal and eucalyptus micro-fibres was calculated for the same fibre volume fraction (Vf = 1.28%) whereas for cellulose nano-pulps a dosage of 0.2% by volume was employed, also to guarantee target workability requirements.

For sisal fibres, the effect of a pre-saturation was also checked, soaking them in water for one hour and then wiping on absorbent paper before adding into the mix.

Six 160 × 40 × 40 mm specimens were cast with each for drying and autogenous shrinkage tests. The environment where storing and testing was done was held at a temperature of 20 ± 1 °C and relative humidity of 50 ± 5%. During production all samples were embedded with 25 mm steel pins in the middle of the two 40 × 40 mm faces that protruded the faces by approximately 5 mm. The specimens were demoulded 24 h after casting and, for autogenous shrinkage tests they were sealed by four layers of plastic film and a layer of aluminium tape. During the first four days, measures were taken twice per hour, and then the measuring interval was progressively relaxed up to two measures per days taken from seven days onward. The mass change of the samples was also daily measured in a balance scale of 0.01 g accuracy. Thus it was possible to verify the efficacy in sealing per-formed and if there was any portion of the retraction as referring to shrinkage, assuming that the perfect seal occurs when the mass loss of the specimen with respect to their water mass reaches the maximum value of 0.5%.

In Fig. 9.76, the results of the measured total and autogenous shrinkage strains are shown. All types of natural fibres were able to affect the total and autogenous shrinkage strain development. As a matter of fact a reduction in the autogenous shrinkage was generally observed, because of the extra amount of internal (curing) water that natural fibres could absorb because of their hydrophilic nature, accompanied by a complementary increase in the total shrinkage, because of the additional porous path for moisture escape that the porous structure of natural fibres.

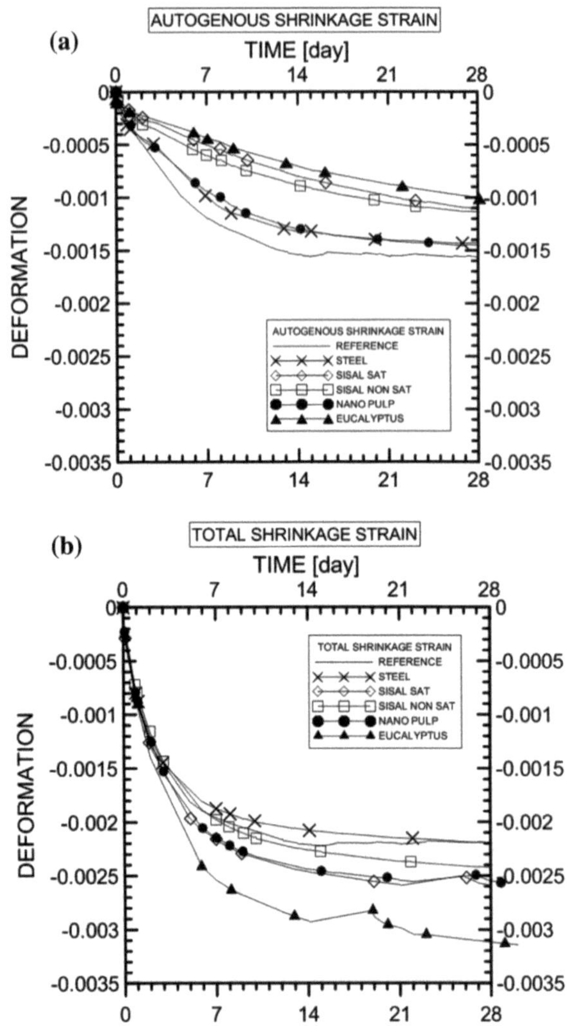

Fig. 9.76 Autogenous and total shrinkage strain for different investigated HPFRCC mixes (Ferrara et al. 2015a)

The cellulose nano-pulp, in the employed quantity, is likely to provide a slight reduction to the autogenous shrinkage, which could be beneficial when used in synergy with other kinds of natural and artificial fibre reinforcement in HPFRCCs.

9.7 Conclusions

In this section the most significant results of the research performed on cementitious composites reinforced with natural fibres have been presented, in the framework of the current state of the art knowledge on the topic.

Fundamental issues related to the processing of the fibres for the production of Natural Fibre Reinforced Cementitious Composites (NFRCCs) have been first of all reviewed, together with the fundamentals of the mechanical behaviour of the same individual fibres and of the fibre reinforced composites, also with reference to the basic mechanisms of cementitious matrix reinforcement provided by natural fibres. The durability of NFRCCs has also been addressed, with main focus on the stability of the long-term performance of cellulosic fibres in the highly alkaline cementitious matrix environment, together with current approaches adopted to improve it, including the so-called hornification process.

The one-of-kind signature of natural fibres as facilitators of crack sealing and material healing processes in fibre reinforced cementitious composites has been deeply characterized, also through quantitative comparison of analogous functionalities in cementitious composites with similar matrix composition and reinforced with industrial (steel) fibres only. It has been in fact clearly demonstrated that natural fibres, thanks to their porous hierarchical structure, can effectively vehiculate water, e.g. absorbed at a cracked site, throughout the surrounding matrix. This is effective on the one hand at promoting the delayed hydration and/or carbonation reaction, whose products seal the cracks and reconstruct through crack matrix continuity, thus contribution to the healing of the material mechanical properties. On the other hand, healing reaction products also deposit at the fibre-matrix interface improving the fibre-matrix bond and thus further contributing to the recovery of the mechanical performance of the composite.

Finally, emerging technologies in the use of advanced cellulosic constituents, such as micro- and nano- cellulose crystals and fibrils have been reviewed, also through dedicated experience on their effectiveness in mitigating autogenous shrinkage in high performance cement based matrices to be employed in cutting edge applications, such as oil well cementing.

Appendix: Natural Fibres in Cementitious Composites

Jute Fibres

Plant: obtained from the stem of *Corchorus capsularis* (Fig. 9.77)
Extraction process: combination of processes which comprise:

Cutting of stems with knives
Retting, i.e. a process which employs the action of microorganisms and moisture to dissolve or rot away much of the cellulose tissues and pectins surrounding fibre bundles and so facilitating the separation of the fibres from the stem. Water retting is generally employed, in stagnant or slowly moving water or in tanks; dew wetting is performed where climate allows
Shredding

Fig. 9.77 Plant of *Corchorus capsularis* (white jute). *Source* Wikipedia

Fig. 9.78 Jute fibres drying after water retting. *Source* Wikipedia

Drying (Fig. 9.78)
Sorting, packing and classification

Fibre yield: depends on sowing time and related plant growth; published data report between about 90 and 150 g/10 plant (Saha et al. 2015)
Chemical composition: (*as from* jute.org/composition.htm)

Cellulose 60–63%
Hemicellulose 21–24%
Lignin 12–13%
Wax and fat 0.4–1.0%
Pectin 0.2–1.5%
Proteins and nitrogenous substances 0.8–1.9%
Ashes 0.7–1.2%

Geometrical and morphological properties:

diameter of individual fibres in the order of fifty microns (±20%) has been reported (Defoirdt et al. 2010) and cross section areas in the range of 0.003–0.003 mm^2. Morphological microstructure, as in Fig. 9.79, shows each fibre consists of a few tens of microcells (20–30) with cell wall thickness of about 2.5 μm and lumen diameter of about 7 μm (Alves Fidelis et al. 2013).

Mechanical properties

Density (*unless specified differently data from* jute.org/composition.htm):

true density 1.46 Mg/m^3
apparent density: 1.1–1.34 Mg/m^3—(Defoirdt et al. 1.39 Mg/m^3)
bulk density: 0.4–0.5 Mg/m^3

Tensile strength:

Defoirdt et al. (2010): from 249 ± 89 N/mm^2 to 314 ± 13 N/mm^2
gauge lengths increasing from 10 *to* 60 mm; *strain rate equal to* 0.2 mm/min
Alves Fidelis et al. (2013): from 307 ± 84 N/mm^2 to 399 ± 100 N/mm^2
Gauge lengths increasing from 5 *to* 35 mm; *strain rate equal to* 0.1 mm/min
Lower gauge lengths generally correspond to higher strength values.
IIT Kharagpur (2011): from 297 to 610 N/mm^2
Depending on type (with alkali or latex at different concentrations) and duration of surface treatment performed to stabilize long term resistance when embedded in cement paste. No information and gauge length and strain rate.

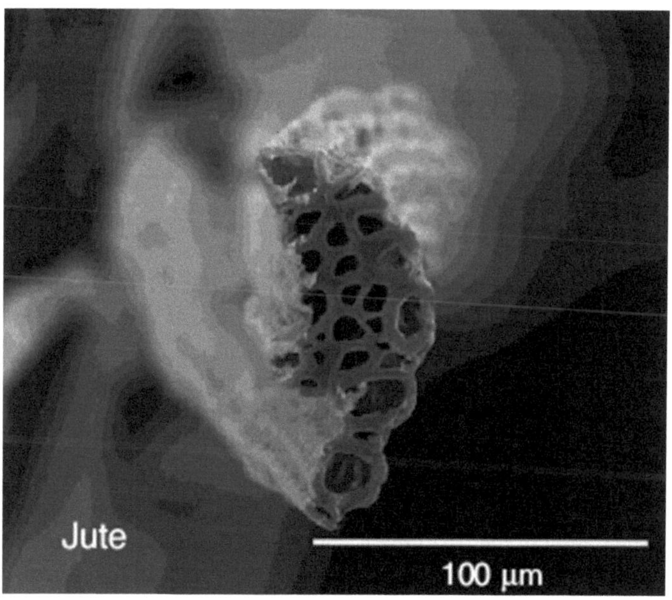

Fig. 9.79 Micrograph of a jute fibre from SEM analysis (Alves Fidelis et al. 2013)

Young modulus:

Defoirdt et al. (2010): 26.25 GPa
Alves Fidelis et al. (2013): from 35.2 ± 15.1 GPa to 48.4 ± 13.3 GPa
Lower gauge lengths generally correspond to lower stiffness values.

Strain to failure:

Defoirdt et al. (2010): from 1.4 ± 0.2 to 1.6 ± 0.2%
Alves Fidelis et al. (2013): from 0.6 ± 0.2 to 0.9 ± 0.3%
Lower gauge lengths generally correspond to slightly higher strain values.
IIT Kharagpur (2011): from 1.34 to 1.7% (*surface treatments as above*)

Applications as reinforcement in cement based materials:

Mansur and Aziz (1982a, b) studied jute fibre reinforced cementitious composites, incorporating different volume fractions of fibres with different lengths. Results of specimens tested in direct tension, flexure, axial compression and impact showed, for the first time, to the authors' knowledge, the feasibility of using jute fibres in developing a low-cost construction material particularly for roofing, wall panels and boards in countries where jute fibres are readily available.

IIT Kharagpur (2011) investigated compressive and flexural strength of concrete reinforced with up to 4% by vol. chopped jute fibres. Applications related to pre-cast non-pressure sewerage pipes, prestressed electric poles, pavers blocks and roofing sheets were proposed.

Sen and Jagannatha Reddy (2013) used jute textile reinforced cementitious composite, as alternative to CFRP and GFRP, to retrofit reinforced concrete beams.

Ozawa et al. (2013) used 0.19% by vol. jute fibres (12 mm long and 10–30 μm in diameter) to prevent explosive spalling in High Performance Concrete subjected to high temperatures; observed performance was better than with similar volume fractions of polypropylene and water soluble poly-vinyl-alcohol fibres.

Hemp Fibres

Plant: obtained from the stem of *Cannabis sativa* (Fig. 9.80). The hemp stem is composed by a woody hollow core surrounded by a soft skin containing long and short fibres (Fig. 9.81).

Extraction process: for high quality textile-production plants are grown up to 4 m, stalks are maintained in bundles during collection; and, finally maceration of plants in water and extraction of fibres occur. In case of low quality production, plants are harvested in traditional or round bales and then fibres are extracted via mechanical and physical-chemical treatments without soaking plants in water. After water or dew retting the separation of the bast fibre is carried out through scratching (breaking the woody core of the stems into short pieces) and decortication (the separation of bast fibre from the hurds) using specialised machineries (Ingrao et al. 2014).

Fig. 9.80 Plant of *Cannabis sativa*. *Source* Wikipedia

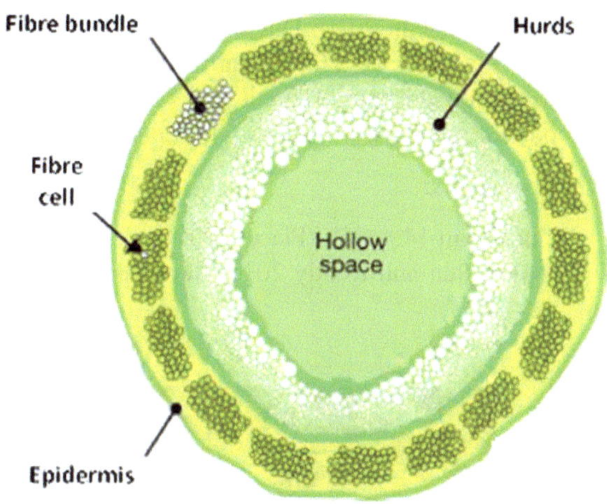

Fig. 9.81 Hemp stem cross section (from Ingrao et al. 2014)

Fibre yield: the amount of fibres contained in a stem is about 30% of its weight, approximately 2/3 of which consist of long fibres, generally employed for weaving and high quality textile products, and the remaining 1/3 of short fibres, employed for other uses. The formers, generally from 20 to 50 mm long, are also called primary fibres, whereas the latters, about 2 mm long, are called also secondary fibres and are most likely to be found at the bottom of the stems.

Chemical composition: (*as from* Marrot et al. 2013)

Cellulose 80.2%
Hemicellulose 12%
Lignin 2.6%
Pectin 3.0%
Ashes 0.5%
Water soluble substances 1.7%

Geometrical and morphological properties:

Hemp fibres have a characteristic polygonal shape (Fig. 9.82). As from morphological measurements, average fibre area is 179.5 ± 63.8 μm^2, to which an average diameter 18.35 ± 4.56 μm corresponds, distribution of fibre diameters in a single stem following a bimodal distribution with two central values for the equivalent diameters around 15 and 19 μm, roughly corresponding to mean diameters of external and internal fibres respectively.

Mechanical properties

Density: 1.4–1.5 Mg/m^3 (Li et al. 2006; Awwad et al. 2012)

Tensile strength:

A wide variety of results has been reported, from as low as 270 ± 40 N/mm^2 (Eichhorn and Young 2004) to as high as 1735 N/mm^2 (Thygesen et al. 2006). Most of the investigations reported values between 400 and 800 N/mm^2 (Saheb and Jog 1999; Eichhorn and Young 2004; Li et al. 2006; Thygesen et al. 2006; Beckermann and Pickering 2008; Sedan et al. 2008; Placet 2009; Bourmaud and Baley 2009; Fan 2010; Duval et al. 2011; Sen and Reddy 2011; Awwad et al. 2012; Placet et al. 2012; Marrot et al. 2013; Berzins et al. 2014). Limited information on gauge length and on strain rate is available; the latter, when reported, ranged from 0.1 to 1 mm/min.

Young modulus:

Values from 14.4 GPa (Placet 2009) and 19.6 ± 14.8 GPa (Duval et al. 2011) to 70 GPa (Sun and Reddy 2011) have been reported.

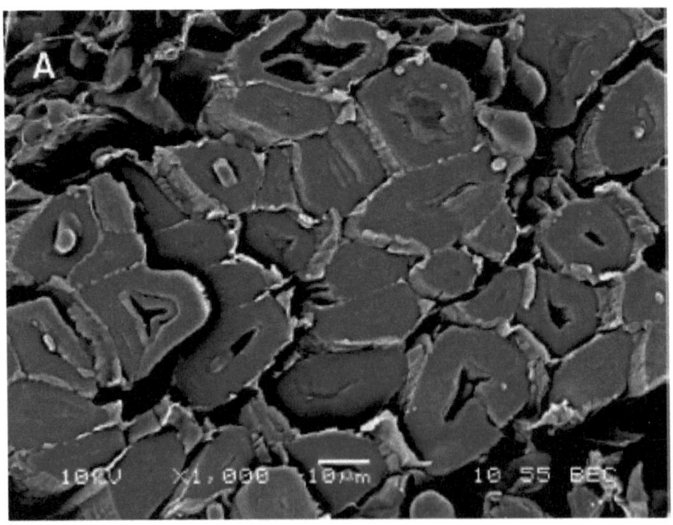

Fig. 9.82 Micrograph of hemp fibres from stem thin sections (from Marrot et al. 2013)

Strain to failure:

0.8 ± 0.1% (Eichhorn and Young 2004); 3.0 ± 1.5% (Duval et al. 2011).

Other mechanical properties

Berzins et al. (2014) tested longitudinal and transverse compressive strength, ranging from 11.7 to 26.4 and from 0.55 to 1.55 N/mm^2, respectively.

Applications as reinforcement in cement based materials:

Hemp is used to produce rope, cloth, food, lighting oil and medicine. Currently hemp fibres are used to manufacture bank notes and were valued hugely before the development of plastic fibres from petrochemicals (Sen and Reddy 2011). Hemp fibres have been mixed in lime-based mortars, for plastering, flooring and non-load bearing wall applications (Murphy et al. 2010); chemical treatment of fibres, e.g. in alkaline solution (NaOH or $ALCl_3$ based) is required to enhance the matrix/fibre interface (Sedan et al. 2008; Troëdec et al. 2011). Composites with hemp fibres and hurds bonded by a binder based on MgO and a reactive vegetable protein have been produced as insulating mats (Sassoni et al. 2014).

Li et al. (2006), investigated compressive and flexural strength of hemp-FRC with fibre lengths ranging from 10 to 30 mm, dosages from 0.18 to 1.06% by vol.

Awwad et al. (2012) tested fresh and hardened state properties (compression and splitting tensile strength, modulus of elasticity, flexural performance) of hemp- FRC with 30 mm long fibres in volume percentages up to 1%. Thermal properties were also studied, which were greatly improved by hemp fibre addition.

Awwad et al. (2013) proposed to use hemp FRC (fibres up to 2% by vol.) for concrete masonry blocks; hemp fibres negatively affect compressive strength.

Awwad et al. (2014) tested beams, cast with previously investigated hemp FRC mixes, finding similar performance as for conventional r/c beams; lower modulus of elasticity of hemp-FRC results in more deformable load-deflection behaviour.

Flax Fibres

Plant: obtained from the stem of *Linum usitatissimum* (Fig. 9.83).

Extraction process: the mature plant is cut with mowing equipment, similar to hay harvesting, and raked into windrows, or harvested manually, pulled up with the roots, so as to increase the fibre length. When dried sufficiently, seeds are separated through a threshing process similar to that used for grains.

The usable flax fibres are then separated from other components, which is done pulling the stems through a hackle and/or beating the plants to break them. Flax processing is divided into two parts: the first part is generally done by the farmer, to bring the flax fibre into a fit state for general or common purposes. This can be performed by three machines: one for threshing out the seed, one for breaking and separating the straw (stem) from the fibre, and one for further separating the broken straw and matter from the fibre. The second part of the process brings the flax into a

Fig. 9.83 Plants of *Linum usitatissimum*. *Source* Wikipedia

state for the very finest purposes, such as lace, cambric, damask, and very fine linen. This second part is performed by a refining machine.

Before the flax fibres can be fit for their purpose (spun into linen or any other use), they must be separated from the rest of the stalk. The first step in this process is retting, which is the process of rotting away the inner stalk, leaving the outer parts intact. At this point, straw, or coarse outer stem (cortex and epidermis), is still remaining. To remove this, the flax is "broken", the straw is broken up into small, short bits, while the actual fibre is left unharmed. Scutching scrapes the outer straw from the fibre. The stems are then pulled through "hackles", which act like combs to remove the straw and some shorter fibres out of the long fibre.

This process is also called, in textile industry, "dressing the flax" and consists of:

- breaking breaks up the straw into short segments;
- scutching removes some of the straw from the fibre;
- heckling is pulling the fibre through various sizes of heckling combs or heckles. A heckle is a bed of "nails"—sharp, long-tapered, tempered, polished steel pins driven into wooden blocks at regular spacing.

Fibre yield (https://www.richters.com/show.cgi?page=InfoSheets/d2701.html)

Typical dry straw yields are 1600–2800 kg per acre. About 15–20% of that, or 240–560 kg/acre, is extractable fibre (1acre ≅ 0.4 ha; info accessed on June 22, 2016).

Chemical composition: (*values from various sources reported by* Baley 2002).

Cellulose 60–74%
Hemicellulose 11–17%
Lignin 2–2.9%
Wax and fat 1.5%
Pectin 1.8%
Water soluble substances 4.9%
Water 8–10%

Geometrical and morphological properties:

The structure of the flax stem, consisting of a woody hollow core surrounded by a soft fibrous matter skin, is shown in Fig. 9.84. The flax stem is made up of fibres of varying diameters and lengths. The lengths of the elementary fibre are in the range of 20–50 mm and have a polyhedral cross-section, with diameters ranging from 10 to 20 μm. The grouping of the elementary fibres (single plant cells) make up the technical fibre, which is of a size that can easily be incorporated into concrete (Baley 2002; Boghossian and Wegner 2008; Yaremko 2012). Single fibres show a concentric layer structure (Fig. 9.85).

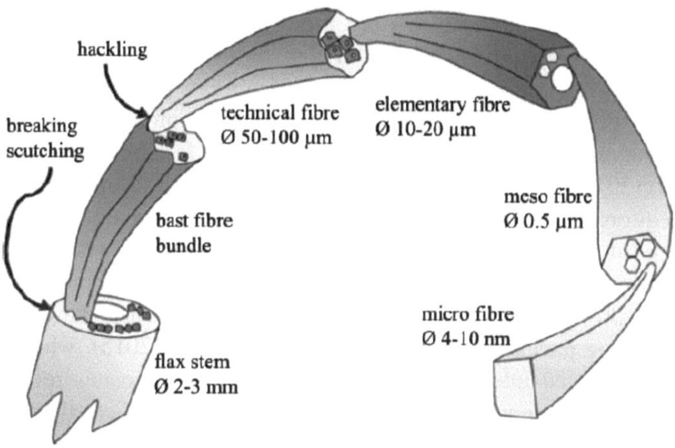

Fig. 9.84 Structure of flax stem (Baley 2002)

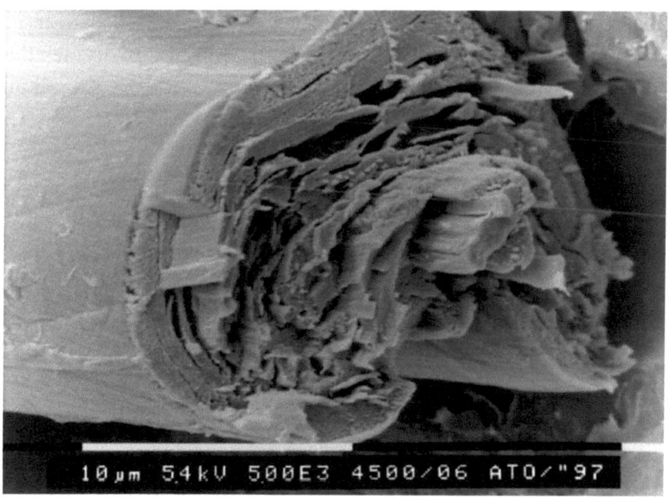

Fig. 9.85 SEM image of a flax fibre. Scale bar = 10 μm (from Bos et al. 2002)

Mechanical properties (*values from various sources reported by* Baley 2002).

Density (*from* Baley 2002): true density 1.54 Mg/m^3
apparent density: 1.38 Mg/m^3

Tensile strength: from 600 to 2000 N/mm^2
Young modulus: from 12 to 85 GPa
Strain to failure: from 1 to 4%

Bos et al. (2002) measured, through a methodology called "*elastic loop test*" the compressive strength of flax fibres, which was in the range of 1200 ± 370 N/mm^2.

Applications as reinforcement in cement based materials:

Coutts (1983) employed short (a few mm long) flax fibres in mass percentages from 2 to 12% in cement-based mortars, highlighting positive effects on MOR.

Boghossian and Wegner (2008) used flax fibres from 10 to 38 mm long in volume fractions ranging from 0.05 to 0.30% to mitigate plastic shrinkage cracking in concrete. Performance similar to that of monofilament and fibrillated Polypropylene and Glass (except for medium volume fractions) fibres was found.

Yaremko (2012) investigated durability under wet/dry and freeze/thaw cycles of concrete reinforced with 0.3% by volume 38 mm long flax fibres, finding stable performance of compressive and flexural strength. Fibres were pre-treated in autoclave or in NaOH and ethanol-water-silane solutions for impurity removal.

Use of 10–20 mm long flax fibres in lime and lime-cement based mortars for thermal insulating panels has been reported by Fic et al. (2015), whereas Assaedi et al. (2014) reported interesting use of geopolymer composites reinforced with different layers of woven flax fibres (from 2.4 to 4.1%), fabricated using a lay-up technique. The use for structural retrofitting can be foreseen, also in analogy with flax fibre reinforced polymer composites, e.g. in concrete filled tubes (Xia et al. 2016).

Snoeck and de Belie (2012) investigated mechanical and self-healing properties of cementitious composites reinforced with flax and cottonised flax, in comparison with polyvinyl alcohol fibres (Fig. 9.86). Cottonisation of flax divides the technical fibre into bundles of elementary fibres and partially removes the alkali-sensitive

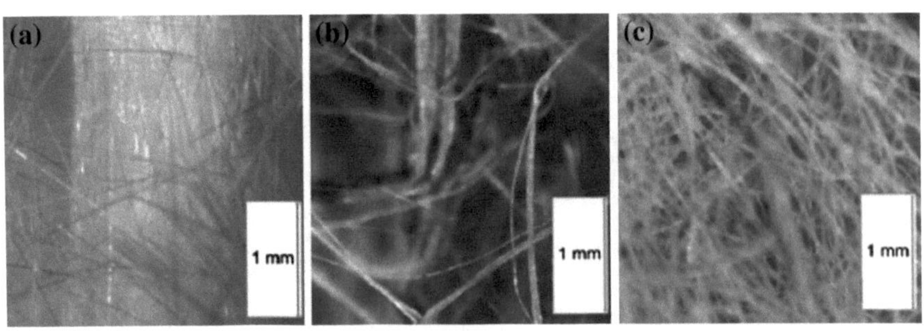

Fig. 9.86 Typical view of PVA-fibres (**a**), technical flax fibres (**b**) and cottonised flax fibres (**c**). The scale bars are 1 mm in height (from Snoeck and de Belie 2012)

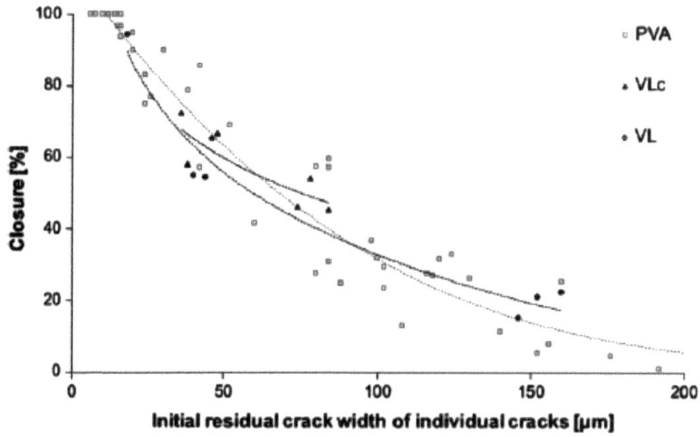

Fig. 9.87 Visual measurement of the crack healing for individual cracks after wet/dry cycles, independent of the volume percent of fibres added (Snoeck and de Belie 2012)

pectin and hemicellulose. Cottonisation of flax enhanced the modulus of elasticity, the peak stress and the strength at first crack formation, in comparison to technical flax fibres. The healing of cracks was independent of the fibre type (Fig. 9.87).

Kenaf Fibres

Plant: obtained from the stem of *Hibiscus cannabinus* (Fig. 9.88).
Extraction process
Similar to the one reported for other bast fibres
Fibre yield:
No precise information was found. As a matter of fact the yield and composition of these plant components can be affected by many factors, including cultivar, planting date, photosensitivity, length of growing season, plant populations, and plant maturity (Akil et al. 2011).
Chemical composition: (*values reported by* Akil et al. 2011).

Cellulose 45–57%
Hemicellulose 21.5%
Lignin 8–13%
Pectin 3–5%

Geometrical and morphological properties:
The stem is straight and unbranched and is composed of an outer layer (bark) and a core. It is easy to separate the stem into bark and core, either by chemicals and/or by enzymatic retting. The bark constitutes 30–40% of the stem dry weight and shows a rather dense structure. On the other hand, the core is wood-like and

Fig. 9.88 Plants of *Hibiscus cannabinus* (from Akil et al. 2011)

makes up the remaining 60–70% of the stem. The core reveals an isotropic and almost amorphous pattern. However, the bark shows an orientated high crystalline fibre pattern. Scanning electron micrographs (SEM's) indicate that the core of the kenaf plant is made up of a complex porous structure with a primary pore structure and minor pores which form an interconnected pore structure. This is reflected in reported strongly different values of core and bark fibres density.

Figure 9.89a, b show SEM's of the cross-section of an untreated kenaf fibre at magnifications of 1000× and 500× respectively (from Elsaid et al. 2011). The fibre cross-section is oval in shape rather than being approximately circular as initially assumed. The ovalized shape of the fibre cross-section explains the helical spiral of the fibre which can be clearly seen in Fig. 9.90. Essentially no porosity along the fibre cross-section exist at this scale.

Mechanical properties

Density (Elsaid et al. 2011: bark fibres 1.29 Mg/m^3 (bulk density 1.0 Mg/m^3) core fibres 0.09–0.11 Mg/m^3

Tensile strength: from 157 to 600 N/mm^2 (Elsaid et al. 2011) 930 N/mm^2 (Akil et al. 2011)

Young modulus: from 12.8 to 34.2 GPa (Elsaid et al. 2011) 53 Goa (Akil et al. 2011)

Strain to failure: from 1.5 to 1.9% (Akil et al. 2011; Elsaid et al. 2011)

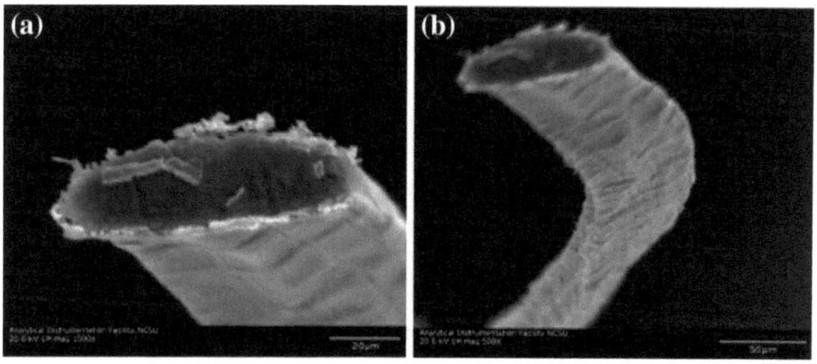

Fig. 9.89 Scanning electron micrograph of kenaf fibre cross-section **a** 1000× magnification **b** 500× magnification (from Elsaid et al. 2011)

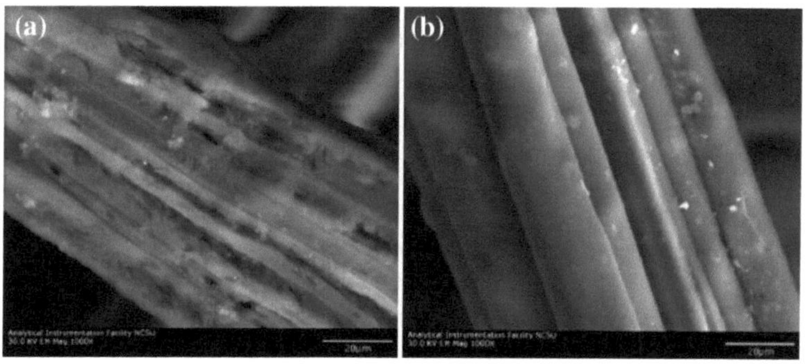

Fig. 9.90 Scanning electron micrograph of surface of **a** untreated and **b** chemically pretreated kenaf fibres. Scale bar at 20 mm (from Elsaid et al. 2011)

The mechanical properties of the fibres depend on the location along the stalk from which the fibres are taken. Fibres from the middle of the stalk tend to be stronger and stiffer while fibres taken from the ends of the stalk tend to be weaker and more flexible possibly due to exposure to environmental conditions, decay, and damage caused by insects and other pests.

Applications as reinforcement in cement based materials:

A review of the main challenges for kenaf fibre as reinforcement in polymer composites, contributing towards the development of eco-friendly assets for the automotive, sports industries, food packaging and furniture industries has been provided by Akil et al. (2011). Motivation has also come from the introduction of kenaf crops as potential replacement for the diminishing tobacco farming industry in the south-eastern United States (Elsaid et al. 2011).

As for the use of kenaf fibres in concrete and cement based composites only very few studies are available to date. Elsaid et al. (2011) characterize the mechanical properties of KFRC including the compressive strength, compressive elastic modulus, splitting tension strength, modulus of rupture, and toughness of Kenaf

Fibre Reinforced Concrete, employing fibres from 25 to 38 mm long and in volume percentages up to 2.4% (24 kg/m^3). Fibres were chemically pretreated to enhance the bond with the inorganic concrete matrix, with a propriety, urethane based surface coating was prepared in a dilute aqueous solution. The fibres were thoroughly coated by immersion in the solution and subsequently left to air dry to allow cross-linking of the polymer on the fibre surface. Images of untreated and pretreated fibres (Fig. 9.90a, b) show that the surface of the fibre is not smooth but rather has longitudinal ridges which could possibly enhance the mechanical interlock with the cement matrix. Measured mechanical performance confirmed the possibility of using KFRC, in the investigated compositions, for applications such as slabs on grade.

Awang et al. (2015) investigated the use of kenaf fibres, in comparison with polypropylene ones, in the production of Lightweight Foamed Concrete, fibres serving to mitigate the early shrinkage cracking potential. Bond issues were called to justify the somewhat lower performance of, in this case, untreated kenaf fibres as compared to that of polypropylene fibre reinforced concrete.

Ramie Fibres

Plant: obtained from the stem of *Boehmeria nivea.*
Extraction process: The extraction of the fibre occurs in three stages. First the cortex or bark is removed; this can be done by hand or by machine (decortication). Second the cortex is scraped to remove most of the outer bark, the parenchyma in the bast layer and some of the gums and pectins. Finally the residual cortex material is washed, dried, and de-gummed to extract the spinnable fibre. Spinning the fibre is made difficult by its brittle quality and low elasticity; and weaving is complicated by the hairy surface of the yarn, resulting from lack of cohesion between the fibres. The greater utilization of ramie depends upon the development of improved processing methods.
Fibre yield: The dry weight of harvested stem from crops ranges from 3.4 to 4.5 t/ha/year, yielding 1600 kg/ha/year of dry non-de-gummed fibre. The weight loss during de-gumming can be up to 25%.
Chemical composition (in Faruk and Sain 2015):

Cellulose 68.6–76.2%
Hemicellulose 13.1–16.7%
Lignin 0.6–0.7%
Pectin 1.9%
Wax 0.3%

Mechanical properties (in Faruk and Sain 2015):

Density: 1.5 Mg/m^3
Tensile strength: from 333 to 1250 N/mm^2
Young modulus: from 15.5 to 65 GPa
Strain to failure: from 1.6 to 3.5%

Applications: Ramie is used to make such products as industrial sewing thread, packing materials, fishing nets, and filter cloths. It is also made into fabrics for household furnishings (upholstery, canvas) and clothing, frequently in blends with other textile fibres (for instance when used in admixture with wool, shrinkage is reported to be greatly reduced when compared with pure wool.) Shorter fibres and waste are used in paper manufacture. Ramie ribbon is used in fine bookbinding as a substitute for traditional linen tape.

Use of ramie fibres in ecological bioplastics in automotive industry and as reinforcement in ramie fibre-reinforced polymer composites has been reported, targeted to applications as bullet-proof armors, socket prosthesis and laminates for civil engineering applications. No information is available on their use as reinforcement in cement based materials.

Sisal Fibres

Plant: obtained from the leaves of *Agave sisalana* (Fig. 9.91).
Extraction process:

Decortication: leaves are crushed by a rotating wheel with blunt knives, then dried—generally under sunlight, then combed by a machine and sorted into several groups, largely on the basis of the in-field different sizes of the leaves.

Fig. 9.91 Plant of *Agave sisalana* (from de Andrade Silva et al. 2008)

Fibre yield:

A sisal plant produces from 200 to 250 leaves before flowering. Each of them contains approximately 700–1400 fibre bundles, 0.5–1 m long. From 100 kg leaves 3.5 kg extractable fibres are obtained.

Three types of fibres reinforce the leaf (Fig. 9.92): structural fibres, arch fibres and xylem fibres (Fig. 9.93). Structural fibres give the sisal leaf its stiffness and are found in the periphery of the leaf. They present a rarely circular and usually horse shoe shaped cross-section and a rough surface. Arch fibres occur in association with the conducting tissues and are usually developed in the median line of the leaf. The xylem occurs opposite to the arch fibres.

Chemical composition:

Cellulose 54–66%
Hemicellulose 12–17%
Lignine 7–14%
Pectine 1%
Ashes 1–7%.

Geometrical and morphological properties:

Every technical fibre contains numerous (between 100 and 200) elongated individual fibre cells which are about 1.8 mm in length and 6–30 μm in diameter. Cell wall thickness of a few microns have been reported, the remaining part of each fibre cell being occupied by the lumen. Cross section areas of individual technical fibre in the range 0.02–0.05 mm^2 have been reported.

Mechanical properties

Density:	1.33 Mg/m^3
Tensile strength:	from 350 to about 600 N/mm^2
Young modulus:	from 8–9 to 19 GPa (with lower variation if corrected for compliance)
Strain to failure:	from 2.5 to about 5%

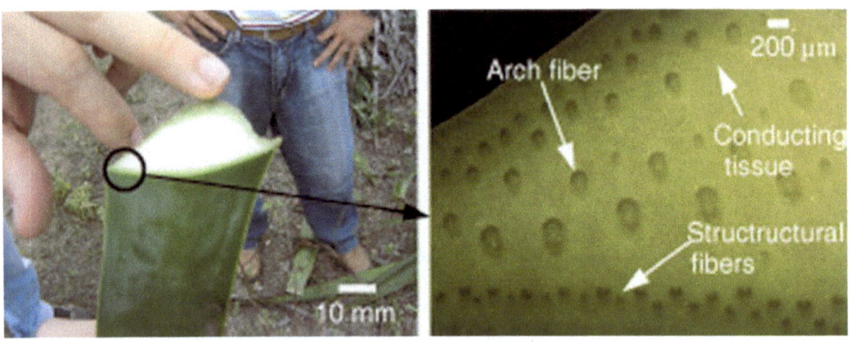

Fig. 9.92 Cross-section view and optical microscopy of the leaf (de Andrade Silva et al. 2008)

Fig. 9.93 Different sisal fibre types: (*top*) structural fibre with a horse shoe shape geometry and (*bottom*) arch fibre (from de Andrade Silva et al. 2008)

Applications as reinforcement in cement based materials:

The good mechanical, environmental and also economical properties of sisal fibres make them ideal candidate for reinforcement in composites in a series of applications encompassing different fields of engineering. It is worth remarking that the selection of the type of natural fibre for manufacturing products is strongly influenced by their availability. Because of this, for example, products from India and Asia mainly contain jute, ramie and kenaf fibre, products manufactured in Europe tend to use flax or hemp fibre, while products from South America tend to use sisal, curaua and ramie (Faruk and Sain 2015).

Anyway, applications of sisal fibres have been reported in automotive engineering, for both automobile interior and exterior parts due to their low cost, low density, superior sound and heat insulation, and environmentally friendly characteristics, contributing to achieve the goal of lowering the overall weight of a vehicle, thereby increasing fuel efficiency, as well as making the manufacturing process more sustainable. Appliances such as computer casings can be manufactured with injection moulding techniques. Leisure products such as skis, golf clubs and butt stocks can also be manufactured using sisal fibre reinforced composites (Faruk and Sain 2015).

The construction industry constitutes the second largest sector to employing natural materials in a range of products including light structural walls, insulation materials, floor and wall coverings, geotextiles and thatch roofing. Sisal fibre-reinforced cement based products such as roof tiles and building blocks are being produced in countries such as Tanzania and Brazil. The Brazilian research group involved in the EnCoRe project has been world-leader in promoting such kind of applications, which will be reviewed in detail in the course of this chapter.

Pineapple Leaf Fibres (PALF)

Plant: obtained from the leaves of *Ananas comosus* (Fig. 9.94)

Fig. 9.94 Plant of *Ananas comosus* (*source* Wikipedia)

Extraction process:

Manual process: begins with shredding through beating, scraping and husking the leaves. After this process the fibres are soaked in water, in case with addition of chemicals to accelerate the activity of microorganisms that digest the unwanted materials and separate the fibres. The wetting material is washed and cleaned, dried in the sun and combed or carded (Fig. 9.95).

Mechanical process: through a decorticator, as for sisal fibres.

Fibre yield:

On average, about 22 units of pineapple leaf weigh a kilogram. The reported fibre yield is about 2.7–3.5% of fibres (datacited in Faruk and Sain 2015).

Chemical composition (*as from studies surveyed by* Asim et al. 2015)

Cellulose 67–85%
Hemicellulose 0–18.8%
Lignine 4.4–15.4%
Pectine 1.1–1.2%
Wax and fat 3.2–4.2%
Ashes 0.9–4.7%.

Geometrical and morphological properties:

Diameters of individual fibres from a few microns to 300 μm (Asim et al. 2015) and length of individual fibres from 2.5 to 4.5 mm and of the technical fibre from 50 to 120 mm have been reported (Fig. 9.96) (Faruk and Sain 2015).

Mechanical properties (*as from studies surveyed in* Asim et al. 2015)

Density:	1.07–1.53 Mg/m^3
Tensile strength:	from 170 to about 1600 N/mm^2
Young modulus:	from 4.49 to 82.5 GPa
Strain to failure:	from 1.4 to 4%

Fig. 9.95 Manual separation of fibres from the leaf (*source* Wikipedia)

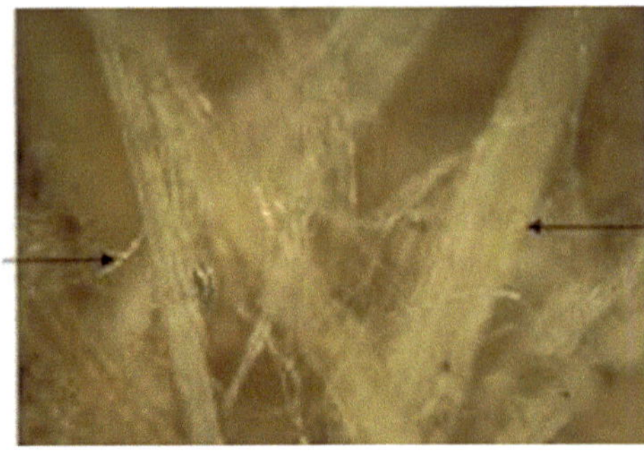

Fig. 9.96 Pineapple leaf fibres (*source* Wikipedia)

Applications as reinforcement in cement based materials:

The outstanding mechanical properties of individual PALF, due to their high cellulose content and low microfibril angle, make them, traditionally employed in textile industry, e.g. in far eastern countries, ideal candidates to being utilised in polymer matrix to develop composites with improved mechanical strength. In Faruk and Sain (2015) and in Asim et al. (2015) applications have been reported of PALFs used, also in hybrid combination with other kinds of natural fibres, in epoxy-, polyethylene- and low density polyethylene-, polypropylene-, vinyl ester- and polyester-based composites. PALFs have been also employed to reinforce thermoset, thermoplastic, biodegradable plastics and natural rubber (Asim et al. 2015). To the authors' knowledge the sole study employing PALFs in concrete is BSc thesis by Elizer et al. (2012), who employed scattered pineapple leaf fibres in FRC beam specimens. Only flexural strength (MOR) values were provided with no proper toughness measurement.

Curaua Fibres

Plant: obtained from the leaf of *Ananas erectifolius* (Fig. 9.97)
Extraction process: *decortication*, which includes the following processes

- Crushing leaves by a rotating wheel with blunt knives;
- Drying, under sunlight
- Combing, by machine, sorting and brushing (Fig. 9.98).
- Cutting, into required length

Fibre yield: Each plant produces 50–60 leaves per year; leaves are hard, flat, erect, 1–1.5 m long, about 40 mm wide and up to 5 mm thick, with an average weight of 150 g each. 100 kg of leaves result into 2 kg of extractable fibres (maximum without irrigation and precipitation higher than 2000 mm/year).

Fig. 9.97 Plant of *Ananus erectifolius* (white jute) *Source* Wikipedia

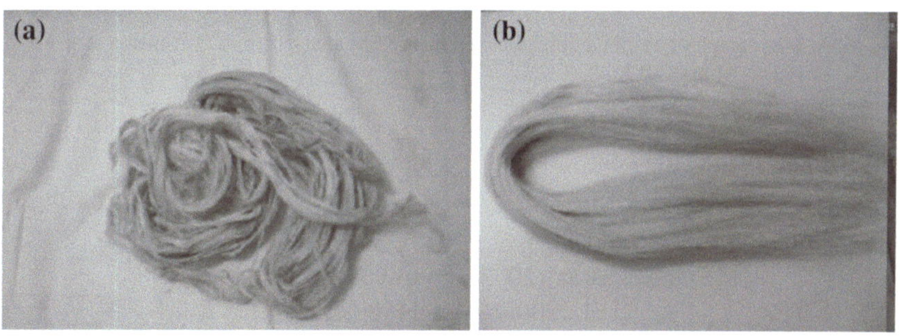

Fig. 9.98 Curaua fibres as obtained after drying (**a**) and after combing and brushing (**b**) (from D'Almeida et al. 2009)

Chemical composition: (*as from* Caraschi and Leão 2001)

- 73.6% cellulose
- 9.9% hemicellulose
- 7.5% lignine

Geometrical and morphological properties:

Cross section of individual technical fibre: $0.004 \pm 50\%\ \text{mm}^2$

Every technical fibre (Fig. 9.99) contains between 10 and 20 elongated individual fibre cells about 10 μm in diameter; average cell wall thickness of 3.5 μm has been reported, which is the thickest among the most commonly used natural fibres; lumen diameter of the same order of magnitude has also been reported, yielding to the highest cell wall thickness to lumen diameter ratio.

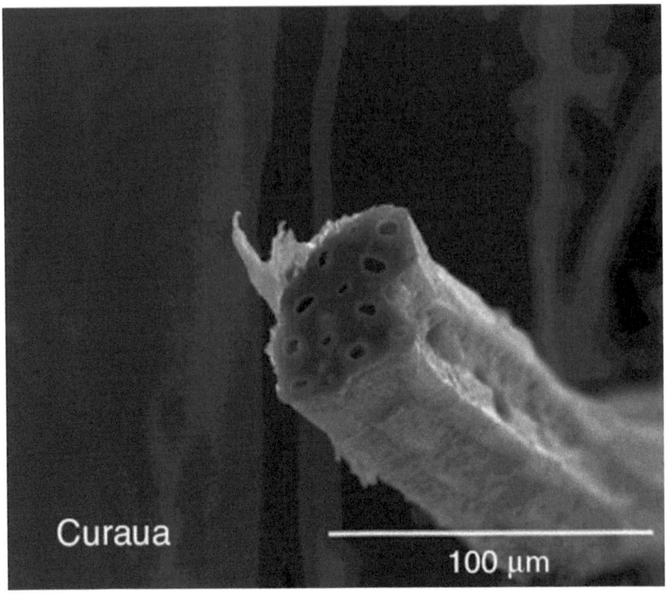

Fig. 9.99 Micrograph of a curaua fibre obtained from SEM analysis from D'Almeida et al. (2009) and Alves Fidelis et al. (2013)

Mechanical properties

Density:	1.38 Mg/m^3
Tensile strength:	Alves Fidelis et al. (2013): 543 ± 100 N/mm^2
	Gomes et al. (2007): up to 900 N/mm^2
	Leao et al. (2009): 500–1150 N/mm^2
	Satyanarayana et al. (2007): 1250–3000 N/mm^2
	Spinacè et al. (2009): 500–1100 N/mm^2
	Souza et al. (2010): 800 N/mm^2
Young modulus:	Alves Fidelis et al. (2013): 63.7 ± 32.5 GPa
	Leao et al. (2009): 11.8 GPa
	Satyanarayana et al. (2007): 30–80 GPa
	Spinacè et al. (2009): 26–46 GPa
	Souza et al. (2010): 10.7 GPa
Strain to failure:	Alves Fidelis et al. (2013): 1 ± 0.2%
	Gomes et al. (2007): 3.9%
	Leao et al. (2009): 3.7–4.3%
	Satyanarayana et al. (2007): 4.5–6%
	Spinacè et al. (2009): 2–4%

Applications as reinforcement in cement based materials:

Several applications of curaua fibres in polymer matrix composites have been reported by Souza et al. (2010, Chap. 22 in Faruk and Sain 2015). Because of their

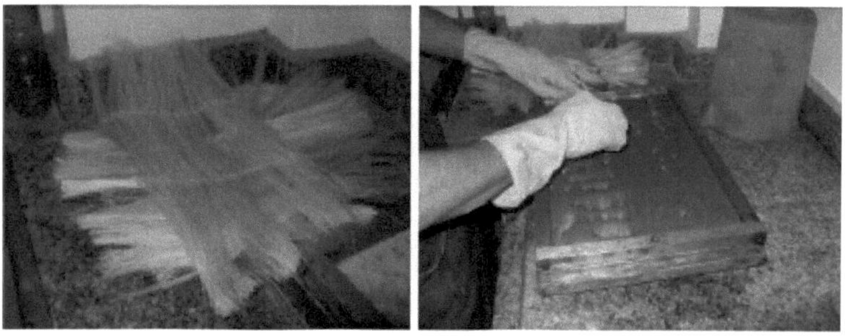

Fig. 9.100 Curaua fibre tissue and production of the laminates (from D'Almeida et al. 2009)

physical properties, which nearly equates those of glass fibres, curaua fibres have attracted the interest of automobile industry, resulting in economic benefits (up to 50% cheaper than fibre glass composites) and social advantages (Zah et al. 2007).

D'Almeida et al. 2010 investigated the use of curaua fibres, both 25 and 50 mm long and in volume percentages ranging from 2 to 6%, for the production of HPFRCCs, highlighting the possibility of achieving deflection hardening response with stable multiple cracking with higher long fibre dosages. The production of compression molded laminates reinforced with multiple layers of long and aligned curaua fibres (Fig. 9.100) was also addressed (D'Almeida et al. 2009).

Because of high cellulose content recently the possibility of obtaining from curaua fibres nanocellulose and/or cellulose nanofibres for composite applications has also been addressed (Souza et al. 2010, Chap. 22 in Faruk and Sain 2015).

Piassava Fibres

Plant: obtained from the petiole of the leave of *Attalea funifera* (Fig. 9.101)
Extraction process:

Stems are cut with knives and fibres are washed, brushed and sorted by length. Longer and more flexible fibres have the highest commercial values, whereas shorter and more rigid fibres are used for brooms.

Fibre yield: The fibres, up to 4 m long with an average diameter of 1.1 mm, are harvested once a year. Each palm tree can yield 8–10 kg of fibres.
Chemical composition: (*as from* d'Almeida et al. 2006)

- 31.6% cellulose
- 48.4% lignine

Geometrical and morphological properties:

Cross section areas of individual technical fibre in the range 0.585 ± 50% mm^2 have been reported (Fig. 9.102—from Alves Fidelis et al. 2013).

Fig. 9.101 Plant of *Attalea funifera*. *Source* Wikipedia

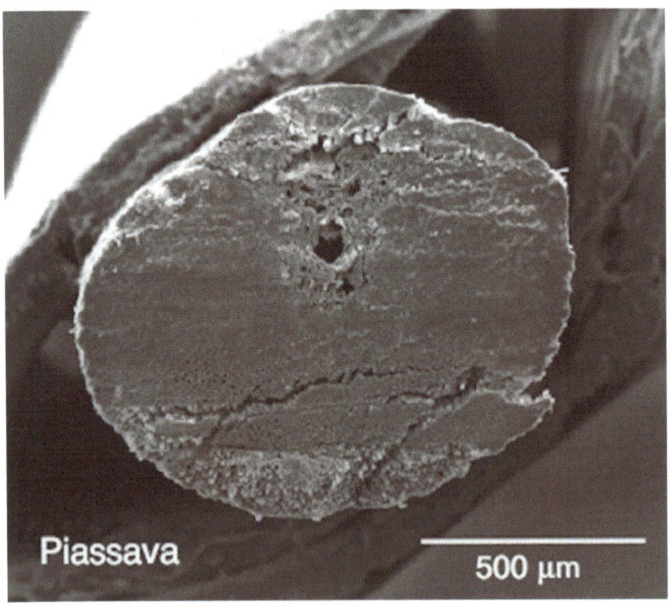

Fig. 9.102 Micrograph of a piassava fibre obtained from SEM analysis (from Alves Fidelis et al. 2013)

Mechanical properties

Density: 1.38 Mg/m^3
Tensile strength:
Alves Fidelis et al. (2013): 131 ± 36 N/mm^2
Gauge lengths 40 mm
D'Almeida et al. (2006): from 108.5 ± 16 to 147.3 ± 25 N/mm^2
gauge lengths from 15 *to* 150 mm—*quite independent of it*
Young modulus:
Alves Fidelis et al. (2013): 3.8 ± 0.9 GPa
D'Almeida et al. (2006): from 1.07 ± 0.27 to 4.59 ± 0.15 GPa
Gauge lengths from 15 *to* 150 mm—*stiffer for shorter lengths*
Strain to failure:
Alves Fidelis et al. (2013): 11.4 ± 3.6%
D'Almeida et al. (2006): from 21.9 ± 4.3 to 6.4 ± 2.3%
Gauge lengths from 15 *to* 150 mm—*shorter length providing higher ultimate strain values*

Applications as reinforcement in cement based materials:

Iglesias et al. (2016) reported the use of piassava fibres in lightweight cementitious composites, to temper the strength decrease due to the addition of waste Ethylene-Vinyl Acetate grains, employed for specific gravity reduction.

Banana and Abaca Fibres

Plant: obtained from the leaf stems of *Musa sepentium and Musa textilis* (Fig. 9.103).

Extraction process:

(*source*—http://www.textileschool.com/articles/373/natural-plant-bast-fibres-banana—*accessed July* 6, 2016)

Banana fibres can be extracted by employing chemical, mechanical or biological methods. Chemical method causes environmental pollution, while mechanical method fails to remove the gummy material from the fibre bundle surface.

Biological procedures yield more fibre bundles than the other two procedures without any harm to the environment. There are basic two types of biological extraction procedures, namely Bacnis method and Loenit method. In the Bacnis method, Banana Fibres are produced from waste stalk of the plant. The outer sheath is tightly covered layers of fibre. The fibre is located primarily adjacent to the outer surface of the sheath and can be peeled-off readily in ribbons of strips of 5–8 cm wide and 2-4 mm thick, the entire length of sheath. This stripping process is known as tuxying the strips being called tuxies. The trunks are pulled apart and the sheath separated according to their position in stalk. They are then flattened and the fibre is stripped from the stem by cutting the pulpy portion and pulling away the tuxy.

Fig. 9.103 Plant of *Musa sepentium*. *Source* Wikipedia

In Loenit method, the tuxies are pulled off the stalk from one sheath at a time. In either of these methods tuxies are tied into bundles of 23–27 kg and brought to the stripping knife for cleaning. In this process tuxies are pulled under a knife blade, which is pressed tightly against the tuxy in order to scrape away the plant tissue between the fibres. The clean fibre is then air dried and made up into bundles for subsequent grading and bailing. In addition to hand stripping, machines are used where the trunks from which the dark outer sheaths have been removed, are cut into sections of 120–180 cm in length. The sections are then crushed between rolls and the pulpy tissues are scraped away, one half the length at a time, by two large revolving drums, the rim of which are fitted with scrapping blade which scrape the sheath while it is pressed against a bed plate, oven dried (replacing the traditional sun-drying process—see Fig. 9.104), graded and baled.

Fibre yield: the amount of fibres obtainable for yarn spinning from a pseudo-stem is about 30% of its weight.

Chemical composition: (*as from* Mukhopadhyay et al. 2008)

Cellulose 31.3%
Hemicellulose 15%
Lignin 15%
Extractives 4.5%
Moisture 9.7%
Ashes 8.7%

Fig. 9.104 Abaca fibres sun-drying in an abaca fibre farm in Costarica (*source* Wikipedia)

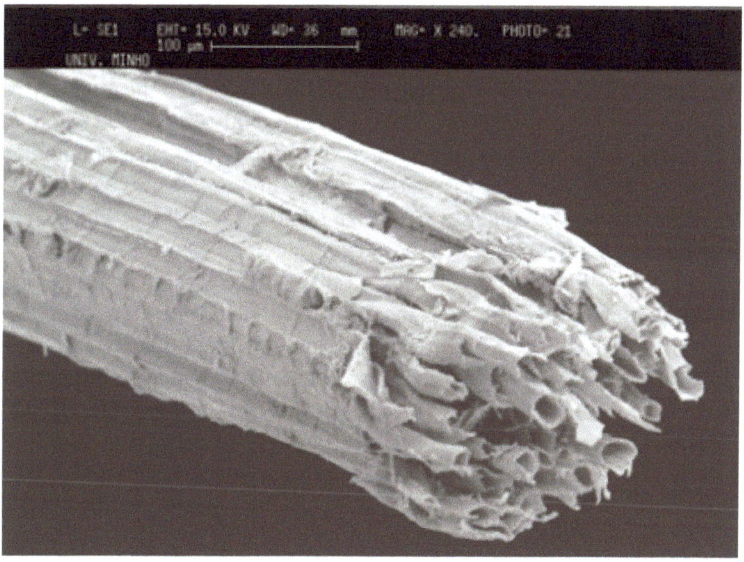

Fig. 9.105 Micrograph of fractured banana fibre (from Mukhopadhyay et al. 2008)

Geometrical and morphological properties:

Banana fibre is a multiple celled structure, with circular cross section (Fig. 9.105), and with large lumens in relation to the wall thickness.

Diameters range from 50–100 to 70–210 μm, according to Mukhopadhyay et al. (2008) and to Sathish and Kesavan (2015), respectively.

Individual fibre length ranges from 2.5 to 13 mm.

Mechanical properties

Density: 1.4-Mg/m^3 (Mukhopadhyay et al. 2008)

Tensile strength:

Sathish and Kesavan (2015) reported tensile strength from 199 to 781 N/mm^2. Mukhopadhyay et al. (2008) reported tensile strengths in a gross range between 150 and 200 N/mm^2, with scant influence of the strain rate (varying between 0.1 and 10 min^{-1}).

Young modulus:

6.6–21.6 GPa (Sathish and Kesavan 2015)

Strain to failure:

1.8–3.3% (Sathish and Kesavan 2015)
from 3.0% (strain rate 0.1 min^{-1}) to 1.2% (strain rate 10 min^{-1})
as reported by Mukhopadhyay et al. (2008)

Applications as reinforcement in cement based materials:

Because of their long staple length, strength and cellulose content, banana and abaca fibres have been traditionally used in the manufacture of specialized papers including tea and coffee bags, sausage casing, electrolytic papers, currency notes, cigarette filters, medical and disposal papers. Innovative applications in polymer matrix composites for automotive industry has been also reported (Anon 2005).

Application of abaca fibres to reinforce concrete panels also containing phosphogypsum waste aggregates was studied by Yetimoglu as early as in 1984.

Systematic studies on the use of banana and abaca fibres, either pulped or chopped, as reinforcement in cement based matrices date back to Coutts and Warden (1987), Zhu et al. (1994), Savastano et al. (2000). All the authors reported increase of flexural strength and toughness with increasing pulp and fibre content up to 8–10% by weight, together with reduction in Young modulus and density and increase in water absorption. Saturation or worsening effects were evident for higher contents.

Positive synergy with steel fibres was also reported (Vaishnav and Titiksh 2016).

Reynaldo (2011) explored hybrid sisal and abaca strand reinforcement to manufacture textile reinforced mortar slabs, also replacing Portland Cement with rice husk.

Sayed (2014) performed comprehensive investigation on the use of abaca fibres in FRC; optimal dosage of 2% by volume was found with reference to mechanical and durability properties, including freeze-and-thaw resistance, which could be negatively affected by the water absorbed by the same fibres, e.g. during the mixing stage. Influence of different chemical treatments to modify surface fibre roughness and improve fibre-matrix bond and composite durability was also investigated.

Etri et al. (2015) have reported the use of mortars reinforced both with discontinuous abaca fibres and with abaca rope meshes for the reinforcement of masonry walls, the latter resulting in a much more effective integrity-keeping and ductile

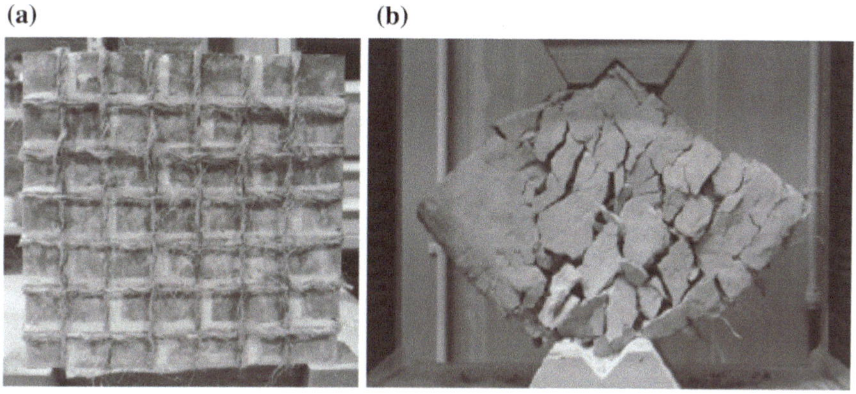

Fig. 9.106 Abaca rope meshes for retrofitting masonry walls (**a**) and retrofitted wall tested in diagonal compression (**b**)

Fig. 9.107 Plants of *Phoenix dactilifera* (**a**), *Elaeis guineensis* (**b**) (*Source* Wikipedia)

retrofitting (Fig. 9.106). Che Muda et al. (2016) reported remarkable improvement in impact resistance of slabs reinforced with volume percentages ranging from 0.5 to 1.5% of banana fibres with assorted lengths (20, 30 and 40 mm).

Mostafa and Uddin (2016) employed banana fibres in compressed earth blocks.

Palm Fibres

Plant: at least 20 different species of palm trees can provide fibres, depending on the zone in which they are grown. Most common are date palm (*Phoenix dactilifera* —Fig. 9.107a), oil palm (*Elaeis guineensis*—Fig. 9.107b *and Elaeis oleifera*) and raffia palm (*Raffia farinifera*).

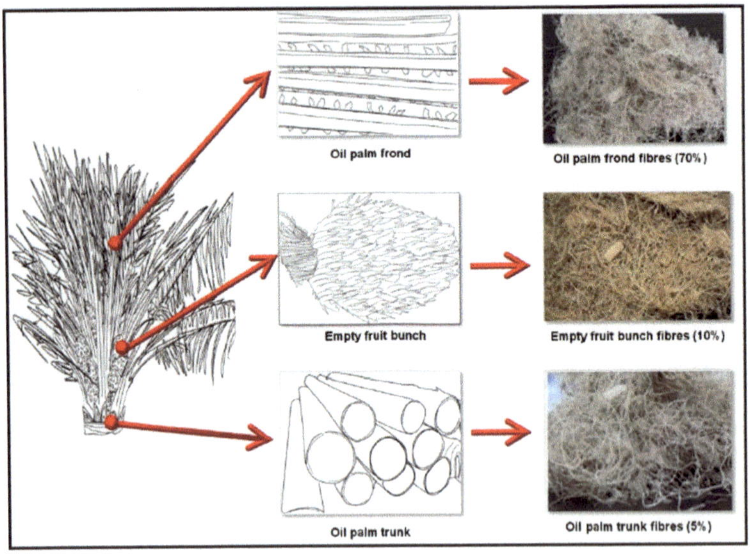

Fig. 9.108 Palm biomass and palm biomass fibres from palm tree (from Khalil et al. 2012)

Extraction process:

Palm-employing industry generates large amounts of biomasses from which not only fibres can be harvested, e.g. from tree trunks, leaves as well as from empty fruit bunches (Fig. 9.108), but which can be as such also utilized as secondary raw materials in cementitious composites production. A retting process is generally utilized to extract fibres from the different parts of the plant.

Fibre yield: The amount of biomass produced by an oil palm tree, inclusive of the oil and lignocellulosic materials, is on the average of 231.5 kg dry weight/year, with oil representing only a small fraction of it. As for the latter oil palm fronds accounts for 70% of the total oil palm biomass produced, while the empty fruit bunches accounts for 10% and the trunk accounts for only about 5% of the total biomass produced.

Chemical composition: (*as from* Khalil et al. 2012—*by %wt.*)

	Oil palm frond	Oil palm fruit bunch	Oil palm trunk
Cellulose	40–50	43–65	29–37
Hemicellulose	34–38	17–33	12–17
Lignin	20–21	13–37	18–23
Ashes	2–3	1–6	2–3

Geometrical and morphological properties:

Individual date palm fibres are reported having diameters in the range between 0.1 and 0.8 mm. As for oil palm fibres the following data are reported by Khalil et al. (2012).

For oil palm fibres Khalil et al. (2012) report individual fibre lengths in the range 0.5–1.5 mm, independently of the source; diameters are reported in the range 10–20 μm for frond fibres, 3–300 μm for fruit bunch fibres and 30–35 μm for trunk fibres. Lumen width of about 8 μm for fruit and frond fibres and of about 18 μm for trunk fibres has been observed.

	Oil palm frond	Oil palm fruit bunch	Oil palm trunk
Cellulose	40–50	43–65	29–37
Hemicellulose	34–38	17–33	12–17
Lignin	20–21	13–37	18–23
Ashes	2–3	1–6	2–3

Mechanical properties

Density:

Date palm (Kriker et al. 2008)

bulk density ranges from 0.5 to 1.1 Mg/m^3

absolute density ranges from 1.3 to 1.45 Mg/m^3

Oil palm (*as reported by* Khalil et al. 2012)

Frond: 0.6–1.2 Mg/m^3

Fruit bunch: 0.7–1.55 Mg/m^3

Trunk: 0.5–1.1 Mg/m^3

Tensile strength:

For date palm fibres properties depend on the plant, classified as male (producing pollen), and female (producing fruits), the latter having better tensile strength. Kriker et al. (2008) reported tensile strength values from 170(dry)–175(wet) N/mm^2 to 290(dry)–300(wet) N/mm^2, for gauge lengths ranging from 100 to 20 mm respectively.

For oil palm fibres in Khalil et al. (2012) values of tensile strength in the range 20–200, 50–400 and 300–600 N/mm^2, respectively for frond, fruit bunch and trunk fibres are reported.

Young modulus:

Date palm (Kriker et al. (2008)

Dry conditions: 4.74–5.75 GPa (gauge length decreasing from 100 to 20 mm)

Wet conditions: 3.78–3.55 GPa (gauge length decreasing from 100 to 20 mm)

Oil palm (Khalil et al. 2012)

Frond: 2–8 GPa

Fruit bunch: 0.6–9 GPa

Trunk: 8–45 GPa

Strain to failure:

Date palm (Kriker et al. 2008)

Dry conditions: 16–11% (gauge length decreasing from 100 to 20 mm)

Wet conditions: 17.4–12% (gauge length decreasing from 100 to 20 mm)

Oil palm (Khalil et al. 2012)

Frond: 3–16%
Fruit bunch: 2.56–18%
Trunk: 5–25%

Applications as reinforcement in cement based materials:
Because of the large palm tree waste biomass available, especially from oil palm industry in the last years, use of (oil) palm fibres as reinforcement in concrete and cementitious composites is becoming very popular, confirming classical advantages linked to the use of dispersed fibre reinforcement in a cementitious matrix (see, e.g., Ahmad et al. 2010a, b). Use as reinforcement of laterite bricks (Mahat et al. 2010), crushed lightweight aggregate concrete blocks (Ramli and Dawood 2010), gypsum boards (Abuh and Umoh 2015) has been also reported. Use of oil palm shells as aggregates (Teo et al. 2006) or of palm stems as reinforcement bars in concrete beams (Ofuyatan and Olutoge 2013) has been also investigated.

Kriker et al. (2005, 2008) and Ozerkan et al. (2013) investigated use of date palm fibres, in percentages up to 3% by volume, as reinforcement in concrete: slight residual tensile strength retainment capacity was observed, together with improvements in restrained shrinkage cracking and resistance to sulphate attacks.

Al Adili et al. (2013) investigated use of date palm fibres, from 1 to 16 mm long and up to 20% by volume, for soil stabilization, finding improvement of soil cohesion and friction angle and decrease of internal voids.

Coconut/Coir Fibres

Plant: obtained from the husk of the fruit of *Cocos nucifera* (Fig. 9.109).
Extraction process
Fibres are obtained from the husk part of the coconut fruit; the separation of the husk from the nut is generally time consuming and laborious. Fruit shells are retted, from three to six months in ponds of brackish waters or in backwaters or lagoons is generally performed. The fruit shells are then beated to separated the fibres, which are then hackled, washed, dried in the shaded, loosened and cleaned. In alternative a mechanical process using defibreing and decorticating is employed, which requires the fruit shells to be retting-processed for just five days; than husks are crushed in a breaker and revolving drums separate the coarse long fibres, which are then washed, cleaned, dried, hackled and combed.

Fibres can be extracted from unripe nuts (“white coir”), while “brown coir” is extracted after ripening of the coconut.
Fibre yield:
The fibres are obtained from the husk part of the coconut fruit. The husk represents about 35% of the fruit weight, and contains about 30% by weight of fibres.

Fig. 9.109 Plant of *Cocos nucifera. Source* Wikipedia

Chemical composition: (*values reported in* Faruk and Sain 2015).

Cellulose 43.4%
Hemicellulose 45.8
Lignin 0.25%
Pectin 3%
Water soluble residues 5.2%

Geometrical and morphological properties:

Diameter of individual fibres in the order of a few hundreds of microns has been reported (Defoirdt et al. 2010)—Fig. 9.109.

Mechanical properties

Density: 1.01 ± 0.05 Mg/m^3 for white coir

1.29 ± 0.07 Mg/m^3 for brown coir (Defoirdt et al. 2010)

Tensile strength:

90 ± 35 MPa reported by Alves Fidelis et al. (2013)—gauge length 40 mm; strain rate 0.2 mm/min

from 192 ± 37 to 162 ± 32 MPa (white coir) and from 343 ± 36 (15 mm gauge length) to 186 ± 55 MPa (brown coir) reported by Defoirdt et al. (2010) for gauge lengths increasing from 5 to 35 mm and 5 mm/min strain rate

Young modulus: 2.6 ± 0.7 GPa (Alves Fidelis et al. 2013)

3.44 (white) and 4.94 (brown) (Defoirdt et al. 2010)

Strain to failure:

18.8 ± 9.1% (Alves Fidelis et al. 2013: gauge lengths 40 mm; 0.2 mm/min)

from 42.4 ± 14% to 26.1 ± 5.6% (white coir) and from 59.5 ± 5 to 24.5 ± 6.8% (brown coir), for increasing gauge lengths and 5 mm/min strain rate (Defoirdt et al. 2010).

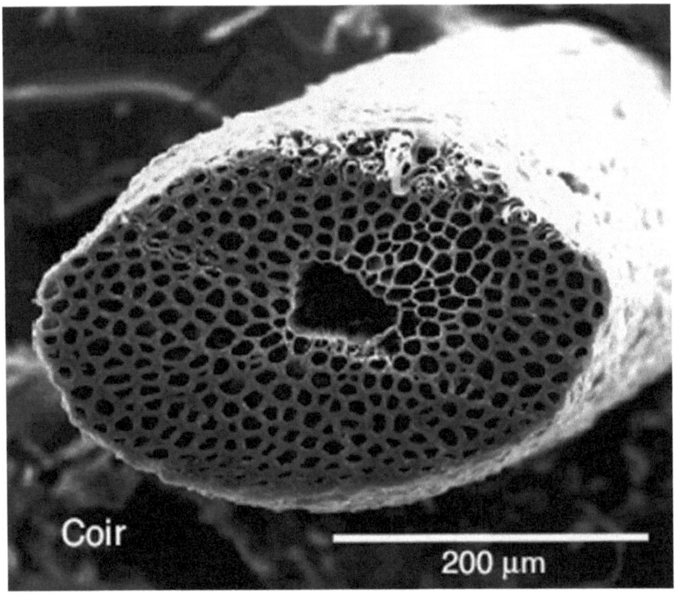

Fig. 9.110 Micrograph of a coir fibre (from Alves Fidelis et al. 2013)

Applications as reinforcement in cement based materials:

Early applications of coconut fibres in cement based matrices for the production of building components, such as corrugated slabs and cement-bonded boards, date back to the mid-eighties (Paramasivam et al. 1984; Aggarwal 1992).

Use of coir fibres to mitigate plastic shrinkage in concrete mixes was investigated by Sanjuàn and Toledo Filho (1998).

Subsequently the durability of coir, and other natural, fibres in alkaline environments was investigated in detail (Tolêdo Filho et al. 2000), also proposing effective methods to mitigate the fibre degradation and subsequent embrittlement of the composite, such as use of pozzolanic cement to reduce the $Ca(OH)_2$ content of the mix (John et al. 2005) or pre-enveloping fibres in a silica fume slurry. Recently, curing for a few days through accelerated carbonation (Almeida et al. 2013) has been found likewise effective to the aforementioned purpose.

Studies identifying the macroscopical mechanical properties of coir FRC have been recently published by Ali et al. (2012) and Al-Masoodi et al. (2013).

Interesting applications in thin ferrocement roof elements have been proposed by Alavez Ramirez et al. (2014) with the aim of improving the thermal insulating characteristics of the element. Similar experience have been also reported by Khedari et al. (2005), with reference to fibre reinforced soil-cement blocks and by Lertwattanaruk and Suntijitto (2015), with reference to coir and oil-palm fibre reinforced concrete mixes.

Ali et al. (2012) and Ali (2014) have proposed to use a reinforcement system made with coir fibre ropes (Fig. 9.111), eventually in combination with the use of

Fig. 9.111 Column reinforcement with coir ropes (Ali 2014)

Fig. 9.112 Cotton plant. *Source* Wikipedia

coir fibres dispersed in the mix, as alternative to conventional steel reinforcement and have effectively tested it in columns subjected to earthquake shake-table actions. Chen and Chouw (2016) studied the behaviour of coir FRC columns confined with flax fibre reinforced polymers.

Use is soil consolidation has been also reported (Ghavami et al. 1999).

Cotton fibres

Plant: fluffy staple fibre that grows in a boll, or protective case, around the seeds of cotton plants of the genus *Gossypium* in the family of *Malvaceae* (Fig. 9.112).

Chemical composition: (*source Wikipedia*)

cellulose 91.00%
water 7.85%
protoplasm, pectins 0.55%
waxes, fatty substances 0.40%
mineral salts 0.20%

Geometrical and morphological properties:
Individual cotton fibres are Fairly uniform in width, 12–20 μm; length varies from 10 to 60 mm.
Mechanical properties

Density	1.54–1.56 Mg/m^3
Tensile strength	400 N/mm^2 (Alomayri and Low 2013)
Young modulus	4.8 GPa (Alomayri and Low 2013)
Strain to failure	no information available

Applications as reinforcement in cement based materials:
A few recent studies on the use of cotton fibres to reinforce geopolymer composites have been found; optimum dosage for enhancement of strength and impact resistance appears to be around 0.5% by weight (Alomayri and Low 2013; Alomayri et al. 2013, 2014). Use to reinforce gypsum boards has been reported.

Bamboo Fibres

Plant: obtained from different trees of the family of *Bambusoideae* (Fig. 9.113). The bamboo culm, in general, is a cylindrical shell, which is divided by transversal diaphragms at the nodes. Bamboo shells are orthotropic materials with high strength in the direction parallel to the fibres and low strength perpendicular to the fibres respectively.

Bamboo is a composite material, consisting of long and parallel cellulose fibres embedded in a ligneous matrix. The density of the fibres in the cross-section of a bamboo shell varies along its thickness. This presents a functionally gradient material, evolved according to the state of stress distribution in its natural environment. As seen in Fig. 9.114, the fibres are concentrated in regions closer to the outer skin. This is consistent with the state of stress distribution when the culm is subjected to wind forces (Ghavami 2005).
Extraction process: Bamboo fibres are reported to have been obtained from bamboo pipe through various mechanical methods, such as crushing, heat steaming, and shearing, after swelling the bamboo stems. A combination of chemical and mechanical processes is used in the pulp and paper industries for pulping,

Fig. 9.113 Bamboo plants. *Source* Wikipedia

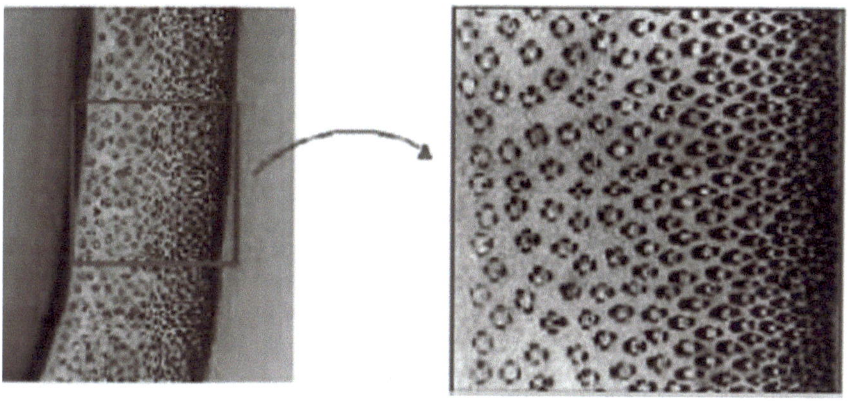

Fig. 9.114 Non-uniform fibre distribution over cross section of a bamboo culm (Ghavami 2005)

and is applicable for the extraction of bamboo fibres. Two conventional mechanical methods, compression molding technique (CMT) and roller mill technique (RMT), were explored prior to alkali treatment. In the CMT, a bed of bamboo strips was placed between two flat platens and subjected to a constant load for a predetermined time. Compression time and the starting bed thickness are important parameters that must be optimized to obtain good quality fibres. In the RMT, the bamboo strips were forced between two rollers, one of which was fixed and the other rotated. The diameter of the rollers, strip thickness, and the speed of the rotating cylinder were chosen to yield flattened strips of bamboo. These alkaline and mechanical-treated strips could be easily separated into individual fibres (Liu et al. 2012).

Chemical composition: (*source* Liu et al. 2012)

cellulose 60.8%
hemicellulose 8%
lignin 32.2%

Geometrical and morphological properties:

Fibres with diameters of some hundred microns (366 ± 74—Defoirdt et al. 2010) and sticks with rectangular cross section from 1 × 5 mm^2 to 3 × 7 mm^2 have been reportedly used (Ahmad et al. 2014).

Parameswaran and Liese (1980) reported that bamboo fibre consists of a multilayered or 'polylamellated' cell wall structure (Fig. 9.115), with alternate broad and narrow layers with different fibrillar orientation. In the thick layers the fibrils are oriented at a slight angle to the fibre axis, whereas the thin ones generally show a more transverse orientation. This cell wall structure is very important because it determines the mechanical properties of the fibres.

Mechanical properties (all data from Defoirdt et al. 2010)

Density:	1.38 Mg/m^3
Tensile strength:	from 639 ± 175 to 813 ± 94 N/mm^2 (slightly scattered according to gauge length, varying from 5 to 35 mm; strain rate 1 mm/min)
Young modulus:	33.37 GPa
Strain to failure:	from 2.9 ± 07% to 2.0 ± 0.6%, decreasing with gauge length

Applications as reinforcement in cement based materials:

Bamboo has been longtime traditionally employed as a construction material in China and South-Eastern Asia countries. Use of bamboo "bars" as reinforcement in concrete structural elements has been reported by Mansur and Aziz (1983) and by Ghavami (1995, 2005), for beams and columns, as well as for slabs, where bamboo permanent shutter forms have been employed (Fig. 9.116). In this framework,

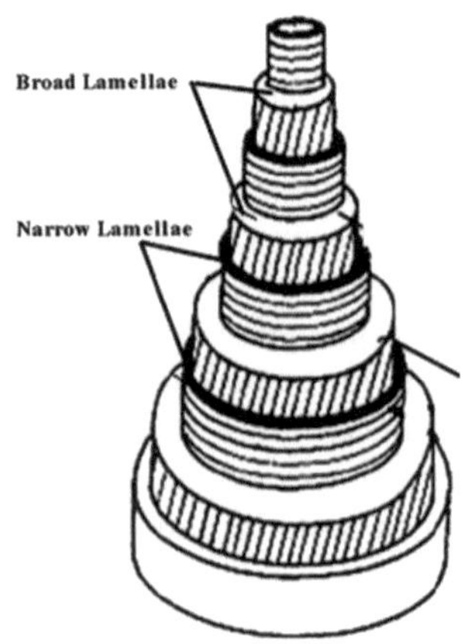

Fig. 9.115 "Polylamellated" bamboo fibre structure (Parameswaran and Liese 1980)

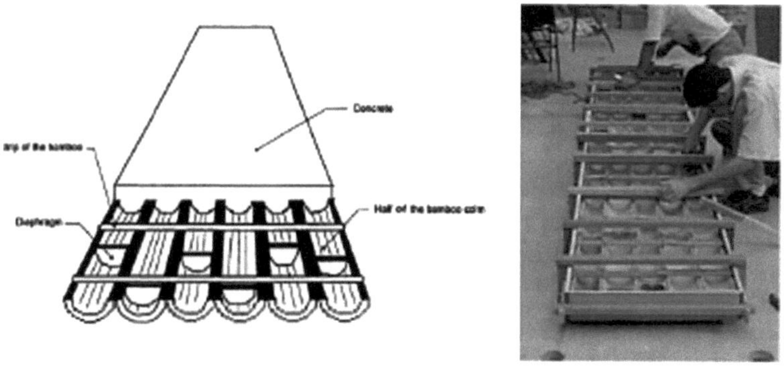

Fig. 9.116 Concrete slab reinforced with bamboo shutter forms (Ghavami 2005)

Fig. 9.117 Bamboo culms (a), air dried bamboo fibres (b), washed bamboo fibres before use if geopolymer composites (c) and reinforcing bamboo strips (d) (Sà Ribeiro et al. 2016)

Agarwal et al. (2014) have investigated the efficacy of different adhesive surface treatment to improve the bamboo-matrix bond.

Moroz et al. (2014) used bamboo rods for strengthening masonry walls.

Frias et al. (2012) and Vilar-Cociña et al. (2014) investigated effectiveness of bamboo leaf ashes as pozzolanic cement replacement in concrete, because of high silica content (about 78% SiO_2).

Sá Ribeiro et al. (2016) has recently investigated the production of geopolymer composites reinforced with either bamboo fibres (5% by weight) as well as with bamboo strips of suitable geometry (Fig. 9.117).

Straw and Cane Fibres

Plant: obtained from *Triticum aestivum* (wheat); *Oryza sativa* (rice); *Zea mais* (maize/corn); *Avena sativa* (oat); *Hordeum vulgare* (barley); *Secale cereale* (rye); *Saccharum officinarum* (sugarcane)
Extraction process: different mechanical and/or chemical methods, depending on the plant and on the geographical location

Chemical composition: (*by wt.—data reported in* Faruk and Sain 2015)

	Cellulose	Hemicell.	Lignin	Starch	Fat	Ashes	Silica
Corn stalk	38–40	7–21	28				
Corn pith	24.6	19.1	12.3				
Oat straw	44–53	27–38	16–19			6–8	4–6.5
Barley straw	47–48	24–29	14–15			5–7	3–6
Barley husk	39	12	22	11	4	2	
Rye straw	50–54	27–30	16–19			2–5	1–4
Rye husk	26	16	13	17	7	11	1
Wheat straw	33–52%	20–36	10–23				
Rice straw	38.3	28	14.9			18.8	
Rice husk	54	19	19.5			15	
Sugarcane	50	25	25				

Geometrical and morphological properties:

Large variety, depending whether straw fibres or husk and on processing

Mechanical properties (all data reported in Faruk and Sain 2015)

Wheat

Tensile strength: from 58 ± 51 to 146.3 ± 53 to 139.9 ± 53 N/mm^2 (respectively for mechanical, chemical processed and microbial retted fibres)

Young modulus: from 3.7 ± 2.6 to 7.9.3 ± 3.7 to 4.89 ± 1.5 GPa (respectively for mechanical, chemical processed and microbial retted fibres)

Sugarcane

Tensile strength: from 170 to 290 N/mm^2

Young modulus: from 15 to 19 and up to 27.1 GPa

Strain to failure: 1.1%

Applications as reinforcement in cement based materials:

Rice husk and rice husk ashes have been longtime (Al-Khalaf and Yousif 1984; Salas et al. 1986, 1987) utilized as a pozzolanic cement replacement in concrete and cementitious composites (Bui et al. 2005; Rodríguez de Sensale 2006; Giaccio et al. 2007; Safiuddin et al. 2010; Cordeiro et al. 2011; Gholizadeh Vayghan et al. 2013; Kunchariyakun et al. 2015; Venkatanarayanan and Rangaraju 2015), as well as, more recently, even in the production of blended cements (Bie et al. 2015; Prasittisopin and Trejo 2015) and of geopolymer composites (He et al. 2013). Use of internal curing agents in UHPC has been also reported (Van et al. 2014). Benefits also included improvement in chloride penetration resistance and hence in delaying corrosion of reinforcement (Gastaldini et al. 2007, 2010; Chalee et al. 2013), as well as improved resistance versus sulphate and acid attacks as well as alkali-silica reaction (Rodríguez de Sensale 2010).

Use of ashes from other sources, such as wheat (Al-Akhras and Abu-Alfoul 2002), sugarcane straw (Frias et al. 2005; Martirena et al. 2006; Villar-Cociña et al. 2003; Calligaris et al. 2015; Moraes et al. 2015), corn stalk and sunflower (Aksogan

Fig. 9.118 Example of plant stalks and their ashes (from Aksogan et al. 2016)

et al. 2016—Fig. 9.118) and elephant grass (Nakanishi et al. 2014; Chagas Cordeiro and Pessôa Sales 2015a, b) has been also reported.

Moraes et al. (2016) studied use of sugarcane bagasse ashes in alkali activated binders. Ashes are obtained by burning in controlled heat furnaces the plant straw; rapid ash cooling results in crystallinity, resulting into pozzolanic activity.

Application of wheat straw fibres in cement bonded boards has been reported by Soroushian et al. (2004). Biricik et al. (2000) found better resistance to sulphate attack in concrete reinforced with wheat straw.

Bouhicha et al. (2005) reported use of barley straw fibres for soil consolidation.

Wang and Wu (2013) proposed to use rice straw coke, obtained from rice straw fibres dehydrated at 110 °C and then thermally treated at 450 °C, instead of rice straw and husk ashes, to reduce CO_2 emissions.

Thomson and Walker (2014), Lawrence et al. (2009), Rahim et al. (2016) and Belhadj et al. (2016) reported the use of barley and rape straws to reduce the thermal conductivity and improve the moisture buffer capacity of cementitious composites. Bederina et al. (2016) investigated the use of different treatments to improve the interaction of barley fibres with surrounding cementitious matrix, including coating the fibres in cement, lime, bitumen and pozzolan, immersing the fibres in water, oil-gas solution, waste oil or varnish.

Qiang et al. (2013a, b) reported the use of straw fibres and bentonite in pavements.

References

Abe H, Ryo F (2005) Review—the orientation of cellulose microfibrils in the cell walls of tracheids in conifers. A model based on observations by field emission scanning electron microscopy. IAWA J 26(2):161–174

Aggarwal LK (1992) Studies on cement-bonded coir fibre boards. Cem Concr Compos 14(1):63–69

Abuh NM, Umoh AA (2015) Palm kernel fruit fibre reinforced gypsum-cement based wall panels: it's physical and mechanical characteristics. Pollution 1(2):117–126

Agarwal A, Nanda B, Maity D (2014) Experimental investigation on chemically treated bamboo reinforced concrete beams and columns. Constr Build Mater 71:610–617

Ahmad Z, Ibrahim A, Tahir PM (2010a) Drying shrinkage characteristics of concrete reinforced with oil palm trunk fibre. Int J Eng Sci Technol 2(5):699–708

Ahmad Z, Ibrahim A, Tahir PM (2010b) Oil palm trunk fibre as a bio-waste resource for concrete reinforcement. Int J Mech Mater Eng (IJMME) 5(2):199–207

Ahmad S, Raza A, Gupta H (2014) Mechanical properties of bamboo fibre reinforced concrete. In: Proceedings ICRSET'2014, 2nd international conference on research in science, engineering and technology, 21–22 March 2014, Dubai, UAE, pp 162–166

Akil HM, Omar MF, Mazuki AAM, Safiee S, Ishak ZAM, Abu Bakar A (2011) Kenaf fibre reinforced composites. A review. Mater Design 32:4107–4121

Aksogan O, Binici H, Ortlek E (2016) Durability of concrete made by partial replacement of fine aggregate by colemanite and barite and cement by ashes of corn stalk, wheat straw and sunflower stalk ashes. Constr Build Mater 106:253–263

Al Adili A, Al-Soudany K, Azzam R (2013) Effect of random inclusion of date palm leaf (fibres) on some properties of soil. Sci J Archit (SJA) 3(4):56–64

Al-Akhras N, Abu-Alfoul B (2002) Effect of wheat straw ash on mechanical properties of autoclaved mortar. Cem Concr Res 32(6):859–869

Alavez-Ramirez R, Chiñas-Castillo F, Morales-Dominguez V (2014) Effect of wheat straw ash on mechanical properties of autoclaved mortar. Cem Concr Res 32:859–863

Al-Darbi MM, Saeed NO, Aijolaiya LO, Islam MR (2006) A novel oil well cementing technology using natural fibres. Pet Sci Technol 24(11):1267–1282

Al-Khalaf MN, Yousif HA (1984) Use of rice husk ash in concrete. Int J Cem Compos Lightweight Concr 6(4):241–248

Ali M (2014) Seismic performance of coconut-fibre-reinforced-concrete columns with different reinforcement configurations of coconut-fibre ropes. Constr Build Mater 70:226–230

Ali M, Liu A, Sou H, Chouw N (2012) Mechanical and dynamic properties of coconut fibre reinforced concrete. Constr Build Mater 30:814–825

Al-Masoodi AHH, Kawan A, Kasmuri M, Hamid R, Khan MNN (2013) Static and dynamic properties of concrete with different types and shapes of fibrous reinforcement. Constr Build Mater 104:247–262

Almeida EFS, Tonoli GHD, Santos SF, Savastano H (2013) Improved durability of vegetable fibre reinforced cement composite subject to accelerated carbonation at early age. Cem Concr Compos 42:49–58

Alomayri T, Low IM (2013) Synthesis and characterization of mechanical properties in cotton fibre-reinforced geopolymer composites. J Asian Ceram Soc 1:30–34

Alomayri T, Vickers L, Shaikh FUA, Low IT (2013) Mechanical properties of cotton fabric reinforced geopolymer composites at 200–1000 °C. J Adv Ceram 3(3):184–193

Alomayri T, Shaikh FUA, Low IT (2014) Characterisation of cotton fibre-reinforced geopolymer composites. Compos Part B Eng 50:1–6

Alves Fidelis ME, Castro Pereira TV, de Martins Gomes OF, de Andrade Silva F, Toledo Filho RD (2013) The effect of fibre morphology on the tensile strength of natural fibres. J Mater Res Technol 2(2):149–157

Anon (2005) DaimlerChrysler awarded for banana fibre use in Mercedes A class. (http://www.netcomposites.com/news), Germany

Ardauny M, Claramunt J, Toledo Filho RD (2015) Cellulosic fiber reinforced cement-based composites: a review of recent research. Constr Build Mater 79:115–128
Asim M, Abdan K, Jawaid M, Nasir M, Dashtizadeh Z, Ishak MR, Enamul Hoque M (2015) A review on pineapple leaves fibre and its composites. Int J Polym Sci. doi:10.1155/2015/950567
ASTM C496/C496M (2011) Standard test method for splitting tensile strength of cylindrical concrete specimens. Am Soc Test Mater
ASTM C1609/C1609M (2012) Standard Test Method for Flexural Performance of Fiber-Reinforced Concrete (Using Beam With Third-Point Loading). Am Soc Test Mater
ASTM C1399/C1399M (2015) Standard Test Method for Obtaining Average Residual-Strength of Fiber-Reinforced Concrete. Am Soc Test Mater
Assaedi HS, Alomayri TS, Shaikh FUA, Low I-M (2014) Synthesis and mechanical properties of flax fabric reinforced geopolymer composites. Adv Mater Res 3(3):151–161
Aulin C, Gallstedt M, Lindstrom T (2010) Oxygen and oil barrier properties of microfibrillated cellulose films and coatings. Cellulose 17:559–574
Awang H, Ahmad MH, Al-Mulali MZ (2015) Influence of kenaf and polypropylene fibres on mechanical and durability properties of fibre reinforced lightweight foamed concrete. J Eng Sci Technol 10(4):496–508
Awwad E, Mabsout M, Hamad B, Talal Farran M, Khatib H (2012) Studies on fibre-reinforced concrete using industrial hemp fibres. Constr Build Mater 35:710–717
Awwad E, Chouetier D, Khatib H (2013) Concrete masonry blocks reinforced with local industrial hemp fibres and hurds. In: Claisse P et al (eds) Proceedings 3rd international conference on sustainable construction materials and technology, Kyoto, Japan, 18–21 Aug 2013, 11 pp
Awwad E, Hamad B, Mabsout M, Khatib H (2014) Structural behaviour of simply supported beams cast with hemp-reinforced concrete. ACI Struct J 111:1307–1316
Azizi Samir MAS, Alloin F, Paillet M, Dufresne A (2004) Tangling effect in fibrillated cellulose reinforced nanocomposites. Macromolecules 37(11):4313–4316
Azizi Samir MAS, Alloin F, Dufresne A (2005) Review of recent research into cellulosic whiskers, their properties and their application in nanocomposite field. Biomacromol 6(2):612–626
Azubuike CP, Esiaba J (2012) Investigation into some physico-technical and tableting properties of low-crystallinity powdered cellulose prepared from corn residues. J Pharm Res Opin 8:94–98
Azwa ZN, Yousif BF, Manalo AC, Karunasena W (2013) A review on the degradability of polymeric composites based on natural fibres. Mater Design 47:424–442
Baley C (2002) Analysis of the Flax fibres tensile behaviour and analysis of the tensile stiffness increase. Compos Part A 33:939–948
Banfill PFG, Starrs G, Derruau G, Mc Carter WJ, Chrisp TM (2006) Rheology of low carbon fiber content reinforced cement mortar. Cem Concr Compos 28(9):773–780
Baril MA, Sorelli L, Rethore J, Baby F, Toutlemonde F, Ferrara L, Bernardi S, Fafard M (2016) Effect of casting flow defects on the crack propagation in UHPFRC thin slabs by means of stereovision digital image correlation. Constr Build Mater 129:182–192
Beck-Candanedo S, Roman M, Gray DG (2005) Effect of reaction conditions on the properties and behaviour of wood cellulose nanocrystal suspensions. Biomacromol 6(2):1048–1054
Beckermann GW, Pickering KL (2008) Engineering and evaluation of hemp fibre reinforced polypropylene composites: fibre treatment and matrix modification. Compos A Appl Sci Manuf 39:979–988
Bederina M, Belhadj B, Ammari MS, Gouilleux A, Makhloufi Z, Montrelay N, Quéneudéc M (2016) Improvement of the properties of a sand concrete containing barley straws—treatment of the barley straws. Constr Build Mater 115:464–477
Belgacem MN, Gandini A (2005) Surface modification of cellulose fibres. Compos Interfaces 12:41
Belgacem MN, Gandini A (2008a) Monomers, polymers and composites from renewable resources. Elsevier, Amsterdam

Belgacem MN, Gandini A (2008b) Surface modification of cellulose fibres. In: Belgacem MN, Gandini A (eds) Monomers, polymers and composites from renewable resources. Elsevier, Amsterdam

Belhadj B, Bederina M, Makhloufi Z, Dheilly RM, Montrelay N, Quéneudec M (2016) Contribution to the development of a sand concrete lightened by the addition of barley straws. Constr Build Mater 113:513–522

Berhane Z (1999) Performance of natural fibre reinforced mortar roofing tiles. Materials and Structures 27:347–352

Berndt ML, Philippacopoulos AJ (2002) Incorporation of fibres in geothermal well cements. Geothermics 31:643–656

Berzins R, Kakitis A, Berzins U, Cukurs J (2014) Evaluation of hemp straw and fibre strength. In: Proceedings engineering for rural development, Jelgava, 29–30 May 2014, pp 198–203

Bie RS, Song XF, Liu QQ, Ji XY, Chen P (2015) Studies on effects of burning conditions and rice husk ash (RHA) blending amount on the mechanical behaviour of cement. Cem Concr Compos 55:162–168

Biricik H, Aköz F, Turker F, Berktay I (2000) Resistance to magnesium sulfate and sodium sulfate attack of mortar containing wheat straw ash. Cem Concr Res 30:1189–1197

Bledzki AK, Gassan J (1999) Composites reinforced with cellulose based fibres. Prog Polym Sci 24:221–274

Boghossian E, Wegner LD (2008) Use of flax fibres to reduce plastic shrinkage cracking in concrete. Cem Concr Compos 30(10):929–937

Boluk Y, Lahiji R, Zhao L, McDermott MT (2011) Suspension viscosities and shape parameter of cellulose nanocrystals (CNC). Colloids Surf A 377(1–3):297–303

Bondeson D, Mathew A, Oksman K (2006) Optimization of the isolation of nanocrystals from microcrystalline cellulose by acid hydrolysis. Cellulose 13(2):171–180

Bos H, Van den Oever M, Peters O (2002) Tensile and compressive properties of flax fibres for natural fibre reinforced composites. Mater Sci 37:1683–1692

Bouhicha M, Aouissi F, Kenai S (2005) Performance of composite soil reinforced with barley straw. Cem Concr Compos 27:617–621

Boulekbache B, Hamrat M, Chemrouk M, Amziane S (2010) Flowability of fibre-reinforced concrete and its effect on the mechanical properties of the material. Constr Build Mater 24(9):1664–1671

Bourmaud A, Baley C (2009) Rigidity analysis of polypropylene/vegetal fibre composites after recycling. Polym Degrad Stab 94:297–305

Bui VK, Geiker MR, Shah SP (2003) Rheology of fiber-reinforced cementitious materials. In: Naaman AE, Reinhardt HW (eds) Proceedings of the 4th international Rilem symposium on high performance fiber-reinforced cement composites (HPFRCC4), RILEM Pubs., pp 221–231

Bui DD, Hu J, Stroeven P (2005) Particle size effect on the strength of rice husk ash blended gap-graded portland cement concrete. Cem Concr Compos 27:357–366

Calligaris GA, Franco MKKD, Aldrige LP, Rodrigues MS, Beraldo AL, Yokaichiya F, Turrillas X, Cardoso LP (2015) Assessing the pozzolanic activity of cements with added sugar cane straw ash by synchrotron X-ray diffraction and Rietveld analysis. Constr Build Mater 98:44–50

Canovas MF, Selva NH, Kawiche GM (1992) New economical solutions for improvement of durability of portland cement mortars reinforced with sisal fibres. Mater Struct 25:417–422

Cao Y, Shibata S, Fukumoto I (2006) Mechanical properties of biodegradable composites reinforced with bagasse fibre before and after alkali treatment. Compos A 37:423–429

Caraschi JC, Leão AL (2001) Characterization of curaua fibre. Mol Cryst Liq Cryst 353:149–215

Carvalho MA, Calil CJ, Savastano HJ, Tubino R, Tereza Carvalho M (2008) Microstructure and mechanical properties of gypsum composites reinforced with recycled cellulose pulp. Mater Res 11(4)

Chagas Cordeiro G, Pessôa Sales C (2015a) Pozzolanic activity of elephant grass ash and its influence on the mechanical properties of concrete. Cem Concr Compos 55:331–336

Chagas Cordeiro G, Pessôa Sales C (2015b) Influence of calcining temperature on the pozzolanic characteristics of elephant grass ash. Cem Concr Compos 73:98–104
Chalee W, Sasakul T, Suwanmaneechot P, Jaturapitakkul C (2013) Utilization of rice husk–bark ash to improve the corrosion resistance of concrete under 5-year exposure in a marine environment. Cem Concr Compos 37:47–53
Chawla KK (1993) Ceramic matrix composites. Chapman and Hall, London
Chawla N, Kerr M, Chawla KK (2005) Monotonic and cyclic fatigue behaviour of high-performance ceramic fibres. J Am Ceram Soc 88:101–108
Che Muda A, Syamsir A, Nasharuddin Mustapha K, Rifdy Samsudin M, Thiruchelvam S, Usman F, Salmia Beddu N (2016) Impact resistance behaviour of banana fibre reinforced slabs. In: Proceedings international conference on advances in renewable energy and technologies (ICARET 2016) IOP IOP conference series: earth and environmental science vol 32, 012017, 5 pp
Chen J, Chouw N (2016) Compressive behaviour of flax FRP double tube confined coconut fibre reinforced concrete. Constr Build Mater 112:666–673
Chung DDL (2005) Dispersion of short fibres in cement. ASCE J Mater Civil Eng 17(4):379–383
Claramunt J, Arganuy M, Arevalo R, Pares F, Toledo Filho RD (2011a) Mechanical performance of ductile cement mortar composites reinforced with nanofibrillated cellulose. In: Proceedings 2nd international Rilem conference on SHCC-2, Rio de Janeiro, Brazil, RILEM Pubs, pp 131–138
Claramunt J, Ardauny M, Garcia-Hortal JA, Toldeo Filho RD (2011b) The hornification of vegetable fibres to improve the durability of cement mortar composites. Cem Concr Compos 33:586–595
CNR-DT 204 (2006) Istruzioni per la Progettazione, l'Esecuzione ed il Controllo di Strutture di Calcestruzzo Fibrorinforzato. Consiglio Nazionale delle Ricerche, Roma (in Italian)
Coutts RSP (1983) Flax fibres as a reinforcement in cement mortars. Int J Cem Compos Lightweight Concr 5(4):257–262
Coutts RSP, Warden PG (1987) Air-cured Abaca reinforced cement composite. Int J Cem Compos Lightweight Concr 9:69–73
Cordeiro GC, Toledo Filho RD, Tavares LM, de Rego Fairbairn FM, Hempel S (2011) Influence of particle size and specific surface area on the pozzolanic activity of residual rice husk ash. Cem Concr Compos 33:529–534
d'Almeida JRM, Aquino RCMP, Monteiro SN (2006) Tensile mechanical properties, morphological aspects and chemical characterization of piassava (*Attalea funifera*) fibres. Compos A 37(9):1473–1479
D'Almeida ALFS, Melo Filho JA, Toledo Filho RD (2009) Use of curaua fibres as reinforcement in cement composites. Mech Chem Eng Trans 17:1717–1722
D'Almeida A, Toledo Fliho R, Melo Filho J (2010) Cement composites reinforced by short curaua fibres. Rev Mater 15(2):151–156
Defoirdt N, Biswas S, De Vriese L, Ngoc Tran LQ, Van Acker J, Ahsan W, Gorbatikh L, Van Vuure A, Verpoest I (2010) Assessment of the tensile properties of coir, bamboo and jute fibre. Compos A 41:588–595
de Andrade Silva F, Chawla N, Toledo Filho RD (2008) Tensile behaviour of high performance natural (sisal) fibres. Compos Sci Technol 68:3438–3443
de Andrade Silva F, Mobasher B, Toledo Filho RD (2009) Cracking mechanisms in durable sisal fibre reinforced cement composites. Cem Concr Compos 31:721–730
de Andrade Silva F, Chawla N, Toledo Filho RD (2010a) Mechanical behaviour of natural sisal fibres. J Biobased Mater Bioenergy 4:1–8
de Andrade Silva F, Mobasher B, Toledo Filho RD (2010b) Fatigue behaviour of sisal fibre reinforced cement composites. Mater Sci Eng A 527:5507–5513
de Andrade Silva F, Toledo Filho RD, Melo Filho JDA, Fairbairn EDMR (2010c) Physical and mechanical properties of durable sisal fibre–cement composites. Constr Build Mater 24:777–85
de Andrade Silva F, Zhu D, Mobasher B, Soranakom C, Toledo Filho RD (2010d) High speed tensile behaviour of sisal fibre cement composites. Mater Sci Eng A 527:544–552

de Andrade Silva F, Toledo Filho RD, Mobasher B, Chawla N (2010e) A multi-scale investigation of the mechanical behaviour of durable sisal fibre cement composites. Matéria (UFRJ) 15:338–344
de Melo Filho JA, de Andrade Silva F, Toledo Filho RD (2013) Degradation kinetics and aging mechanisms on sisal fibre cement composite systems. Cem Concr Compos 40:30–39
di Prisco M, Plizzari G, Vandewalle L (2009) Fibre reinforced concrete. New design perspectives. Mater Struct 42:1261–1281
Dhonde HB, Mo YL, Hsu TTC, Vogel J (2007) Fresh and hardened properties of self-consolidating fiber-reinforced concrete. ACI Mater J 104(5):491–500
Dong XM, Revol JF, Gray DG (1998) Effect of microcrystallite preparation conditions on the formation of colloid crystals of cellulose. Cellulose 5(1):19–32
Dufresne A (2010) Polymer nanocomposites from biological sources. In: Nalwa HS (ed) Encyclopedia of nanoscience and nanotechnology. American scientific Publisher, Valencia
Duval A, Bourmaud A, Augier L, Baley C (2011) Influence of the sampling area of the stem on the mechanical properties of hemp fibres. Mater Lett 65:797–800
Eichhorn SJ, Young RJ (2004) Composite micromechanics of hemp fibres and epoxy resin microdroplets. Compos Sci Technol 64(5):767–772
Eichhorn SJ, Dufresne A, Aranguren M, Marcovich NE, Capadona JR, Rowan SR, Weder C, Thielemans W, Roman M, Renneckar S, Gindl W, Veigel S, Yano H, Abe K, Nogi M, Nakagaito AN, Mangalam A, Simonsen J, Benight AS, Bismarck A, Berglund LA, Pejis T (2010) Review: current international research into cellulose nanofibres and nanocomposites. J Mater Sci 45:1
de Aro EF, Fajardo BHJC, Perez MA (2012) Properties of pineapple leaf fiber-reinforced concrete beams. B.Sc thesis, Mapua Institute of Technology October 2012. 143 pp
Elsaid A, Dawood M, Seracino R, Bobko C (2011) Mechanical properties of kenaf fibre reinforced concrete. Constr Build Mater 25:1991–2001
EN 12390-6 (2000) Testing hardened concrete—part 6: tensile splitting strength of test specimens. European Committee for Standardization, Brussels, BE
EN 14651 (2005) Test method for metallic fibre concrete. Measuring the flexural tensile strength (limit of proportionality (LOP), residual). European Committee for Standardization, Brussels, BE
Eriksen K, Syberud Ø, Gregersen (2008) The use of microfibrillated cellulose produced from kraft pulp as a strength enhancer in TMP paper. Nordic Pulp Pap Res J 23(3):299–304
Etri S, Numada M, Meguro K (2015) URM wallets retrofitted with fibre reinforced mortar (FRM) and abaca rope mesh (ARM) subjected to diagonal compression test. Bulletin of ERS, 48, Institute of Industrial Science, University of Tokyo, pp 103–111
European Concrete Societies Network, European Concrete Award 2014, 13 pp
Fan M (2010) Characterization and performance of elementary hemp fibres: factors influencing tensile strength. BioResources 5:2307–2322
Faruk O, Mohini S (eds) (2014) Biofiber reinforcements in composite materials, 1st edn. Woodhead Publishing, 772 pp, eBook ISBN:9781782421276, Hardcover ISBN:9781782421221
Faruk O, Sain M (2015) Biofibre reinforcement in composite materials. Woodhead Publishing, 772 pp, ISBN-9781782421221
Fengel D, Wegener G (1984) Wood—chemistry, ultrastructure, reactions, 1st edn. Wiley, Berlin, 613 pp
Ferrara L, Park YD, Shah SP (2008) A method for mixdesign of fiber-reinforced self-compacting concrete. Cem Concr Res 37:957–971
Ferrara L, Ozyurt N, di Prisco M (2011) High mechanical performance of fibre reinforced cementitious composites: the role of "casting-flow" induced fibre orientation. Mater Struct 44 (1):109–128
Ferrara L, Faifer M, Toscani S (2012a) A magnetic method for non destructive monitoring of fibre dispersion and orientation in steel fibre reinforced cementitious composites—part 1: method calibration. Mater Struct 45:575–589

Ferrara L, Faifer M, Muhaxheri M, Toscani S (2012b) A magnetic method for non destructive monitoring of fibre dispersion and orientation in steel fibre reinforced cementitious composites —part 2: correlation to tensile fracture toughness. Mater Struct 45:591–598

Ferrara L, Krelani V, Carsana M (2014) A fracture testing based approach to assess crack healing of concrete with and without crystalline admixtures. Const Build Mater 68:515–531

Ferrara L (2015a) Tailoring the orientation of fibres in high performance fibre reinforced cementitious composites: part 1—experimental evidence, monitoring and prediction. J Mater Struct Integrity 9:72–91

Ferrara L (2015b) Tailoring the orientation of fibres in high performance fibre reinforced cementitious composites: part 2—correlation to mechanical properties and design implications. J Mater Struct Integrity 9:92–107

Ferrara L, Park YD, Shah SP (2007) A method for mix-design of fiber reinforced self compacting concrete. Cem Concr Res 37:957–971

Ferrara L, Ferreira SR, Della Torre M, Krelani V, de Silva FA, Toledo Filho RD (2015a) Effect of cellulose nanopulp on autogenous and drying shrinkage of cement based composites. In: Sobolev K, Shah SP (eds) Nanotechnology in construction, proceedings Nicom, vol 5. Springer, Chicago, 26–28 May 2015, pp 325–330, ISBN 9783319170879

Ferrara L, Ferreira SR, Krelani V, Della Torre M, de Andrade Silva F, Toledo Filho RD (2015b) Natural fibres as promoters of autogenous healing in HPFRCCs: results from an on-going Italy-Brasil cooperation. In Chiorino MA et al (eds) ACI special publication 305. Durability and sustainability of concrete structures—workshop proceedings, Bologna, Italy, 1–3 Oct 2015, pp. 11.1–11.10, ISBN-13: 978-1-942727-44-6

Ferrara L, Krelani V, Moretti F (2016a) On the use of crystalline admixtures in cement based construction materials: from porosity reducers to promoters of self healing. Smart Mater Struct 25:084002 (17 pp). doi:10.1088/0964-1726/25/8/084002

Ferrara L, Krelani V, Moretti F, Roig Flores M, Serna Ros P (2016b) Effects of autogenous healing on the recovery of mechanical performance of high performance fibre reinforced cementitious composites (HPFRCCs): part 1, submitted for publication to cement and concrete composites, 13 Dec 2015, resubmitted with revisions 2 Sept 2016

Ferrara L, Krelani V, Moretti F (2016c) Autogenous healing on the recovery of mechanical performance of high performance fibre reinforced cementitious composites (HPFRCCs): part 2 —correlation between healing of mechanical performance and crack sealing. Submitted for publication to Cem Concr Compos 73:299–315

Ferrara L, Ferreira SR, da Cunha Moreira TN, Silva F de A, Toledo Filho RD (2017) Self-healing capacity of hybrid sisal-steel fiber reinforced cementitious composites and comparison with steel-only HPFRCC. Submitted for publication to Cement and Concrete Composites, October 13, 2016, resubmitted with revisions, April 15, 2017

Ferreira SR, Lima PRL, Silva F de A, Toledo Filho RD (2014) Effect of sisal fibre hornification on the fibre matrix bonding characteristics and bending behaviour of cement based composites. Key Engineering Materials 600:421–432

Fic S, Brzyski P, Szelag M (2015) Composite based on foam lime mortar with flax fibres for use in the building industry. In: Proceedings ECOPole, pp 26–31. doi:10.2429/proc.2015.9(1)003

FLEXPAKRENEW (2007) FP7 program. (NMP-2007-2.4-3): renewable materials for functional packaging applications

Frias M, Savastano H, Villar E, Sanchez de Rojas MI, Santos S (2012) Characterization and properties of blended cement matrices containing activated bamboo leaf wastes. Constr Build Mater 34:1019–1023

Frıas M, Villar-Cocina E, Sanchez de Rojas MI, Valencia-Morales E (2005) The effect that different pozzolanic activity methods has on the kinetic constants of the pozzolanic reaction in sugar cane straw-clay ash/lime systems: application of a kinetic–diffusive model. Cem Concr Res 35:2137–2142

Garas V, Kahn L, Kurtis K (2009) Short-term tensile creep and shrinkage of ultra-high performance concrete. Cem Concr Compos 31:147–152

Gastaldini ALG, Isaia GC, Gomes NS, Sperb JEK (2007) Chloride penetration and carbonation in concrete with rice husk ash and chemical activators. Cem Concr Compos 29:176–180

Gastaldini ALG, Isaia GC, Saciloto AP, Missau F, Hoppe TF (2010) Influence of curing time on the chloride penetration resistance of concrete containing rice husk ash: a technical and economical feasibility study. Cem Concr Compos 32:783–793

Ghavami K (1995) Ultimate load behaviour of bamboo-reinforced lightweight concrete beams. Cem Concr Compos 17:281–288

Ghavami K (2005) Bamboo as reinforcement in structural concrete elements. Cem Concr Compos 27:637–649

Ghavami K, Toledo Filho RD, Barbosac NM (1999) Behaviour of composite soil reinforced with natural fibres. Cem Concr Compos 21:39–48

Gholizadeh Vayghan A, Khaloo AR, Rajabipour F (2013) The effects of a hydrochloric acid pre-treatment on the physicochemical properties and pozzolanic performance of rice husk ash. Cem Concr Compos 39:131–140

Giaccio G, Rodrıguez de Sensale G, Zerbino R (2007) Failure mechanism of normal and high-strength concrete with rice-husk ash. Cem Concr Compos 29:566–574

Gomes A, Matsuo T, Goda K, Ohgi J (2007) Development and effect of alkali treatment on tensile properties of curaua fiber green composites. Compos: Part A 38:1811–1820

Grunewald S (2004) Performance-based design of selfcompacting fiber reinforced concrete. Ph.D. thesis. Section of Structuring and Building Engineering, Delft University of Technology, The Netherlands

Habibi Y, Lucia LA, Rojas OJ (2010) Cellulose nanocrystals: chemistry, self-assembly, and applications. Chem Rev 110(6):3479–3500

He J, Jie Y, Zhang J, Yu Y, Zhang G (2013) Synthesis and characterization of red mud and rice husk ash-based geopolymer composites. Cem Concr Compos 37:108–118

Henriksson M, Berglund LA (2007) Structure and properties of cellulose nanocomposite films containing melamine formaldehyde. J Appl Polym Sci 106:2817–2824

Herrick FW, Wash S (1984a) Process for preparing microfibrillated cellulose. US Patent Application No. US 4,481,077

Herrick FW, Wash S (1984b) Redispersible microfibrillated cellulose. US Patent Application No. US 4,481,076

Herrick FW, Casebier RL, Hamilton JK, Sanberg KR (1983) Microfibrillated cellulose: morphology and accessibility. In: Proceedings conference: IX—cellulose conference, vol 37. Syracuse, NY, USA, pp 797–813

Hoyos CG, Cristia E, Vazquez A (2013) Effects of cellulose microcrystalline particles on properties of cement based composites. Mater Des 51:810–818

Hughes BP, Fattuhi NI (1976) The workability of steel-fiber-reinforced concrete. Mag Concr Res 28(9):157–161

Hult EL, Iotti M, Lenes M (2010) Efficient approach to high barrier packaging using microfibrillar cellulose and shellac. Cellulose 17:575–586

Iglesias SM, Almeida HC, Dominguez DS, Santos JFL (2016) Three-dimensional image processing applied to the characterization of lightweight mortar reinforced with piassaba fibres. In: Marcal PV, Yamagata N (eds) Design and analysis of reinforced fibre composites. Springer, pp 19–24, ISBN 978-3-319-20006-4

Ingrao C, Lo Giudice A, Bacenetti J, Tricase C, Dotelli G, Fiala M, Siracusa V, Mbohwa C (2014) Energy and environmental assessment of industrial hemp for building applications: a review. Renew Sustain Energy Rev 51:29–42

IIT Kharagpur, Materials Science Center (2011) Development of jute fibre reinforced cement concrete composites. Final report Project No: JMDC/JTM/MM-IV/7.1/2008, sponsored by National Jute Board, Ministry of Textile, Govt. of India: 78 pp. Available online at http://www.jute.com/documents/10437/0/JFR_FINAL_PROJECT_REPORT.pdf/971388b6-640a-44db-8465-bea71f9244a4. Accessed 14 June 2016

Jarabo R, Fuente E, Monte MC, Savastano H Jr, Mutié P, Negro C (2012) Use of cellulose fibres from hemp core in fibre-cement production. Effect on flocculation, retention, drainage and product properties. Ind Crops Prod 39:89–96
John VM, Cincotto MA, Sjöström C, Agopyan V, Oliveira CTA (2005) Durability of slag mortar reinforced with coconut fibre. Cem Concr Compos 27:565–574
Julien B, David V, Cécile B, Alain D (2011) Correlation between stiffness of sheets prepared from cellulose whiskers and nanoparticles dimensions. Carbohydr Polym 84:211–215ù
Kan L, Shi H (2012) Investigation of self-healing behaviour of engineered cementitious composites (ECC) materials. Constr Build Mater 29:356–358
Kan L, Shi H, Sakulich A, Li V (2010) Self-healing characterization of engineered cementitious composite materials. ACI Mater J 107(6):617–624
Khalil HPSA, Jawaid M, Hassan A, Paridah MT, Zaidon A (2012) Oil palm biomass fibres and recent advancement in oil palm biomass fibres based hybrid biocomposites. In: Composites and their application, pp 187–200. doi:10.5772/48235 (ch. 8)
Khedari J, Watsanasathaporn P, Hirunlabh J (2005) Development of fibre-based soil–cement block with low thermal conductivity. Cem Concr Compos 27:111–116
Kim D, Kang S, Ahn T (2014) Mechanical characterization of high-performance steel-fibre reinforced cement composites with self-healing effect. Materials 7(1):508–526
Klemm D, Schumann D, Kramer F, Hessler N, Hornung M, Schmauder HP et al (2006) Nanocelluloses as innovative polymers in research and application. Adv Polym Sci 205:49–96
Klemm D, Kramer F, Moritz S, Lindstrom T, Ankerfors M, Gray D et al (2011) Nanocelluloses: a new family of nature-based materials. Angew Chem Int Ed 50(24):5438–5466
Kriker A, Debicki G, Bali A, Khenfera MM, Chabannet M (2005) Mechanical properties of date palm fibres and concrete reinforced with date palm fibres in hot-dry climate. Cem Concr Compos 27:554–564
Kriker A, Bali A, Debicki G, Bouziane M, Chabannet M (2008) Durability of date palm fibres and their use as reinforcement in hot dry climates. Cem Concr Compos 30:639–648
Kuder KG, Ozyurt N, Mu EB, Shah SP (2007) Rheology of fiber-reinforced cementitious composites. Cem Concr Res 37:191–199
Kunchariyakun K, Asavapisit S, Sombatsompop K (2015) Properties of autoclaved aerated concrete incorporating rice husk ash as partial replacement for fine aggregate. Cem Concr Compos 55:11–16
Lavouine N, Desloges I, Dufresne A, Bras J (2012) Microfibrillated celulose—its barrier properties and applications in cellulosic materials: a review. Carbohydr Polym 90(2):735–764
Lawrence M, Heath A, Walker P (2009) Determining moisture levels in straw bale construction. Constr Build Mater 23:2763–2768
Leao AL, Machado IS, Souza SF, Soriano L (2009) Production of curaua fibers for industrial applications: characterization and micropropagation. Acta Hortic 822:227–238
Lertwattanaruk P, Suntijitto A (2015) Properties of natural fibre cement materials containing coconut coir and oil palm fibres for residential building applications. Constr Build Mater 94:664–669
Li M, Li VC (2011) Cracking and healing of engineered cementitious composites under chloride environment. ACI Mater J 108:333–340
Li M, Li VC (2013) Rheology, fibre dispersion, and robust properties of engineered cementitious composites. Mater Struct 46(3):405
Li VC, Stang H, Krenchel H (1993) Micromechanics of crack bridging in fibre reinforced concrete. Mater Struct 20:486–494
Li VC, Lim Y, Chan Y (1998) Feasibility of a passive smart self-healing cementitious composite. Compos B Eng 29:819–827
Li Z, Wnag X, Wang L (2006) Properties of hemp fibre reinforced concrete composites. Compos A 37:497–505
Liu D, Song J, Anderson DP, Chang PR, Hua Y (2012) Bamboo fibre and its reinforced composites: structure and properties. Cellulose 19(5):1449–1480

Lu J, Askeland P, Drzal LT (2008) Surface modification of microfibrillated cellulose for epoxy composite applications. Polymer 49(5):1285–1296
Lv Z, Chen D (2014) Overview of recent work on self-healing in cementitious materials. Materiales de Construcción 64(316):1–12
Mahat N, Yaacob Z, Mastan NF, Abd Rashid AF, Zainordin Z, Mohamed Rashid MR, Nizam Husin H, Khalil N, Najib Mat Noor M, Iskandar Wan Abdullah WF, Abd Rahman NA, Ahmad S (2010) Mod Appl Sci 4(7):119–129
Majeed K, Jawaid M, Hassan A, Abu Bakar A, Abdul Khalil HPS, Salema AA, Inuwa I (2013) Potential materials for food packaging from nanoclay/natural fibres filled hybrid composites. Mater Design 46:391–410
Maji AK, Shah SP (1988) Process zone and acoustic emission measurements in concrete. Exp Mech 28:27–33
Mansur MA, Aziz MA (1982a) A study of jute fibre reinforced cement composites. Int J Cem Compos Lightweight Concr 4(29):75–82
Mansur MA, Aziz MA (1982b) Study of bamboo-mesh reinforced cement composites. J Cem Compos Lightweight Concr 5(3):165–171
Mansur MA, Aziz MA (1983) Study of bamboo-mesh reinforced cement composites. Int J Cem Compos Lightweight Concr 5(3):165–171
Markovic I (2006) High-performance hybrid-fiber concrete: development and utilization. PhD Thesis, Department of Underground Infrastructure, Delft University of Technology
Marrot L, Lefeuvre A, Pontoire B, Bourmauda A, Baleya C (2013) Analysis of the hemp fibre mechanical properties and their scattering (Fedora 17). Ind Crops Prod 61:317–327
Martinie L, Rossi P, Roussel N (2010) Rheology of fiber reinforced cementitious materials: classification and prediction. Cem Concr Res 40:226–234
Martirena F, Middendorf B, Day RL, Gehrke M, Roque P, Martínez L, Betancourt S (2006) Rudimentary, low tech incinerators as a means to produce reactive pozzolan out of sugar cane straw. Cem Concr Res 36:1056–1061
Mihashi H, Nishiwaki T (2012) Development of engineered self-healing and self-repairing concrete: state-of-the-art report. J Adv Concr Technol 10:170–184
Miller RA, Shah SP, Bjelkhagen HI (1988) Crack profiles in mortar measured by hollographic interferrometry. Exp Mech 28:388–394
Minelli M, Baschetti MG, Doghieri F, Ankefors M, Lindstrom T, Siro I et al (2010) Investigation of mass transport properties of microfibrillated cellulose (MFC) films. J Membr Sci 358:67–75
Mobasher B, Castro-Montero A, Shah SP (1990) A study of fracture in fiber-reinforced cement-based composites using laser holographic interferometry. Exp Mech 30(3):286–294
Mohamed MAS, Ghorbel E, Wardeh G (2010) Valorization of micro-cellulose fibres in self-compacting concrete. Constr Build Mater 4(12):2473–2480
Mohr BJ, Nanko H, Kurtis KE (2005a) Durability of kraft pulp fibre–cement composites to wet/dry cycling. Cem Concr Compos 27:435–448
Mohr BJ, Nanko H, Kurtis KE (2005b) Durability of thermomechanical pulp fibre-cement composites to wet/dry cycling. Cem Concr Res 35:1646–1649
Mohr BJ, Biernacki JJ, Kurtis KE (2006a) Microstructural and chemical effects of wet/ dry cycling on pulp fibre–cement composites. Cem Concr Res 36:1240–1251
Mohr BJ, Nanko H, Kurtis KE (2006b) Aligned kraft pulp fibre sheets for reinforcing mortar. Cem Concr Compos 28:161–172
Mohr BJ, Biernacki JJ, Kurtis KE (2007) Supplementary cementitious materials for mitigating degradation of kraft pulp fibre-cement composites. Cem Concr Res 37:1531–1543
Mohr BJ, Hood KL, Kurtis KE (2009) Mitigation of alkali–silica expansion in pulp fibre–mortar composites. Cem Concr Compos 31:677–681
Moon RJ, Martini A, Nairn J, Simonsen J, Youngblood J (2011) Cellulose nanomaterials review: structure, properties and nanocomposites. Chem Soc Rev 40:3941–3994
Moraes JCB, Akasaki JL, Melges JLP, Monzó J, Borracherob MV, Soriano L, Payá J, Tashima MM (2015) Assessment of sugar cane straw ash (SCSA) as pozzolanic material in

blended portland cement: microstructural characterization of pastes and mechanical strength of mortars. Constr Build Mater 94:670–677

Moraes JCB, Tashima MM, Akasaki JL, Melges JLP, Monzó J, Borrachero MV, Soriano L, Payá J (2016) Increasing the sustainability of alkali-activated binders: the use of sugar cane straw ash (SCSA). Constr Build Mater 124:148–154

Moroz JG, Lissel SL, Hagel MD (2014) Performance of bamboo reinforced concrete masonry shear walls. Constr Build Mater 61:125–137

Morseburg K, Chinga-Carrasco G (2009) Assessing the combined benefits of clay and nanofibrillated cellulose in layered TMP-based sheets. Cellulose 16(5):795–806

Mostafa M, Uddin N (2016) Experimental analysis of Compressed Earth Block (CEB) with banana fibers resisting flexural and compression forces. Case Stud Constr Mater 5:53–63

Mukhopadhyay S, Fangueiro R, Arpaç Y, Şentürk Ü (2008) Banana fibres—variability and fracture behaviour. J Eng Fibres Fabr 3(2):39–45

Murphy F, Pavia S, Walker R (2010) An assessment of the physical properties of lime-hemp concrete. In: Ní Nualláin NA, Walsh D, West R, Cannon E, Caprani C, McCabe B (eds) Proceedings of BRI/CRI. Cork, pp 431–439

Naaman AE, Reinhardt HW (2006) Proposed classification of HPFRC composites based on their tensile response. Mater Struct 39:547–555

Nakanishi EY, Frías M, Martínez-Ramírez S, Santos SF, Rodrigues MS, Rodríguez O, Savastano H Jr (2014) haracterization and properties of elephant grass ashes as supplementary cementing material in pozzolan/$Ca(OH)_2$ pastes. Constr Build Mater 73:391–398

Ofuyatan O, Olutoge F (2013) Flexural characteristics and potentials of oil palm stem as reinforcement in concrete beams. J Emerg Trends Eng Appl Sci (JETEAS) 4(4):642–647

Ozawa M, Bo Z, Kawaguchi J, Uchida Y (2013) Preventive effect on spalling of UFC using jute fibre at high temperature. In: MATEC web of conferences, vol 6, p 02006 (1–6)

Özbay E, Sahmaran M, Yücel H, Erdem T, Lachemi M, Li V (2013) Effect of sustained flexural loading on self-healing of engineered cementitious composites. J Adv Concr Technol 11:167–179

Ozerkan NG, Ahsan B, Mansour S, Iyengar SR (2013) Mechanical performance and durability of treated palm fibre reinforced mortars. Int J Sustain Built Environ 2:131–142

Ozyurt N, Mason TO, Shah SP (2007) Correlation of fiber dispersion, rheology and mechanical performance of FRCs. Cem Concr Compos 29(2):70–79

Parameswaran N, Liese W (1980) Ultrastructure aspects of bamboo cells. Cellul Chem Technol 14:587

Paramasivam P, Nathan GK, Das Gupta NC (1984) Coconut fibre reinforced corrugated slabs. Int J Cem Compos Lightweight Concr 6(1):19–27

Payen A (1838) Memoir on the composition of the tissue of plants and of woody [material]. Comptes Rendus 7:1052–1056

Pereira ENB, Fischer G, Barros JAO (2012) Direct assessment of tensile stress-crack opening behaviour of strain hardening cementitious composites (SHCC). Cem Concr Res 42:834–846

Peters SJ (2009) Fracture toughness investigations of micro and nano cellulose fibre reinforced ultra high performance concrete. MSc thesis, University of Maine

Placet V (2009) Characterization of the thermo-mechanical behaviour of hemp fibres intended for the manufacturing of high performance composites. Compos A Appl Sci Manuf 40(8):1111–1118

Placet V, Trivaudey F, Cisse O, Gucheret-Retel V, Boubakar ML (2012) Diameter dependence of the apparent tensile modulus of hemp fibres: a morphological, structural or ultrastructural effect? Compos A Appl Sci Manuf 43:275–287

Prasittisopin L, Trejo D (2015) Hydration and phase formation of blended cementitious systems incorporating chemically transformed rice husk ash. Cement Concr Compos 59:100–106

Qian S, Zhou J, de Rooij MR, Schlangen E, Ye G, van Breugel K (2009) Self-healing behaviour of strain hardening cementitious composites incorporating local waste materials. Cem Concr Compos 31:613–621

Qian SZ, Zhou J, Schlangen E (2010) Influence of curing condition and precracking time on the self-healing behaviour of engineered cementitious composites. Cem Concr Compos 32:686–693

Qiang X, Lei L, Yi-jun C (2013a) Study on the action effect of pavement straw composite fibrematerial in asphalt mixture. Constr Build Mater 43:293–299

Qiang X, Xia-ting F, Lei L, Yi-jun C, Xiao L (2013b) Evaluation of pavement straw composite fibre on SMA pavement performances. Constr Build Mater 41:834–843

Rahim M, Douzane O, Le Tran AD, Promis G, Langlet T (2016) Characterization and comparison of hygric properties of rape straw concrete and hemp concrete. Constr Build Mater 102:679–687

Rahimirad M (2012) Properties of oil well cement reinforced by carbon nanotubes. National Iranina South Oil Company, Javad Dehghani Baghbadorani, Islamica Azad University, Omidiyeh, Iran. Society of petroleum engineer

Rahman MM, Khan MA (2007) Surface treatment of coir (*Cocos nucifera*) fibres and its influence on the fibres' physico-mechanical properties. Compos Sci Technol 67:2369–2376

Ramli M, Dawood ET (2010) Effects of palm fibre on the mechanical properties of lightweight concrete crushed brick. Am J Eng Appl Sci 3(2):489–493

Rånby BG, Ribi E (1950) Experientia 6:12

Reier GE, Shangraw R (1966) Microcrystalline cellulose in tableting. J Pharm Sci 55(5):510–514

Revol JF, Gobdout L, Dong XM, Gray DG, Chanzy H, MARET G (1994) Chiralnematic suspensions of cellulose crystallites; phase separation and magnetic field orientation. Liq Cryst 16(1):127–134

Reynaldo MP Jr (2011) Natural organic fibre meshes as reinforcements in cement mortar matrix. In: Proceedings of the 2011 IAJC-ASEE international conference. ISBN 978-1-60643-379-9, Paper 108, INT vol 302, 12 pp

Rodríguez de Sensale G (2006) Strength development of concrete with rice-husk ash. Cem Concr Compos 28:158–160

Rodríguez de Sensale G (2010) Effect of rice-husk ash on durability of cementitious materials. Cem Concr Compos 32:718–725

Roig-Flores M, Moscato S, Serna P, Ferrara L (2015a) Self-healing capability of concrete with crystalline admixtures in different environments. Constr Build Mater 86:1–11

Roig-Flores M, Pirritano F, Serna Ros P, Ferrara L (2015b) Effect of crystalline admixtures on the self-healing capability of early-age concrete studied by means of permeability and crack closing tests. Constr Build Mater 114:447–457

Roig-Flores M, Pirritano F, Serna Ros P, Ferrara L (2016) Effect of crystalline admixtures on the self-healing capability of early-age concrete studied by means of permeability and crack closing tests. Const Build Mater 114:447–457

Rossi P (1992) Mechanical behaviour of metal-fibre reinforced concretes. Cem Concr Compos 14(1):3–16

Safiuddin M, West JS, Soudki KA (2010) Hardened properties of self-consolidating high performance concrete including rice husk ash. Cem Concr Compos 32:708–717

Saha BC, Iqbal Hossein AJM, Rahman T, Hossain MJ, Zakaria KM (2015) Influence of sowing time and variety on growth and yield of jute (*Corchorus capsularis* L) at Southern Region of Bangladesh. Eur Acad Res 3(4):4391–4402

Saheb DN, Jog JP (1999) Natural fibre polymer composites: a review. Adv Polym Technol 18:351–363

Sahmaran M, Li V (2008) Durability of mechanically loaded engineered cementitious composites under highly alkaline environments. Cem Concr Compos 30:72–81

Sahmaran M, Yildirim G, Noori R, Ozbay E, Lachemi M (2015) Repeatability and pervasiveness of self-healing in engineered cementitious composites. ACI Mater J 112(4):513–522

Salas J, Alvarez M, Veras J (1986) Lightweight insulating concretes with rice husk. J Cem Compos Lightweight Concr 3:171–180

Salas J, Alvarez M, Veras J (1987) Rice husk and fly ash concrete blocks. Int J Cem Compos Lightweight Concr 9(3):177–182

Salehian H, Barros JAO, Taheri M (2014) Evaluation of the influence of post-cracking response of steel fibre reinforced concrete (SFRC) on load carrying capacity of SFRC panels. Constr Build Mater 73:289–304

Sanjuàn MA, Toledo Filho RD (1998) Effectiveness of crack control at early age on the corrosion of steel bars in low modulus sisal and coconut fibre reinforced mortars. Cem Concr Res 28 (4):555–565

Sá Ribeiro RA, Sá Ribeiro MG, Sankar K, Kriven WM (2016) Geopolymer-bamboo composite—a novel sustainable construction material. Const Build Mater 123:501–507

Sassoni E, Manzia S, Motoria A, Montecchi M, Canti M (2014) Novel sustainable hemp-based composites for application in the building industry: physical, thermal and mechanical characterization. Energy Build 77:219–226

Sathish P, Kesavan R (2015) Banana fibre reinforced composites: a review. Int J Adv Res Sci Eng Technol 2(10):872–874

Satyanarayana KG, Guimaraes JL, Wypych F (2007) Studies on lignocellulosic fibers of Brazil. Composites 38:1694–1709

Savastano H, Warden PG, Coutts RSP (2000) Brazilian waste fibres as reinforcement for cement-based composites. Cem Concr Compos 22(5):379–384

Savastano H Jr, Warden P, Coutts R (2001) Ground iron blast furnace slag as a matrix for cellulose-cement materials. Cem Concr Compos 23:389–97

Savastano H, Warden PG, Coutts RSP (2003) Mechanically pulped sisal as reinforcement in cementitious matrices. Cem Concr Compos 25:311–319

Savastano H, Warden PG, Coutts RSP (2005) Microstructure and mechanical properties of waste fibre–cement composites. Cement Concr Compos 27:583–592

Savastano H, Turner A, Mercer C, Soboyejo WO (2006) Mechanical behaviour of cement-based materials reinforced with sisal fibres. J Mater Sci 41:6938–6948

Savastano H, Santos SF, Radonjic M, Soboyejo WO (2009) Fracture and fatigue of natural fibre-reinforced cementitious composites. Cem Concr Compos 31:232–243

Sayed M (2014) Plant fibre reinforced composites. Abaca fibres in concrete. Investigation on the possible application of Abaca fibre as reinforcement in concrete to create ductility. PhD thesis. Department of Materials and Environment, Delft University of Technology, 392 + xvii pp

Schlangen E, Sangadji S (2013) Addressing infrastructure durability and sustainability by self healing mechanisms—recent advances in self healing concrete and asphalt. Proc Eng 54:39–57

Schlangen E, Jonkers H, Qian S, Garcia A (2010) Recent advances on self healing of concrete. In: Oh B et al (eds) Fracture mechanics of concrete and concrete structures—recent advances in fracture mechanics of concrete. Proceedings of FraMCoS-7, Jeju, South Korea, pp 291–298

Schrader EK (1978) Formulating guidance for testing of fibre concrete in ACI committee 544. In: Proceedings RILEM symposium on testing and test methods of fibre cement composites. Construction Press Ltd., Lancaster, pp 9–21

Sedan D, Pagnoux C, Smith A, Chotard T (2008) Mechanical properties of hemp fibre reinforced cement: influence of the fibre/matrix interaction. J Eur Ceram Soc 28:183–192

Sen T, Jagannatha Reddy H (2011) Various industrial applications of hemp, kinaf, flax and ramie natural fibres. Int J Innov Manag Technol 2(3):192–198

Sen T, Jagannatha Reddy H (2013) Strengthening of RC beams in flexure using natural jute fibre textile reinforced composite system and its comparative study with CFRP and GFRP strengthening systems. Int J Sustain Built Environ 2:41–55

Simon J, Muller HP, Koch R, Muller V (1998) Thermoplastic and biodegradable polymers of cellulose. Polym Degrad Stab 59:107–115

Siqueira G, Bras J, Dufresne A (2010) Cellulosic bionanocomposites: a review of preparation, properties and applications. Polymer 2(4):728–765

Siro I, Plackett D (2010) Microfibrillated cellulose and new nanocompositematerials: a review. Cellulose 17(3):459–494

Snoeck D, de Belie N (2012) Mechanical and self-healing properties of cementitious composites reinforced with flax and cottonised flax, and compared with polyvinyl alcohol fibres. Biosyst Eng 111:325–335

Snoeck D, de Belie N (2015) From straw in bricks to modern use of microfibres in cementitious composites for improved autogenous healing—a review. Constr Build Mater 95:774–787
Snoek D, De Belie N (2016) Repeated autogenous healing in strain-hardening cementitious composites by using superabsorbent polymers. ASCE J Mater Civil Eng 8. doi:10.1061/(ASCE)MT.1943-5533.0001360
Soranankom C, Mobasher B (2008) Correlation of tensile and flexural responses of strain softening and strain hardening cement composites. Cem Concr Compos 30(6):465–477
Soroushian P, Aouadi F, Chowdhury H, Nossoni A, Sarwar G (2004) Cement-bonded straw board subjected to accelerated processing. Cem Concr Compos 26:797–802
Soroushian P, Elzafraney M, Nossoni A, Chowdhury H (2006) Evaluation of normal-weight and light-weight fillers in extruded cellulose fiber cement products. Cem Concr Compos 28(1):69–76
Soroushian P, Won J-P, Hassan M (2012) Durability characteristics of CO_2 cured cellulse fibre reinforced cement composites. Constr Build Mater 34:44–53
Souza SF, Leao AL, Cai JH, Crystal W, Sain M, Cherian BM (2010) Nanocellulose from Curaua fibers and their nanocomposites. Mol Cryst Liq Cryst 522:42–52
Spinacè MAS, Lambert CS, Fermoselli KKG, De Paoli MA (2009) Characterization of lignocellulosic curaua fibers. Carbohydr Polym 77:47–53
Sturcova A, Davies GR, Eichhorn SJ (2005) Elastic modulus and stress-transfer properties of tunicate cellulose whiskers. Biomacromol 6(2):1055–1061
Sugiyama J, Chanzy H, Revol JF (1994) On the polarity of cellulose in the cell wall of Valonia. Planta 193(2):260–265
SUNPAP (2009) http://sunpap.vtt.fi/
Tang W, Kardani O, Cui H (2015) Robust evaluation of self-healing efficiency in cementitious materials—a review. Constr Build Mater 81:233–247
Teo DCL, Mannan MA, Kurian JV (2006) Flexural behaviour of reinforced lightweight concrete beams made with oil palm shell (OPS). J Adv Concr Technol 4(3):459–468
Tashiro K, Kobayashi M (1991) Theoretical evaluation of three-dimensional elastic constants of native and regenerated celluloses: role of hydrogen bonds. Polymer 32(8):1516–1526
Thomson A, Walker P (2014) Durability characteristics of straw bales in building envelopes. Constr Build Mater 68:135–141
Thygesen A, Daniel G, Lilholt H, Thomsen AB (2006) Hemp fibre microstructure and use of fungal defibration to obtain fibres for composite materials. J Nat Fibres 2:19–37
Tolêdo Filho RD, Scrivener K, England GL, Ghavami K (2000) Durability of alkali-sensitive sisal and coconut fibres in cement mortar composites. Cem Concr Compos 22(2):127–143
Tolêdo-Filho RD, Ghavami K, England GL, Scrivener K (2003) Development of vegetable fibre-mortar composites of improved durability. Cem Concr Compos 25:185–196
Toledo Filho RD, Ghavami K, Ma Sanjuán, England GL (2005) Free, restrained and drying shrinkage of cement mortar composites reinforced with vegetable fibres. Cem Concr Compos 27:537–546
Toledo Filho RD, de Andrade Silva F, Fairbairn EMR, Melo Filho JDA (2009) Durability of compression molded sisal fibre reinforced mortar laminates. Constr Build Mater 23:2409–2420
Tonoli GHD, Fuente E, Monte C, Savastano H, Lahr FAR, Blanco A (2009a) Effect of fibre morphology on flocculation of fibre–cement suspensions. Cem Concr Res 39:1017–1022
Tonoli GHD, Rodrigues Filho UP, Savastano H, Bras J, Belgacem MN, Lahr Fa R (2009b) Cellulose modified fibres in cement based composites. Compos A Appl Sci Manuf 40:2046–2053
Tonoli GHD, Savastano H, Fuente E, Negro C, Blanco A, Lahr FAR (2010a) Eucalyptus pulp fibres as alternative reinforcement to engineered cement based composites. Ind Crops Prod 31:225–232
Tonoli GHD, Santos SF, Joaquim AP, Savastano H (2010b) Effect of accelerated carbonation on cementitious roofing tiles reinforced with lignocellulosic fibre. Constr Build Mater 24:193–201

Tonoli GHD, Santos SF, Savastano H, Delvasto S, Mejía de Gutiérrez R, Lopez de Murphy MDM (2011) Effects of natural weathering on microstructure and mineral composition of cementitious roofing tiles reinforced with fique fibre. Cem Concr Compos 33:225–232

Troëdec M Le, Dalmay P, Patapy C, Peyratout C, Smith A, Chotard T (2011) Mechanical properties of hemp-lime reinforced mortars: influence of the chemical treatment of fibres. J Compos Mater 45(22):2347–2357

Turbak AF, Snyder FW, Sandberg KR (1985) Micro-fibrillated cellulose and process for producing it. Patent no CH 648071, A5

Vaishnav P, Titiksh A (2016) Evaluating the performance of hybrid fibre reinforced self compacted concrete (Hfrscc) using steel and banana fibres. J Ceram Concr Sci 1(1):1–10

Van VTA, Rößler C, Bui DD, Ludwig H-M (2014) Rice husk ash as both pozzolanic admixture and internal curing agent in ultra-high performance concrete. Cem Concr Compos 53:270–278

Van Tittelboom K, De Belie N (2013) Self-healing in cementitious materials—a review. Materials 6:2182–2217

Venkatanarayanan HK, Rangaraju PR (2015) Effect of grinding of low-carbon rice husk ash on the microstructure and performance properties of blended cement concrete. Cem Concr Compos 55:348–363

Villar-Cociña E, Valencia Morales E, Gonzalez-Rodrıguez R, Hernandez-Ruız J (2003) Kinetics of the pozzolanic reaction between lime and sugar cane straw ash by electrical conductivity measurement: a kinetic–diffusive model. Cem Concr Res 33:517–524

Villar-Cociña E, Valencia Morales E, Santos SF, Savastano H, Frias M (2014) Pozzolanic behaviour of bamboo leaf ash: characterization and determination of the kinetic parameters. Constr Build Mater 33:68–73

Wang W-J, Wu C-H (2013) Benefits of adding rice straw coke powder to cement mortar and the subsequent reduction of carbon emissions. Constr Build Mater 47:616–622

Wu M, Johannesson B, Geiker M (2012) A review: self-healing in cementitious materials and engineered cementitious composite as a self-healing material. Constr Build Mater 28:571–583

Xia Y, Xian G, Wang Z, Li H (2016) Static and cyclic compressive properties of self-compacting concrete-filled flax fibre–reinforced polymer tubes. ASCE J Compos Constr. DOI 10.1061/(ASCE)CC.1943-5614.0000706

Yang Y, Lepech MD, Yang En-H, Li VC (2009a) Autogenous healing of engineered cementitious composites under wet–dry cycles. Cem Concr Res 39:382–390

Yang EH, Sahmaran M, Yang Y, Li VC (2009b) Rheological control in the production of engineered cementitious composites. ACI Mater J 106(04):357–366

Yaremko C (2012) Durability of flax fibre reinforced concrete. MSc thesis, Department of Civil Engineering, University of Saskatchewan, Saskatoon, Canada, pp 129 + ix

Yetimoglu S (1984) Abaca fibre reinforced phosphogypsum concrete panels (modulus of rupture, toughness, splitting tensile strength). DA Thesis, UMiami. Dissertations from ProQuest. Paper 1426. http://scholarlyrepository.miami.edu/dissertations/1426

Yildirim G, Alyousif A, Sahmaran M, Lachemi M (2015a) Assessing the self-healing capability of cementitious composites under increasing sustained loading. Adv Cem Res 27(10):581–592

Yildirim G, Keskin O, Keskin S, Sahmaran M, Lachemi M (2015b) A review of intrinsic self-healing capability of engineered cementitious composites: recovery of transport and mechanical properties. Constr Build Mater 101:10–21

Yousif B, EL-Taayeb N (2007) Tribological evaluations of polyester composites considering three orientations of CSM glass fibres using BOR machine. Appl Compos Mater 14:105–116

Zah R, Hischier R, Leao AL, Braun I (2007) Curaua fibers in the automotive industry and the sustainability assessment. J Clean Prod 15:1032–1040

Zimmermann T, Pohler E, Geiger T (2004) mCellulose fibrils for polymer reinforcement. Adv Eng Mater 6(9):754–761

Zhu WH, Tobias BC, Coutts RSP, Langfors G (1994) Air-cured banana-fibre-reinforced cement composites. Cem Concr Compos 16:3–8

Chapter 10
Approaches for the Design of Structures Made by Concrete Reinforced with Sustainable Fibres

Joaquim A.O. Barros and Liberato Ferrara

Abstract In this section design recommendations for Fibre Reinforced Concrete and Cementitious Composites (FRCC) are presented, with main reference to the ones proposed by RILEM TC 162-TDF (2003) for SFRC and the ones proposed by CEB-FIP Model Code 2010 (MC2010). As a matter of fact, no specific reliable design recommendations are available so far for concrete reinforced with sustainable, either recycled or natural, fibres due to the relative novelty of these composite materials. However, when short discrete natural or recycled fibres, herein encompassed within the common wording of sustainable fibres, are used as a randomly distributed reinforcement in a concrete matrix, the design recommendations of the MC2010 can, in general, be adopted for the design of sustainable FRCC elements. Nevertheless, to the aforementioned purpose, the fibre reinforced composite materials have to be experimentally characterized according to the procedures described in Sect. 8.2.5, in order to obtain the residual flexural tensile strength parameters (f_{Rj}) to model the post-cracking behaviour. Some design tools are also proposed to design specific aspects of sustainable FRCC, such is the case of composites reinforced with continuous (long) vegetable fibres.

10.1 Introduction

The energy consumed in concrete cracking is the property most benefited from adding fibres to concrete. To characterize toughness enhancement provided by fibre reinforcement mechanisms, ASTM C 1018 (1990) recommended the use of toughness indexes, I_N, and residual strength factors, while the Japanese Society of

J.A.O. Barros (✉)
University of Minho, Braga, Portugal
e-mail: barros@civil.uminho.pt

L. Ferrara
Politecnico di Milano, Milan, Italy

J.A.O. Barros et al. (eds.), *Recent Advances on Green Concrete for Structural Purposes*, Research for Development, DOI 10.1007/978-3-319-56797-6_10

Civil Engineers (JSCE 1984) proposed the concept of the flexural toughness factor, *FT*. Gopalaratnam et al. (1991) and Banthia and Trottier (1995a, b) pointed out the fragilities of the aforementioned two approaches, namely: the I_N indexes of ASTM C 1018 (1990) are susceptible to human judgment errors, since they require an accurate assessment of the crack initiation, which is almost impossible to guarantee; the *FT* of JSCE is dependent on specimen geometry, and the limit deflection of span/150 (mm) used in its calculation is not based on serviceability considerations. Using a test set-up identical to JSCE (1984), Banthia and Trottier (1995a) proposed the concept of post-crack strength, *PCS*. The main contributions of this technique are the exclusion of the energy up to peak load (energy not influenced by fibre reinforcement) from the total energy up to target deflections, and the consideration of various deflection limits to cover distinct requirements on serviceability limit state analysis.

The recommendations of RILEM TC 162-TDF (2003) to characterize the post-cracking behaviour of SFRC (Vandewalle et al. 2000a, b) can be regarded as an improvement of the technique proposed by Banthia and Trottier (1995a). To decrease the scatter in the values generally reported when using four point un-notched beam tests (like JSCE 1984 and Banthia and Trottier 1995a, b test set up), RILEM TC 162-TDF (2003) proposed a three-point notched beam test. Since plain concrete shows a strain-softening phase, RILEM TC 162-TDF (2003) recommended a more precise procedure for excluding the parcel of energy due to matrix cracking from the total energy absorbed by a SFRC. Furthermore, two deflection limits were used for the evaluation of the equivalent flexural tensile strength parameters, one to be used on the design at serviceability limit states, $f_{eq,2}$, and the other on the design at ultimate limit states, $f_{eq,3}$ (Vandewalle et al. 2002b). The f_{eq} concept corresponds to *FT* of JSCE (1984) and PCS of Banthia and Trottier (1995a, b) in as much as it is a function of the energy dissipated up to a given deflection. It deviates from *FT* and PCS, however, since f_{eq} is presumed to only take the energy absorption capacity provided by fibre reinforcement mechanisms. Later, RILEM TC 162-TDF (2003) proposed the replacement of f_{eq} for the concept of residual flexural tensile strength, f_R, which provides the stress for distinct deflections or crack mouth opening displacements, (CMOD) (Vandewalle et al. 2002). Although this last concept has the advantage of being easier to evaluate, it is more susceptible to the irregularities of the force-deflection relationships registered in the tests. The f_{eq} and f_R parameters were also used to define the stress-strain constitutive law proposed for modelling the post cracking behaviour of SFRC (Vandewalle et al. 2002b, 2003). The concept of f_R parameters have been adopted by MC2010 for the toughness classification of FRC, as well as for the definition of the constitutive laws under the framework of the design of structures made by this composite material.

10.2 Constitutive Relationships

10.2.1 RILEM TC 162-TDF Approach

Figure 10.1 shows the σ–ε diagram proposed by RILEM TC 162-TDF (2003) (Vandewalle et al. 2003) to model the uniaxial behaviour of SFRC. While the diagram in the compression stress regime is the same as for a plain concrete with the same compressive strength class, for the part of the diagram defining the behaviour in the tensile stress regime the relevant points are determined from the following relations:

$$\sigma_1 = C_1 f_{ctm,fl}(1.6 - d); \quad \sigma_2 = C_2 f_{R,1}\kappa_h; \quad \sigma_3 = C_3 f_{R,4}\kappa_h \tag{10.1}$$

$$\varepsilon_1 = \sigma_1/E_c; \quad \varepsilon_2 = \varepsilon_1 + C_4(‰); \quad \varepsilon_3 = C_5\,(‰); \quad E_c = 9500(f_{cm})^{1/3} \tag{10.2}$$

where $f_{ctm,fl}$ and E_c are the FRC average flexural tensile strength and Young's modulus (in MPa), respectively, d is the effective beam depth (in m), with $1.6 - d \geq 1.0$, and κ_h is a parameter that intends to simulate the influence of the depth of the beam's cross section on the post peak behaviour of FRC. RILEM TC 162-TDF (2003) proposed for the C_i parameters the following values: $C_1 = 0.7$, $C_2 = 0.45$, $C_3 = 0.37$, $C_4 = 0.1$ and $C_5 = 25$. As long as more advanced models are not available to evaluate κ_h, RILEM TC 162-TDF (2003) recommended the graph represented in Fig. 10.2. A procedure to determine the values of Ci parameters through inverse analysis will be described in the next section. (Barros et al. 2005).

10.2.2 CEB-FIP Model Code 2010 Approach

Using the values of f_{Rj} determined according to the approach described in Sect. 8.2.5, the stress-strain constitutive laws for ultimate limit states (ULS) and for serviceability limit states (SLS) can be derived.

For ULS the two models schematically represented in Fig. 10.3 are recommended, where f_{Fts} represents the serviceability residual strength, defined as the post-cracking strength for serviceability crack openings, while f_{Ftu} represents the

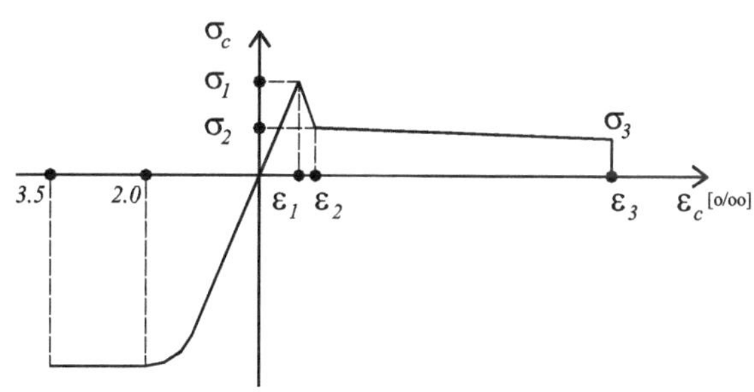

Fig. 10.1 σ–ε diagram for SFRC, according to RILEM TC 162-TDF (Vandewalle et al. 2003)

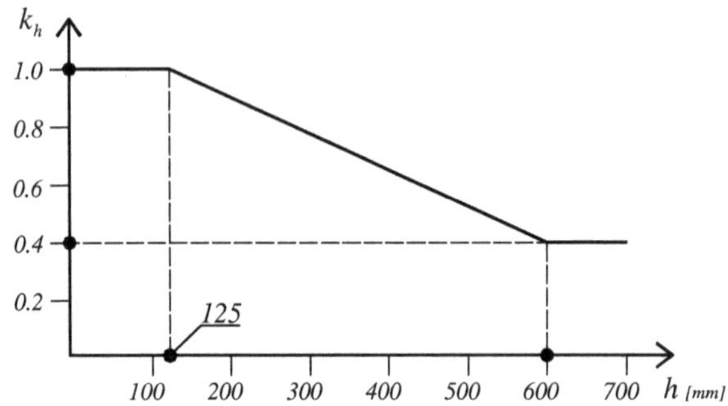

Fig. 10.2 Size factor according to RILEM TC 162-TDF (Vandewalle et al. 2003)

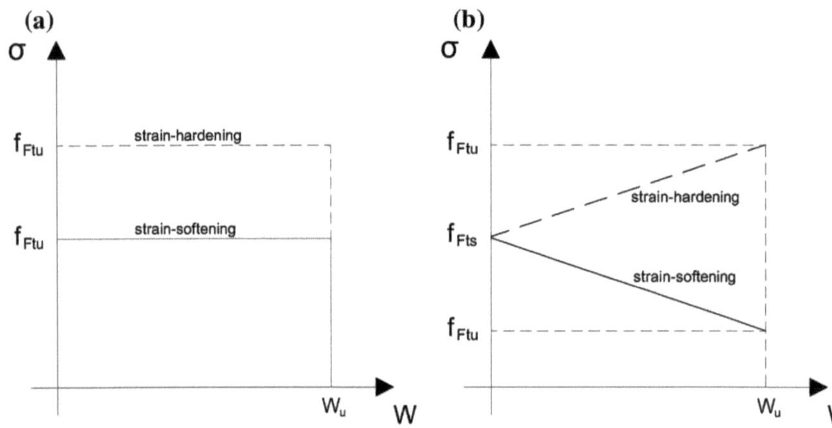

Fig. 10.3 Simplified stress-crack opening constitutive laws: **a** rigid-plastic model, **b** linear model (*continuous* and *dashed lines* refer to softening and hardening post-cracking behaviour, respectively)

ultimate residual strength. These two parameters are calculated from the following equations:

$$f_{Fts} = 0.45 f_{R1} \tag{10.3}$$

$$f_{Ftu} = f_{Fts} - \frac{w_u}{CMOD_3}(f_{Fts} - 0.5 f_{R3} + 0.2 f_{R1}) \geq 0 \tag{10.4}$$

where CMOD3 = 2.5 mm and fR1 and fR3 are the residual nominal flexural strength values at a CMOD equal to 0.5 and 2.5 mm respectively, obtained in 3-point notched beam bending test according to the EN 14651 (2005) recommendations.

If a strain-softening FRC is used, the strain can be derived from the crack width by using a characteristic length, l_{cs}, of the structural element:

$$\varepsilon = w/l_{cs} \tag{10.5}$$

If strain softening FRC is used with tensile steel rebars, the structural characteristic length l_{cs} can be estimated from:

$$l_{cs} = \min\{s_{rm}, y\} \tag{10.6}$$

where s_{rm} is the mean distance value between cracks, y is the distance between the neutral axis and the tensile surface of the cross section, evaluated in the elastic cracked phase by neglecting the residual tensile strength of FRC, and for a load configuration corresponding to the SLS of crack opening and crack spacing.

In sections without traditional reinforcement under bending or under combined tensile-flexural and compressive-flexural forces, with resulting force external to the section, $y = h$ is assumed. The same assumption can be taken for slabs.

The ultimate crack width can be calculated as $w_u = l_{cs}\, \varepsilon_{Fu}$, by assuming ε_{Fu} equal to 2 and 1% for, respectively, variable and constant strain distribution along the cross section. In any case, the maximum crack width shall not exceed 2.5 mm.

Figure 10.4a (CASE I) represents the stress-strain diagram recommended for the SLS analysis when a strain softening FRC is used. Up to ε_B the same σ–ε diagram adopted for the parent plain concrete matrix is used (MC2010 2011), while in the post-peak phase the bilinear diagram is defined from the following equations:

$$\frac{\sigma - f_{ct}}{0.2f_{ct} - f_{ct}} = \frac{\varepsilon - \varepsilon_B}{\varepsilon_Q - \varepsilon_B}, \quad \text{for} \quad \varepsilon_B \le \varepsilon \le \varepsilon_C \tag{10.7}$$

with

$$\varepsilon_Q = \frac{G_f}{f_{ct} \cdot l_{cs}} + \left(\varepsilon_B - \frac{0.8f_{ct}}{E_c}\right) \tag{10.8}$$

where G_f represents the fracture energy of the parent plain concrete matrix. According to MC 2010, $\sigma_A = 0.9f_{ct}$ and $\varepsilon_B = 0.15‰$. For strain softening FRC:

$$\varepsilon_{SLS} = CMOD_1/l_{cs} \tag{10.9}$$

$$\varepsilon_{ULS} = w_u/l_{cs} = \min\left(\varepsilon_{Fu}, 2.5/l_{cs}\right), \quad \text{with } l_{cs} \text{ in mm} \tag{10.10}$$

with $\text{CMOD}_1 = 0.5$ mm and $\varepsilon_{Fu} = 2\%$ for variable strain distribution along the cross section, and 1% for only tensile strain distribution along the cross section.

For materials characterized by a stable crack propagation up to ε_{SLS} with a tensile strength f_{Fts} larger than f_{ct}, two cases can be considered:

CASE II, Fig. 10.4b: the cracking process becomes stable up to the SLS strain, ε_{SLS}, and again four branches define the σ–ε constitutive law. The first two branches

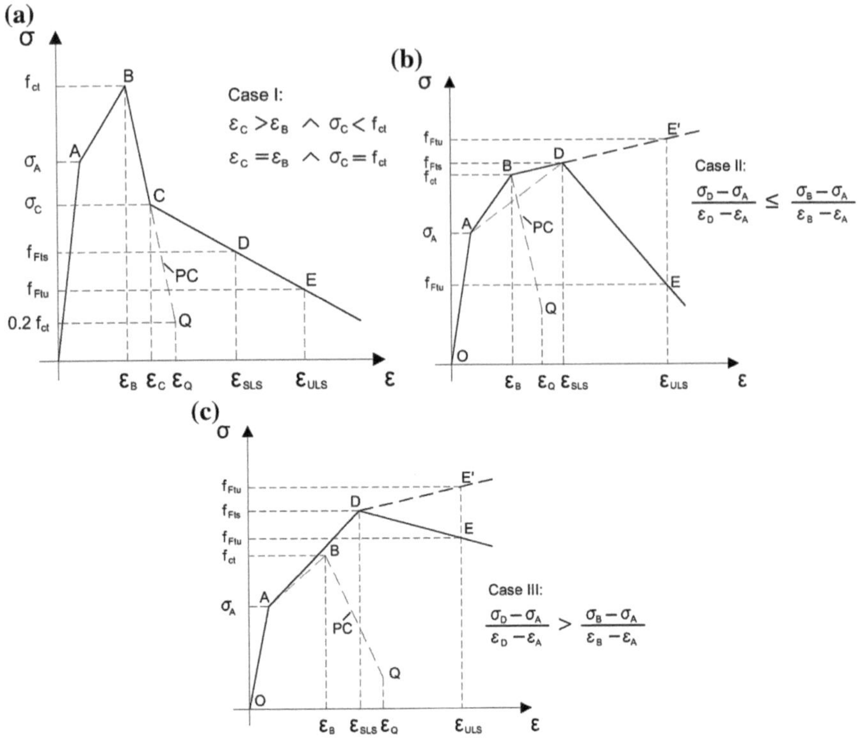

Fig. 10.4 Stress-strain diagrams at SLS for: **a** softening; and **b** and **c** softening or hardening behaviour of FRC (MC2010)

remain those corresponding to PC, while the third branch $(\overline{BD})$ is analytically described as:

$$\frac{\sigma - f_{ct}}{f_{Ftsd} - f_{ct}} = \frac{\varepsilon - \varepsilon_B}{\varepsilon_{SLS} - \varepsilon_B}, \quad \text{for} \quad \varepsilon_B \leq \varepsilon \leq \varepsilon_{SLS} \tag{10.11}$$

CASE III, Fig. 10.4c: the cracking remains stable up to the SLS strain, ε_{SLS}, but now three branches define the $\sigma - \varepsilon$ constitutive law. The second branch $(\overline{AD})$ is defined as:

$$\frac{\sigma - \sigma_A}{f_{Ftsd} - \sigma_A} = \frac{\varepsilon - \varepsilon_A}{\varepsilon_{SLS} - \varepsilon_A}, \quad \text{for} \quad \varepsilon_A \leq \varepsilon \leq \varepsilon_{SLS} \tag{10.12}$$

For Cases II and III, the material can have softening $(\overline{DE})$ or hardening $(\overline{DE'})$ behaviour, depending on the slope of the last branch.

Design values for the post-cracking strength parameter at ULS can be determined as:

Table 10.1 Partial safety factor

FCR in	Partial safety factors
Compression	As plain concrete
Tension (limit of linearity)	As plain concrete
Tension (residual strength)	$\gamma_F = 1.5$

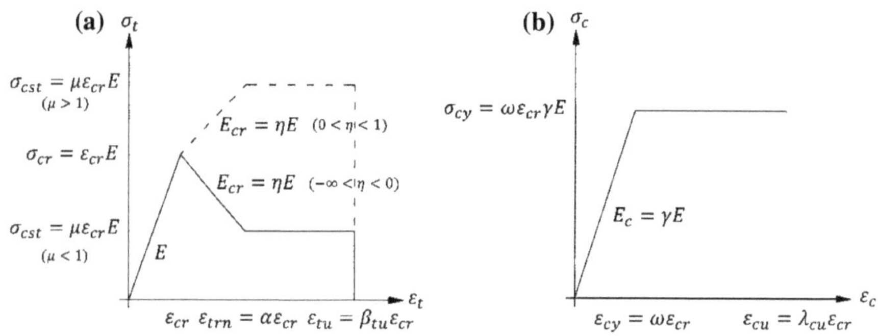

Fig. 10.5 Idealized stress-strain response the behaviour of FRCC in: **a** tension; **b** compression (based on Soranakom and Mobasher 2008)

$$f_{Ftsd} = f_{Ftsk}/\gamma_F \tag{10.13}$$

$$f_{Ftud} = f_{Ftuk}/\gamma_F \tag{10.14}$$

where the f_{Xd} and f_{Xk} are the design and characteristic values of the parameter f_X, and γ_F is the material safety factor, whose values are indicated in Table 10.1. For serviceability limit states (SLS), the partial factors should be taken as 1.0.

10.3 Flexural Reinforcement

The flexural capacity of a cross section of a FRC element (M_P) can be determined according to the closed formulations proposed by Soranakom and Mobasher (2008) or Taheri et al. (2012):

$$M_P = M'_u \times M_{cr} = \frac{3\mu\omega}{\mu + \omega} M_{cr} \tag{10.15}$$

where (Fig. 10.5)

$$\mu = \frac{\sigma_{cst}}{E\varepsilon_{cr}} = \frac{\sigma_{cst}}{\sigma_{cr}} \tag{10.16}$$

$$\omega = \frac{\varepsilon_{cy}}{\varepsilon_{cr}} = \frac{\sigma_{cy}}{E\varepsilon_{cr}} = \frac{\sigma_{cy}}{\sigma_{cr}} \quad (10.17)$$

In Eq. (10.15) M_{cr} is the bending moment at crack initiation of a section:

$$M_{cr} = \frac{h^2 \sigma_{cr}}{6} \quad (10.18)$$

being h the height of the cross section of the FRCC element.

The parameters defining the constitutive law represented in Fig. 10.5a can be obtained by inverse analysis, as described in Sect. 11, or from the recommendations of RILEM TC 162 TDF (2003) or MC2010 (2011), as detailed in the Sect. 10.2.

These type approaches have bee implemented in cross section layered models (Salehian et al. 2014) or in closed form solutions (Soranakom and Mobasher 2008; Taheri et al. 2012). The parametric studies carried out by Taheri et al. (2012) demonstrated the η parameter (the one that defines the strain softening/hardening character, see Fig. 10.5) has a determinant effect on the load carrying capacity of FRC structures at serviceability limit state conditions, mainly when the structure has a minimum conventional flexural reinforcement that assures a stabilized crack propagation.

10.4 Shear Reinforcement

10.4.1 RILEM TC 162-TDF Approach

In accordance with RILEM TC 162-TDF (2003) approach (Vandewalle et al. 2003), the shear resistance of SFRC beams, V_{Rd}, is calculated as follow:

$$V_{Rd} = (V_{cd} + V_{fd}) + V_{wd} \quad (10.19)$$

where V_{cd}, V_{fd} and V_{wd}, are the contribution of concrete, fiber reinforcement, and steel stirrups, respectively. The contribute of concrete, V_{cd}, is obtained from the following equation:

$$V_{cd} = \left[\frac{0.18}{\gamma_c} k \left(100\, \rho_{sl} f_{ck}\right)^{1/3} + 0.15\sigma_{cp} \right] b_w d \quad (10.20)$$

where

$$k = 1 + \sqrt{200/d} \leq 2.0 \quad (10.21)$$

is a size factor with d being the effective depth of the flexural reinforcement, in mm, $\rho_{sl} = A_{sl}/b_w d$ is the reinforcement ratio of the longitudinal reinforcement, being the A_{sl} the cross sectional area of the reinforcement which extends $\geq l_{bd} + d$ beyond the considered section (l_{bd} is the design anchorage length), and b_w is the smallest width of the cross-section in the tensile area.

The shear resistance related to the contribution of steel fiber reinforcement, V_{fd}, is obtained from the following equation:

$$V_{fd} = 0.7 k_f \, k \, \tau_{fd} \, b_w d \tag{10.22}$$

where

$$k_f = 1 + n \left(\frac{h_f}{b_w}\right) \left(\frac{h_f}{d}\right) \leq 1.5 \tag{10.23}$$

$$n = \frac{b_f - b_w}{h_f} \leq 3 \tag{10.24}$$

$$\tau_{fd} = \frac{0.18}{\gamma_c} f_{R,4k} \tag{10.25}$$

k_f is the factor for taking into account the contribution for the shear resistance of the flange in a T cross section beam, and τ_{fd} is the design value of the shear strength provided by the fibre reinforcement. In Eq. (10.23), h_f is the height of the flange, while in Eq. (10.24), b_f is the width of the flange. In Eq. (10.25) $f_{R,4k}$ is the characteristic value of the residual tensile strength parameter at a deflection of 3.0 mm obtained in the 3PNBBT according to the recommendations of Vandewalle et al. (2002), see Sect. 8.2.5.

10.4.2 CEB-FIP Model Code 2010 Approach

To determine the shear resistance of FRC beams, the MC2010 (2011) merges the contribution of fiber reinforcement, V_{fd}, and concrete, V_{cd}, in an unique term, $V_{Rd,F}$:

$$V_{Rd} = V_{Rd,F} + V_{wd} \leq V_{Rd,\max} \tag{10.26}$$

where the equations for the evaluation of the contribution of the steel stirrups V_{wd} and of the maximum value of the shear force $V_{Rd,\max}$ to avoid crushing of the compression struts are given in the prEN 1992-1-1 (2010). The term $V_{Rd,F}$ represents the contribution of the FRC for the shear resistance and is obtained from equation:

$$V_{\mathrm{Rd,F}} = \left\{ \frac{0.18}{\gamma_{\mathrm{c}}} \cdot k \cdot \left[100 \cdot \rho_{sl} \cdot \left(1 + 7.5 \cdot \frac{f_{\mathrm{Ftuk}}}{f_{\mathrm{ctk}}} \right) \cdot f_{\mathrm{ck}} \right]^{1/3} + 0.15 \cdot \sigma_{\mathrm{cp}} \right\} \cdot b_{\mathrm{w}} \cdot d \tag{10.27}$$

where γ_c is the partial safety factor for the concrete (1.5), f_{ctk} and f_{ck} are characteristic values of the, respectively, the tensile and compressive strength for the FRC, and $\sigma_{cp} = N_{Ed}/A_c < 0.2 f_{cd}$ is the average stress acting on the concrete cross section, A_c, for an axial force N_{Ed}, due to loading or pre-stressing actions ($N_{Ed} > 0$ for compression).

Both RILEM TC 162-TDF (2003) and MC2010 (2011) guidelines address the contribution of the transversal reinforcement, V_{wd}, in the same way (as prEN 1992-1-1 2010):

$$V_{wd} = \frac{A_{sw}}{s} 0.9 d\, f_{ywd} (1 + \cot \alpha) \sin \alpha \tag{10.28}$$

where f_{ywd} is the design value of the yield stress of shear reinforcement, and α is angle formed by this reinforcement with the longitudinal axis of the beams.

10.4.3 Assessment of the Potentialities of RSFRC for the Shear Reinforcement of RC Beams

Zamanzadeh et al. (2015) explored the use of RSFRC for the shear reinforcement of RC beams, by using the RSFRC with the mix composition presented in Sect. 8.2.5 (Table 10.2), whose relevant properties are also indicated in that section. Figure 10.6 shows the geometry and reinforcement details of the beams produced for this experimental program, as well as the loading and support conditions. Two specimens were tested per each of the three series, where the difference is resumed to the width of the web of the I type cross section adopted for the beams (Fig. 10.6). All the beams were flexurally reinforced with a relatively high reinforcing ratio in order to promote shear failure for the beams.

The relationship between the applied load and the deflection at the loaded section obtained in the tested series of beams is represented in Fig. 10.7.

Table 10.2 Shear capacity according to analytical formulations and experimental tests (Zamanzadeh et al. 2015)

Beam's designation	V_{exp} (kN)	$V_{\mathrm{Rd,RILEM}}$ (kN)	$V_{\mathrm{exp}}/V_{\mathrm{Rd,RILEM}}$	$V_{\mathrm{Rd,FIB}}$ (kN)	$V_{exp}/V_{\mathrm{Rd,FIB}}$
S_W70	81.290	29.806	2.72	26.042	3.121
S_W110	95.810	45.356	2.11	41.690	2.298
S_W150	109.172	56.497	1.93	51.266	2.129

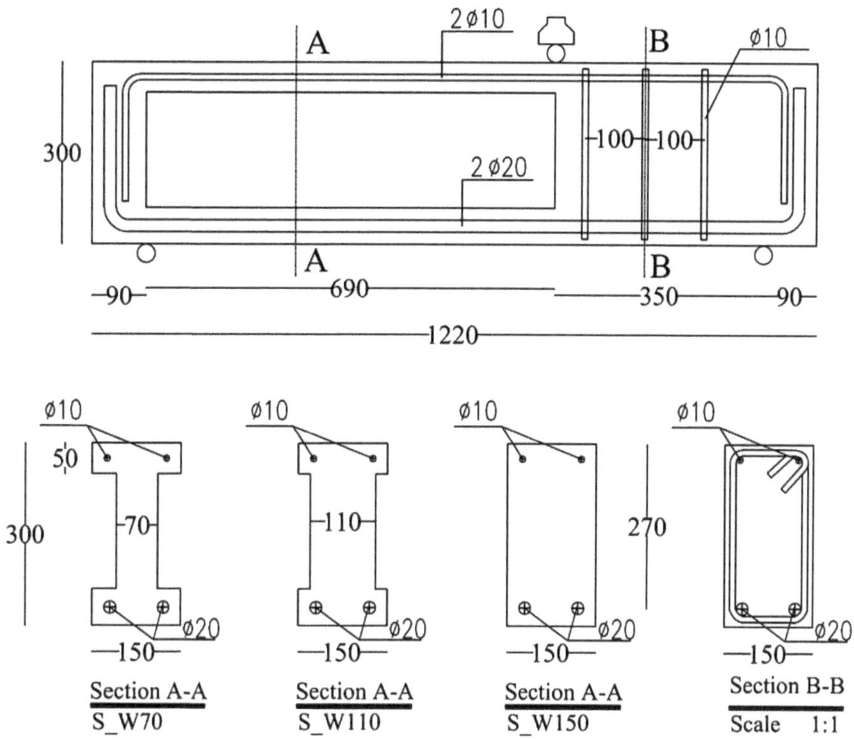

Fig. 10.6 Geometry of the beams (dimensions in mm) (Zamanzadeh et al. 2015)

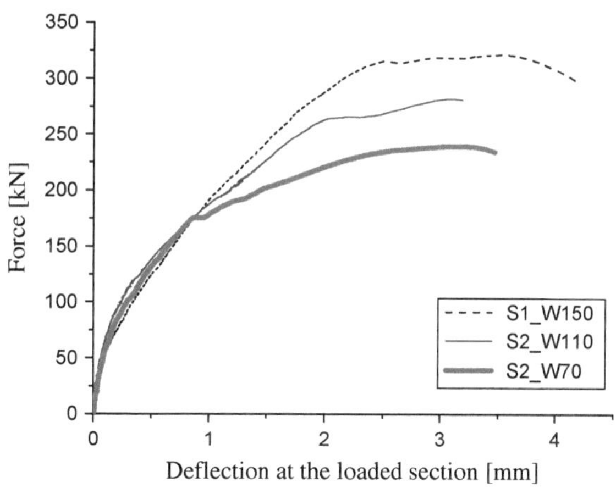

Fig. 10.7 Load—deflection relationship at the loaded section for the tested series of beams (Zamanzadeh et al. 2015)

Table 10.2 presents the shear capacity of the tested beams predicted by applying the formulation proposed by MC2010 (2011) and RILEM TC 162-TDF (2003), where characteristic values were adopted for the material properties (according to the equations of both formulations), and $\gamma_c = 1.5$. In Table 10.2 the label S_Wj was used to differentiate the tested beams, where "j" identifies the web's cross-section thickness (in mm) of the part of the beam without shear reinforcement.

From the obtained results it is verified that by increasing the beam's web thickness the load carrying capacity has increased without affecting significantly the deflection at maximum load. It is also verified that the ratio between the shear capacity obtained experimentally ($\boldsymbol{V_{exp}}$) and applying the analytical formulations ($\boldsymbol{V_{ana}}$) has decreased with the increase of the beam web thickness. Since design values are being used for the properties of the intervening materials, the $\boldsymbol{V_{exp}/V_{Rd,ana}}$ should be higher than 1.5 in order to guarantee safe predictions by the analytical approach. However, the decrease of the $\boldsymbol{V_{exp}/V_{Rd,ana}}$ with the increase of the width of the web's beam cross section (b_w) indicates that the formulations do not consider properly the favourable effect of the fibre orientation when b_w decreases. In fact, fibres become more preferentially aligned with the axis of the beam when b_w decreases due to a more pronounced wall effect, leading to more effective fibre reinforcement mechanisms in terms of arresting the crack propagation (Barros 2011). The irregular shape of the RSF indicates that a higher tendency for this effect is expected when using ISF, due to their higher aspect ratio.

10.5 Punching Reinforcement

10.5.1 Introduction

Moraes-Neto et al. 2013 has proposed a model for predicting the contribution of industrial steel fibres for the punching resistance of concrete slabs flexurally strengthened with conventional reinforcement. The model is conceptually applicable also to concrete reinforced with recycled steel fibres, as long as some of the model material parameters are conveniently characterized experimentally.

This model determines the punching resistance of a SFRC flat slab by defining two curves, one corresponding to the relationship between the applied load and the rotation of the slab ($V - \psi$), and the other that represents a failure criterion. The interception of these curves provides the punching failure load. For the establishment of the $V - \psi$ relationship, the column-slab connection is assumed in axisymmetric conditions (Fig. 10.8a). The crack pattern of a slab failing in punching can be assumed as forming radial segments (Fig. 10.8a, b). Each radial segment is delimited by a tangential crack formed close to the column, by two radial cracks, and by the edge considered as a free boarder of the slab (Fig. 10.8b).

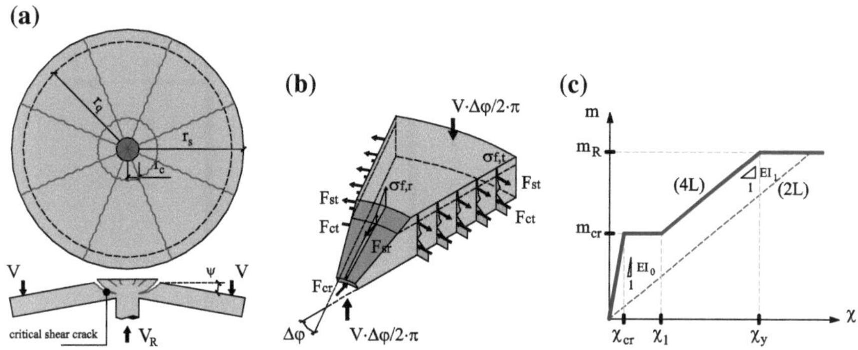

Fig. 10.8 **a** Assumed crack pattern in a column-slab connection; **b** stresses and resultant forces in a radial segment of the slab; and **c** 2L and 4L moment curvature ($m - \chi$) diagrams

10.5.2 Load-Rotation Approach

According to Guandalini (2005) the $V - \psi$ relationship of an axisymmetric slab can be obtained by using the 2L bilinear moment-curvature diagram ($m - \chi$) represented in Fig. 10.8c.

To derive the load versus rotation relationship supported on the 2L $m - \chi$ diagram, it is necessary to assume the slab decomposed in two regions: elastic ($r_y < r_0$) and elasto-plastic ($r_0 < r_y < r_s$). By virtue of equilibrium of bending moments of the force components installed in the radial segment represented in Fig. 10.8b, the following two equations are obtained, after some mathematical manipulations described elsewhere (Moraes-Neto et al. 2013):

$$V(\psi) = \frac{2 \cdot \pi}{(r_q - r_c)} \cdot E \cdot I_1 \cdot \psi \cdot \left[1 + \ln\left(\frac{r_s}{r_0}\right)\right] \quad for\, r_y \le r_0 (elastic\, regime) \tag{10.29}$$

$$V(\psi) = \frac{2 \cdot \pi}{(r_q - r_c)} \cdot E \cdot I_1 \cdot \psi \cdot \left[1 + \ln\left(\frac{r_s}{r_y}\right)\right] \\ for\ r_0 < r_y < r_s (elasto - plastic\, regime) \tag{10.30}$$

In these equations the variables r_c, r_0, r_q and r_s represent, respectively, the radius of column's cross section, the distance to the axis of the column to the punching failure surface ($r_0 = r_c + d/2$), the distance of the circumferential loading line, and the radius of the slab. The EI_1 represents the flexural stiffness of the slab's cross section after crack initiation (Fig. 10.8c).

The evaluation of EI_1 is done following the procedures adopted for RC members (Barros et al. 2012), and assuming a stabilized cracking phase:

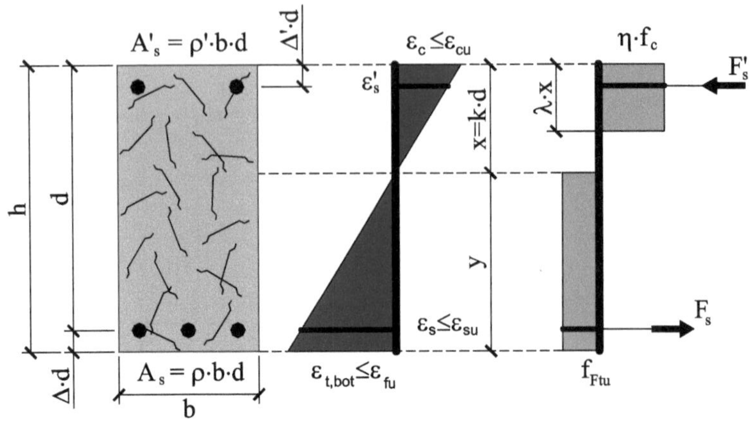

Fig. 10.9 Adopted approach to evaluate the ultimate bending moment, m_R (adapted from the MC2010)

$$E \cdot I_1 = \rho \cdot \beta \cdot E_s \cdot d^3 \cdot \left(1 - \frac{x}{d}\right) \cdot \left(1 - \frac{x}{3 \cdot d}\right) \quad (10.31)$$

The contribution of fibre reinforcement to EI_1 is only indirectly taken in the evaluation of the neutral axis, x, Fig. 10.9 (η and λ parameters are evaluated according to MC2010 (2011), and f_{Ftu} according to Eq. (10.4)).

In Eq. (10.31) β is a factor intending to take into account the arrangement of the reinforcement, since the deduction of Eq. (10.29) was supported on the principle of axisymmetric structural conditions, but the majority of the built and tested RC flat slabs have orthogonal arrangement of the reinforcement (Guandalini 2005). According to Muttoni (2008), $\beta = 0.6$ yields satisfactory results. The evaluation of the position of the neutral axis, x, can be made according to the recommendations of MC2010 (2011), see Fig. 10.9.

To evaluate f_{Ftu} of concrete reinforced with RSF, the experimental tests presented in Sect. 8.2.5, following the recommendations of MC2010 (2011), should be performed.

The load corresponding to the flexural failure of the slab (V_{flex}) can be determined from the following equation (Moraes-Neto et al. 2013).

$$V_{flex} = 2 \cdot \pi \cdot m_R \cdot \frac{r_s}{(r_q - r_c)} \quad for\ r_y = r_s\ (flexural\,failure\ load) \quad (10.32)$$

where m_R represents the resisting bending moment (plastic bending moment) of the slab's cross section.

Combining Eqs. (10.29) and (10.31) and assuming for the r_y/r_s the value of 0.35 obtained by Barros et al. (2015), and after executing some mathematical manipulations, described in Moraes-Neto et al. (2013), the following equation is obtained:

$$\psi = \Delta \cdot \frac{m_R \cdot r_s}{E \cdot I_1} \cdot \left(\frac{V}{V_{flex}}\right)^{3/2} \tag{10.33}$$

where $\Delta = 0.65$ for regular concrete and $\Delta = 1.625$ for concrete of lightweight aggregates. For slabs in axisymmetric structural conditions, V_{flex} is obtained from Eq. (10.32), while for square slabs the yield line theory leads $V_{flex} = 8 \cdot m_R$.

To evaluate the plastic bending moment, m_R, the recommendations of MC2010 (2011) for the simulation of the contribution of fibre reinforcement can be adopted (see Eq. (10.4) and Fig. 10.9).

10.5.3 Failure Criterion

Following the recommendations of ACI 318 (2008), Muttoni (2008), and Muttoni and Schwartz (1991), and taking the main achievements of Walraven (1981), and Vecchio and Collins (1986) on the contribution of the aggregate interlock for the concrete shear resistance, and the relevant results of Moraes-Neto (2013) on the contribution of fibre reinforcement for the concrete shear resistance, the following equation was determined that defines the punching failure criterion of *SFRC* slabs:

$$\frac{V}{b_0 \cdot d \cdot \sqrt{f_c} \cdot \left[\lambda_f + k_f^{1/3}\right]} = \frac{1}{1.33 + 20 \cdot \mu_f \cdot \psi \cdot d \cdot k_{dg}} \quad [\text{MPa, mm}] \tag{10.34}$$

where b_0 is the perimeter of the punching failure surface ($b_0 = 4 \cdot e + \pi \cdot d$ for column of square cross section), $k_{dg} = 1/(d_{g0} + d_g)$, being $d_{g0} = 16$ mm the reference diameter, and d_g the maximum diameter of the aggregates, and

$$\lambda_f = \begin{cases} 0 & if\ C_f \neq 0 \\ 1 & if\ C_f = 0 \end{cases}; \quad \mu_f = \begin{cases} (1/11) & if\ C_f \neq 0 \\ 1 & if\ C_f = 0 \end{cases} \tag{10.35}$$

where C_f is the content of fibres (Kg/m^3). The k_f must be calibrated for the adopted RSFRC. Based on a database where hooked ends steel fibres were used for the punching reinforcement of RC slabs, Barros et al. (2015) proposed an equation for the determination of this parameter.

10.6 Minimum Flexural Reinforcement, Crack Width and Crack Spacing

10.6.1 Minimum Flexural Reinforcement

To control the crack width in FRC elements under bending, the MC2010 (2011) recommends the use of a cross sectional area of longitudinal steel bars given by the following equation:

$$A_{s,\min} = k_c\, k\, (f_{ctm} \; - f_{Ftsm})\, \frac{A_{ct}}{\sigma_s} \tag{10.36}$$

where

- k_c is a coefficient that takes into account the nature of the stress distribution within the section immediately prior to cracking and of the change of the lever arm. The equations for the evaluation of k_c are given in the prEN 1992-1-1 (2010);
- k is the coefficient that considers the effect of non-uniform self-equilibrating stresses, which lead to a reduction of restraint forces. The recommending values are indicated in the prEN 1992-1-1 (2010);
- f_{ctm} is the average value of the tensile strength of the FRC;
- A_{ct} is the area of concrete within the tensile zone. The tensile zone is that part of the section which is calculated to be in tension just before the formation of the first crack;
- σ_s is the absolute value of the maximum stress permitted in the reinforcement immediately after formation of the crack. This may be taken as the yield strength of the reinforcement, f_{yk};
- f_{Ftsm} is the average value of the residual strength of the FRC. The equations for the evaluation of this parameter are given in the MC2010 (2011)

To apply this equation to concrete reinforced with sustainable fibres, the evaluation of the f_{Ftsm} of these composites should be performed by means of the experimental tests according to the recommendations of MC2010 (see Sect. 8.2.5).

10.6.2 Crack Width and Crack Spacing

To estimate the design value of the crack width (w_d) for members of fiber reinforced concrete (FRC) including longitudinal steel bars (R/FRC) and subjected mainly to flexure or tension, the MC2010 (2011) proposed the following equation:

$$w_d = 2\{k_6 \cdot c + \frac{1}{4} \cdot \frac{\phi}{\rho_{s,ef}} \cdot \frac{(f_{ctm} - f_{Ftsm})}{\tau_{bm}}\} \cdot \frac{1}{E_s} \cdot (\sigma_{st} - k_7\sigma_{sr} + k_8\varepsilon_s E_s) \quad (10.37)$$

where ϕ is the bar diameter, $\rho_{s,ef}$ is the effective reinforcement ratio:

$$\rho_{s,ef} = \frac{A_s}{2.5(c + \phi/2)b} \quad (10.38)$$

being A_s the cross sectional area of the longitudinal reinforcement contained within the effective FRC area in tension $A_{c,ef} = 2.5(\mathrm{c} + /2)b$, where c is the concrete cover and b is the width of the cross section. τ_{bm} in Eq. (10.37) is the average bond strength between reinforcing bars and surrounding concrete:

$$\tau_{bm} = 1.8 f_{ctm} \quad (10.39)$$

while σ_{st} is the stress in the reinforcement, E_s is modulus of elasticity of steel bars, and σ_{sr} is the maximum steel stress at crack section of the crack formation stage (MC2010):

$$\sigma_{sr} = \frac{(f_{ctm} - f_{Ftsm})}{\rho_{s,ef}} \cdot (1 + \gamma_s\rho) \quad (10.40)$$

where f_{ctm} and f_{Ftsm} are, respectively, the average value of the tensile strength, and the average value of the residual flexural tensile strength of FRC (f_{Fts}):

$$f_{Fts} = 0.45 f_{R1} \quad (10.41)$$

being f_{R1} the residual flexural tensile strength at a crack mouth opening displacement (CMOD) of 0.5 mm.

In Eq. (10.37) ε_s is the strain of rebar at the onset of cracking. Since the tensile strength is not significantly affected when using content and type of fibers used in FRC applied in the majority of structural applications, Eq. (10.39) can be assumed also applicable for RSFRC. In the evaluation of the σ_{st}, the effect of the fiber reinforcement should be taken into account. To evaluate the maximum crack spacing in R/FRC elements, the MC2010 (2011) proposes the following equation:

$$L_{cs,\max} = k_6 c + \frac{1}{4} \frac{(f_{ctm} - f_{Ftsm})}{\tau_{bm}} \cdot \frac{\phi_s}{\rho_{s,ef}} \quad (10.42)$$

In Eqs. (10.36) and (10.41) k_6 to k_8 are non-dimensional coefficients, whose values can be found in MC2010 (2011). These equations can, conceptually, be applied to concrete reinforced with RSF, as long as the involved material properties are conveniently characterized experimentally according to the recommendations of MC2010. However, their predictive performance must be assessed by experimental evidence, so specific research in this topic is recommended.

10.7 Approaches of General Applicability for Composite Materials

More sophisticated approaches, based on a cross sectional layered model, capable of considering directly the stress-crack width relationship of FRCC and the bond properties between the flexural reinforcements and surrounding FRCC, have been proposed for the evaluation of the flexural capacity of beams type elements (Barros et al. 2015). This type of model was extended to slab type structures (Salehian et al. 2014) by determining the stress-crack width relationship of FRCC from panel tests, like the ones described in Sect. 8.2.7. The authors demonstrated that this strategy is more appropriate for 2D randomly distributed fibers, such is the case of slabs and shells, since the occurrence of several cracks in the experimental tests capture more correctly the fibre reinforcement mechanisms developed in this type of structures, where in general a large number of cracks are formed due to their statically indeterminate character (Espion 2004; Mandl 2008; Salehian et al. 2015). These type of approaches have been considered in the design recommendations for elevated SFRC slabs (steel fibre reinforced concrete slabs supported on columns) recently published by ACI (Mobasher et al. 2015).

References

ACI 318 (2008) Building code requirements for structural concrete. American Concrete Institute, Farmington Hills, Michigan

ASTM C 1018 (1990) Standard test method for flexural toughness and first crack strength of fiber reinforced concrete (using beam with third-point loading (4.02). Am Soc Test Mater Philadelphia, 637–644

Banthia N, Trottier JF (1995a) Test methods for flexural toughness characterization of fiber reinforced concrete: some concerns and a proposition. ACI Mater J 92(1), 48–57

Banthia N, Trottier JF (1995b) Concrete reinforced with deformed steel fibres, Part II: Toughness characterization. ACI Mater J 92(2), pp 146–154

Barros JAO, Taheri M, Salehian H (2015a) A model to simulate the moment-rotation and crack width of FRC members reinforced with longitudinal bars. Eng Struct 100:43–56

Barros JAO, Moraes-Neto BN, Melo GSSA, Frazão CMV (2015b) Assessment of the effectiveness of steel fibre reinforcement for the punching resistance of flat slabs by experimental research and design approach. Compos Part B J 78:8–25

Barros JAO, Salehian H, Pires NMMA, Gonçalves DMF (2012) Design and testing elevated steel fibre reinforced self-compacting concrete slabs. In: Barros J et al. (ed) 8th RILEM international symposium on fibre reinforced concrete: challenges and opportunities, p 12, September, 2012

Barros JAO (2011) Technology, design and applications of steel fibre reinforced self compacting concrete. In: 6th international conference fibre concrete 2011 technology, design, application, CTU in Prague, Masarykova kolej, 8 and 9 Sept 2011 (Invited Keynote Lecturer)

Barros JAO, Cunha VMCF, Ribeiro AF, Antunes JAB (2005) Post-cracking behaviour of steel fibre reinforced concrete. RILEM Mater Struct J 38(275):47–56

EN 14651 (2005) Test method for metallic fibre concrete. Measuring the flexural tensile strength (limit of proportionality (LOP), residual). European Committee for Standardization, Brussels, BE

Espion B (2004) Test report n°33396. University of Brussels, Belgium
Gopalaratnam VS, Shah SP, Batson GB, Criswell ME, Ramakrishnan V, Wecharatana M (1991) Fracture toughness of fiber reinforced concrete. ACI Mater J 88(4):339–353, July–August
Guandalini S (2005) Poinçonnement symétrique des dalles en béton armé. PhD Thesis 3380, Ecole Polytechnique Fédérale de Lausanne, Switzerland, 2005 (in French)
Japan Society of Civil Engineers (JSCE) (1984) Method of test for flexural strength and flexural toughness of fiber reinforced concrete. Standard, SF-4, 1984, pp 58–66
Mandl J (2008) Flat slabs made of steel fibre reinforced concrete (SFRC). CPI worldwide, 1
MC2010, CEB fib Model Code 2010 – Final Draft, 2011
Mobasher B, Barros JAO, Naaman AE, Destree X et al (2015) Report on design and construction of steel fiber-reinforced concrete elevated slabs. ACI 544.6R-15, Reported by ACI Committee 544, 2015
Moraes-Neto BN (2013) Punching behaviour of steel fibre reinforced concrete slabs submitted to symmetric loading. PhD in Civil Engineering, Department of Civil and Environmental Engineering, University of Brasília, Brasília, DF, January 2013 (in Portuguese)
Moraes-Neto BN, Barros JAO, Melo GSSA (2013) A model for the prediction of the punching resistance of steel fibre reinforced concrete slabs centrically loaded. Constr Build Mater J 46:211–223
Muttoni A (2008) Punching shear strength of reinforced concrete slabs without transverse reinforcement. ACI Struct J 105(4), 440–450, Jul/Aug 2008
Muttoni A, Schwartz J (1991) Behaviour of beams and punching in slabs without shear reinforcement. IABSE Colloquium, vol 62, pp 703–708, Zurich, Switzerland, 1991
prEN 1992-1-1 (2010) Eurocode 2: design of concrete structures—Part 1-1. General rules and rules for buildings
RILEM TC 162-TDF (2003) Test and design methods for steel fibre reinforced concrete. 1327 r-e-design method. Final Recommendation. Mater Struct 36:560–567
Soranakom C, Mobasher B (2008) Moment-curvature response of strain softening and strain hardening cement based composites. Cem Concr Compos 30(6):465–477
Salehian H, Barros JAO, Taheri M (2014) Evaluation of the influence of post-cracking response of steel fibre reinforced concrete (SFRC) on load carrying capacity of SFRC panels. Constr Build Mater J 73:289–304
Salehian H, Barros JAO (2015) Assessment of the performance of steel fibre reinforced self-compacting concrete in elevated slabs. Cem Concr Compos 55:268–280
Taheri M, Barros JAO, Salehian HR (2012) A parametric study on the use of strain softening/hardening FRC for RC elements failing in bending. ASCE Mater Civ Eng J 24 (3):259–274
Vandewalle L et al (2003) Test and design methods for steel fibre reinforced concrete—σ–ε design method—final recommendation. Mater Struct 36(262):560–567
Vandewalle L et al (2002) Test and design methods for steel fibre reinforced concrete—final recommendation. Mater Struct 35(253):579–582
Vandewalle L et al (2000a) Test and design methods for steel fiber reinforced concrete. Recommendations for bending test. Mater Struct 33(225), 3–5
Vandewalle L et al (2000b) Test and design methods for steel fiber reinforced concrete. Recommendations for σ−ε design method. Mater Struct 33(226), 75–81
Vecchio FJ, Collins MP (1986) The modified compression-field theory for reinforced concrete elements subjected to shear. ACI J Proc 83(2):219–231
Walraven JC (1981) Fundamental analysis of aggregate interlock. J Struct Eng ASCE 107 (11):2245–2270
Zamanzadeh Z, Lourenço LAP, Barros JAO (2015) Recycled steel fibre reinforced concrete failing in bending and in shear. J Constr Build Mater 85:195–207

Chapter 11
Inverse Analysis for Deriving the Fracture Properties of RSFRC

Joaquim A.O. Barros

Abstract This section is dedicated to the description of a numerical strategy to derive the parameters that define the fracture mode I and fracture mode II of fibre reinforced concrete (FRC). This strategy is herein designated by inverse analysis (IA), and it was used to determine the mode I and II fracture parameters of Recycled Steel Fibre Reinforced Concrete (RSFRC) developed in Chap. 8. From the theoretical point of view the proposed IA is applicable to any type of FRC, as long as the setup of the experimental tests, whose results are adopted in the IA, assures the fracture is limited to a unique crack.

11.1 Fracture Mode I Parameters

The results from the experimental flexural tests introduced in Sect. 8.2.5 can be used to derive, by inverse analysis (IA), the mode I fracture parameters of Recycled Steel Fibre Reinforced Concrete (RSFRC). In fact the 3-point notched beam bending test (3PNBBT) is specially used for the evaluation of the fracture mode I parameters, namely the stress at crack initiation, the mode I fracture energy and the shape of the stress-crack width. This strategy was also adopted by Neocleous et al. (2006), who pointed out that the recommendations of the RILEM TC 162-TDF (2003) for determining the tensile constitutive law of RSFRC provided unsafe predictions.

Figure 11.1 shows the schematic representation of the geometry of a 3PNBBT and a typical finite element mesh adopted in the IA. 2D line interface finite elements (IFE), of four or six nodes, were located in the specimen's symmetry axis. In the remaining parts of the specimen four- or eight-nodes Serendipity plane-stress finite elements (SPSFE) are generally used. Details about these finite elements and FEM-analysis can be found elsewhere (Sena-Cruz et al. 2004). Gauss-Lobatto integration scheme (Schellekens 1990) with two (in case of IFE of four nodes) or three integration points, IP, (in case of IFE of six nodes) is used for the 2D line IFE,

J.A.O. Barros (✉)
University of Minho, Braga, Portugal
e-mail: barros@civil.uminho.pt

J.A.O. Barros et al. (eds.), *Recent Advances on Green Concrete for Structural Purposes*, Research for Development, DOI 10.1007/978-3-319-56797-6_11

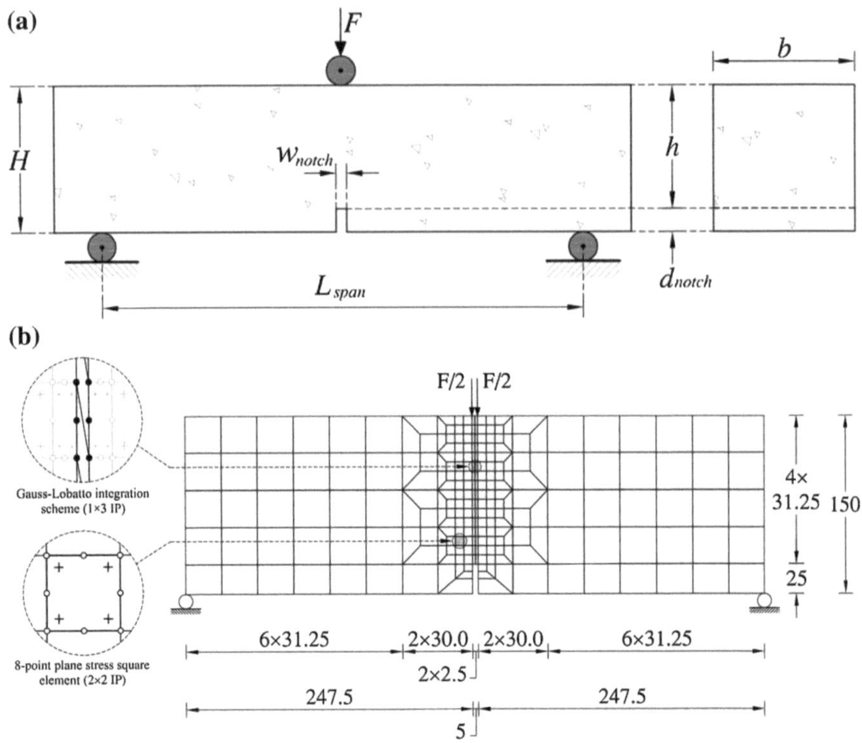

Fig. 11.1 Inverse analysis based on the results from 3PNBBT: **a** geometry; **b** finite element mesh (the dimensions, in mm, correspond to the specimen proposed by MC2010 (2011))

while Gauss-Legendre integration scheme with 2×2 IP is adopted for the SPSFE. To avoid undesired spurious oscillations of the stress field, a value close to 1.0×10^4 N/mm^3 is recommended for the initial stiffness of the component of the constitutive law of the IFE defining the crack opening process (Schellekens 1990). Since in this problem sliding does not occur in the interface elements, the analysis is independent of the values assigned to the stiffness of the component of the constitutive law of the IFE defining the crack sliding process. The component of the constitutive law of the IFE defining the crack opening process can be simulated by the stress-crack width multi-linear diagram represented in Fig. 11.2a.

Instead of using IFE for modelling the fracture initiation and propagation, the column of material above the notch can be simulated by SPSFE with a smeared crack model (Fig. 11.3). The Gauss-Legendre integration scheme with 2×2 IP is used in all elements, with the exception of the elements at the specimen symmetry axis, where 1×2 IP are adopted. This has the purpose of forcing the propagation of the crack along the specimen symmetry axis, over the aligned integration points, which is generally observed in the experimental tests.

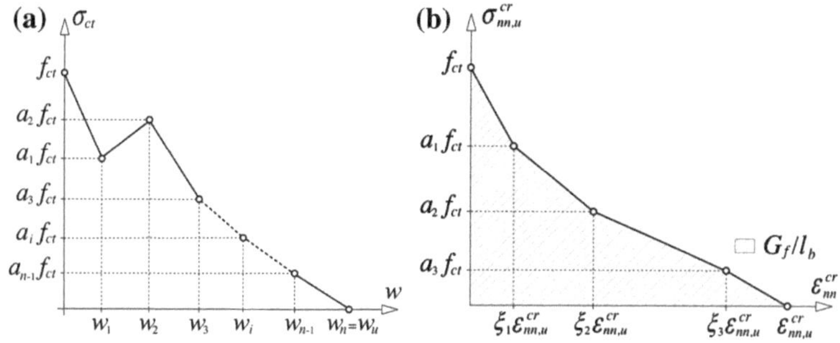

Fig. 11.2 Multi-linear normal stress versus crack width adopted to simulate the crack opening component of the constitutive law of the IFE

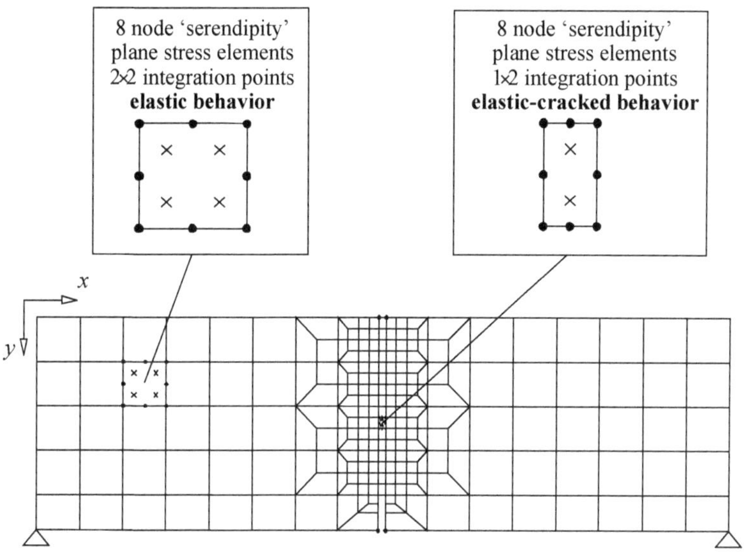

Fig. 11.3 Finite element mesh with a smeared crack approach for modelling the initiation and propagation of crack along the notched plane (Pereira et al. 2008)

Linear elastic material behaviour is assigned to all elements, with the exception of the elements above the notch, where elastic-cracked behaviour in tension is assumed (Pereira et al. 2008). A multi-linear crack normal stress versus crack normal strain is adopted, identical to the one represented in Fig. 11.2b, where the crack width is the product of the crack normal strain to the crack band width, l_b. The l_b is assumed to be equal to the width of the notch, which is the same of the elements above the notch $\left(w_i = \xi_i \cdot \varepsilon_{nn,u}^{cr} \cdot l_b\right)$ (l_b=5 mm in the referred study).

The Inverse Analysis consists of a trial and error process for obtaining the points defining the diagram adopted to characterize the fracture mode I parameters, by

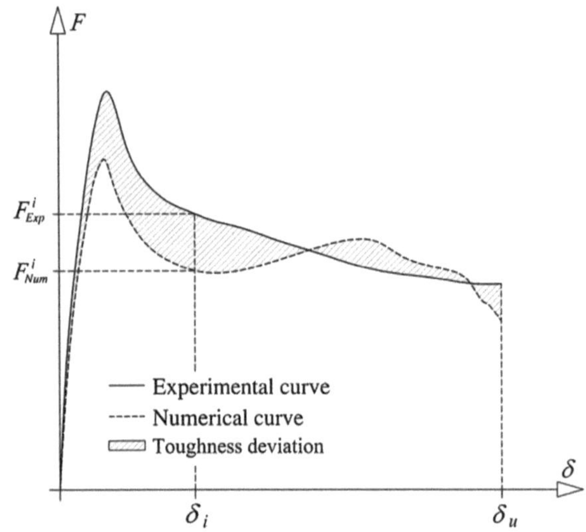

Fig. 11.4 Assessment of the numerical force determined by the inverse analysis by comparing to the experimental force

fitting with the minimum error the force-deflection (or force-crack mouth opening displacement) relationship recorded in the experimental tests. The two convergence criteria schematically represented in Fig. 11.4, and obtained from the evaluation of the following two error parameters, are generally used to end this trial error process:

$$err_F = \frac{\left| F^k_{Exp} - F^k_{Num} \right|}{F^k_{Exp}} \tag{2.5.1}$$

$$err_A = \frac{\left| A^{(F-\delta)_k}_{Exp} - A^{(F-\delta)_k}_{Num} \right|}{A^{(F-\delta)_k}_{Exp}} \tag{2.5.2}$$

where $A^{(F-\delta)_k}_{Exp}$ and $A^{(F-\delta)_k}_{Num}$ are the area beneath, respectively, the experimental and numerical force-deflection (or CMOD) curves up to the central deflection of δ^k.

By using the smeared crack approach with the inverse analysis (the one corresponding to the finite element mesh indicated in Fig. 11.5), Zamanzadeh et al. (2015a) have obtained the fracture parameters indicated in Table 11.1 for RSFRC reinforced with 45, 60 and 90 kg/m^3 of RSF (designated in this table by, RSFRC45, RSFRC60 and RSFRC90, respectively, see also Table 8.3). A trilinear crack normal stress versus crack normal strain diagram was adopted: therefore the parameters determined by inverse analysis were the stress at crack initiation, f_{ct}, the mode I fracture energy, G_{fI}, and the parameters defining the transition points of the trilinear diagram, α_i and ξ_i (i = 1, 2). These RSFRCs were developed for exploring their use as alternative/complementary shear reinforcement for reinforced concrete beams.

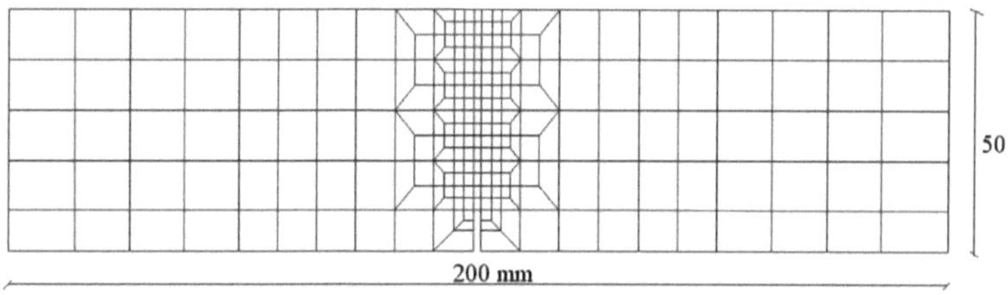

Fig. 11.5 Geometry and finite element mesh of the specimens adopted in the inverse analysis for the determination of the fracture mode I parameters of the RSFRC of M2_3.8n composition (span length = 200 mm; width of the cross section = 50 mm; width and depth of the notch = 4 and 15 mm, respectively)

The relevant results characterizing the flexural behaviour of these RSFRC are resumed in Figs. 8.29 and 8.30. The values of mode I fracture energy are close to the ones predicted by the equations proposed by Barros and Figueiras (1999) for concrete reinforced with hooked ends steel fibres of relatively small aspect ratio (ratio between the length and the diameter of the fibre). It should be noted that, since the IA determines the complete stress-crack width relationship of the material (up to the stage of null stress transfer capacity in mode I), the fracture energy is directly evaluated. Therefore, in spite of the non-null load capacity at the ultimate deflection measured in the 3PNBBT, δ_u, (or CMODu), the IA has the capacity of determining directly the G_{fI}. However, the obtained value is in general an upper bond of the values of G_{fI} measured in direct tensile tests, since in the IA it is assumed the propagation of only one plane crack (along the notched plane), while in the experimental tests the fracture surface is not perfectly plane, and in several situations the failure crack is followed with some micro-cracks where fracture energy is also dissipated.

Zamanzadeh (2017) has also determined the fracture mode I parameters of the RSFRC developed according to the technology shown in Fig. 8.19. These thin panels were used for the shear strengthening of RC beams susceptible to shear failure (Zamanzadeh et al. 2015b). The relevant properties of the RSFRC employed in the cited study are summarized in Sects. 8.2.3 (direct tensile behaviour in Fig. 8.21) and 8.2.6 (direct shear behaviour in Fig. 8.38). Since this RSFRC has a mortar type matrix due to the relatively high content of RSF used, the geometry of the specimens and the finite element mesh adopted in the IA, represented in Fig. 11.5, was downscaled from the beam's geometry recommended by MC2010 (2011) for current FRC. It was used 244 plane stress elements of 8 nodes, and Gauss-Legendre integration scheme with 2 × 2 Integration Points.

The results obtained with inverse analysis strategy for the specimens of the M2_3.8n RSFRC are presented in the last line of Table 11.1. It is verified that by using 3.8% in volume of RSF applied according to the SIFCON type technology has assured a RSFRC with tensile strain hardening nature (α_1 and α_2 are higher than 1.0, Fig. 11.2b), with a quite high mode I fracture energy (12.0 N/mm). However, it should be aware that the inverse analysis based on the experimental results with

Table 11.1 Values defining the fracture mode I parameters of RSFRC, obtained from inverse analysis (Zamanzadeh et al. 2015a)

Series	f_{ct} (N/mm^2)	ξ_1	α_1	ξ_2	α_2	G_{fI} (N/mm)
RSFRC45	2.250	0.012	0.650	0.280	0.520	6.000
RSFRC60	2.300	0.032	0.750	0.350	0.730	6.300
RSFRC90	2.620	0.100	0.930	0.600	0.730	7.700
M2_3.8n	3.000	0.070	1.200	0.600	1.130	12.000

3PNBBT provides an upper bound values of mode I fracture energy, mainly in FRC of tensile strain hardening character, since in the IA it is assumed the formation of an unique fracture surface (coinciding with the specimen symmetry axis), while in the experimental tests more than one crack is formed during the loading process of this type of specimen. Even executing very thin notches for promoting the localization of cracking process in the notch, micro- and meso-cracks are always formed away from the notched plane (Pereira et al. 2010), and the evaluation of the real area of the fracture surface is, in the current state-of-the-art, still not possible to be reliability assessed.

11.2 Fracture Mode II Parameters

Baghi and Barros (2016) have also used the inverse analysis to determine the fracture mode II parameters of SHCC. These fracture parameters are the shear retention parameter, β, that together with the crack shear strength, $\tau_{t,p}^{cr}$, defines the first branch of the crack shear stress versus crack shear strain diagram represented in Fig. 11.6, $\tau_t^{cr} - \gamma_t^{cr}$, ($\gamma_{t,p}^{cr} = \tau_{t,p}^{cr} / D_{t,1}^{cr}$; $D_{t,1}^{cr} = \beta/(1-\beta) \cdot G_C$ being G_c the transverse elastic modulus), and the mode II fracture energy $(G_{f,s})$. This diagram simulates the crack shear stress transfer during the cracking process of concrete structures. The relevance of using this type of diagram for capturing accurately the load carrying capacity and crack pattern of RC beams failing in shear was demonstrated by Barros (2016).

Zamanzadeh (2017) following equal strategy of the one proposed by Baghi and Barros (2016) has determined the fracture mode II parameters of the RSFRC developed according to the technology shown in Fig. 8.19, whose direct tensile and direct shear properties are described in Sects. 8.2.3 and 8.2.6, respectively.

Figure 11.7 represents the finite element mesh used for the simulation of the specimen, which is composed of 2015 nodes and 1920 serendipity 4 nodes plain stress elements with 2 × 2 Gauss-Legendre integration scheme. This figure also shows the support and load conditions. The RSFRC specimens can be considered as isotropic material in its plane due to random orientation nature of the fibers. In fact, due to the relatively small thickness of the panels from which the Iosipescu specimens were extracted, it was assumed that fibres were oriented primarily in the plane.

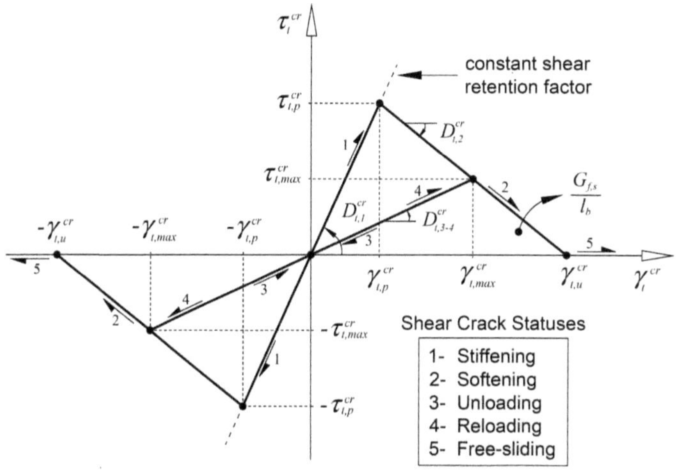

Fig. 11.6 Diagram to simulate the relationship between the crack shear stress and crack shear strain component, and possible shear crack statuses (Ventura-Gouveia 2011)

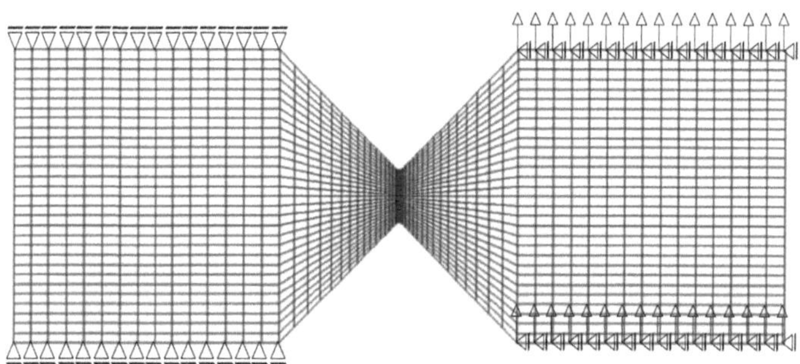

Fig. 11.7 Finite element mesh of the Iosipescu specimen for determining by inverse analysis the mode II fracture parameters of the RSFRC of M2_3.8n

The force is the sum of all vertical forces in each node in contact with the movable part of the testing fixture. This force represents the uniform load imposed on the specimens and was measured by the load cell in the experimental tests (Fig. 11.8).

By fitting as much as possible the experimental results presented in Fig. 8.38, the following fracture mode II parameters for the developed RSFRC were obtained (Fig. 11.6): $\beta = 0.5$, $\tau_{t,p}^{cr} = 3.5$ MPa, $G_{f,s} = 2.5$ N/mm. Comparing these results to the ones obtained by Baghi and Barros (2016) for SHCC ($\beta = 0.15$, $\tau_{t,p}^{cr} = 1.0$ MPa, $G_{f,s} = 0.5$ N/mm) it is concluded that the RSFRC developed by Zamanzadeh (2017) has higher crack shear stress transfer performance.

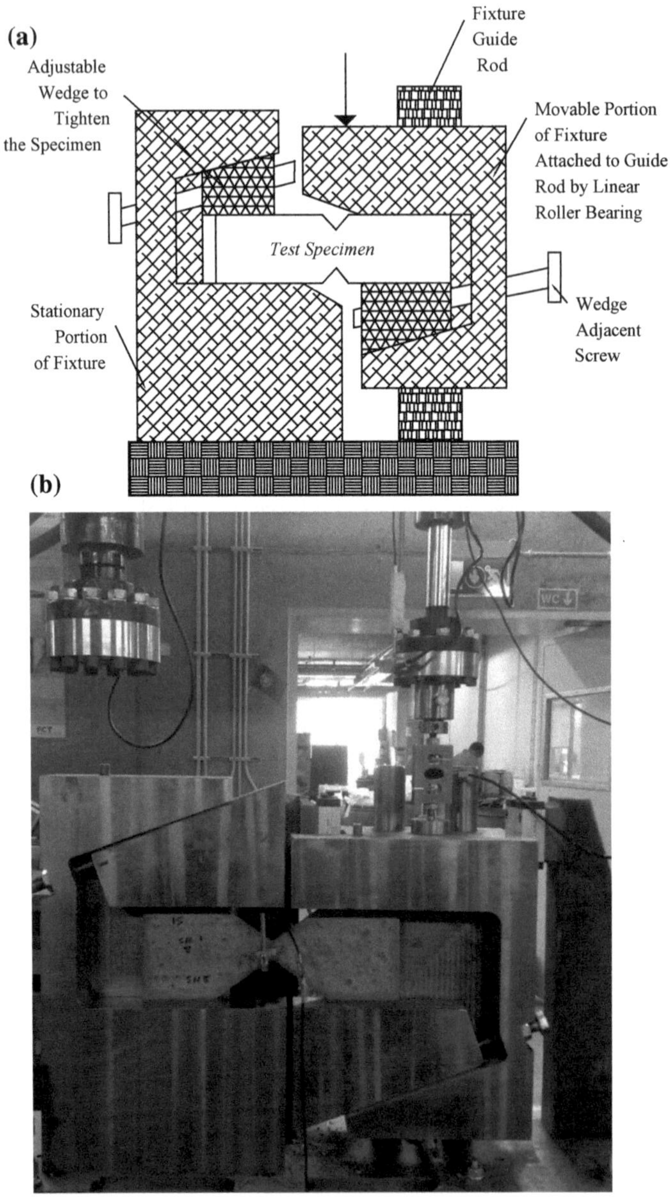

Fig. 11.8 Iosipescu test setup (Baghi and Barros 2016)

References

Baghi H, Barros JAO (2016) Shear properties of the strain hardening cementitious composite materials. ASCE J Mater Civ Eng 28(10)

Barros JAO (2016) Debilities and strengths of FEM-based constitutive models for the material nonlinear analysis of steel fiber reinforced concrete structures. In: Saouma V, Bolander J, Landis E (eds) 9th International conference on fracture mechanics of concrete and CONCRETE structures, FraMCoS-9, California, 29 May–1 June 2016

Barros JAO, Figueiras JA (1999) Flexural behavior of steel fibre reinforced concrete: testing and modelling. ASCE Mater Civ Eng J 11(4):331–339

MC2010 (2011) CEB fib Model Code 2010—Final Draft

Neocleous K, Tlemat H, Pilakoutas K (2006) Design issues for concrete reinforced with steel fibres, including fibres recovered from used tyres. Mater Civ Eng 18(5):677–685

Pereira EB, Fischer G, Barros JAO, Lepech M (2010) Crack formation and tensile stress-crack opening behavior of fiber reinforced cementitious composites (FRCC). In: Oh BH, Choi OC, Chung L (eds) Proceedings of FraMCoS-7, Jeju, Korea, 23–28 May 2010.

Pereira EB, Barros JAO, Camões AFFL (2008) Steel fibre reinforced self-compacting concrete—experimental research and numerical simulation. ASCE Struct Eng J 134(8):1310–1321

RILEM TC 162-TDF (2003) Test and design methods for steel fibre reinforced concrete. 1327 r-e-design method. Final Recommendation. Mater Struct 36:560–567

Schellekens JCJ (1990) Interface elements in finite element analysis. TU-Delft report 25.2-90-5-17/TNO-IBBC report BI-90-165

Sena-Cruz JM, Barros JAO, Ribeiro AF, Azevedo AFM, Camões AFFL (2004) Stress-crack opening relationship of enhanced performance concrete. In: 9th Portuguese conference on fracture, ESTSetúbal, Portugal, 395–403, 18–20 Feb 2004

Ventura-Gouveia A (2011) Constitutive models for the material nonlinear analysis of concrete structures including time dependent effects. PhD Thesis, Department of Civil Engineering, University of Minho

Zamanzadeh Z (2017) Cement based materials reinforced with recycled steel fibres for the shear reinforcement and strengthening of RC beams. PhD Thesis, University of Minho (in press)

Zamanzadeh Z, Lourenço LAP, Barros JAO (2015a) Recycled steel fibre reinforced concrete failing in bending and in shear. J Constr Build Mater 85:195–207

Zamanzadeh Z, Lourenço LAP, Barros JAO (2015b) Thin panels of cement composites reinforced with recycled fibres for the shear strengthening of reinforced concrete elements. Fibre concrete 2015, FC2015, Prague, Czech Republic, 10–11 Sept 2015

Chapter 12
Advanced Numerical Models for the Analysis of Sustainable FRC Structures

Antonio Caggiano, Joaquim A.O. Barros and Guillermo Etse

Abstract This chapter introduces some of the relevant modelling approaches that are being used to simulate the behaviour of cementitious materials reinforced with discrete fibres. The major part of these approaches were originally proposed for modelling non-fibrous reinforced cement based materials, therefore the main focus herein given is related to the aspects how fibre reinforcement mechanisms have been considered. These approaches were grouped in two classes, one where fibres are explicitly considered in the finite element mesh (FEMesh), herein designated as Discrete Fibre Reinforcement Approaches (DFRAs). The other class is designated by smeared fibre reinforcement approaches (SFRAs), where fibres are not part of the FEMesh and their contribution is basically considered attributing a constitutive law to the FRC that simulates the fibre reinforcement mechanisms in terms of the fracture modes of this composite material.

12.1 Introduction

Modeling of Fiber Reinforced Concrete (FRC) can be performed according to two main strategies: the first one considers the fibers individually in the finite element mesh with their reinforcement mechanisms directly (through interface elements) or indirectly (through analytical models) simulated, and by adopting a strategy for estimating the fiber distribution and orientation; the second one simulates the fibre reinforcement mechanisms by means of the fracture mechanics properties attributed

A. Caggiano
University of Buenos Aires, Intecin - CONICET, Buenos Aires, Argentina

J.A.O. Barros (✉)
University of Minho, Braga, Portugal
e-mail: barros@civil.uminho.pt

G. Etse
National University of Tucumán, Tucumán, Argentina

G. Etse
CONICET, Buenos Aires, Argentina

J.A.O. Barros et al. (eds.), *Recent Advances on Green Concrete for Structural Purposes*, Research for Development, DOI 10.1007/978-3-319-56797-6_12

to the cracked FRC. Approaches following the first strategy will be herein designated by "Discrete Fibre Reinforcement Approaches" (DFRAs), whereas for those developed in the framework of the second one the wording "Smeared Fibre Reinforcement Approaches" (SFRAs) will be adopted, in order to reflect how the fibre reinforcement mechanisms and cracking are taken into account.

It is well known the dependence of the mechanical properties of FRC on its rheological properties, mixing method, casting technology and geometry of the element to be produced in consequence of the fibre distribution and orientation (Dupont and Vandewalle 2005; Ferrara and Meda 2006; Ozyurt et al. 2007; Laranjeira et al. 2010; Torrijos et al. 2010; Abrishambaf et al. 2013, 2016a). Therefore, more recently, research has been done for integrating, into modelling of FRC, fluid mechanics in viscous medium (concrete flowability), maturation of cement based materials (hardening process), and physics of particles (impact and friction of aggregates during casting process), in order to better capture the aforementioned phenomena and, consequently, get a more accurate prediction of fibre distribution and orientation in the final hardened FRC (Deeb et al. 2014a, b).

For modeling concrete reinforced with sustainable fibres (i.e., recycled and vegetable), the previous approaches can be adopted, as long as the corresponding model parameters are conveniently identified by means of dedicated experimental investigation. In case of continuous vegetable fibres (relatively long fibres), they can be regarded as discrete or smeared reinforcement, and due to the pronounced weakest bond to the surround matrix, interface elements are currently used, whose properties are calibrated on fibre pullout tests (Barros et al. 2016).

12.2 Discrete Fibre Reinforcement Approaches (DFRAs)

Discrete Fibre Reinforcement Approaches (DFRAs) aim to simulate in an explicit way the fibre reinforcement mechanisms by considering each fibre in the finite element mesh. DFRAs are being mainly used for modeling the behavior of steel fibre reinforced concrete (SFRC) with relatively low content of fibres.

12.2.1 Bi-phase Type Models

The bi-phase model developed by Cunha et al. (2011) regards fibre reinforced concrete like a two-phase material consisting of plain concrete and discrete fibres. The plain concrete is simulated by a 3D smeared fixed crack model, while the fibre reinforcement mechanisms are modelled using micro-mechanical behaviour laws determined from experimental fibre pull-out tests. Other researchers have followed this approach with some minor modifications (Soetens 2015). The fixed smeared crack model adopted for the plain concrete phase is based on the strain decomposition concept and nonlinear fracture mechanics (mode I) proposed by de Borst and Nauta (1985). This approach was later generalized for a multidirectional fixed and rotating smeared crack version applicable to 2D plain stress and shell type elements for the analysis of concrete and FRC structures (Barros 1995). The algorithms for implementing rigorously the constitutive laws adopted to simulate both the cracks and the concrete between cracks were latter improved by Sena-Cruz (2004), and extended to 3D concrete structures by Ventura-Gouveia (2011).

The fibre structure phase consists of a random distribution of the discrete fibres in the bulk concrete according by the following procedures (Abrishambaf et al. 2016b):

(a) The initial parameters are defined, namely, the content and geometric data of fibres (fibre length, cross sectional area and mass weight of a single fibre) and geometry of specimen. Since recycled steel fibres (RSF) can have quite different geometric characteristics, as indicated in Chap. 8, they should be simulated by an algorithm capable of reproducing the statistical distribution of these characteristics, representative of the RSF applied in the studied application of RSFRC.
(b) The random number generation engine is initialized using the Mersenne-Twister (MT) procedure, which generates uniform pseudo-random values.
(c) The total number of fibres contained in the specimen is computed:

$$N_f^{vol.} = V_{sp} \cdot \frac{C_f}{m_f} \tag{12.1}$$

where, V_{sp} is the volume of specimen, C_f is the fibre content (kg/m^3) and m_f (kg) is the mass weight of a single fibre. In case of RSF, due to the dispersion of fibre mass weight, m_f should be a representative value of the type of RSF adopted in the application being simulated. The following steps are executed up to $N_f^{vol.}$ is attained.
(d) The fibre's gravity centre coordinates, as well as its orientation are randomly generated. For laminar type structures reinforced with ISF it was demonstrated that a Gaussian distribution provides accurate predictions of the fibre orientation profile (Abrishambaf et al. 2013, 2016a), but for sustainable fibres specific research should be executed in this respect.

(i) A set of three uniform random numbers (between 0 and 1), (ξ_x, ξ_y, ξ_z), are generated and the centre of gravity coordinates are determined, (x_i, y_i, z_i), for the ith fibre:

$$\begin{aligned} x_i &= \xi_x b_{rct} \\ y_i &= \xi_y l_{rct} \\ z_i &= \xi_x h_{rct} \end{aligned} \tag{12.2}$$

where, b_{rct}, l_{rct} and h_{rct} are width, length and height of the representative geometry of the element to be simulated.

(ii) According to the specified values of mean and standard deviation provided by the Gaussian distribution (or the one that best represents the fibre orientation profile of the sustainable fibres), a random value between 0 and 1 is generated for the dependent variable of the cumulative distribution function (Cunha 2010). The fibre orientation is set the independent variable, $x = \cos\theta_i$, and is obtained by the inverse distribution function method (Cunha 2010).

(iii) The fibre orientation versor $\hat{u}_i = (u_i, v_i, m_i)$ (see Fig. 12.1) is computed by randomly generating the first two components and determining the other component by solving $\cos(\theta_i) = \hat{u} \otimes \hat{\upsilon}$ with the Newton-Raphson method, where $\cos(\theta_i)$ indicates the fibre orientation towards the crack surface crossed by the fibre, based on the cumulative distribution function calculated in the previous step, and $\hat{\upsilon}$ is the versor orthogonal to this crack plane (known from the crack initiation process of the fixed smeared crack model representing the plain concrete phase).

(iv) The end-nodes' coordinates for the *i*th fibre (j = 1, 2) are computed by (in RSF l_f should be a representative value of the type of RSF adopted in the application being simulated):

$$\begin{aligned} x_i^j &= x_i \pm l_f u_i/2 \\ y_i^j &= y_i \pm l_f \upsilon_i/2 \\ z_i^j &= z_i \pm l_f m_i/2 \end{aligned} \tag{12.3}$$

(v) The end-nodes coordinates of *i*th fibre must accomplish the specimen's boundary conditions. For the fibres do not satisfying these conditions, a new orientation is randomly generated up to their accomplishment. The fibres intersecting the moulds are removed from the fibre structure.

Fig. 12.1 Definition of fibre orientation versors

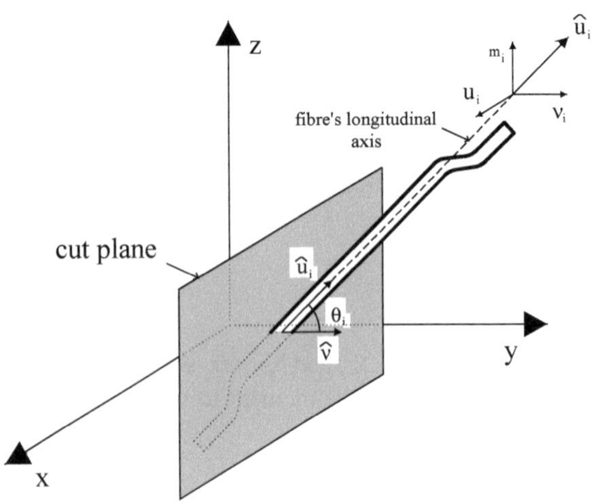

Figure 12.2 shows the fibre distribution meshes applying this algorithm, in the case of the simulations of the three point notched beam bending tests introduced in Sect. 8.2.5, and for the series β = [0°–15°], [15°–45°], [45°–75°] and [75°–90°], where β is the angle between the expected concrete flow and notched plane direction of the specimen (Fig. 12.3). β = 0° corresponds to a concrete flow parallel to the notched plan, promoting a preferential orientation of ISF fibres orthogonal to the notched plane (Cunha 2010; Abrishambaf et al. 2016a).

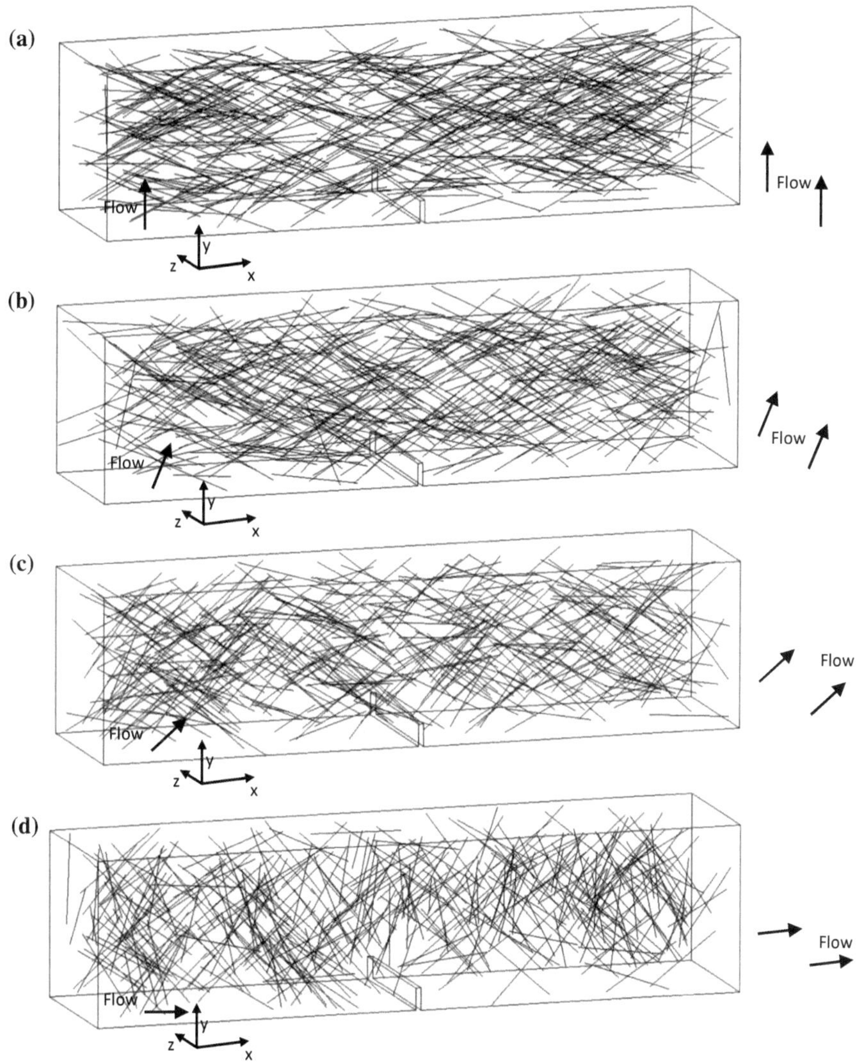

Fig. 12.2 Fibre distribution in a prismatic specimen extracted from a panel (the specimen's lateral face coincides with the *bottom* surface of the panel, i.e. in contact with the mould): β = **a** [0°–15°], **b** [15°–45°], **c** [45°–75°] and **d** [75°–90°] (Abrishambaf et al. 2016b)

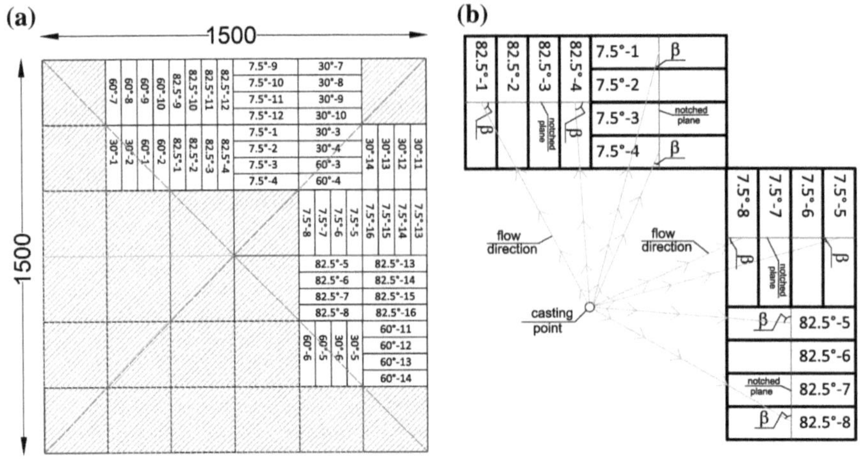

Fig. 12.3 **a** Specimen's extracting plane and **b** definition of β angle (Abrishambaf et al. 2016b)

The fibre distribution mesh, consisting of 3D embedded cable type 2 node elements, is coupled to the 3D solid finite element type mesh simulating the concrete phase, and the stiffness matrix of a FRC solid finite element, $\underline{K}^{rc}$, is determined from (Cunha et al. 2012):

$$\underline{K}^{rc} = \underline{K}^{crco} + \sum_{i=1}^{n_f} \underline{K}_i^f \tag{12.4}$$

where $\underline{K}^{crco}, \underline{K}_i^f$ and n_f are the stiffness matrix of the cracked concrete, the stiffness matrix of the *i*th fibre embedded in the concrete mother-element, and the total number of the embedded fibres into the mother-element, respectively. The embedded elements are modelled assuming a perfect bond between fibre and matrix. Therefore, the fibre resisting mechanisms, captured by fibre pullout tests (Cunha et al. 2010), are indirectly simulated, since a direct approach requires the adoption of fibre-matrix interface finite elements leading to quite high computing time. The fibre load–slip, *F*–*s*, relationship obtained in fibre pullout tests is, therefore, converted into a fibre stress–strain, σ_f–ε_f, relationship, by adopting the approach schematically represented in Fig. 12.4, where ε_f, *s* and l_b are the fibre's strain, fibre's slip and the crack band width of the solid finite element (which guarantees that the results of the simulations are not dependent of the finite element mesh refinement). The fibre stress, σ_f, is calculated from the pull-out force, *F*, divided by the fibre's cross section area, A_f.

In order to adopt this approach for sustainable fibres, the results from pullout tests should be collected and organized in order to define representative clusters of pullout resisting mechanism for this type of fibres. This strategy was already adopted by Cunha et al. (2011), but limited to the fibre orientation. For the case of sustainable fibres other geometric characteristics of these fibres must be also considered, like representative diameter, surface and shape attributes.

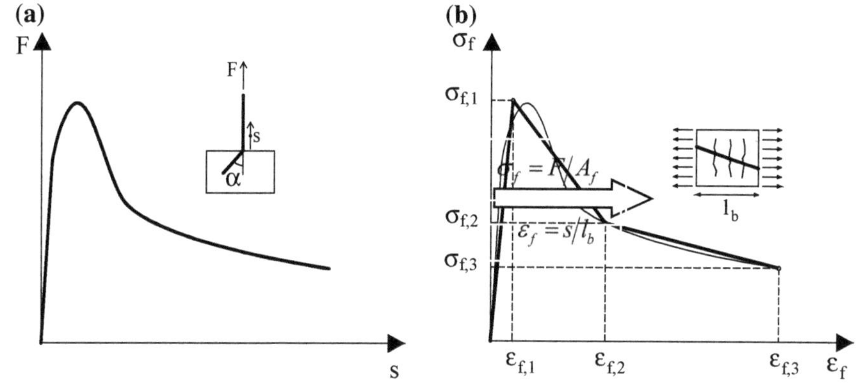

Fig. 12.4 Determination of the embedded cable's stress–strain diagram based on the experimental pull-out force–slip relation (Cunha et al. 2011)

12.2.2 Lattice Type Models

In the framework of DFRA and fracture behavior of both concrete and FRC, the lattice-based model is a simple and effective tool for understanding the physics of fracture processes in concrete members (Yip et al. 2006). Lattice type models can be based on either truss (Schorn and Rode 1987) or beam elements as highlighted in Fig. 12.5 (Lilliu and van Mier 2003). Within the lattice framework, fiber effects on FRC can be easily modeled since each beam (or truss) can be considered as the short reinforcement to be modelled (Bolander et al. 2008).

Fracture processes in lattice models are typically reproduced by means of sequential removal of failed elements from the mesh. Particularly it is assumed that each element behaves with a linear-elastic mode until failure: once failure occurs the element should be removed. It is worth to mention that despite the elastic-brittle character of each element, globally the structural response of the specimen can be softening or even hardening in cases of a great amount of fibers.

Failure can be detected in several ways. The most used is the one that evaluates the effective stress, "computed" at the ends of the lattice beam/truss, according to the following expression:

$$\bar{\sigma} = \frac{F}{A} + \alpha \frac{\sqrt{M_x^2 + M_y^2}}{W} \tag{12.5}$$

being A and W the area and modulus of the cross-section; F, M_x and M_y are the internal axial force and bending moments (in case of truss M_x and M_y are null). The coefficient α is adopted to adjust the contribution of the bending moment in a beam element (Lilliu and van Mier 2003).

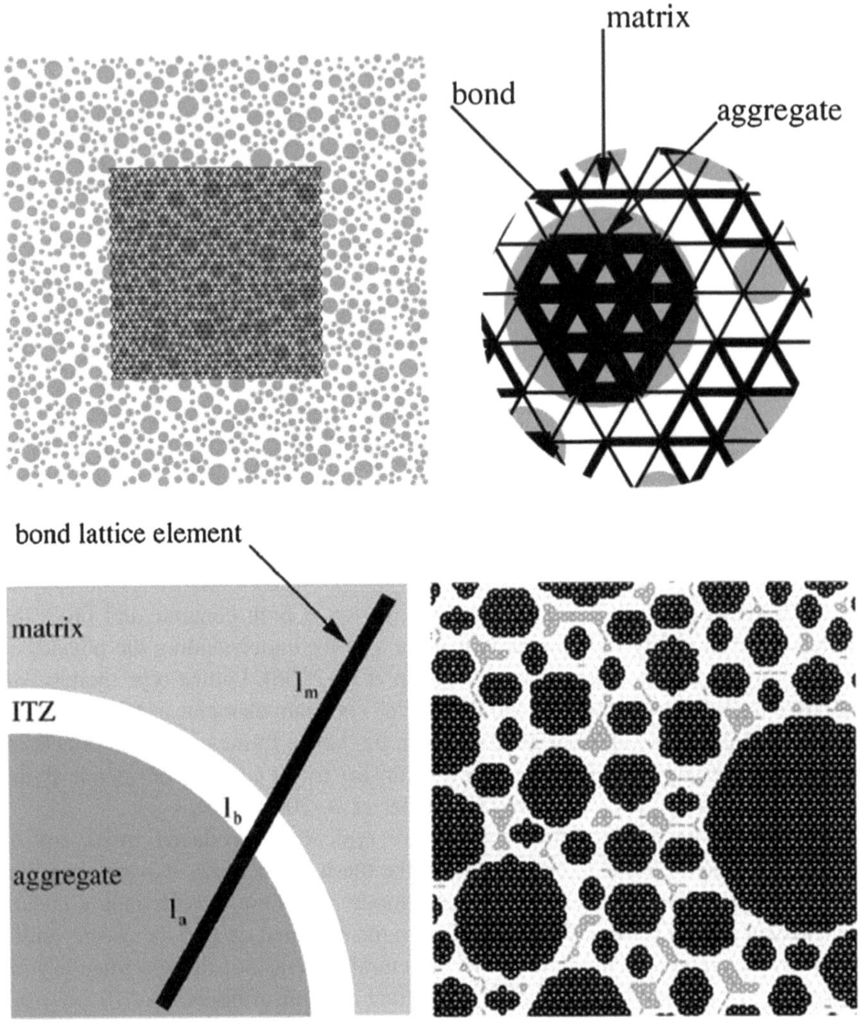

Fig. 12.5 Meso-mechanical simulation by means of the lattice model by Lilliu and van Mier (2003)

The element to be removed, which in a FRC meso-structure can be represented by a coarse aggregate, cementitious matrix or a fiber element, is that in which the effective stress $\overline{\sigma}$ reaches a threshold value (namely failure strength).

Let's consider a short fiber embedded within a lattice model of a cement-based composite. It contributes to the elastic stiffness of the crossed Voronoi cells (associated a certain set of points, which are then connected between them throughout lattice elements) but also has an explicit influence on the post-cracking response of opening cracks.

Particularly, in order to take into account the enhancement in the lattice stiffness, a "zero-length spring", as proposed by Bolander and Sukumar (2005) and highlighted in Fig. 12.6 can be employed. This spring is naturally aligned in the fiber direction, connects the two Voronoi cells at the boundary crossing, and its stiffness depends on the fiber type/geometry as follows:

$$k_f = \frac{A_f \sigma_f}{(h_v/\cos\psi)\varepsilon_m} \tag{12.6}$$

where A_f is the fiber cross-section area, σ_f is the fiber axial stress at the boundary crossing ε_m represents the surrounding matrix strain in the fiber direction, ψ deals with the fiber inclination and finally h_v is the projected distance between the centroids of the lattice Voronoi diagrams (Fig. 12.6).

Besides the fiber contributions on the overall composite stiffness before the crack onset as above described, lattice models are particularly suitable for modeling the fracture responses and cracking phenomena of FRC. As a matter of fact, discontinuous short reinforcements can be easily and explicitly added by means of lattice elements (truss or beams), suitable for capturing the toughness effects provided by fibers crossing cracks in cement-based systems. The process is mainly due to debonding along the fiber-matrix interface. At the onset of matrix fracture, one or more fibers that cross the developing crack can be identified and the associated stiffness of the so called "zero-length spring" (Eq. 12.6) has to be modified to take into account the bond-slip mechanisms. Particularly by combining equilibrium, compatibility and constitutive (of fiber, matrix and bond-slip) rules (see Fig. 12.7) for the embedded fiber lengths to each side of the spring, the axial stiffness of a modified spring can be derived.

Thus, pullout relations for each embedded length can be derived from the constitutive properties of the $\tau - s$ (bond stress vs. slip) interface constitutive law. These formulations are all based on adopting a certain analytical expression or more complex constitutive models for describing the interface bond-slip law. A bi-linear

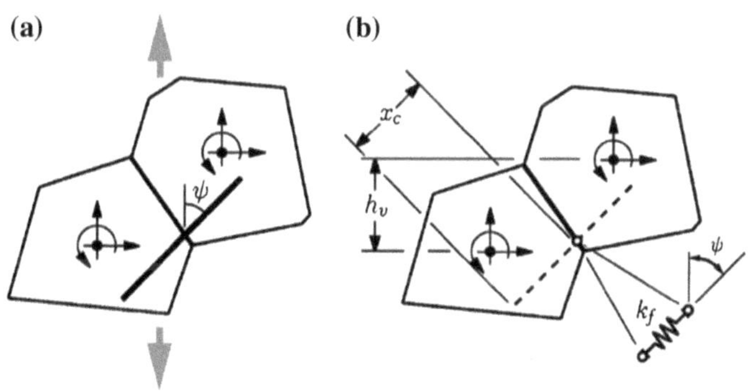

Fig. 12.6 **a** Scheme of two lattice (Voronoi) elements with a fiber inclusion and **b** stiffness fiber contribution by Bolander and Sukumar (2005)

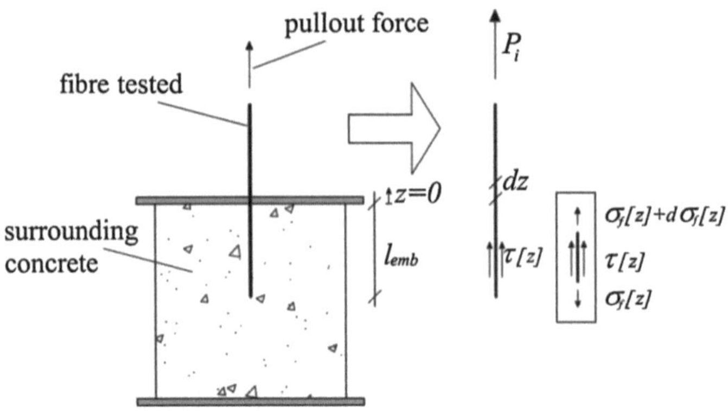

Fig. 12.7 Schematic components of a fiber pull-out analysis (Caggiano and Martinelli 2012)

elastic-softening bond-slip relationship was adopted by Caggiano et al. (2012), whereas other proposals are based on exponential negative type (Caggiano and Martinelli 2012) or multilinear (Cunha et al. 2010) bond–slip laws. Furthermore, it is worth mentioning that these formulations are generally based on the assumption of a fracture process occurring in pure debonding mode and neglecting the effect of the interface normal (peeling) stresses and the occurrence of out-of-plane (uplift) displacements.

12.2.3 Strong Discontinuity Approaches

Strong discontinuity approaches incorporate displacement discontinuities in the finite element formulations for capturing arbitrary crack propagations within fixed FE mesh without losing the mesh objectivity. Particularly, new types of finite elements were proposed in the literature, essentially formulated by enriching the (continuous) displacement fields with additional discontinuities.

Two families of strong discontinuity approaches can be distinguished, depending on the technique employed to enrich the displacement field:

- the so-called Embedded strong discontinuity Finite Element Method (E-FEM), proposed among others by Dvorkin et al. (1990), Oliver (1996), Oliver et al. (2002) and Armero and Linder (2009), which are able to reproduce displacement "jumps" through elemental discontinuity enrichments (Fig. 12.8a), and
- the extended Finite Element Method (X-FEM) in which the discontinuity of the displacement field is captured by means of nodal enrichments of the kinematic field (8b) by Wells and Sluys (2001) and Liu et al. (2011).

Typical fracture problems in solid mechanics exhibit cracks or displacement jumps in the considered spatial domain Ω as highlighted in Fig. 12.8.

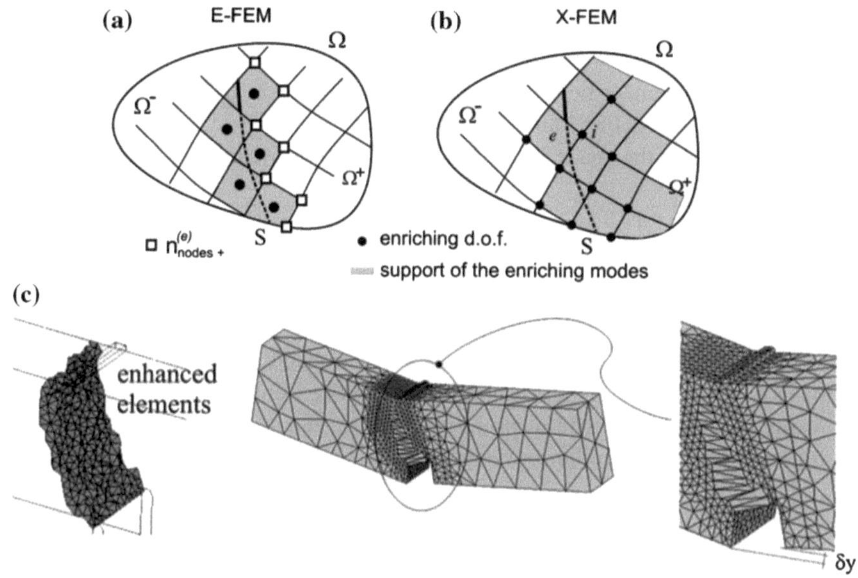

Fig. 12.8 **a** Elemental (E-FEM) and **b** extended (X-FEM) enrichment approaches for **c** the 3-point bending scheme by Oliver et al. (2006)

The following discontinuous displacement field needs to be introduced in strong discontinuity approaches (Oliver et al. 2006):

$$\mathbf{u}(\mathbf{x}) = \bar{\mathbf{u}}(\mathbf{x}) + \mathrm{H}_s(\mathbf{x})\boldsymbol{\beta}(\mathbf{x}); \quad \mathrm{H}_s(\mathbf{x}) = \begin{cases} 1 & \forall\, \mathbf{x} \in \Omega^+ \\ 0 & \forall\, \mathbf{x} \in \Omega^- \end{cases} \tag{12.7}$$

being $\mathbf{u}(\mathbf{x})$ the displacement field, while $\bar{\mathbf{u}}(\mathbf{x})$ and $\boldsymbol{\beta}(\mathbf{x})$ are the so-called regular displacement field and the displacement jump, respectively. $\mathrm{H}_s(\mathbf{x})$ is the Heaviside (step) function, Ω^+ and Ω^- are the positive and negative sides of the cracked domain.

Under the hypothesis of infinitesimal strain, the strain field $\boldsymbol{\varepsilon}(\mathbf{x})$ results:

$$\boldsymbol{\varepsilon}(\mathbf{x}) = \nabla^{sym}\mathbf{u}(\mathbf{x}) = \nabla^{sym}\bar{\mathbf{u}}(\mathbf{x}) + \mathrm{H}_s\nabla^{sym}\boldsymbol{\beta}(\mathbf{x}) + \delta_s[\mathbf{n}(\mathbf{x}) \otimes \boldsymbol{\beta}(\mathbf{x})]^{sym} \tag{12.8}$$

where $\mathbf{n}(\mathbf{x})$ is the normal vector to the fracture line S while δ_s deals with the Dirac's delta-function.

Being Ω^h a discretized body, the governing variational equation of equilibrium in the standard form and in absence of body forces can be written as follows:

$$\int_{\Omega^h} \nabla^{sym}\delta\mathbf{u}^h \cdot \boldsymbol{\sigma}\, d\Omega = \int_{\Gamma_\sigma^h} \delta\mathbf{u}^h \cdot \bar{\mathbf{t}}\, d\Gamma_\sigma \quad \forall\, \mathbf{u}^h \in \Upsilon_0^h \tag{12.9}$$

In Eq. (12.9) $\Gamma_{\sigma}^{h} \subset \partial\Omega^{h}$ is the boundary with the prescribed traction $\bar{\mathbf{t}}$, Υ_{0}^{h} represents the space of the admissible displacements.

Both the displacement field of Eq. (12.7) and the space Υ_{0}^{h} are represented in a different way by the E-FEM and the X-FEM.

On the one hand the space of interpolation functions defined for the X-FEM is:

$$\Upsilon_{X-FEM}^{h} = \left\{ \mathbf{u}^{h}(\mathbf{x}) \middle| \mathbf{u}^{h}(\mathbf{x}) = \sum_{i=1}^{n_{node}} \left(N_i(\mathbf{x})\mathbf{d_i} + \mathrm{H}_s N_i(\mathbf{x})\boldsymbol{\beta}_i \right) \right\} \tag{12.10}$$

where $N_i(\mathbf{x})$ accounts for the standard interpolation FE shape functions, $\mathbf{d_i}$ are the nodal regular displacement vector, $\boldsymbol{\beta}_i$ is the vector collecting the nodal displacement jumps and n_{node} is the number of nodes of the finite element mesh.

Finally the variations, with respect to parameters $\mathbf{d_i}$ and $\boldsymbol{\beta}_i$ of Eq. (12.10) lead to the following discrete equilibrium equations:

$$\delta\mathbf{d_i} \cdot \left(\int_{\Omega^h} \nabla N_i \cdot \boldsymbol{\sigma}\, d\Omega - \int_{\Gamma_\sigma^h} N_i \cdot \bar{\mathbf{t}}\, d\Gamma \right) = 0 \quad \forall\, \delta\mathbf{d_i};\ i = 1, n_{node} \tag{12.11a}$$

$$\delta\boldsymbol{\beta}_i \cdot \left(\int_{\Omega^h} \mathrm{H}_s \nabla N_i \cdot \boldsymbol{\sigma}\, d\Omega + \int_{S} N_i \cdot (\boldsymbol{\sigma}_S \cdot \mathbf{n})\, dS \right) = 0 \quad \forall\, \delta\boldsymbol{\beta}_i;\ i = 1, n_{node} \tag{2.6.11b}$$

On the other hand the space of interpolation functions defined for the E-FEM is:

$$\begin{aligned} \Upsilon_{E-FEM}^{h} &= \left\{ \mathbf{u}^{h}(\mathbf{x}) \middle| \mathbf{u}^{h}(\mathbf{x}) = \sum_{i=1}^{n_{node}} \left(N_i(\mathbf{x})\mathbf{d_i} \right) + \sum_{e=1}^{n_{elem}} \left(M_S^{(e)} \boldsymbol{\beta}_e \right) \right\}, \\ M_S^{(e)} &= \mathrm{H}_s - \varphi^{(e)};\ \varphi^{(e)} = \sum_{i=1}^{n_{node+}^{(e)}} N_i^{(e)} \end{aligned} \tag{12.12}$$

where n_{elem} is the number of elements while $n_{node+}^{(e)}$ refers to those nodes placed in Ω^{+}, $\boldsymbol{\beta}_e$ are degrees of freedom describing the elemental displacement jumps and $M_S^{(e)}$ represents the elemental unit jump function.

Thus, by calculating the variations with respect to parameters $\mathbf{d_i}$ and $\boldsymbol{\beta}_e$ of Eq. (12.12), the following discrete equilibrium equations can be obtained:

$$\delta\mathbf{d_i} \cdot \left(\int_{\Omega^h} \nabla N_i \cdot \boldsymbol{\sigma}\, d\Omega - \int_{\Gamma_\sigma^h} N_i \cdot \bar{\mathbf{t}}\, d\Gamma \right) = 0 \quad \forall\, \delta\mathbf{d_i};\ i = 1, n_{node} \tag{12.13a}$$

$$\delta\boldsymbol{\beta}_e \cdot \left(\int_{\Omega^{(e)}} \nabla\varphi^{(e)} \cdot \boldsymbol{\sigma}\, d\Omega + \int_{S^{(e)}} (\boldsymbol{\sigma}_S \cdot \mathbf{n})\, dS \right) = 0 \quad \forall\, \delta\boldsymbol{\beta}_e;\ e = 1, n_{elem} \qquad (12.13b)$$

The stress vector $\boldsymbol{\sigma}_S \cdot \mathbf{n} = \mathbf{t}_s$ in Eqs. (2.6.11b) and (12.13b) represents the interface cohesive traction in the crack front that can be substituted with each class of stress-crack opening constitutive rule.

An interesting comparison between the E-FEM and X-FEM approaches, employed to analyze strong discontinuities in concrete materials, has been reported in Oliver et al. (2006).

Strong discontinuity approaches employed for modeling the fracture behavior of FRC have been employed by Bouhala et al. (2013) and Ramm et al. (2006) for the X-FEM, while Oliver et al. (2008) and Ibrahimbegovic et al. (2010) explicitly modeled through the E-FEM approach the concrete, steel and bond-slip behaviors in RC frames.

In the approach proposed by Oliver et al. (2008), FRC is assumed to be a composite material constituted by a matrix plus embedded fibers (Fig. 12.9). According to the basic hypothesis of the well-known Mixture Theory (Trusdell et al. 1960), the composite can be considered as a continuum in which each infinitesimal volume is simultaneously occupied by each one of the constituents (i.e., matrix and fibers).

Assuming this parallel mechanical system, the strain field of the matrix, $\boldsymbol{\varepsilon}^m$, and the composite strain field, $\boldsymbol{\varepsilon}$, can be considered as equal. The axial strain of the fiber r is $\varepsilon^{fr} = \mathbf{r} \cdot \boldsymbol{\varepsilon} \cdot \mathbf{r}$ (being $\mathbf{r}$ its direction) while $\varepsilon^{fs} = \mathbf{s} \cdot \boldsymbol{\varepsilon} \cdot \mathbf{s}$ for fiber s (in the $\mathbf{s}$ direction). Then, transversal to the fiber direction the dowel strain can be finally considered as $\gamma^f = 2\,\mathbf{r} \cdot \boldsymbol{\varepsilon} \cdot \mathbf{s}$.

According to these assumptions, the composite material stress tensor can be computed as the sum of the stresses of each constituent, weighted by the corresponding volumetric fraction k^m (matrix), k^{fr} (r fiber) and k^{fs} (s fiber) as figures out in Fig. 12.10. Particularly, $\boldsymbol{\sigma}^m, \sigma^{fr}, \sigma^{fs}$ and τ^f represent the stresses for matrix and fibers and are based on the adopted constitutive model for each one of them.

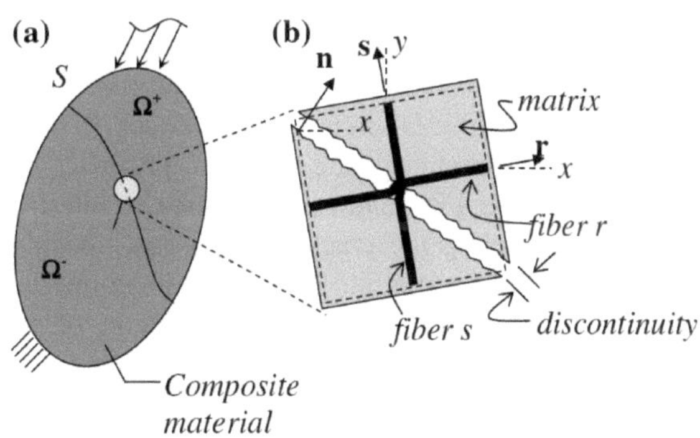

Fig. 12.9 **a** FRC with a discontinuity surface S and **b** representative material point of the composite material by Oliver et al. (2008)

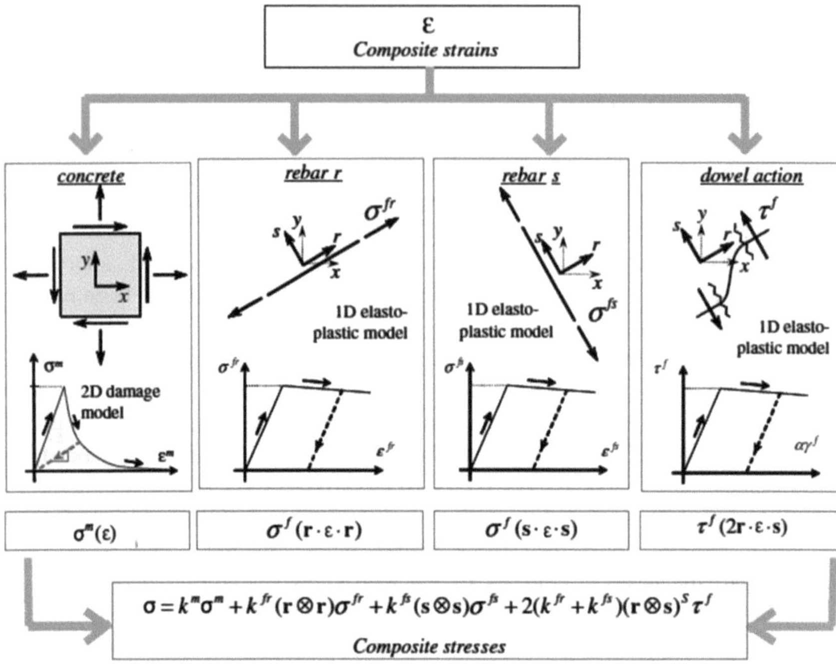

Fig. 12.10 Stresses composition scheme for the composite material under consideration

12.2.4 Particle Models

Particle models for quasi-brittle materials are based on the formulation of the microscopic inter-particle contact layers of the matrix particles (Jirasek and Bazant 1994). The pioneer proposals of particle simulations for granular media and rock movements have been reported in the works by Cundall (1971), Rodriguez (1974) and Kawai (1980). These works mainly modelled the behavior of granular solids (such as sand) considering rigid particles that interact by friction. Furthermore, a particle model for brittle composite materials was proposed by Zubelewicz and Bazant (1987), and later adapter by Bazant et al. (1990) for simulating cracking localization in concrete elements and fiber composites.

In this framework, a mesoscopic approach, based on a further development of the so-called Lattice Discrete Particle Model (LDPM) (Cusatis et al. 2011a, b), was proposed to simulate the failure processes possibly developing at the mortar-to-mortar FRC and mortar-aggregate interfaces (Fig. 12.11). This proposal aimed at modeling the crack-bridging mechanisms of fibers in FRC (Schauffert et al. 2012; El-Helou et al. 2014) combining the development of two independent research efforts that led to the formulation of the Confinement Shear Lattice (CSL) model (Cusatis et al. 2006) and the Discrete Particle Model (DPM) (Pelessone 2005) for cement-based composites.

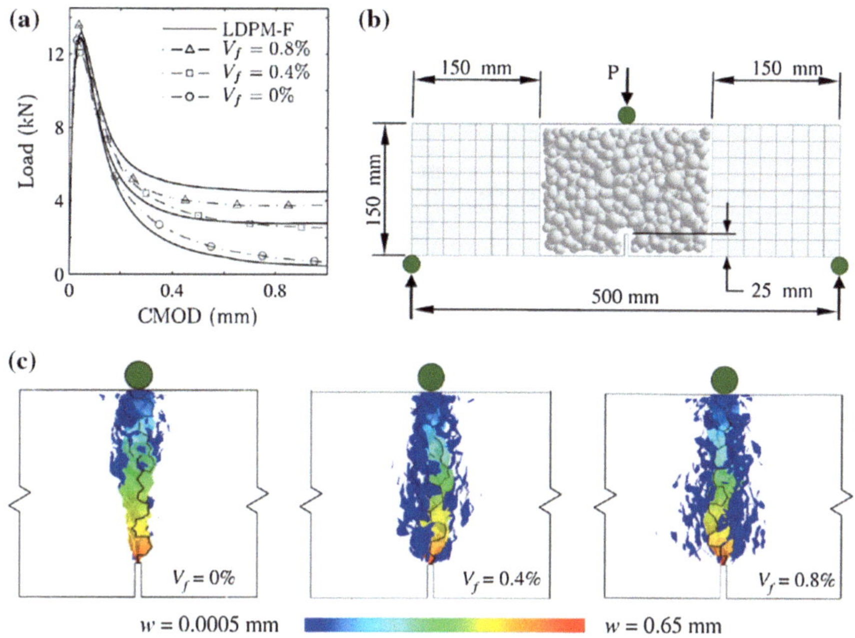

Fig. 12.11 3-point bending: **a** numerical versus experimental results, **b** particles and **c** mesoscale crack patterns (Schauffert et al. 2012)

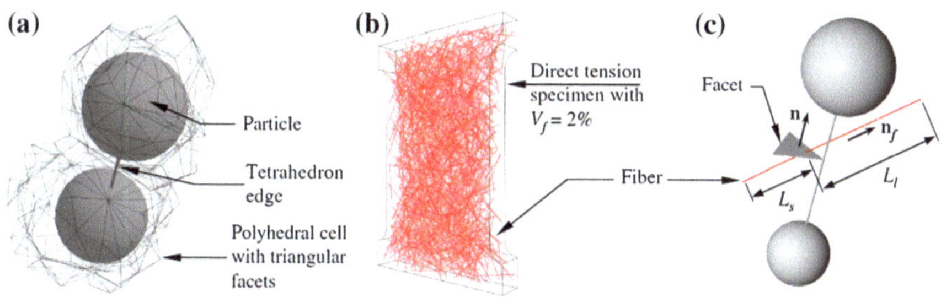

Fig. 12.12 **a** Tetrahedralization of adjacent particles, **b** specimen with random distributed fibers and **c** facet with intersecting fiber (Schauffert et al. 2012)

LPDM model deals with a mesoscale representation procedure for heterogeneous cementitious materials. The model is capable to simulate material mesostructure trough modeling coarse aggregates and surrounding mortar as polyhedral cells and fibers explicitly. The so-called tetrahedralization of the particle centers generates the lattice framework where each lattice member is associated with a triangular shape ("facet") which represents the plane of contact between two cells (Fig. 12.12).

LDPM describes concrete deformation and crack mechanisms through the adoption of a rigid-body kinematics (Cusatis et al. 2011a). Thus, for a given displacement/rotation field relating to the particles associated with a given facet, the relative displacement $\mathbf{u}_C$ at the facet centroid can be used to calculate the strains:

$$\varepsilon_N = \frac{\mathbf{n} \cdot \mathbf{u}_C}{l} \quad \varepsilon_M = \frac{\mathbf{m} \cdot \mathbf{u}_C}{l} \quad \varepsilon_L = \frac{\mathbf{l} \cdot \mathbf{u}_C}{l} \tag{12.14}$$

being l the tetrahedron length associated with the considered facet, $\mathbf{m}$ and $\mathbf{l}$ two mutually orthogonal unit vectors tangential to the facet and $\mathbf{n}$ the normal unit vector (Fig. 12.12).

Strain definitions highlighted in Eq. (12.14) are full consistent to define the strain field in classical continuum mechanics (Cusatis and Schauffert 2010) and have been used in the LDPM model to compute the following facet stress vector:

$$\boldsymbol{\sigma}_C = \sigma_N \mathbf{n} + \sigma_M \mathbf{m} + \sigma_L \mathbf{l} \tag{12.15}$$

which depends on the constitutive laws employed in the normal and tangential direction of the facet:

$$\sigma_N = \sigma_N(\varepsilon_N) \quad \sigma_M = \sigma_M(\varepsilon_M) \quad \sigma_L = \sigma_L(\varepsilon_L) \tag{12.16}$$

When the facet strain components reach the yield elastic limit crack openings can be calculated as:

$$\mathbf{w} = w_N \mathbf{n} + w_M \mathbf{m} + w_L \mathbf{l} \tag{12.17}$$

where:

$$\begin{aligned} w_N &= l(\varepsilon_N - \sigma_N/E_N) \\ w_M &= l(\varepsilon_M - \sigma_M/E_T) \\ w_L &= l(\varepsilon_L - \sigma_L/E_T) \end{aligned} \tag{12.18}$$

being E_N and E_T the elastic normal and tangential LDPM stiffness, respectively.

As result of coupling between fibers and the surrounding concrete matrix, total stress vector $\boldsymbol{\sigma}$ on each LDPM facet for FRC can be computed as:

$$\boldsymbol{\sigma} = \boldsymbol{\sigma}_C + \frac{1}{A_c} \sum_{f \in A_c} \mathbf{P}_f \tag{12.19}$$

where A_c represents the facet area and $\mathbf{P}_f$ collects the crack-bridging force for each fiber crossing the LDPM facet in the **m, l** and **n** direction:

$$\mathbf{P}_f = \mathbf{P}_{f,N}\mathbf{n} + \mathbf{P}_{f,M}\mathbf{m} + \mathbf{P}_{f,L}\mathbf{l} \tag{12.20}$$

where $\mathbf{P}_{f,N}$, $\mathbf{P}_{f,M}$ and $\mathbf{P}_{f,L}$ are the fiber effects in the normal and tangential directions of the facet.

Fiber contribution of Eqs. (12.19) and (12.20) is assumed to be negligible for all elastic responses (pre-cracking) and also in case of inelastic behaviors in which the normal facet stress component becomes negative (compressive case). Based on this, it can be assumed that $\mathbf{P}_f = \mathbf{0}$ when $\varepsilon_N \leq 0$. Furthermore, the fiber bridging effect depends on the meso/micro mechanical crack-opening mechanisms at facet level. The resulting stress-crack opening force $\mathbf{P}_f = \mathbf{P}_f(\mathbf{w})$ can be analyzed individually for each fiber intersecting the facet, this latter aimed at representing a possible crack. The calculation is based on a pullout model similarly to that described in the lattice-based approach of Sect. 12.2.2 and followed by other proposals as that based on the zero-thickness interface proposal presented in Sect. 12.2.5.

12.2.5 Zero-Thickness Based Interface Models

The discrete approach based on interface elements has been followed by several authors and researchers, as an alternative to the continuous method. This approach is based on the use of zero-thickness joints that connect continuum solid elements with the possibility of simulating the initiation and propagation of cracks.

Zero-thickness interface elements, formulated in terms of contact stresses versus opening relative displacements, have been historically employed for modelling the response behavior of both material discontinuities, such as mechanical contacts (Lei 2001) or bonds (Caggiano and Martinelli 2012), and crack evolutions in quasi-brittle materials like concrete (Ciancio et al. 2013). Several plasticity-based interface formulations have been proposed to predict failure behaviors of discontinuities in soil/rock mechanisms (D'Aguiar et al. 2011).

However, one of the most frequent use of interface elements in computational concrete mechanics is related to mesoscopic failure simulations. Particularly, the use of non-linear interfaces can be applied to model the aggregate-mortar joint (Stankowski et al. 1993). This strategy can be also combined with the use of nonlinear continuum models for the mortar in between aggregates. Alternatively, the use of nonlinear interfaces may involve both the aggregate-mortar and mortar-mortar joints, while linear elastic models are considered for the continuum (mortar and coarse aggregate) elements. This approach was proposed by Lopez et al. (2008a, b) for rate independent failure behavior analysis of concrete and by Lorefice et al. (2008) for time dependent simulations of concrete failure cracks.

Further constitutive models available in literature for interface proposals have been proposed by Hillerborg et al. (1976), Carol et al. (1997), and Pandolfi and Ortiz (2002) between others. Sometimes, interface formulations may only include traction-separation laws (Olesen 2001; Oh et al. 2007; Buratti et al. 2011;

Pereira et al. 2012) or even constitutive relations based also on more complex mixed-modes of fracture (Carol et al. 1997; Hillerborg et al. 1976; Pandolfi et al. 2000; Park et al. 2010; Lopez et al. 2008a, b).

Recently, the approach proposed by Lopez et al. (2008a, b) and the related interface proposals was extended by Caggiano et al. (2011, 2012) to model the cracking behavior of systems made of FRC. The proposed constitutive model for FRC interfaces is based on an application of the "Mixture Theory" and combines three internal stress-crack opening relationships, listed below:

(1) A *fracture energy-based cracking* formulation for plain mortar/concrete interfaces that relates normal and tangential stress components with the corresponding relative displacements in the plane of the interface. Its maximum strength criterion is defined according to the three-parameter hyperbolic failure surface by Carol et al. (1997);
(2) A *fiber bond-slip* formulation to describe the uniaxial inelastic behavior of fibers crossing the interfaces by means of an elasto-plastic model;
(3) A formulation for *fiber dowel action* based on elastic foundation concepts to obtain the dowel force-displacement relationship.

More specifically, according to the fundamental assumptions of the Mixture Theory (Trusdell et al. 1960), each infinitesimal volume of the mixture is ideally occupied by all mixture constituents (i.e., "phases") that are subjected to the same displacement fields. Consequently, the composite stress is obtained from the addition of the mixture stresses weighted by the volume fraction of each mixing constituent.

The composite schematization of a considered FRC specimen takes into account three main phases: (i) coarse aggregates, (ii) cementitious matrix or mortar and (iii) fibers (Fig. 12.13). It follows that the mechanical properties of the considered composite material can be defined according to the "Mixture Theory" above mentioned.

For instance, a generic mechanical parameter of the composite can be derived as:

$$p_c = \rho_a p_a + \rho_m p_m + \rho_f p_f \tag{12.21}$$

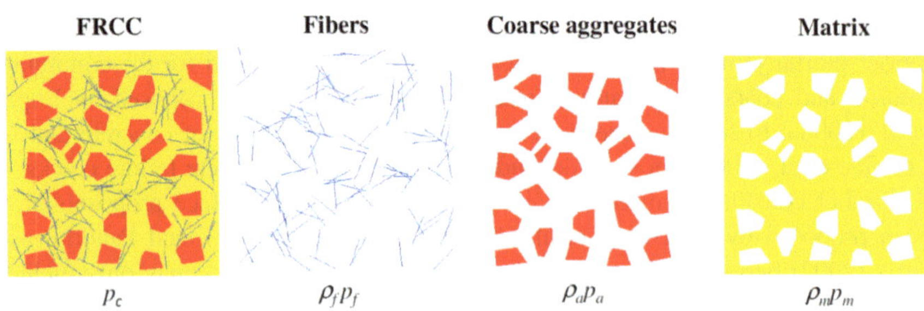

Fig. 12.13 Mixture components of the FRCC material (Caggiano 2013)

being $p_{\#}$ the considered mechanical parameter (i.e., the elastic modulus or Poisson's ratio) while $\rho_{\#}$ refers to the volume fraction of each $_{\#}$ component: i.e., # = c for the considered composite, # = a for coarse aggregates, # = m for the matrix and # = f for fibers.

In a similar way, it can be considered a generic relative displacement vector **u** at the interface (Fig. 12.14), where the axial displacement of the single fiber is given by $u_N = \mathbf{u} \cdot \mathbf{n}_N$ (being $\mathbf{n}_N$ the fiber direction), while in the transversal direction $u_T = \mathbf{u} \cdot \mathbf{n}_T$. Thereby, $\mathbf{n}_T$ is a unitary vector orthogonal to $\mathbf{n}_N$. Consequently, the axial and tangential fiber strains can be derived as $\varepsilon_N = u_N / l_f$ and $\gamma_T = u_T / l_f$ respectively, being l_f the fiber length (each fiber is ideally assumed to cross the interface fracture surface at its mid-length, i.e. at $l_f/2$).

The interface constitutive model is formulated in incremental form, as generally assumed for the flow theory of plasticity. According to the hypotheses of the composite model, the rate of the stress vector at the interface $\dot{\mathbf{t}} = [\dot{\sigma}, \dot{\tau}]^t$ (being $\dot{\sigma}$ and $\dot{\tau}$ the normal and tangential rate interfacial composite stresses, respectively) is calculated by means of the following $\Box_{\#}$-weighted sum:

$$\dot{\mathbf{t}} = w[\rho_i]\dot{\mathbf{t}}^i + \sum_{f=0}^{n_f} w[\rho_f]\left(\dot{\sigma}_f[\dot{\varepsilon}_N]\,\mathbf{n}_N + \dot{\tau}_f[\dot{\gamma}_T]\mathbf{n}_T\right). \tag{12.22}$$

The indices i and f in Eq. (12.22) refer to interface and fiber, respectively, σ_f and τ_f mean the bond-slip force and dowel effect of the single considered fiber, n_f is the number of fibers crossing the interface, while $w[\rho_{\#}]$ are weighting functions of the Mixture Theory.

The fiber bridging effect, formulated herein in terms of bond-slip and dowel mechanisms, are schematically represented in the interface plane as shown in Fig. 12.15.

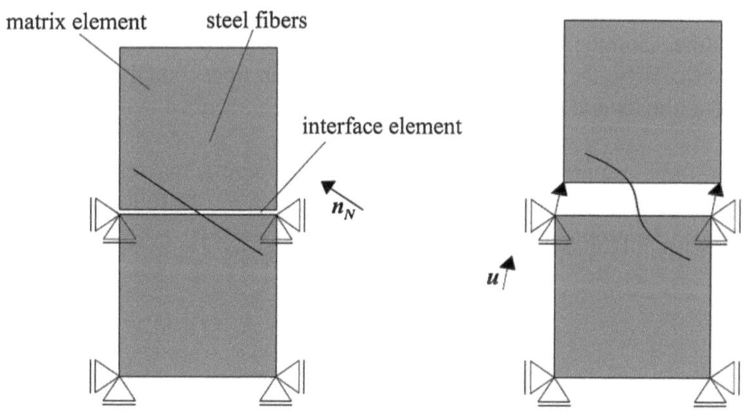

Fig. 12.14 Schematic configuration of an interface element crossed by one fiber

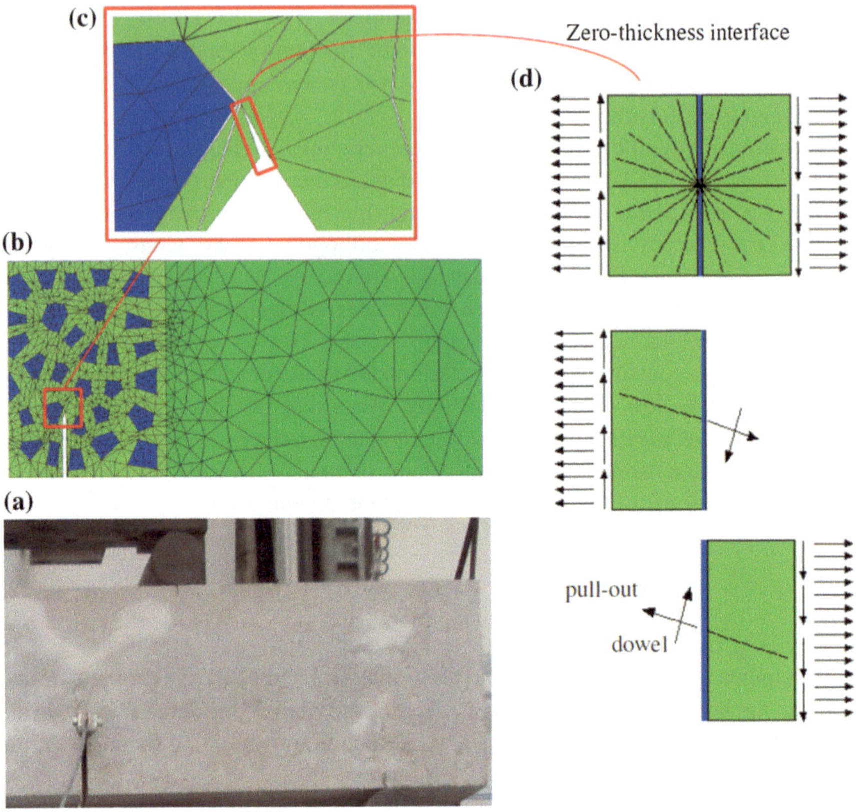

Fig. 12.15 FE discretisation including fiber reinforced concrete and mortar, coarse aggregates and interfaces (Etse et al. 2012)

On the one hand, fracture opening processes in concrete passing through fibers activate bridging effects on both crack sides due to fiber axial forces. The bond between fibers and the surrounding concrete matrix actually controls those bridging effect. Particularly, the axial (tensile) stresses on the fibers equal the shear stresses throughout the lateral contact surfaces of fibers embedded within the concrete matrix.

Under these simplified assumptions and considering that each generic fiber crosses the fracture line at its mid-length, i.e. $l_{emb} = l_f/2$, the following equilibrium equation can be proposed:

$$\frac{d\sigma_f[x]}{dx} = -\frac{4\tau_a[x]}{d_f} \tag{12.23}$$

where σ_f is the fiber axial tensile stress, τ_a the local bond stress between fiber and matrix and d_f the fiber diameter. Furthermore, a bilinear shear-slip law was proposed to model the fiber-concrete debonding process as follows:

$$\tau_a[x] = \begin{cases} -k_E s[x] & s[x] \leq s_e \\ -\tau_{y,a} + k_S(s[x] - s_e) & s_e < s[x] \leq s_u \\ 0 & s[x] > s_u \end{cases} \tag{12.24}$$

where $s[x]$ defines the tangential displacement between the fiber and concrete, at the point of the abscissa x. The positive constants k_E and k_S represent the elastic and softening slopes of such bond-slip relationships, respectively; $\tau_{y,a}$ is the shear bond strength while s_e and s_u are the elastic and ultimate slips, respectively. The complete derivation of the whole bond-slip numerical model and its validation against pull-out experimental tests, were proposed in a Caggiano and Martinelli (2012).

On the other hand, the dowel action resulting in a shear transfer mechanism across cracks represents an important component on the overall bridging effect of steel fibers (e.g. recycled steel fibers recovered from waste tires) in fracture processes of FRC. A simple analytical model which accounts for the dowel action of fibers crossing cracks was developed, based on the definition of a lateral force-displacement law on a generic fiber, embedded in a cracked concrete matrix.

Thus, the well-known Winkler beam theory is used to describe the dowel force, V_d, corresponding to the transversal displacement, Δ. Its analytical solution is obtained as:

$$V_d = E_s J_s \lambda_f^3 \Delta \tag{12.25}$$

where E_s is the steel elastic modulus and J_s the fiber moment of inertia, while the Winkler parameter, λ_f can analytically derived as:

$$\lambda_f = \sqrt[4]{\frac{k_c d_f}{4 E_s J_s}} = \sqrt[4]{\frac{16 k_c}{E_s \pi d_f^3}} \tag{12.26}$$

where k_c is the foundation (herein, the surrounding mortar) stiffness.

Finally, the empirical expression proposed by Dulacska (1972) for RC-structures is taken as reference for the maximum dowel strength $V_{d,u}$:

$$V_{d,u} = k_{dow} d_f^2 \sqrt{|f_c \,||\, \sigma_{y,s}|} \tag{12.27}$$

being k_{dow} a non-dimensional empirical coefficient based on the experimental results. A typical value $k_{dow} = 1.27$ can be assumed as reference for RC-structures. Finally, f_c and $\sigma_{y,s}$ are the compressive concrete strength and the steel yield stress value, respectively.

The predictive capacity of the interface model outlined in this subsection is demonstrated by comparing the corresponding numerical simulations with several experimental results. Mesoscopic numerical predictions, considering fracture processes in pure Mode I and mixed modes of fracture and developed into the explicit meso-structure of the composite, have been analyzed and discussed. These comparisons clearly demonstrated the advantages and shortcomings of the propose model (Fig. 12.16).

With reference to the modelling of transport problems in concrete through interfaces, it is here worth citing the work by Segura and Carol (2004), who took into account moisture diffusion problems and coupled analysis; the fluid flow through discontinuities by Idiart et al. (2011a), which was related to coupled hygro-mechanical analysis of concrete drying shrinkage; the proposal by Idiart et al. (2011b) whereby a chemo-mechanical analysis of concrete cracking and degradation due to external sulphate attack was performed; and the work by Liaudat et al. (2013) who outline a diffusion–reaction model for Alkali–Silica Reaction (ASR) processes.

12.3 Smeared Fibre Reinforcement Approaches (SFRAs)

12.3.1 Bibliographic Overview on the Smeared Crack Approaches (SCAs)

Concrete cracks have been traditionally treated by means of classical continuum or Smeared-Crack Approaches (SCAs) in which the fracture zone is considered to be

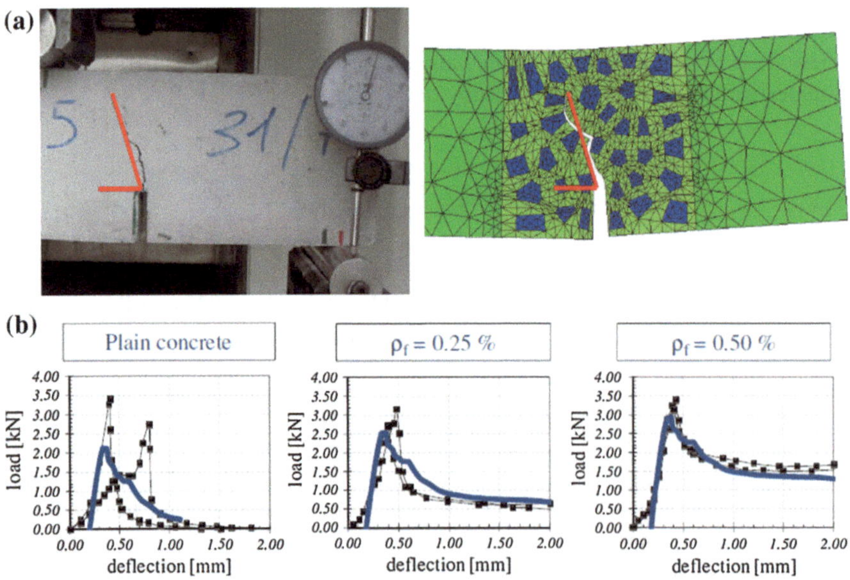

Fig. 12.16 Cracking simulations against experimental results of SFRC beams under three-point notched bending tests in mixed-modes of fracture (Etse et al. 2012)

distributed in a certain region of the solid (de Borst and Gutierrez 1999). Despite their advantages from the computational point of view, classical concrete models based on the smeared crack approach suffer from a strong FE-size dependence of the localization band width (Oliver 1989; Rots et al. 1985). To avoid this severe deficiency of the smeared-crack approach different regularization procedures have been proposed (Willam et al. 1984).

On the one hand, several continuum SCA models are based on fracture mechanics concepts leading to fracture energy regularization, but still suffering from loss of objectivity of the deformation pattern (Etse and Willam 1994). In this field, fracture energy-based concrete models are, among others, due to Bazant and Oh (1983), Shah (1990), Carpinteri et al. (1997), Comi and Perego (2001), Duan et al. (2007) and Meschke and Dumstorff (2007). On the other hand, more sophisticated constitutive theories have been proposed to solve the strong mesh dependency that appears when the governing equations turn ill posed (Carosio et al. 2000). They are based on rate dependency, higher strain gradients, micropolar theory, etc. Among others, it can be here referred the contributions by Vardoulakis and Aifantis (1991), de Borst et al. (1995), Peerlings et al. (2004), Lee and Fenves (1998) and Etse et al. (2003).

In this field, interesting contributions for FRC have been done by Hu et al. (2003) who proposed a single smooth biaxial failure surface for FRC, Seow and Swaddiwudhipong (2005) that used a five parameter failure criterion for FRC with both straight and hooked-end steel fibers, and by Minelli and Vecchio (2006), who used a model based on a modification of the compression field theory. Other relevant contributions worthy of mention can be found in Beghini et al. (2007) and Guttema (2003).

The non-local continuum theory, as developed in the milestone study by Pijaudier Cabot and Bazant (1987), is also worth being cited here, in whose framework several "Continuum Damage" based constitutive models for concrete and quasi brittle materials have been formulated (e.g., Mazars 1984; di Prisco and Mazars 1996). Due to the crack bridging effects provided by the steel reinforcements and/or fibers, when applying non-local continuum based models for modelling both conventional reinforced concrete or fiber reinforced concretes, an appropriate calibration is needed of the non-local integration length scale parameters, as related to the material physical heterogeneity scales, and of its evolution along the different stages of the crack formation and propagation processes (Ferrara and di Prisco 2001, 2002). This so far stands as a major research needs in this field.

Smeared crack models are the most used for simulating FRC structures due to the capacity of fibre reinforcement mechanisms promote the formation of diffuse crack patterns, mainly when using FRC of tensile strain hardening behaviour, or combined with conventional reinforcement. Even using FRC of tensile strain hardening nature, but of deflection hardening capacity, a relatively large number of cracks can be formed in statically indeterminate FRC structures (Salehian and Barros 2015; Barros et al. 2009).

12.3.2 Fixed, Rotating and Multidirectional Fixed Smeared Crack Models

The fixed, rotating and multidirectional fixed smeared crack models are commonly employed for modeling the behaviour of FRC structures that can also include conventional steel or/and even fiber reinforced polymer (FRP) reinforcements.

In the fixed approach no more than two sets of orthogonal smeared cracks are considered per integration point (IP) of the finite element (FE), while in the multidirectional approach several sets of smeared cracks can be formed, depending of the criteria adopted to trigger the formation of a new crack (in the scope of the present section, crack should be regarded as smeared cracks on the area/volume of the FRC representative of the IP). The rotating model allows a maximum of two orthogonal sets of smeared cracks to continuous rotate in order to assure coaxiality between principal strains and principal stresses. Barros (1995) demonstrated that the multidirectional smeared crack model can be regarded as an anisotropic damage model, due to the possibility of directly simulate in the same IP the occurrence of several cracks of different orientation and in different stage of fracture energy dissipation. De Borst (2002) have also proved that fixed and rotating smeared crack models, as well as micro-plane models, "may be considered as a special case of (anisotropic) damage models".

All of these versions of smeared crack models are based on the nonlinear fracture mechanics. Therefore, the fracture properties of the material must be determined. Chapters 8 and 11 have detailed how to determine the fracture mode I and II from experimental tests and inverse analysis, respectively. The contribution of the fibre reinforcement mechanisms to the stiffness and load carrying capacity of a FRC structures is mainly caused by the mode I fracture energy and the shape of the stress-crack width relationship that governs the opening process of a crack.

In a smeared approach the crack width is transformed in a tensile strain normal to the crack by using the concept of crack band width that is dependent of some geometric characteristic of the finite element or IP in order to assure the results of the simulation are not dependent of the finite element mesh refinement. Several approaches are being proposed to define this crack band width (Bažant and Oh 1983; Rots 1988; Oñate et al. 1987; Dahlblom and Ottosen 1990; Cervenka et al. 1990; Oliver 1989), some of them capable of simulating the tensile stress relieve in the concrete surrounding a crack during its softening process (Oliver 1989, 1996), and consequently avoiding spurious strain and intrinsic directional bias that occur when more simple strategies are adopted, like the ones based on a characteristic length that represents the area or volume of the IP (or FE).

Previous sections have demonstrated that the mode I fracture energy and the tensile stress capacity after crack initiation of a FRC materials increase with the content of fibres, and this increase also depend on their geometric and material properties, fibre-matrix bond conditions, and mechanical performance of the matrix. Due to the influence of these several variables, generally also interacting with each other a relatively large scatter is obtained for the fracture mode I parameters of concrete reinforced with industrially produced fibers. Due to the much higher variability of the variables affecting the reinforcement mechanisms when sustainable fibres are used for the reinforcement of cement based materials, such is the case of steel fibres from the recycling process of tyres (RSF—Chap. 8) and vegetable fibres (see Chap. 9), the scatter of the corresponding values of the fracture mode I parameters is expected to increase significantly. Therefore this scatter should be considered in order to obtain the corresponding characteristic values for assuring safety predictions when smeared crack models are adopted in the simulations.

The values of the mode I fracture parameters have a pronounced effect at the structural level, since the crack initiation up to the stage where cracking process is stabilized. This means that these parameters have a significant favorable effect, mainly, in serviceability limit state conditions. In fact, the larger is the fracture energy and the tensile capacity in the cracked stage of a FRC, the higher is the load carrying capacity of the structure made by this FRC at equal crack opening. This beneficial effect increases with the support redundancy of the structure due to stress redistribution this structural character promotes, combined with the higher internal indeterminacy of the FRC caused by the fibre reinforcement mechanisms. These beneficial effects increase with the decrease of the conventional steel reinforcement (Taheri et al. 2012).

FRC is being used in an attempt of decreasing, or even eliminating, the conventional shear reinforcement. In the context of smeared crack models, the crack shear stress transfer is generally simulated by a shear retention factor β, that varies between 1 (at the onset of cracking) and 0 (crack completely open, which means the mode I fracture energy was completely exhausted). The β multiplies the transverse elasticity modulus of uncracked FRC, G_c, in order to simulate the shear stiffness degradation during cracking process. A constant value for β was assumed in the oldest versions of the fixed smeared crack models, while in the rotating crack model the β is directly obtained by the condition of imposing coaxiality between the principal tensile strains and principal tensile stresses (Rots 1988). All these approaches predict unsafe load carrying capacity, uncharacteristic crack patterns and incorrect failure modes for elements failing in shear (deficiently reinforced in shear) (Ventura-Gouveia 2011).

In an attempt of overcoming this deficiencies, a constitutive law for defining the crack shear stress versus crack shear sliding has been used (Rots and de Borst 1987; Breveglieri et al. 2016). The definition of this relationship requires the knowledge of the mode II fracture parameters, namely the stiffness of the branch up to the crack shear strength, the crack shear strength, and the mode II fracture energy. Furthermore, due to the smeared character of this type of crack models, a characteristic length needs to be used to convert crack sliding in crack shear strain. Baghi and Barros (2016) have demonstrated these fracture parameters can be derived by executing inverse analysis with the results obtained from the Iosipescu tests. The same approach was adopted by Zamanzadeh (2016) for RSFRC, as described in Sect. 10.2. Zamanzadeh et al. (2015) have demonstrated that this approach is capable of simulating with reasonable accuracy the behaviour of RSFRC beams failing in shear, see Fig. 12.17.

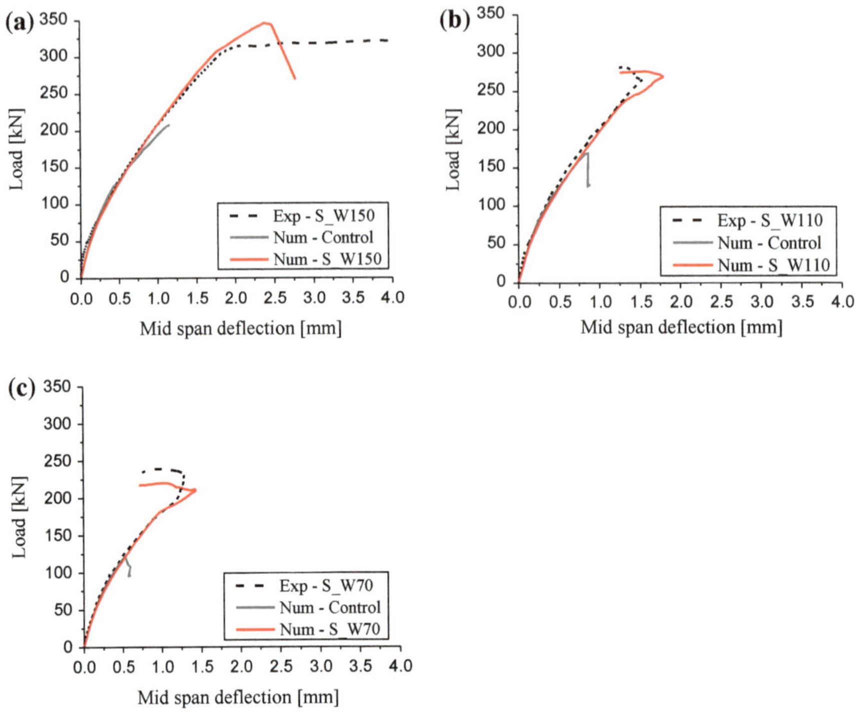

Fig. 12.17 Comparison of the experimental and numerical load-deflection curves of the bending tests with T-shape beams failing in shear: **a** S_W150, **b** S_W110 and **c** S_W70 (Zamanzadeh et al. 2015)

However, more tests tailored at the identification of mode II fracture parameters, such as Iosipescu tests with the corresponding inverse analysis, should be performed and the obtained mode II fracture parameters should be used in the numerical simulations of RSFRC structures failing in shear in order to have a deeper understanding on the applicability and reliability of this approach. Furthermore, the generally accepted assumption of using the same crack band width for mode I and mode II fracture propagation in the smeared crack context is arguable and requires specific research on this topic.

Fibres have also been demonstrated very effective for punching reinforcement (Barros et al. 2015; Ventura-Gouveia et al. 2011). As demonstrated by Teixeira et al. (2015), the load carrying capacity and the deflection performance of SFRC slabs flexurally reinforced with conventional steel reinforcement increase with the content of fibres, in consequence of the increase of the mode I fracture energy (Fig. 12.18).

Ventura-Gouveia et al. (2011) have also demonstrated the relevance of considering a shear softening diagram for simulating the degradation of the out-of-plane shear stress components with the crack opening process when using

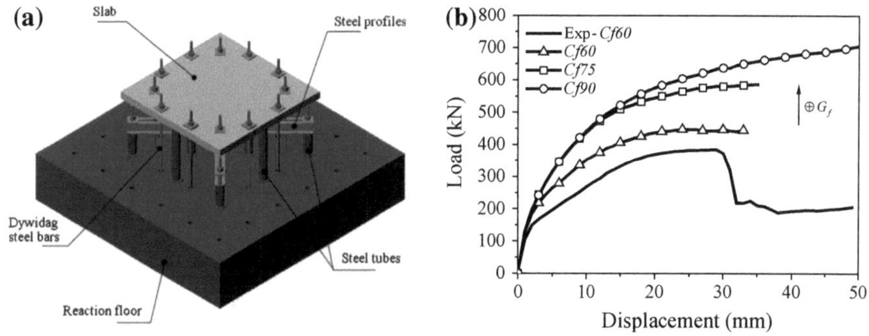

Fig. 12.18 Punching shear tests with steel fibre reinforced self-compacting concrete (SFRSCC) slabs: **a** test setup and **b** influence of the mode I fracture energy on the load carrying and deflection capacity of this type of slabs Teixeira et al. (2015)

the Reissner-Mindlin layered shell theory with a multidirectional fixed smeared crack model. Figure 12.19 shows that the punching failure mode was only capable of being captured when this approach was implemented and applied to the executed type of experimental tests (Barros et al. 2007).

As demonstrated in Sect. 8.2.2, in the compression behaviour, the contribution of the fibre reinforcement is mainly reflected in the post-peak softening stage, where the energy absorption per unit volume of the concrete ($G_{f,c}$ [FL^{-2}]), which is the area under the compression stress–strain diagram in this post-peak phase, is the parameter most used to simulate this beneficial effect. This parameter also depends on the geometric and material characteristics of the fibres, as well as on the properties of the matrix and fibre-matrix bond conditions.

Therefore the influence of the reinforcement provided by sustainable fibres on the compression softening stage of cement based materials requires the evaluation of the aforementioned parameter. However, besides the difficulties intrinsic of the variability of the geometric and material characteristics of this type of fibres, the softening stage in a direct compression tests also depends on the geometry of the specimen, stiffness of the equipment, interface conditions between loading platens of the equipment and the extremities of the specimen (frictional level introduced) and monitoring setup. The evaluation of $G_{f,c}$ is hence affected by uncertainties that cast several doubts over its reliability for modeling the post peak stage of a cement based material in compression.

Apart from these considerations, the smeared crack models capable of simulating the inelastic deformation of FRC in compression demonstrate that the load carrying capacity of a concrete structure, and mainly its deformation capacity, increases with $G_{f,c}$, and this effect is as pronounced as larger is the relative volume of the concrete experience inelastic deformation in compression, as demonstrated in Fig. 12.20 (Behbahani et al. 2016). The simulations in this figure were performed employing a

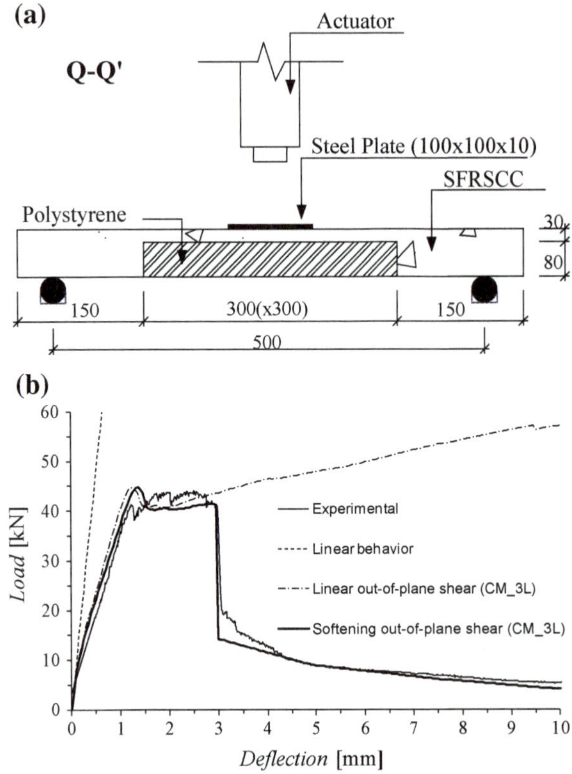

Fig. 12.19 **a** Test setup of a SFRSCC panel submitted to punching and **b** load-deflection at the center of the panel (Ventura et al. 2011)

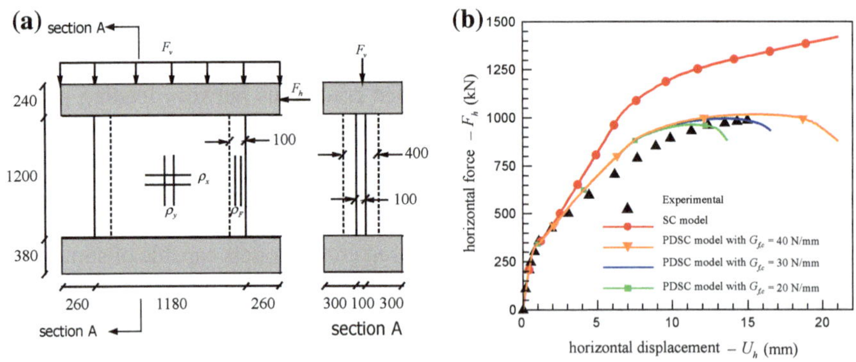

Fig. 12.20 **a** Geometry and loading configurations of the shear walls tested by Maier and Thürlimann (dimensions in mm); horizontal load versus horizontal displacement relationship, F_h–U_h from the numerical simulations (Behbahani et al. 2016)

plasticity-damage multidirectional fixed smeared cracking (PDSC) that combines a multidirectional fixed smeared crack model to simulate the crack initiation and propagation with a plastic-damage model to account for the inelastic compressive behaviour of concrete between cracks (Behbahani et al. 2015). In the same figure the curve is also represented (denoted as SC) corresponding to the simulation where the inelastic deformation of concrete in compression is not taken into account (the concrete between cracks is assumed linear-elastic in compression), being visible that this approach conducts to unsafe predictions.

When long vegetable fibers are used in a stacking layered approach of a sequence of cement mortar layer and a layer of unidirectional fibres for producing thin panels with relatively high fiber volume percentages (4–6% by volume), the bond performance between fibres and surrounding medium is relatively weak (Fig. 12.21). In this case it is fundamental to simulate the sliding between the fibre layer and surrounding mortar, by adopting a modeling strategy similar to the one schematically represented in Fig. 12.22 (Barros et al. 2016), otherwise resulting into unsafe and over-stiff predictions.

12.3.3 Micro-plane Models

The microplane model can be regarded as pertaining to the class of smeared crack approaches. This model has largely been used for predicting the mechanical

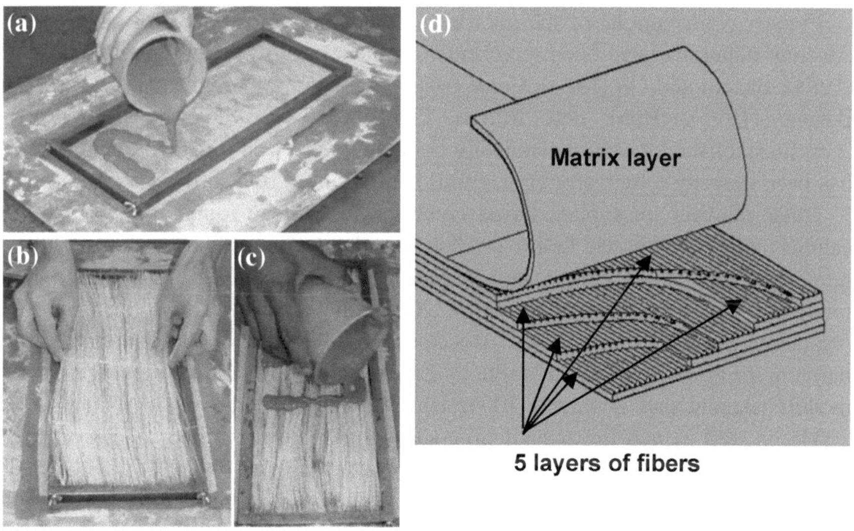

Fig. 12.21 The composite molding procedure: **a** the first layer of matrix being poured in the mold, **b** placement of the first fiber layer, **c** placement of the second matrix layer and **d** an overview of the composite layers (Barros et al. 2016)

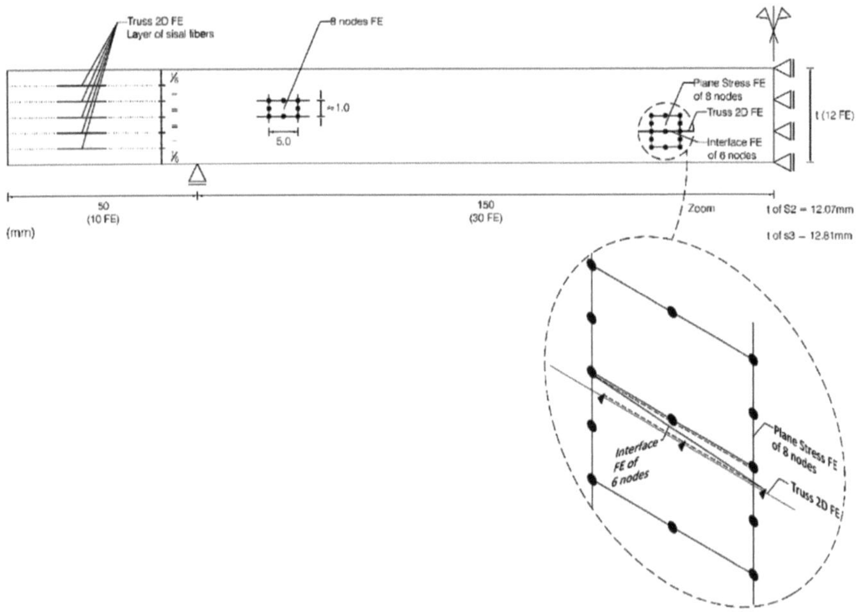

Fig. 12.22 Simulated panels: dimensions (in mm), finite element mesh, and support and loading conditions (Barros et al. 2016)

behavior of quasi-brittle materials, such as concrete and rocks, but its applicability to FRC has also been explored.

Pioneer contributions of the microplane theory in constitutive formulations for concrete materials have been represented in the works by Bazant and Gambarova (1984), Bazant and Oh (1985), Carol et al. (1992), and more recently by Carol and Bazant (1997), Kuhl and Ramm (2000) and Cervenka et al. (2005). A well-established thermodynamically consistent approach for microplane theories has been described by Carol et al. (2001) and Kuhl et al. (2001).

Other relevant microplane-based contributions can be found in several applications including concrete failure prediction under cyclic loads (Ozbolt et al. 2001), numerical analysis of compressed concrete columns confined with carbon fiber reinforced polymer (Gambarelli et al. 2014), the mechanical response of polycrystalline shape memory alloys (Brocca et al. 2002), micropolar continua formulation in the spirit of Cosserat Media by Etse et al. (2003), strain-softening nonlocal models (Bazant and Di Luzio 2004; Di Luzio 2007), large strains (Carol et al. 2004), as well as non-linear hardening-softening behavior of fiber reinforced concretes by Beghini et al. (2007) and Caner et al. (2013).

Recently, Vrech et al. (2016) have formulated a novel thermodynamically consistent fracture-based microplane model for simulating the failure behavior of SFRC. The constitutive formulation at the microplane level was described in terms of normal and shear stresses versus related micro-strains, while fiber effect on the composite failure behavior is taken into account through both a bond-slip formulation and a dowel model depending on the relative orientations between fibers and microplanes.

Kinematic assumptions, as well as constitutive equations of the Vrech et al. (2016) model, are briefly presented in the following. The microplane approach, originally proposed by Bazant and Oh (1985), consists in the formulation of constitutive laws at microplane level defining the mechanical behavior of planes (i.e., the microplanes) generically orientated. Then, the macroscopic response shall be achieved through the consideration of appropriated homogenization process over the responses in all microplanes.

Assuming kinematic constraints, the normal and tangential strains at microplane level, ε_N and $\boldsymbol{\varepsilon}_T$ respectively, are computed by means of the following relationships:

$$\varepsilon_N = \mathbf{N} : \boldsymbol{\varepsilon}^{mac}, \quad \boldsymbol{\varepsilon}_T = \mathbf{T} : \boldsymbol{\varepsilon}^{mac} \tag{12.28}$$

being $\boldsymbol{\varepsilon}^{mac}$ the macroscopic strain tensor projected on a microplane characterized by its normal direction $\mathbf{n}$ (Fig. 12.23). The projection tensors are defined as:

$$\mathbf{N} = \mathbf{n} \otimes \mathbf{n}, \quad \mathbf{T} = \mathbf{n} \cdot \mathbf{I}^{sym} - \mathbf{n} \otimes \mathbf{n} \otimes \mathbf{n} \tag{12.29}$$

being $\mathbf{I}^{sym}$ the symmetric part of the fourth-order identity tensor.

In the elasto-plastic regime and assuming small strains, both macro- and microscopic strains are computed according the Prandtl-Reuss additive decomposition. Particularly, at microplane level, normal and tangential strain rates are obtained as

$$\dot{\varepsilon}_N = \dot{\varepsilon}_N^e + \dot{\varepsilon}_N^p, \quad \dot{\varepsilon}_T = \dot{\varepsilon}_T^e + \dot{\varepsilon}_T^p \tag{12.30}$$

where the superscripts e and p denote elastic and plastic components, respectively.

The incremental rate stress vector at microplane level, $\mathbf{t}_\sigma$, and the rate of the microplane elastic strains, $\dot{\varepsilon}^{el,mic}$, are connected by means of the following elastic stiffness operator:

$$\mathbf{C}^{mic} = \begin{pmatrix} E_N & 0 & 0 \\ 0 & E_T & 0 \\ 0 & 0 & E_T \end{pmatrix} \tag{12.31}$$

Then

$$\dot{\mathbf{t}}_\sigma = \mathbf{C}^{mic} \cdot \dot{\varepsilon}^{el,mic}; \quad \dot{\mathbf{t}}_\sigma = \mathbf{C}^{mic} \cdot \left(\dot{\varepsilon}^{mic} - \dot{\varepsilon}^{p,mic}\right) \tag{12.32}$$

The microscopic elastic moduli are related to the macroscopic ones as demonstrated in Bazant and Prat (1988a, b), $E_N = 3K + 2G$ and $E_T = 2G$, being K and G the bulk and shear macroscopic moduli, respectively.

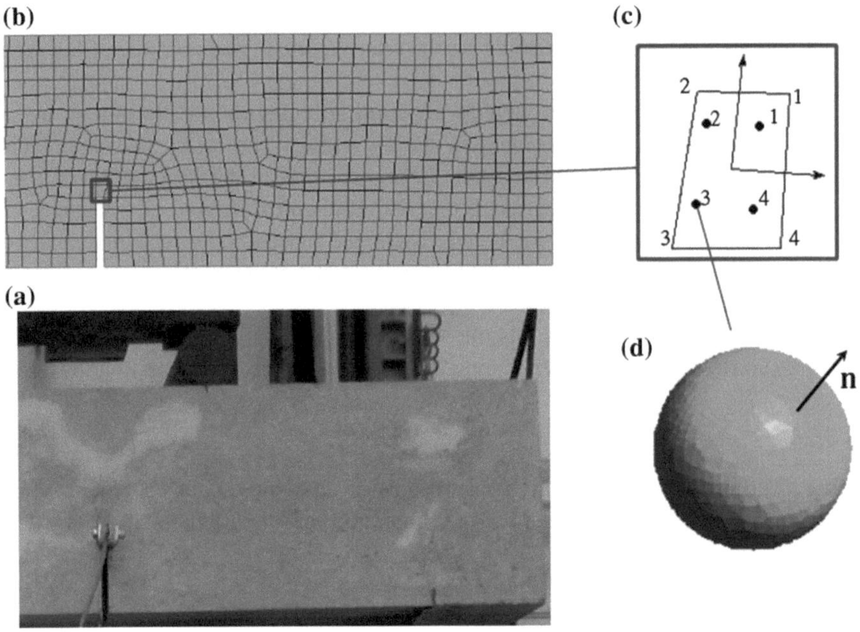

Fig. 12.23 **a** FRC specimen, **b** continuum discretization scale, **c** 4-node continuum FE and **d** spherical region at gauss-point with a generalized microplane normal direction (Caggiano 2013)

The microplane constitutive formulation for SFRC is thus based on the Mixture Theory by Trusdell et al. (1960). The main assumption of this theory is that in every infinitesimal volume the kinematic field of the equivalent continuum and that one of each mixture constituent is the same. Thus, the stress vector of the mixture is defined as:

$$\mathbf{t}_\sigma = \omega^m \boldsymbol{\sigma}^m + \omega^f \left[\sigma_N^f \mathbf{n} + \boldsymbol{\sigma}_T^f \cdot \mathbf{n}_T\right] \tag{12.33}$$

being ω^m and ω^f the weighting functions depending on the volume fraction of each constituent, with m and f referring to concrete matrix, and fibers, respectively.

The concrete matrix stress vector is represented by the constitutive rule for plain concrete $\boldsymbol{\sigma}^m = [\sigma_N^m(\varepsilon_N), \boldsymbol{\sigma}_T^m(\boldsymbol{\varepsilon}_T)]$; while $\sigma_N^f(\varepsilon_N)$ and $\boldsymbol{\sigma}_T^f(\boldsymbol{\varepsilon}_T)$ deals with the bond-slip and dowel stresses due to the post-cracking interaction between fibers and mortar (Fig. 12.24). These stress components are defined in the normal and tangential directions of the fibers, respectively, same as the vectors $\mathbf{n}$ and $\mathbf{n}_T$.

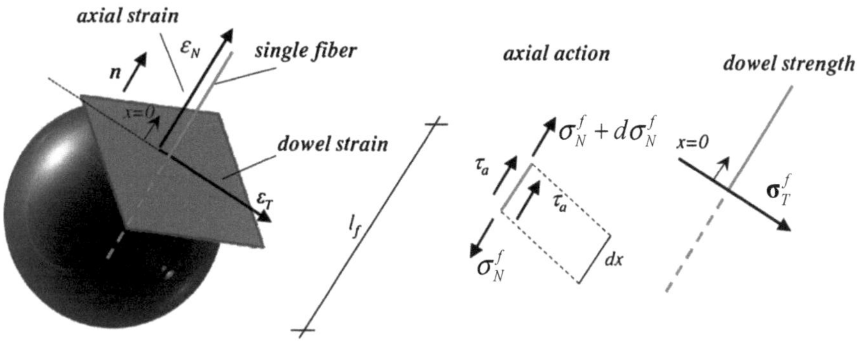

Fig. 12.24 Microplane strain activating the fiber bridging effects (Caggiano 2013)

References

Abrishambaf A, Cunha V, Barros JAO (2016a) The influence of fibre orientation on the post-cracking tensile behaviour of steel fibre reinforced self-compacting concrete. Fract Struct Integrity J 31(1):38–53

Abrishambaf A, Barros JAO, Cunha VMCF (2016b) A two-phase material approach to model steel fibre reinforced self-compacting concrete panels. Eng Fract Mech J 162:1–20

Abrishambaf A, Barros JAO, Cunha VMCF (2013) Relation between fibre distribution and post-cracking behaviour in steel fibre reinforced self-compacting concrete panels. Cem Concr Res 51:57–66

Armero F, Linder C (2009) Numerical simulation of dynamic fracture using finite elements with embedded discontinuities. Int J Fract 160:119–141

Baghi H, Barros JAO (2016) Shear properties of the strain hardening cementitious composite materials. ASCE J Mater Civ Eng 28(10)

Barros JAO, Silva FA, Toledo-Filho RD (2016) Experimental and numerical research on the potentialities of layered reinforcement configuration of continuous sisal fibers for thin mortar panels. Constr Build Mater 102:792–801

Barros JAO, Moraes-Neto BN, Melo GSSA, Frazão CMV (2015) Assessment of the effectiveness of steel fibre reinforcement for the punching resistance of flat slabs by experimental research and design approach. Compos Part B J 78:8–25

Barros JAO, di Prisco M, di Prisco C (2009) Modelling FRC infrastructures taking into account the soil-structure interaction. Conference on numerical methods in engineering 2009, Barcelona, 29 June–2 July 2009

Barros JAO, Pereira EB, Santos SPF (2007) Lightweight panels of steel fiber reinforced self-compacting concrete. ASCE Mater Civ Eng J 19(4):295–304

Barros JAO (1995) Behaviour of fibre reinforced concrete—experimental analysis and numerical simulation. PhD Thesis, Dep. Civil Eng., Faculty of Eng. of Oporto University, 502 p, December 1995 (in Portuguese)

Bazant ZP, Di Luzio G (2004) Nonlocal microplane model with strain softening yield limits. Int J Solids Struct 41(24–25):7209–7240

Bazant ZP, Gambarova PG (1984) Crack shear in concrete: crack band microplane model. J Struct Eng 110:2015–2035

Bažant ZP, Oh BH (1983) Crack band theory for fracture of concrete. Matériaux et Constr 16 (3):155–177

Bazant ZP, Oh BH (1985) Microplane model for progressive fracture of concrete and rock. J Eng Mech 111(4):559–582

Bazant Z, Prat P (1988a) Microplane model for brittle plastic material: I. Theory. J Eng Mech 114 (10):1672–1689
Bazant Z, Prat P (1988b) Microplane model for brittle plastic material: II. Verification. J Eng Mech 114(10):1689–1703
Bazant Z, Tabbara M, Kazemi M, Pijaudier-Cabot G (1990) Random particle model for fracture of aggregate or fiber composites. J Eng Mech, ASCE 116:1686–705
Beghini A, Bazant Z, Zhou Y, Gouirand O, Caner F (2007) Microplane model m5f for multiaxial behavior and fracture of fiber reinforced concrete. ASCE-J Mat Civil Eng 133(1):66–75
Behbahani AE, Barros JAO, Ventura-Gouveia A (2016) Application of plastic-damage multidirectional fixed smeared crack model in the analysis of RC structures. Eng Struct J 374–391. doi:10.1016/j.engstruct.2016.07.013
Behbahani AE, Barros JAO, Ventura-Gouveia A (2015) Plastic-damage smeared crack model to simulate the behaviour of structures made by cement based materials. Int J Solid Struct 73–74:20–40
Bolander JE, Choi S, Duddukuri SR (2008) Fracture of fiber-reinforced cement composites: effects of fiber dispersion. Int J Fract 154(1–2):73–86
Bolander JE, Sukumar N (2005) Irregular lattice model for quasi static crack propagation. Phys Rev B 71(9):094106
Bouhala L, Makradi A, Belouettar S, Kiefer-Kamal H, Fréres P (2013) Modelling of failure in long fibres reinforced composites by X-FEM and cohesive zone model. Compos B Eng 55:352–361
Breveglieri M, Barros JAO, Aprile A, Ventura-Gouveia A (2016) Strategies for numerical modeling the behavior of RC beams strengthened in shear using the ETS technique. Eng Struct J 128:296–315
Brocca M, Brinson LC, Bazant ZP (2002) Three-dimensional constitutive model for shape memory alloys based on microplane model. J Mech Phys Solids 50(5):1051–1077
Buratti N, Mazzotti C, Savoia M (2011) Post-cracking behaviour of steel and macrosynthetic fibre-reinforced concretes. Constr Build Mater 25(5):2713–2722
Caggiano A (2013) Meso-mechanical analysis of steel fiber reinforced cementitious composites. PhD Thesis, Universities of Salerno and Tucuman, Italy/Argentina
Caggiano A, Etse G, Martinelli E (2011) Interface model for fracture behaviour of fiber-reinforced cementitious composites (FRCCs): theoretical formulation and applications. J Environ Civ Eng 15(9):1339–1358
Caggiano A, Etse G, Martinelli E (2012) Zero-thickness interface model formulation for failure behavior of fiber-reinforced cementitious composites. Comput Struct 98–99:23–32
Caggiano A, Martinelli E (2012) A unified formulation for simulating the bond behaviour of fibres in cementitious materials. Mater Des 42:204–213
Caner FC, Bazant ZP, Wendner R (2013) Microplane model M7f for fiber reinforced concrete. Eng Fract Mech 105:41–57
Carol I, Bazant ZP (1997) Damage and plasticity in microplane theory. Int J Solids Struct 34 (29):3807–3835
Carol I, Prat P, Bazant ZP (1992) New explicit microplane model for concrete: theoretical aspects and numerical implementation. Int J Solids Struct 29(9):1173–1191
Carol I, Prat P, Lopez C (1997) Normal/shear cracking model: applications to discrete crack analysis. J Eng Mech—ASCE 123:765–773
Carol I, Jirasek M, Bazant ZP (2001) A thermodynamically consistent approach to microplane theory. Part I. Free-energy and consistent microplane stresses. Int J Solids Struct 38(17):2921–2931
Carol I, Jirasek M, Bazant ZP (2004) A framework for microplane models at large strain, with application to hyperelasticity. Int J Solids Struct 41(2):511–557
Carosio A, Willam K, Etse G (2000) On the consistency of viscoplastic formulations. Int J Solids Struct 37(48–50):7349–7369
Carpinteri A, Chiaia B, Nemati KM (1997) Complex fracture energy dissipation in concrete under different loading conditions. Mech Mater 26(2):93–108

Cervenka J, Bazant ZP, Wierer M (2005) Equivalent localization element for crack band approach to mesh-sensitivity in microplane model. Int J Numer Methods Eng 62(5):700–726
Cervenka V, Pukl H, Eligehausen R (1990) Computer simulation of anchoring technique and design of concrete structures. In: Proc. Second Intern. Conf. on Computer Aided Analysis and Desing of Concrete Structures, Zell am See, Austria, pp 1–19
Ciancio D, Carol I, Cuomo M (2013) A method for the calculation of inter-element stresses in 3D. Comput Methods Appl Mech Eng 254:222–237
Comi C, Perego U (2001) Fracture energy based bi-dissipative damage model for concrete. Int J Solids Struct 38(36–37):6427–6454
Cundall P (1971) A computer model for simulating progressive large scale movements in blocky rock systems. Proc Int Symp Rock Fract, ISRM, Nancy, France, pp 2–8
Cunha VMCF (2010) Steel fibre reinforced self-compacting concrete - from micromechanics to composite behaviour. PhD Thesis, University of Minho, Portugal
Cunha VMCF, Barros JAO, Sena-Cruz JM (2010) Pullout behaviour of steel fibres in self-compacting concrete. ASCE Mater Civ Eng J 22(1):1–9
Cunha VMCF, Barros JAO, Sena-Cruz JM (2011) An integrated approach for modelling the tensile behaviour of steel fibre reinforced self-compacting concrete. Cem Concr Res J 41:64–76
Cunha VMCF, Barros JAO, Sena-Cruz JM (2012) A finite element model with discrete embedded elements for fibre reinforced composites. Comput Struct J 94–95:22–33
Cusatis G, Schauffert EA (2010) Discontinuous cell method (DCM) for cohesive fracture propagation. In: Oh HH, Choi OC, Chung L (eds) Proc., 7th Int. Conf. on Fracture Mechanics of Concrete and Concrete Structures (FraMCoS 7), Electronic Proceedings, Jeju, South Korea, 2010
Cusatis G, Bazant ZP, Cedolin L (2006) Confinement-shear lattice CSL model for fracture propagation in concrete. Comput Methods Appl Mech Eng 195(52):7154–7171
Cusatis G, Pelessone D, Mencarelli A (2011a) Lattice discrete particle model (LDPM) for failure behavior of concrete. I: theory. Cement Concr Compos 33(9):881–890
Cusatis G, Mencarelli A, Pelessone D, Baylot J (2011b) Lattice discrete particle model (LDPM) for failure behavior of concrete. II: calibration and validation. Cement Concr Compos 33 (9):891–905
D'Aguiar SC, Modaressi-Farahmand-Razavi A, dos Santos JA, Lopez-Caballero F (2011) Elastoplastic constitutive modelling of soil–structure interfaces under monotonic and cyclic loading. Comput Geotech 38:430–447
Dahlblom O, Ottosen NS (1990) Smeared crack analysis using generalized fictitious crack model. J Eng Mech, ASCE 116(1):55–76
De Borst R (2002) Fracture in quasi-brittle materials: a review of continuum damage-based approaches. Eng Fract Mech 69:95–112
de Borst R, Gutierrez M (1999) A unified framework for concrete damage and fracture models including size effects. Int J Fract 95:261–277
De Borst R, Nauta P (1985) Non-orthogonal cracks in a smeared finite element model. Eng Comput 2:35–46
de Borst R, Pamin J, Peerlings R, Sluys L (1995) On gradient-enhanced damage and plasticity models for failure in quasi-brittle and frictional materials. Comp Mech 17:130–141
Deeb R, Karihaloo BL, Kulasegaram S (2014a) Reorientation of short steel fibres during the flow of self-compacting concrete mix and determination of the fibre orientation factor. Cem Concr Res 56:112–120
Deeb R, Kulasegaram S, Karihaloo BL (2014b) 3D modelling of the flow of self-compacting concrete with or without steel fibres. Part I: slump flow test. Comput Part Mech 1(4): 1373–1389
Di Luzio G (2007) A symmetric over-nonlocal microplane model M4 for fracture in concrete. Int J Solids Struct 44(13):4418–4441
di Prisco M, Mazars J (1996) Crush-crack': a non-local damage model for concrete. Mech Cohesive-Frictional Mater 1(4):321–347

Duan K, Hu X, Wittmann FH (2007) Size effect on specific fracture energy of concrete. Eng Fract Mech 74(1–2):87–96
Dupont D, Vandewalle L (2005) Distribution of steel fibres in rectangular sections. Cem Concr Compos 27:391–398
Dulacska H (1972) Dowel action of reinforcement crossing cracks in concrete. ACI Struct J 69(12):754–757
Dvorkin E, Cuitino A, Gioia G (1990) Finite elements with displacement embedded localization lines insensitive to mesh size and distortions. Int J Numer Methods Eng 30:541–564
El-Helou RG, Moen CD, Lale E, Cusatis G (2014) Lattice discrete particle modeling of buckling deformation in thin ultra-high-performance fiber-reinforced concrete plates. Comput Model Concr Struct 1:365
Etse G, Willam K (1994) A fracture energy-based constitutive formulation for inelastic behavior of plain concrete. ASCE-JEM 120:1983–2011
Etse G, Nieto M, Steinmann P (2003) A micropolar microplane theory. Int J Eng Sci 41(13–14):1631–1648
Etse G, Caggiano A, Vrech S (2012) Multiscale failure analysis of fiber reinforced concrete based on a discrete crack model. Int J Fract 178(1–2):131–146
Ferrara L, di Prisco M (2001) Mode I fracture behavior in concrete: non-local damage modelling. ASCE J Eng Mech 127(7):678–692
Ferrara L, di Prisco M (2002) A non-local approach with evolutionary internal length for the analysis of mode I fracture processes in concrete. EM2002, 15th ASCE engineering mechanics conference, New York, NY, 2–5 June 2002
Ferrara L, Meda A (2006) Relationships between fibre distribution, workability and the mechanical properties of SFRC applied to precast roof elements. Mater Struct 39:411–420
Gambarelli S, Nistico N, Ozbolt J (2014) Numerical analysis of compressed concrete columns confined with CFRP: microplane-based approach. Compos B Eng 67(5):303–312
Guttema T (2003) Ein beitrag zur realitatsnahen modellierung und analyse von stahlfaserverstarkten stahlbeton- und stahlbetonachentragwerken. Ph.D. Thesis, Universitat Kassel
Hillerborg A, Modeer M, Petersson P (1976) Analysis of crack formation and crack growth in concrete by means of fracture mechanics and finite elements. Cem Concr Compos 6(6): 773–781
Hu X, Day R, Dux P (2003) Biaxial failure model for fiber reinforced concrete. ASCE-J Mat Civil Engrg 15(6):609–615
Ibrahimbegovic A, Boulkertous A, Davenne L, Brancherie D (2010) Modelling of reinforced-concrete structures providing crack-spacing based on X-FEM, ED-FEM and novel operator split solution procedure. Int J Numer Meth Eng 83(4):452–481
Idiart AE, Lopez CM, Carol I (2011a) Modeling of drying shrinkage of concrete specimens at the meso-level. Mater Struct 44:415–435
Idiart AE, Lopez CM, Carol I (2011b) Chemo-mechanical analysis of concrete cracking and degradation due to external sulfate attack: a meso-scale model. Cem Concr Compos 33: 411–423
Jirasek M, Bazant ZP (1994) Macroscopic fracture characteristics of random particle systems. Int J Fract 69:201–228
Kawai T (1980) Some considerations on the finite element method. Int J Numer Meth Eng 16:81–120
Kuhl E, Ramm E (2000) Microplane modelling of cohesive frictional materials. Europ J Mech/A: Solids 19:121–143
Kuhl E, Steinmann P, Carol I (2001) A thermodynamically consistent approach to microplane theory. Part II. Dissipation and inelastic constitutive modelling. Int J Solids Struct 38 (17):2933–2952
Laranjeira F, Grunewald S, Walraven J, Blom C, Molins C, Aguado A (2010) Characterization of the orientation profile of steel fiber reinforced concrete. Mater Struct 14:1093–1111
Lee J, Fenves G (1998) Plastic-damage model for cyclic loading of concrete structures. ASCE—JEM 124(8):892–901

Lei X (2001) Contact friction analysis with a simple interface element. Comput Methods Appl Mech Eng 190:1955–1965

Liaudat J, Rodriguez M, Lopez C, Carol I (2013) Mechanics and physics of creep, shrinkage, and durability of concrete: a tribute to Zdenek P. Bazant. Proceedings of the ninth international conference on creep, shrinkage, and durability mechanics (CONCREEP-9), ASCE Publications, Cambridge, Massachusetts, 22–25 Sept 2013

Lilliu G, van Mier J (2003) 3D lattice type fracture model for concrete. Eng Fract Mech 70: 927–941

Liu Z, Menouillard T, Belytschko T (2011) An XFEM/spectral element method for dynamic crack propagation. Int J Fract 169:183–198

Lopez CM, Carol I, Aguado A (2008a) Meso-structural study of concrete fracture using interface elements. I: numerical model and tensile behavior. Mater Struct 41(3):583–599

Lopez CM, Carol I, Aguado A (2008b) Meso-structural study of concrete fracture using interface elements. II: compression, biaxial and brazilian test. Mater Struct 41(3):601–620

Lorefice R, Etse G, Carol I (2008) Viscoplastic approach for rate-dependent failure analysis of concrete joints and interfaces. Int J Solids Struct 45(9):2686–2705

Mazars J (1984) Application de la mecanique de l'endommagement au comportement non lineaire et a la rupture du beton de structure. These de Doctorat d'Etat es Sciences Physiques, Universitè Pierre et Marie Curie, LMT, Paris (in French)

Meschke G, Dumstorff P (2007) Energy-based modeling of cohesive and cohesionless cracks via X-FEM. Comput Methods Applied Mech Eng 196(21–24):2338–2357

Minelli F, Vecchio F (2006) Compression field modeling of fiber-reinforced concrete members under shear loading. ACI-Struct J 106(2):244–252

Oh B, Kim J, Choi Y (2007) Fracture behavior of concrete members reinforced with structural synthetic fibers. J Eng Mech—ASCE 74:243–257

Olesen J (2001) Fictitious crack propagation in fiber-reinforced concrete beams. J Eng Mech—ASCE 127(3):272–280

Oliver J (1989) Consistent characteristic length for smeared cracking models. Int J Numer Methods Eng 28:461–474

Oliver J (1996) Modelling strong discontinuities in solid mechanics via strain softening constitutive equation", 1. fundamentals. Int J Numer Methods Eng 39:3575–3600

Oliver J, Huespe A, Pulido M, Chaves E (2002) From continuum mechanics to fracture mechanics: the strong discontinuity approach. Eng Fract Mech 69:113–136

Oliver J, Huespe A, Sanchez P (2006) A comparative study on finite elements for capturing strong discontinuities: E-FEM vs X-FEM. Comput Methods Appl Mech Eng 195:4732–4752

Oliver J, Linero DL, Huespe AE, Manzoli OL (2008) Two-dimensional modeling of material failure in reinforced concrete by means of a continuum strong discontinuity approach. Comput Methods Appl Mech Eng 197(5):332–348

Oñate E, Oller S, Oliver J, Lubliner J (1987) A constitutive model for cracking of concrete based on the incremental theory of plasticity. In: Owen DR, Hinton E, Oñate E (eds) Proc. Int. conf. computational Plasticity-models, software and applications, Barcelona, pp 1311–1329

Ozbolt J, Li Y, Kozar I (2001) Microplane model for concrete with relaxed kinematic constraint. Int J Solids Struct 38(16):2683–2711

Ozyurt N, Mason TO, Shah SP (2007) Correlation of fibre dispersion, rheology and mechanical performance of FRCs. Cem Concr Compos 29:70–79

Pandolfi A, Guduru P, Ortiz M, Rosakis A (2000) Three dimensional cohesive-element analysis and experiments of dynamic fracture in C300 steel. Int J Solids Struct 37:3733–3760

Pandolfi A, Ortiz M (2002) An efficient adaptive procedure for three-dimensional fragmentation simulations. Eng Comput 18:148–159

Park K, Paulino G, Roesler J (2010) Cohesive fracture model for functionally graded fiber reinforced concrete. Cem Concr Compos 40:956–965

Peerlings R, Massart T, Geers M (2004) A thermodynamically motivated implicit gradient damage framework and its application to brick masonry cracking. Comput Methods Appl Mech Eng 193(30–32):3403–3417

Pelessone D (2005) Discrete particle method. Technical report", Engineering and Software System Solutions, Inc.
Pereira EB, Fischer G, Barros JA (2012) Direct assessment of tensile stress-crack opening behavior of strain hardening cementitious composites (SHCC). Cem Concr Res 42(6):834–846
Pijaudier-Cabot G, Bazant ZP (1987) Nonlocal damage theory. J Eng Mech 113(10):1512–1533
Ramm E, Hund A, Hettich T (2006) A variational multiscale model for composites with special emphasis on the X-FEM and level sets. In: Computational modelling of concrete structures: proceedings of the EURO-C, pp 27–30
Rodriguez, J., "Study of behavior of granular heterogeneous media by means of analogical mathematical discontinuous models", PhD Thesis, Universidad Politecnica de Madrid, Spain (in Spanish), 1974
Rots JG (1988) Computational modeling of concrete fracture. Dissertation, Delft University of Technology
Rots JG, de Borst R (1987) Analysis of mixed-mode fracture in concrete. J Eng Mech, ASCE 113 (11):1739–1758
Rots JG, Nauta P, Kuster GMA, Blaauwendraad J (1985) Smeared crack approach and fracture localization in concrete. HERON 30(1)
Salehian H, Barros JAO (2015) Assessment of the performance of steel fibre reinforced self-compacting concrete in elevated slabs. Cem Concr Compos 55:268–280
Schauffert EA, Cusatis G, Pelessone D, O'Daniel JL, Baylot JT (2012) Lattice discrete particle model for fiber-reinforced concrete. II: tensile fracture and multiaxial loading behavior. J Eng Mech 138(7):834–841
Schorn H, Rode U (1987) 3-D modelling of process zone in concrete by numerical simulation. In: Shah SP, Swartz SE (eds) Fracture of concrete and rock. Springer-V, New York, pp 220–228
Segura J, Carol I (2004) On zero-thickness interface elements for diffusion problems. Int J Numer Anal Methods Geomech 28:947–962
Sena-Cruz JM (2004) Strengthening of concrete structures with near surface mounted CFRP laminate strips. PhD Thesis, University of Minho, Portugal
Seow PEC, Swaddiwudhipong S (2005) Failure surface for concrete under multiaxial load—a unified approach. ASCE-J Mater Civ Eng 17(2):219–228
Shah S (1990) Size-effect method for determining fracture energy and process zone size of concrete. Mater Struct 23:461–465
Soetens T (2015) Design models for the shear strength of prestressed precast steel fibre reinforced concrete girders. PhD Thesis, Gent University, Belgium
Stankowski T, Runesson K, Sture S (1993) Fracture and slip of interfaces in cementitious composites. I: characteristics, II: implementation. ASCEJ Eng Mech 119:292–327
Taheri M, Barros JAO, Salehian HR (2012) A parametric study on the use of strain softening/hardening FRC for RC elements failing in bending. ASCE Mater Civ Eng J 24 (3):259–274
Teixeira MDE, Barros JAO, Cunha VMCF, Moraes-Neto BN, Ventura-Gouveia A (2015) Numerical simulation of the punching shear behaviour of self-compacting fibre reinforced flat slabs. J Constr Build Mater 74:25–36
Torrijos MC, Barragan BE, Zerbino RL (2010) Placing conditions, mesostructural characteristics and post-cracking response of fibre reinforced self-compacting concretes. Constr Build Mater 24:1078–1085
Trusdell C, Toupin R (1960) The classical field theories. vol III/I. Handbuch der Physik, Springer-Verlag, Berlin
Ventura-Gouveia A (2011) Constitutive models for the material nonlinear analysis of concrete structures including time-dependent effects. PhD Thesis, University of Minho, Portugal
Ventura-Gouveia A, Barros JAO, Azevedo AFM (2011) Crack constitutive model for the prediction of punching failure modes of fiber reinforced concrete laminar structures. Comput Concr 8(6):735–755
Yip M, Li Z, Liao BS, Bolander J (2006) Irregular lattice models of fracture of multiphase particulate materials. Int J Fract 140:113–124

Vardoulakis I, Aifantis EC (1991) A gradient flow theory of plasticity for granular materials. Acta Mech 87:197–217

Vrech S, Etse G, Caggiano A (2016) Thermodynamically consistent elasto-plastic microplane formulation for fiber reinforced concrete. Int J Solids Struct 81:337–349

Wells G, Sluys L (2001) A new method for modelling cohesive cracks using finite elements. Int J Numer Methods Eng 50:2667–2682

Willam K, Bicanic N, Sture S (1984) Constitutive and computational aspects of strain-softening and localization in solids. ASME-WAM84, New Orleans, Symp. Vol. G00274 constitutive equations: micro, macro and computational aspects, pp 233–252

Zamanzadeh Z (2016) Cement based materials reinforced with recycled steel fibres for the shear reinforcement and strengthening of RC beams. PhD Thesis, University of Minho (in press)

Zamanzadeh Z, Lourenço LAP, Barros JAO (2015) Recycled steel fibre reinforced concrete failing in bending and in shear. J Constr Build Mater 85:195–207

Zubelewicz A, Bazant Z (1987) Interface element modeling of fracture in aggregate composites. J Eng Mech—ASCE 113:1619–1630

Chapter 13
Exploring the Use of Cement Based Materials Reinforced with Sustainable Fibres for Structural Applications

Paulo R.L. Lima and Joaquim A.O. Barros

Abstract This section presents some construction applications where sustainable fibres (both natural and from tire recycling industry) are used as fundamental or complementary reinforcement of cement based materials. Recent significant advances have been achieved on the enhancement of durability performance of natural (also currently designated by vegetable) fibres for the reinforcement of cement based matrices, and research efforts have been demonstrating the reinforcement performance of RSF as well. Because of this, and of the increasing concern of the society on the aspects affecting detrimentally the environmental conditions, a new phase seems to appear in the scientific and technical communities dedicated to explore the use of cementitious composites reinforced with different kinds of sustainable fibres for structural applications by demonstrating their economic and technical advantages.

13.1 Natural Fibre Reinforced Concrete

13.1.1 Tiles and Sheets for Roofing

Since 1970s natural fibres are being used as a reinforcement of cement-based products mainly for roofing applications, especially in countries where these fibres are locally produced or obtained at competitive price. Though they do not have the same quality of industrial tiles/sheets (metallic-, ceramic-, cement- or sandwich-type-based), the cost of these products made by natural fibre reinforced cement-based materials, herein designated by the acronym NFRC, is more accessible. The NSFRC products also feature better quality and durability than those traditionally made by lime or clay base matrices reinforced with straw.

P.R.L. Lima
State University of Feira de Santana, Feira de Santana, Brazil

J.A.O. Barros (✉)
University of Minho, Braga, Portugal
e-mail: barros@civil.uminho.pt

J.A.O. Barros et al. (eds.), *Recent Advances on Green Concrete for Structural Purposes*, Research for Development, DOI 10.1007/978-3-319-56797-6_13

A survey conducted by SKAT (1987) identified the regular production of roof elements made by NFRC in sixteen African, Asian and American countries. The matrix is composed of a mortar of cement and sand, in a composition of 1:1 or 1:2, in mass, being reinforced by short vegetable fibres (sisal, coconut, jute or elephant grass) with a length varying between 10 and 30 mm. This fibre reinforcement, used in a volume content generally not exceeding 1%, has the main purpose of contributing for the integrity of the resulting composite material during its manufacturing and application, as well as to avoid plastic shrinkage cracking. In the hardened state of the composite, the contribution of this type of fibres mainly stands for enhancing the resistance of cement based products against impact loadings. For the production of NFRC elements, the mixture is generally cast on a plane plate covered by a plastic film, and vibrated, and then transported still in its fresh state for plastic, timber or concrete formwork that provides the intended geometry and size to the NFRC elements. Initially, the production of NFRC elements was totally manual, but the development of electro-mechanical equipment for mixing and vibrating allowed their production process faster and with better final quality (Fisher and McVay 1993). NFRC products of several sizes and shapes were produced; the tiles and sheets shown in Fig. 13.1 are the most frequent ones (Wells 1994; Derr-Petrossian 1994). Tiles were produced in smaller dimensions and weight than sheets. These NFRC tiles have a length varying between 490 and 500 mm, a width of 250 mm and a thickness ranging from 6 to 8 mm, with a supporting span length of 400 mm: 12 tiles per square meter are hence necessary, as shown in Fig. 13.2. Since the weight of each NFRC tile is approximately 1.62 kg, the total weight per m^2 is about 20 kg. The NFRC tiles have to support their own weight and the impact loads from falling objects like branches or fruits, and those related to the production, stocking and assembling.

With the aim of contributing to the final quality of these products, a standard was proposed for regulating their production and guarantee the resistance requirements during the demoulding process. For this aim a minimum content of 1% of vegetable fibres, by weight, was recommended. Furthermore, a load carrying capacity of

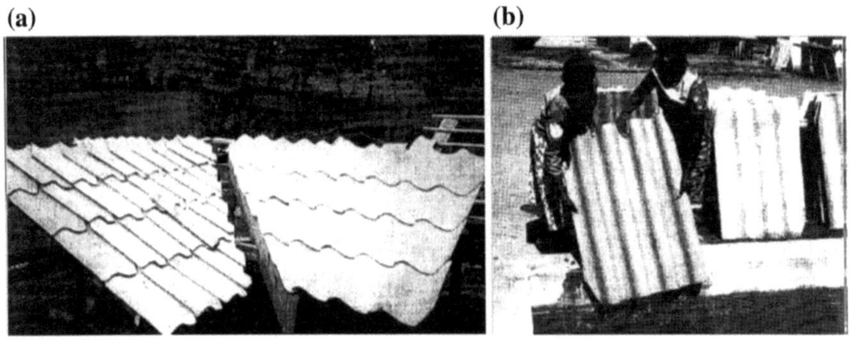

Fig. 13.1 NFRC products for roofing in Africa: **a** tiles and **b** sheets (Derr-Petrossian 1994; Wells 1994)

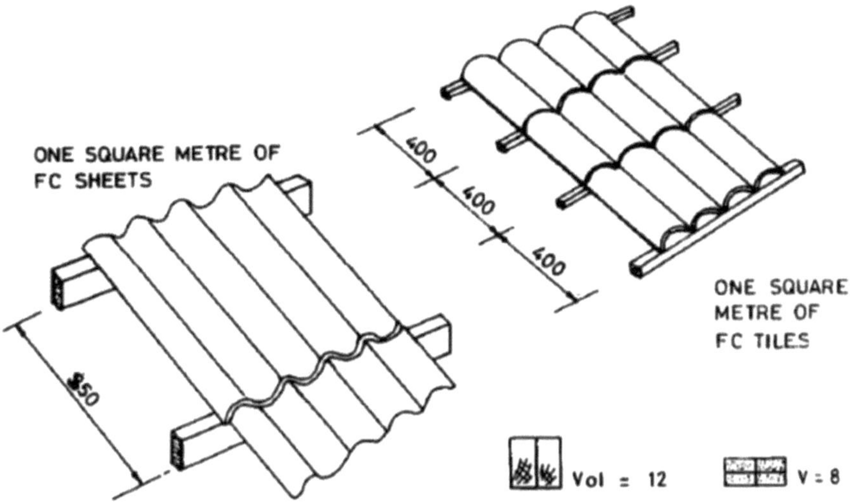

Fig. 13.2 Support systems for sheets (*left*) and tiles (*right*) (Derr-Petrossian 1994)

25 kg, applied in a three-point loading configuration, was required, together with an impact resistance to a steel ball of 250 g falling down from 0.5 m height without cracking occurrence (UNCHS 1989).

NFRC corrugated plates (herein denoted as sheets) are being produced with a length of 1000 mm, a width of 750 mm and a thickness of 10 mm, weighting about 20 kg each one. Since a supporting span length of 850 mm is used, the dead weight per m^2 of NFRC sheets is about 32 kg. When compared to the 500 mm long tiles, these NFRC sheets require less amount of timber supporting structure for the roofing, which decreases its total cost. However, the structural requirements for these NFRC sheets are higher, since apart from resisting to their own weight and impact loadings, they have also to be able to withstand a wind pressure equal to 90 kgf/m^2. Like the tiles made by fibre reinforced cement available in the market, the NFRC sheets are not designed for supporting the dead weight of a person walking on it (Derr-Petrossian 1994).

Tiles of 487 × 263 × 6 mm dimensions and with configuration similar to the ceramic Roman tile were produced in Brazil by Savastano et al. (1999) by using residues of vegetable fibres as reinforcement. These tiles have a load carrying capacity under three point bending equal to 450 N, with a testing span length of 330 mm. Applied in the roofs of cattle farms (Fig. 13.3a), these tiles were able to guarantee a better thermal comfort than the asbestos-cement tiles used so far (Roma et al. 2008). By using automation technology in the production of these type of tiles, and a process for accelerating the carbonation, the flexural capacity and durability of the tiles were increased (better production and quality control, and matrix not so aggressive for the vegetable fibres), according to Tonoli et al. (2010).

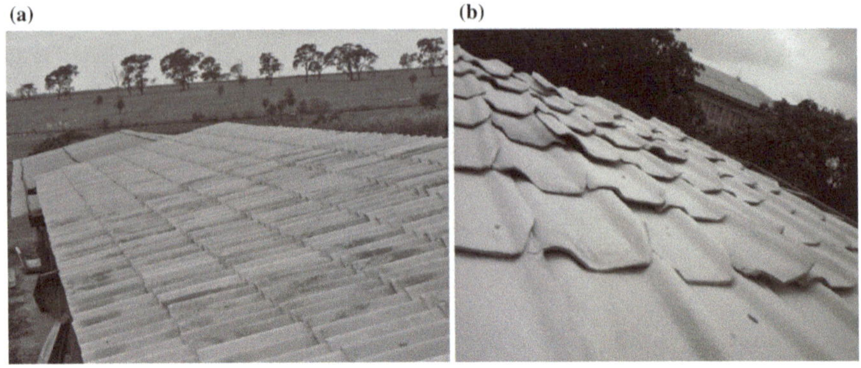

Fig. 13.3 Cement roofing system reinforced with: **a** vegetable waste (Brazil—Roma et al. 2008) and **b** sugarcane bagasse (Nigeria—Omoniyi and Akinyemi 2012)

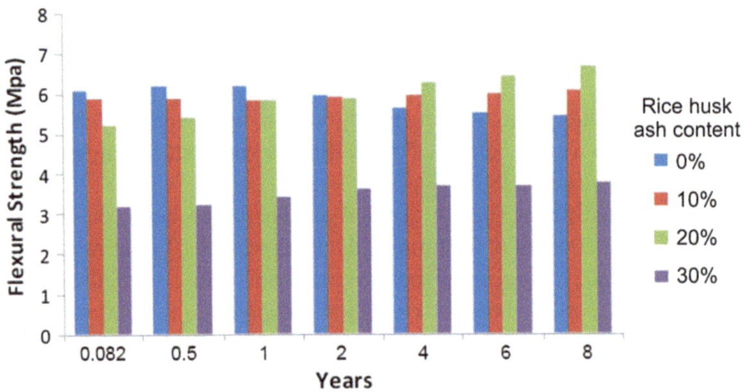

Fig. 13.4 Flexural results of tiles after natural weathering tests (Omoniyi and Akinyemi 2012)

Cement based tiles reinforced with sugarcane (Saccharum officinarum) bagasse in a content that varied between 1 and 4%, by weight of cement, were produced by Omoniyi and Akinyemi (2012). Water cement ratio of 0.4 and 0.5, sand/cement ratio of 1.0, 2.0 and 3.0, and percentage of 10, 20 and 30% of replacement of cement by rice husk ash were adopted. The results have demonstrated an increase of the resistance to the impact loading with fibre contents up to 3%. The flexural resistance of tiles reinforced with 2% of fibres, and cement replaced by rice husk ash in the aforementioned percentages, was evaluated at different ages in samples exposed to natural weathering conditions at the University of Ibadan, in Nigeria, as shown in Fig. 13.3b. The results presented in Fig. 13.4 show that the tiles without rice husk ash featured a flexural strength reduction after 8 years of exposition, while those with 30% of cement replaced by rice husk ash have presented an increase of the flexural capacity when compared to the corresponding values at 28 days.

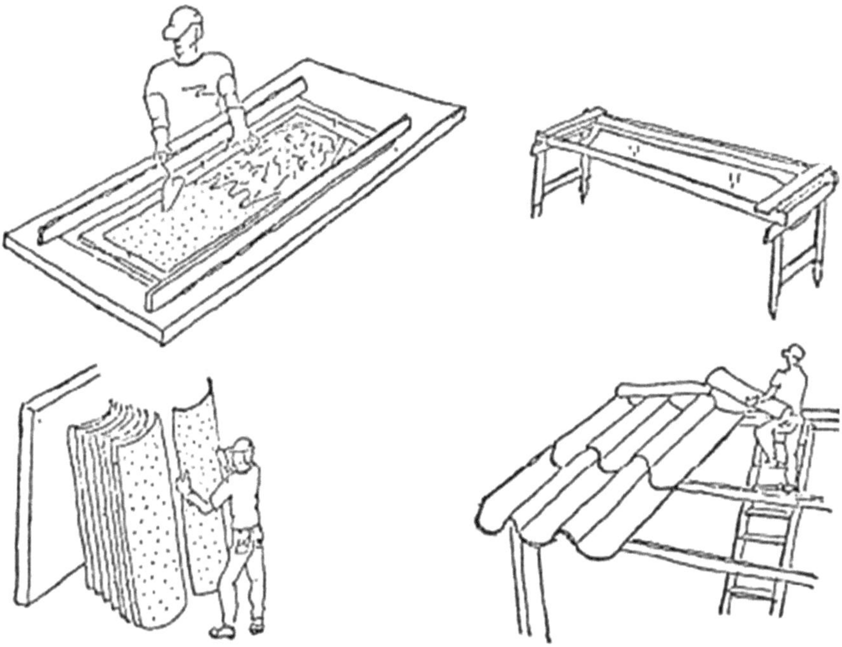

Fig. 13.5 Manufacture and installation of tiles of mortar reinforced with vegetable fibres (Guimarães 1987)

Channel shape type tiles with 1.0 m length and 10 mm thickness were produced in Brazil with a mortar mix of 1:3:0.7 (cement:sand:water/cement ratio, in mass) reinforced with 2% of fibres in volume, Fig. 13.5 (Guimarães 1987). When compared with tiles made of unreinforced mortar, those reinforced with sisal, coconut or piassava (*Attalea funifera* m.) fibres showed an increase of flexural strength of 5, 19 and 26%, respectively.

Paramasivan et al. (1984) have produced corrugated cement-based sheets (length of 915 mm, width of 460 mm and thickness of 10 mm) reinforced with coconut fibres. A mortar mixture of 1:0.5:0.35 (cement:sand:s/c ratio, in mass), three fibre volume percentages (2, 3 and 4%), and three fibre lengths (12.5, 25 and 38 mm) were used. The casting was made in moulds of female-male type. The mortar was initially spread on a plastic film previously placed on a smooth timber mould, and the compacting was then performed on a vibrating table followed by the application of a pressure of 1.5 atm. The tiles were subjected to a three-point bending test, and a flexural strength varying between 19 and 22 MPa was recorded.

Hussein and Zakaria (1990) have also studied the flexural behaviour of thin flat (500 × 100 × 10 mm dimensions) and corrugated (1220 × 630 × 10 mm) sheets made by cement paste (water/cement ratio of 0.35) reinforced with six coconut fibre contents (1–6%, in mass), besides the reference one (unreinforced). The fibre length varied between 37 and 250 mm. A mechanical mixer was used for the production of the cementitious composites. The flat sheets were produced by spreading the mixture on the moulds and applying vibration for three minutes. The corrugated sheets were made from flat sheets that after have been cured for one hour were moved for the corrugated moulds, and subjected to a pressure of 0.4 N/mm^2 for 24 h up to demoulding. The corrugated sheets showed higher flexural strength than the flat sheets when submitted to three and four point bending tests, respectively. Several cracks were formed in the tested sheets, and fibres failed by tensile rupture, demonstrating an effective fibre-matrix bond. The authors have considered the corrugated sheets suitable for being used in affordable constructions, mainly in developing countries.

Corrugated tiles of dimensions 1830 × 920 × 6 mm reinforced with fique fibre (*Furcraea gender*) from Colombia, in a content of 3% by total weight of solids, and with a length of 15 mm, have been developed in the University of Del Valle since 1988, using a low-cost technology, to be used by rural communities, as shown in Fig. 13.6 (Delvasto et al. 2010). The inspection of this type of tiles, by scanning electron microscopy (SEM) after 14 years of exposition to the weathering conditions of Cali (Fig. 13.6), has highlighted that in spite of the fibre-matrix bond conditions have been detrimentally affected by the fibre volume variation, no signal of deterioration was visible in the tiles (Tonoli et al. 2011). The corrugated sheets were also produced using industrial processes, as well as hybrid reinforcement combining fique and asbestos or PVA fibres. Sheet reinforced with 3.3% of fique fibre featured a flexural load bearing capacity of 2875 N/m, as compared to 4250 N/m of commercial tiles made of asbestos-cement (Delvasto et al. 2010).

During the last years, matrices tailored to preserve the long term reinforcement mechanisms of vegetable fibres have been developed for composites reinforced with relatively high content of sisal fibres (up to 6% volume), as by Ferreira et al. (2015). These composites can have a pronounced deflection hardening behaviour accompanied by a multi-cracking process, even using short fibres randomly

(a) (b)

Fig. 13.6 Use of fique fibres: **a** production of corrugated sheets (Delvasto et al. 2010) and **b** prototype for the assessment of the sheets' performance in weathering conditions (Tonoli et al. 2011)

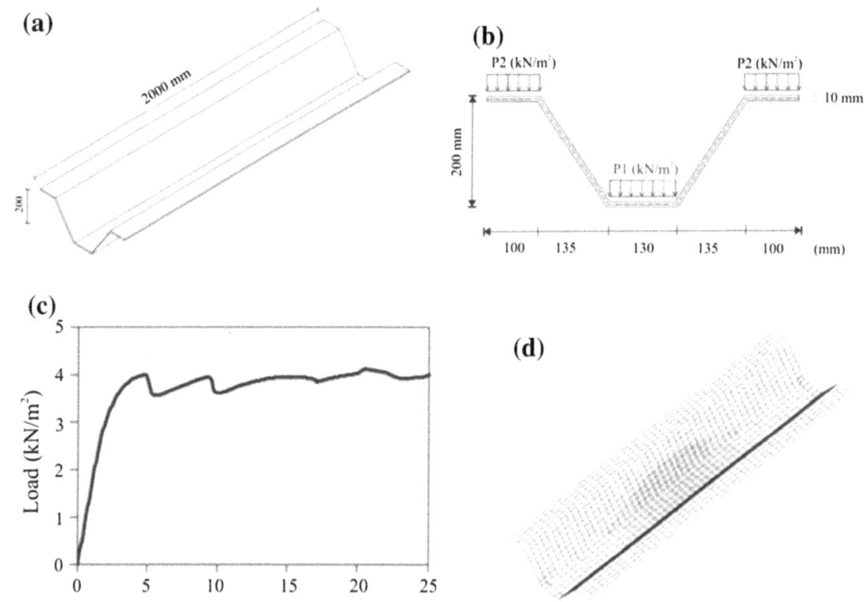

Fig. 13.7 FEM-based material nonlinear analysis of roofing folder type elements of deflection hardening mortar reinforced with vegetable fibres: **a** geometry (dimensions in mm); **b** loading conditions; **c** load-deflection response; **d** cracking pattern

distributed (Lima et al. 2016a). This is opening new perspectives on their use in the construction industry. Using the finite element method (FEM) and nonlinear fracture mechanics capable of simulating the crack formation and propagation in cement based materials, Lima et al. (2016b) demonstrated the possibility of using these composites in prefabricated roof elements (Fig. 13.7).

13.1.2 Formwork Elements

Schafer and Brunssen (1990) have evaluated the flexural performance of permanent moulds made of a cementitious mortar reinforced with long sisal fibres. The arch- and trapezoidal-shaped formworks of 10 and 20 mm thick shown in Fig. 13.8 were produced by using a sequence of mortar and sisal fibre layers. Under bending test configuration, the arch-shaped elements presented a load carrying capacity of 7 kN, which corresponds to a load level two times higher the load at cracking initiation, which is sufficient for resisting the load conditions occurring during casting the fresh reinforced concrete slab. However, according to the authors, the manufacturing procedure is very labour-intensive, thereby only appropriate in countries where labour resources are abundant and at relatively low costs, such is the case of some developing countries.

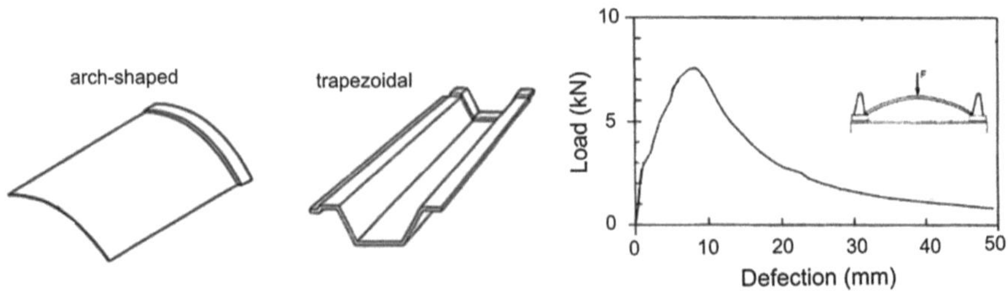

Fig. 13.8 Permanent formwork elements of layered mortar reinforced with long sisal fibres for RC slabs (Schafer and Brunssen 1990)

The trapezoidal thin shell of mortar reinforced with 6% by volume of 40 mm long sisal fibres, shown in Fig. 13.9, was developed by Lima et al. (2016c) for serving as permanent formwork for prefabricated slabs. This lightweight block has a width of 350 mm, a height of 30 mm, a length of 500 mm and a thickness of 10 mm. When submitted to a centre load distributed over an area of 200×75 mm^2, and simply supported (the vertical and horizontal displacements were restricted) in their lateral borders (in order to simulate the real support conditions in the slab, as much as possible), this lightweight block has attained a load carrying capacity of 183 kgf, which is much higher than the load at crack initiation as well as the minimum load required by the standard for this type of elements (70 kgf).

By performing nonlinear finite element analysis with the values for the model material parameters obtained experimentally and by inverse analysis (Fig. 13.10a), new geometric configurations for this type of block were investigated by Lima et al. (2016d), and their respective flexural capacity was estimated (Fig. 13.10b). Figure 13.10 shows that for a block of 700 mm width, length of 1000 mm and height of 100 mm, it is sufficient a thickness of 10 mm in order to guarantee a load carrying capacity and deflection performance higher than the requisites recommended by the NBR 14859 (ABNT 2002).

13.1.3 Wall Elements

Wall panels 2.4 m high, 0.395 m wide and 0.09 m thick were produced with a mortar reinforced with 2%, by volume, of 30 mm long coconut fibres. Besides having demonstrated similar thermal insulation performance to panels made of conventional concrete (thermal resistance equal to 0.14 m^2 K/W), the wall panels attained a compressive strength of 6.5 MPa, which is sufficient for affordable houses. A prototype was built in San Paulo, Brazil, as Fig. 13.11 shows (John et al. 1990), and after 16 years exposed to local weathering conditions no signs of durability concerns were detected (Agopyan et al. 2005).

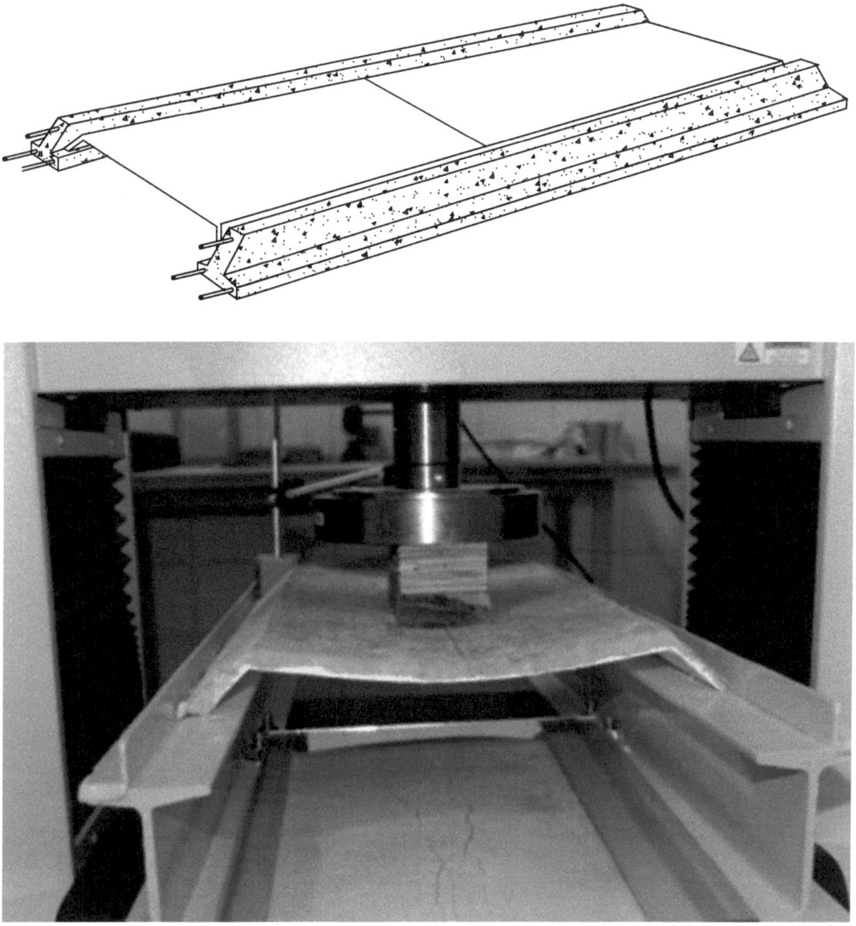

Fig. 13.9 Lightweight block for one-way precast slab of mortar reinforced with sisal fibres: a configuration of the slab and b bending test

Coconut fibres were also used for the reinforcement of cement based matrices (the authors used the acronym CFRC for designating this composite) in the production of the type of interlocking blocks shown in Fig. 13.12a. With this type of blocks the authors (Ali et al. 2012a) had the purpose of producing walls with appropriate performance to resist lateral load conditions representative of seismic events, even without using mortar joints for bonding the blocks. After research on the rheology and mechanical properties of CFRC, it was concluded that 1% by volume of 50 mm long coconut fibres is suitable for producing this type of blocks (Ali et al. 2012a). The blocks were submitted to compression and shear tests (Fig. 13.12b—Tang et al. 2014) at the age of 28 days and after have been stored in a storage room in Auckland, New Zealand for 15 months. The results from these

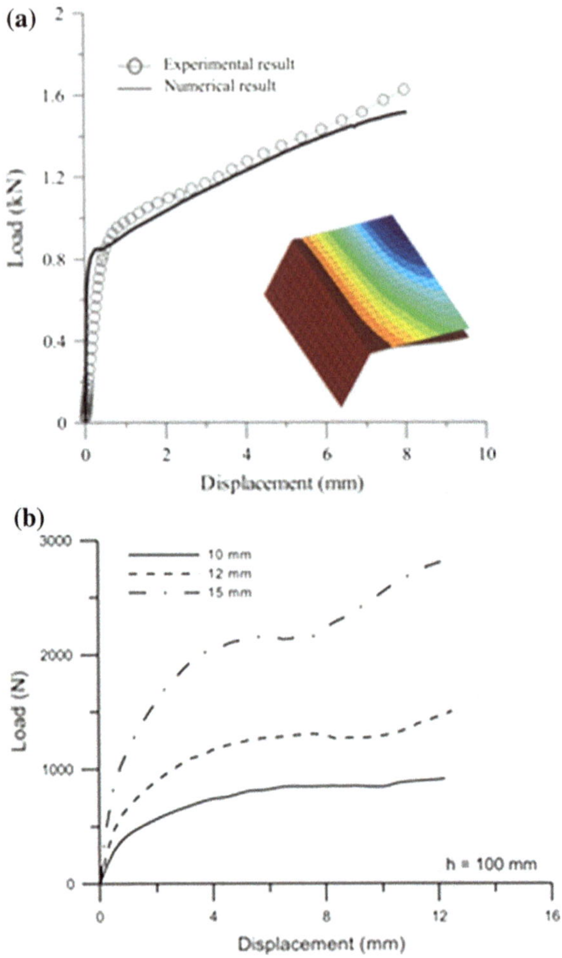

Fig. 13.10 FEM-based material nonlinear analysis of lightweight block for one-way precast slab of mortar reinforced with sisal fibres: **a** simulation of specimens experimentally tested and **b** assessment of block's thickness effect on its load carrying capacity (Lima et al. 2016d)

tests showed a small variation of the compressive strength (an increase of 1.8%) and of the shear strength (increase of 2.7%) when compared to the results obtained at 28 days (Tang et al. 2014).

Pre-fabricated walls were made by Ali et al. (2013) using the CFRC interlocking blocks. In order to increase their resisting capacity to lateral seismic type loading, the walls were post-tensioned by ropes of coconut fibres (Fig. 13.12c). The obtained experimental results indicate that the post-tensioned system has increased the flexural stiffness of this type of walls, but their damping was higher without this system (Ali et al. 2013), which according to the authors is justified by the larger relative movements of the blocks when no post-tension is applied.

Juárez et al. (2010) developed wall-type blocks made by mortar reinforced with agave lechuguilla vegetable fibres (produced in Mexico) in volume content of 0.25–1%. In order to reduce the weight of this type of blocks, post-consumer

Fig. 13.11 Wall panels of affordable house made by mortar reinforced by coconut fibres (John et al. 1990)

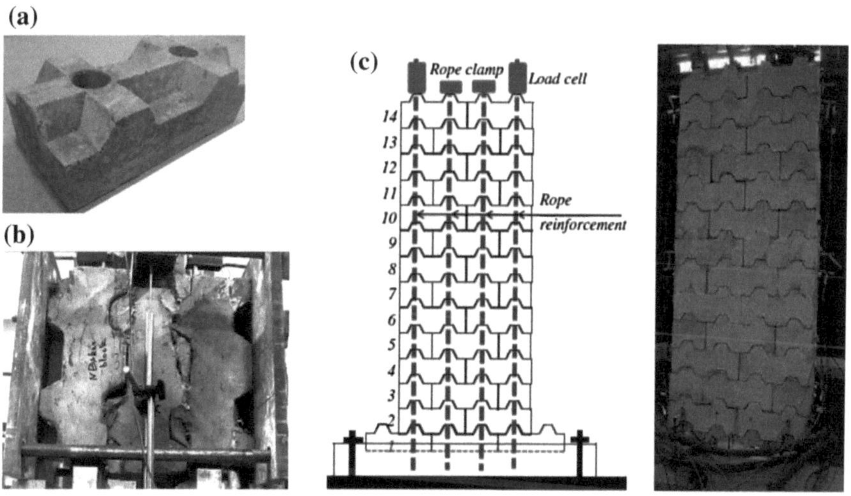

Fig. 13.12 Wall system made by CFRC interlocking blocks for better performance under seismic type loadings: **a** CFRC interlocking block (Ali et al. 2012a), **b** shear test of bricks (Tang et al. 2014), and **c** post-tensioned wall with ropes of coconut fibres (Ali et al. 2013)

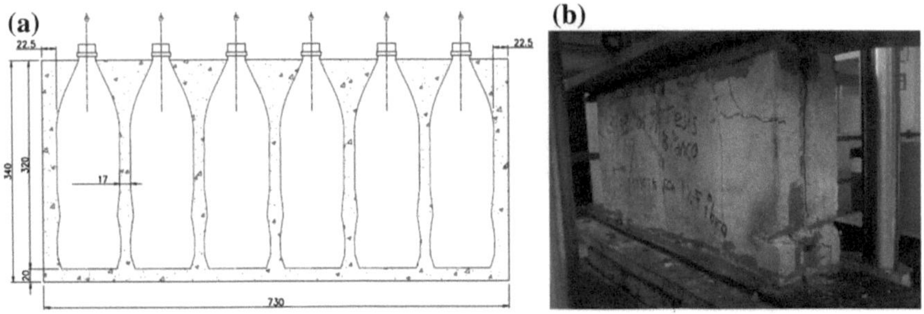

Fig. 13.13 Wall type block made by mortar reinforced with agave lechuguilla vegetable fibres and including post-used PET bottles: **a** configuration and **b** compression test (Juarez et al. 2010)

polyethylene terephthalate (PET) bottles were used according to the configuration represented in Fig. 13.13a. By using 0.75% of 25 mm long fibres, an increase of 10% of the compressive strength was obtained when compared to the unreinforced mortar (Fig. 13.13b).

13.1.4 Beams and Columns

Columns made by thin wall polymer tubes reinforced with flax fibres (flax fibre reinforced polymer—FFRP), filled by concrete reinforced with coconut fibres (CFRC) were produced and tested in compression and bending (Yan and Chouw 2013; Yan et al. 2014; Chen and Chouw 2016), Fig. 13.14. FFRP tubes were fabricated using the hand lay-up process with bidirectional woven flax fabric (550 g/m^2) and epoxy resin. A mass content of 1% of 50 mm long coconut fibres were used in a composition, by weight, of 1:0.68:3.77:2.96 for cement:water:-gravel:sand. This fibre reinforcement has provided an increase of 9.3% in the axial load carrying capacity of the columns, while the axial strain corresponding the peak load has increased from 0.2% (without fibre reinforcement) to 0.54%. At the failure of the columns the unreinforced core concrete has disintegrated, while the CFRC has preserved some integrity thanks to the reinforcement mechanisms of coconut fibres bridging the formed cracks. Under bending, this tubular type composite column has shown an increase of 36.5 and 140% in terms of maximum load and corresponding deflection, respectively, when compared to the column with unreinforced concrete core.

The use of coir fibre as reinforcement of concrete in order to improve the damping capacity of material was evaluated by Ali et al. (2012b) using impact test in beams with dimensions of 100 mm width, 100 mm depth and 915 mm length. A small impact load was applied three times at mid-span of the beam with the help of a calibrated hammer. The response was recorded by accelerometers, located near to mid-span. Weight contents of 1, 2, 3 and 5% of coconut fibres with length equal

Fig. 13.14 FFRP–CFRC tubular composite tubes: **a** casting the CFRC into the FFRC tubes and **b** four-point bending test setup (Chen and Chouw 2016)

Fig. 13.15 Column type elements in concrete reinforced with vegetable fibres (Bragança et al. 2014)

to 25, 50 and 75 mm were used in a composition, in mass, of 1:0.48:2:2 for cement: water:gravel:sand. The increase of fibre content has resulted in a higher damping ratio and a lower fundamental frequency. Beams with 50 mm long fibres had higher damping than the beams reinforced with fibres of other lengths.

Column type elements for supporting cables of electric current distribution (Fig. 13.15) were produced with a concrete reinforced with 0.8% by volume of vegetable fibres (sisal and coconut) by Bragança et al. (2014). As expected, under flexural tests, the behaviour of this type of elements was the same regardless of using conventional or CFRC, since the fibre reinforcement was intended to increase their resistance to the impact type loadings due to the favourable contribution of fibre reinforcement mechanisms in terms of concrete integrity and energy absorption capacity.

13.2 Concrete Reinforced with Recycled Fibres

Polyethylene terephthalate (PET) fibres of aspect ratio (fibre length/fibre diameter) of 25, obtained from manual cutting process of post-used PET bottles, were used in a fibre volume content of 0.5% for producing a material, which was designated by PFRC, that was applied in D-Region (Fig. 13.16a) in an attempt of improving the energy dissipation capacity of reinforced concrete (RC) column-beam joints submitted to seismic loadings (Marthong and Marthong 2016). Three types of conventionally reinforced column-beam joints (without PFRC) with different cross sections and reinforcement were tested (Fig. 13.16b): (a) weak beam in bending (BWF); (b) weak beam in shear (BWS) and (c) weak column in shear (CWS). Figure 13.16b (the suffix "SF" in the legend represents the specimen with PFRC in the D-region) shows the comparisons of envelope curves, as obtained from hysteresis loops. It can be observed that the fibre reinforcement has contributed to increase the lateral load carrying capacity of this type of elements. This increase was more pronounced in the BWF/BWFSF elements, i.e., those with beams deficiently reinforced in bending. According to the authors, the use of PFRC resulted in a smaller damage in the column-beam joint that occurred during the dynamic tests, since fibre reinforcement mechanisms contributed to decrease the crack width and crack spacing when compared to the elements made of ordinary concrete. In consequence, the PFRC has contributed to increase the lateral stiffness (15%) and energy dissipation (24%).

The influence of using 0.5, 0.75 and 1.0%, by volume, of fibres from recycling of post-consumer PET bottles to improve the flexural behaviour of RC beams

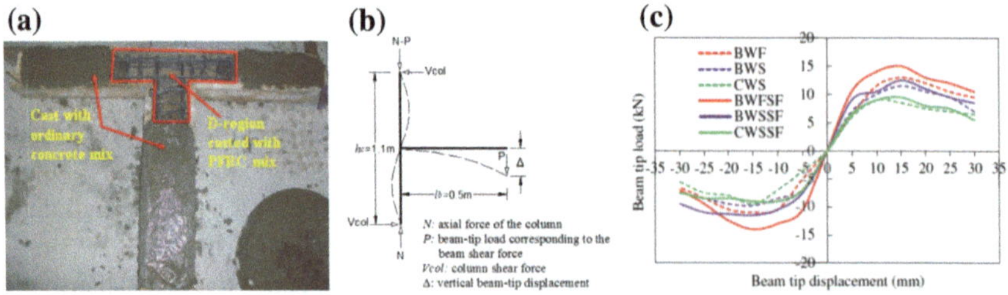

Fig. 13.16 Concrete reinforced with fibres recycled from post-consumed PET bottles (PFRC) to improve the dynamic response of RC column-joints: **a** casting process of a column-joint specimen, **b** configuration of test and **c** beam-tip load-displacement extracted from a hysteretic loop (Marthong and Marthong 2016)

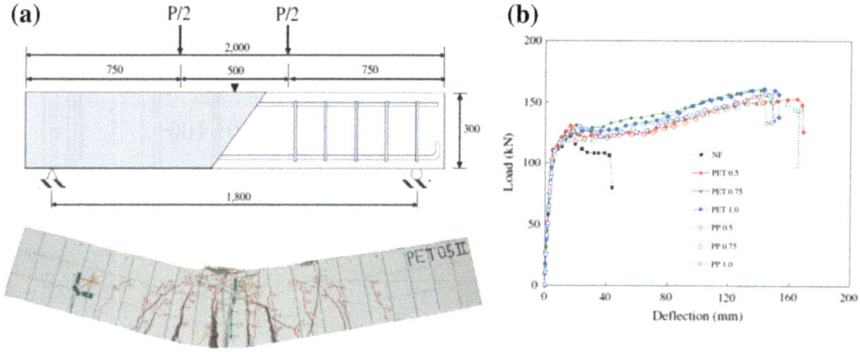

Fig. 13.17 Concrete reinforced with fibres recycled from post-consumed PET bottles for improving the flexural behaviour of RC beams failing in bending: **a** reinforcement arrangement, loading and supporting conditions (dimensions in mm), **b** failure model and respective cracking pattern and **c** load versus midspan deflection (adapted from Kim et al. 2010)

submitted to four point bending tests (Fig. 13.17a) was assessed by Kim et al. (2010). The recycled PET fibres had an embossed geometry of cross section of 0.2 × 1.3 mm, length of 50 mm, elasticity modulus of 10 GPa, tensile strength of 421 MPa and ultimate tensile strain of 11.2%. The RC beams, of cross section of 200 × 300 mm, were flexurally reinforced with three bars of 13 mm diameter in the tension zone (flexural reinforcement ratio of 0.66%) and two bars of 10 mm diameter in the compression zone, and were shear reinforced with steel stirrups of 10 mm diameter at 150 mm spacing.

Despite the beams were designed to fail by yielding of the tensile flexural reinforcement, those with PFRC featured a failure initially controlled by concrete crushing, indicating that fibre reinforcement has delayed the onset of yielding of the flexural reinforcement (Fig. 13.17b). The results of the bending tests, shown in Fig. 13.17c, show that the beams reinforced with 0.5% (PET 0.5), 0.75% (PET 0.75) and 1.0% (PET 1.0) attained a load carrying capacity 25, 31 and 32% higher than the one of the reference beam (made of conventional concrete, NF). Figure 13.17 also shows that the deflection capacity was significantly improved thanks to the reinforcement contribution of PET fibres. When compared to the beams reinforced with polypropylene (PP) fibres in the same volume content, the PET fibre reinforcement provided similar load carrying and deflection capacities (Fig. 13.17c). The PP fibres had a crimped cross section geometry of 0.38 × 0.9 mm, length of 50 mm, elasticity modulus of 6 GPa, tensile strength of 550 MPa and ultimate tensile strain of 15.0%. By assuming, as a ductility index, the ratio between the deflection at ultimate load (just before the abrupt load decay) and the deflection at yield initiation of the tensile flexural reinforcement, the ductility index has increased from 3.53 in the reference beam up to 36.5 in the beam reinforced with 0.5% of PET fibres.

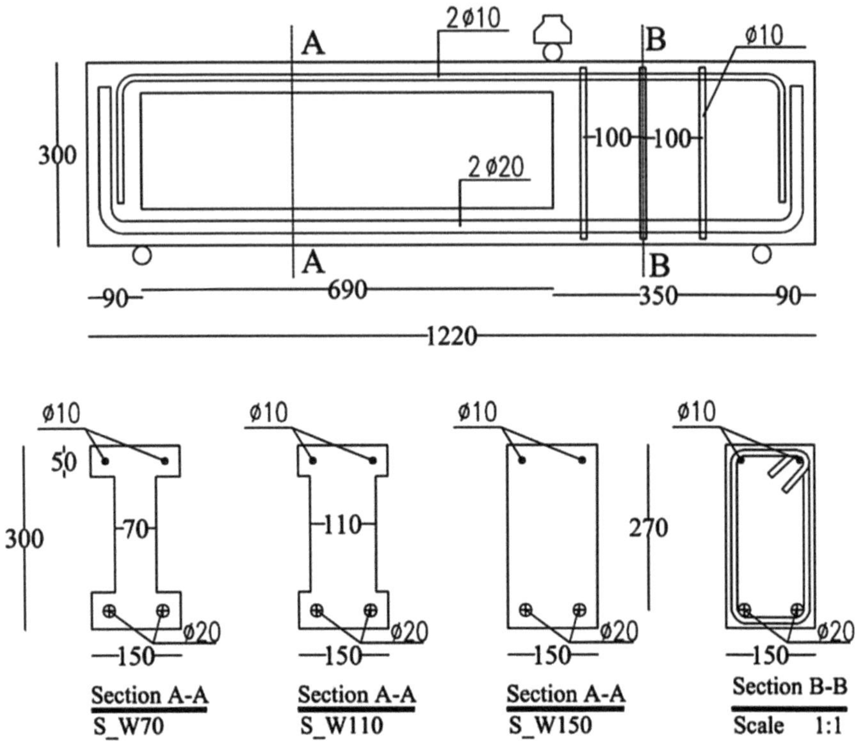

Fig. 13.18 Geometry of the beams, reinforcement configurations, and loading and support conditions (dimensions in mm), (Zamanzadeh et al. 2015a)

Zamanzadeh et al. (2015a, b) explored the potentialities of recycled steel fibres (RSF) for the shear reinforcement of RC beams of different web thickness. These RSF were obtained from tyre recycling industry according to the cryogenic process (described in Sect. 8.1). The relevant information about the tested series of beams is provided in Fig. 13.18. These beams were produced by a concrete reinforced with 60 kg/m^3 of RSF, whose mix composition and characterization of its mechanical properties are provided in Sect. 8.2. The post-cracking equivalent and residual flexural tensile strength parameters, according to the RILEM TC 162 TDF (2003) and Model Code 2010 (2011), respectively, are indicated in Table 8.4 of this section.

The relationship between the applied load and the deflection at the loaded section of the tested series of beams is represented in Fig. 13.19 (in the legend WX, with X = 70, 110 and 150, represents beam's web thickness in mm). It is verified that the beam's load carrying capacity has increased with the web thickness, while the deflection at maximum load was almost the same. After having calibrated the values of the parameters of a constitutive model capable of simulating the material nonlinear behaviour due to crack formation and propagation in this type of materials, the authors performed a parametric study having concluded that 90 kg/m^3 of

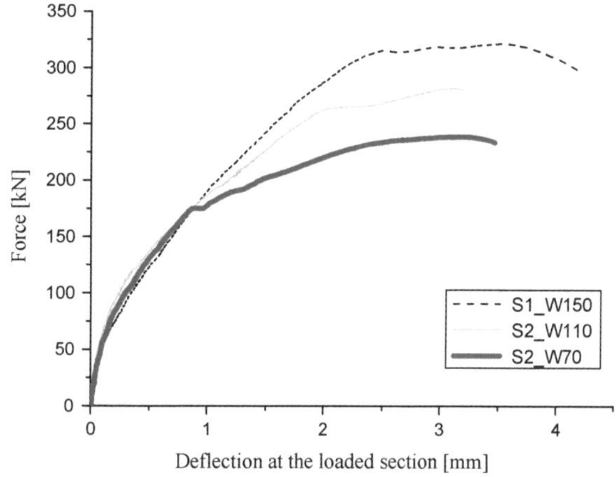

Fig. 13.19 Load–deflection relationship at the loaded section for the tested series of beams (Zamanzadeh et al. 2015a)

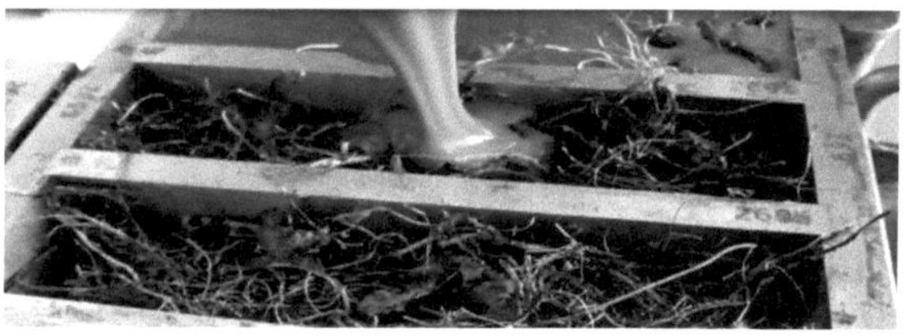

Fig. 13.20 Production of RSFRC thin panels for the shear strengthening of RC beams (Zamanzadeh et al. 2015b)

RSF provide an increase of 95, 81 and 71% in terms of shear capacity of the beams with a web's thickness of 70, 110 and 150 mm, respectively, when the shear capacity of the reference beam (plain concrete with the same flexural reinforcement ratio) is considered for comparison purpose.

Using a technique similar to SIFCON (Fig. 13.20) for producing thin panels of cement based composite reinforced with relatively high content of RSF, Zamanzadeh et al. (2015b) explored the potentiality of using these panels for the shear strengthening of RC beams (Fig. 13.21). The details about the technology for producing these panels, the mechanical characterization of their properties and the experimental programs carried out are detailed elsewhere (Zamanzadeh et al. 2015b). Figure 13.22 shows that attaching two RSFRC panel with a thickness of 20 mm provided an increase of 33% in terms of shear capacity of the beam with a web's thickness of 70 mm when the shear capacity of the reference beam (plain concrete with web's thickness of 110 mm and the same flexural reinforcement ratio) is considered for comparison purposes.

Micelli et al. (2015) have explored the potentialities of using RSF for the reinforcement of slab type elements. The objective is the replacement of conventional steel meshes by RSF in tunnel applications. The replacement of conventional reinforcement with industrial steel fibres (ISF) is already common practice in these applications (Skatun and Spigerverk 1986). For that purpose, the authors performed tests with square panels of 600 × 600 × 100 mm dimensions, supported along

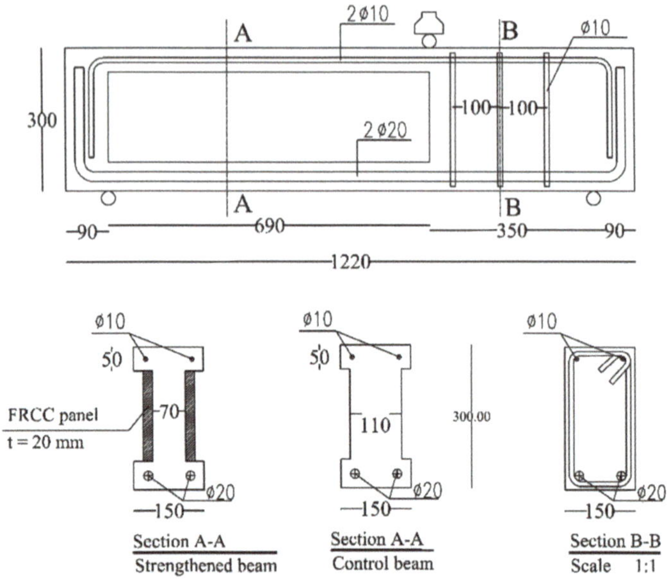

Fig. 13.21 Experimental program for exploring the use of thin RSFRC panels for the shear strengthening of RC beams (dimensions in mm): geometry, support and loading conditions, and shear strengthening arrangement (Zamanzadeh et al. 2015b)

Fig. 13.22 Force versus deflection relationship of a beam with ("strengthened beam") and without ("control beam") thin panels of RSFRC for the shear strengthening (Zamanzadeh et al. 2015b)

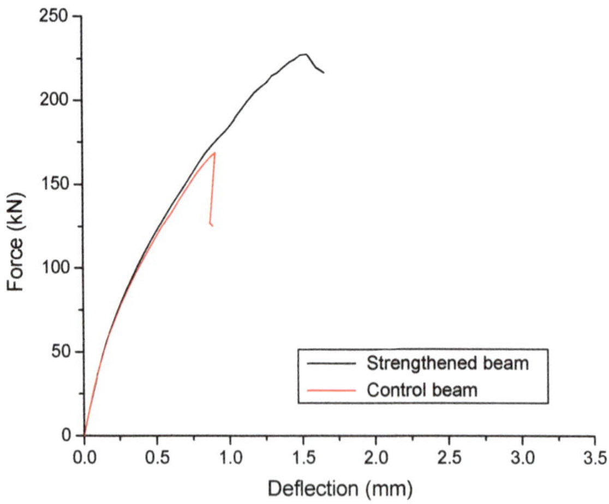

their contour (span length in both directions of 500 mm), following the recommendations of EFNARC (1996), see Fig. 13.23b. FRC panel included 0.28%, by volume, of RSF, while the reference panel was reinforced with a mesh of steel bars, 6 mm in diameter and spaced at 200 mm, with a concrete cover of 35 mm in the tension zone (Fig. 13.23a). Both type of panels showed a similar load-deflection

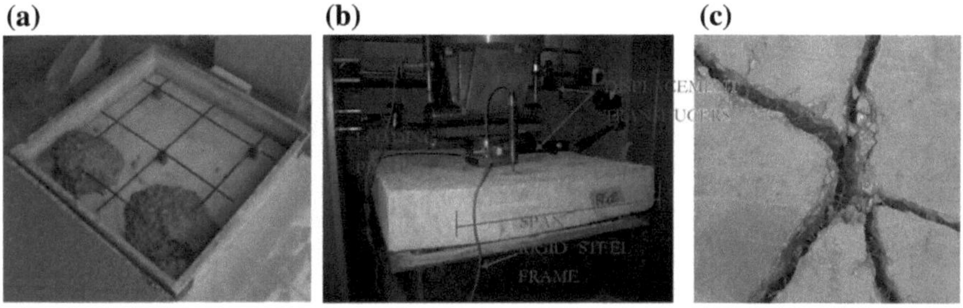

Fig. 13.23 Slab-panel tests according to the EFNARC recommendations (1996) for assessing the potentialities of RSF for tunnelling applications: **a** panels reinforced with conventional steel grid mesh, **b** test setup and **c** RSF bridging the formed macro-cracks (Micelli et al. 2015)

response (post-peak softening), but the panels reinforced with RSF attained a load carrying and an energy absorption capacity (the area under the force-deflection curve), 20 and 33%, respectively, higher than the corresponding ones of the panels reinforced with steel meshes. The RSF failed by fibre pull-out mechanisms (see Fig. 13.23c), while in the panels reinforced with steel mesh the bars ruptured in the section of the macro-cracks. The obtained results indicate the RSF have potential for the tunnelling applications, but deeper studies must be performed for reliable conclusions.

The use of steel fibres from two types of tyre recycling processes was also explored by Tlemat (2004) and Pilakoutas et al. (2004) for the reinforcement of concrete slabs in drainage system of highways and car parking. The conventional reinforcement (consisting of nine ϕ 20 mm steel bars in the investigated case), was replaced by either pyrolysed recycled steel fibres (PRSF) obtained by cutting recycled steel tyre-cord to 15, 25 and 50 mm pieces—the cord was obtained from the microwave induced pyrolysis of whole tyres, or shredded recycled steel fibres (SRSF) obtained from the third stage of mechanical shredding of discarded tyres, thereby featuring different diameter and shape (see Sect. 8.1 for more details on the production of RSF from tyre recycling industry).

Besides the conventional RC slab shown in Fig. 13.24a, other three types of slabs were also produced: concrete slabs reinforced with 6% by weight of PRSF; concrete slabs reinforced with 2% by weight of SRSF; slurry infiltrated concrete slab reinforced with 17.5% by weight of SRSF.

The results from flexural tests under centre point load, as shown in Fig. 13.24b, highlighted that the maximum load decreased by 32.8 and 44.3% when 6% by weight of PRSF and 2% by weight of SRSF were used, respectively, in comparison to the slab conventionally reinforced. However, when using slurry infiltrated concrete reinforced with 17.5% by weight of SRSF, the load carrying capacity increased by 18.2%. While the slab conventionally reinforced has failed in punching (see Fig. 13.24c), the slabs reinforced with recycled fibres have failed by the formation of longitudinal crack, as shown in Fig. 13.24d.

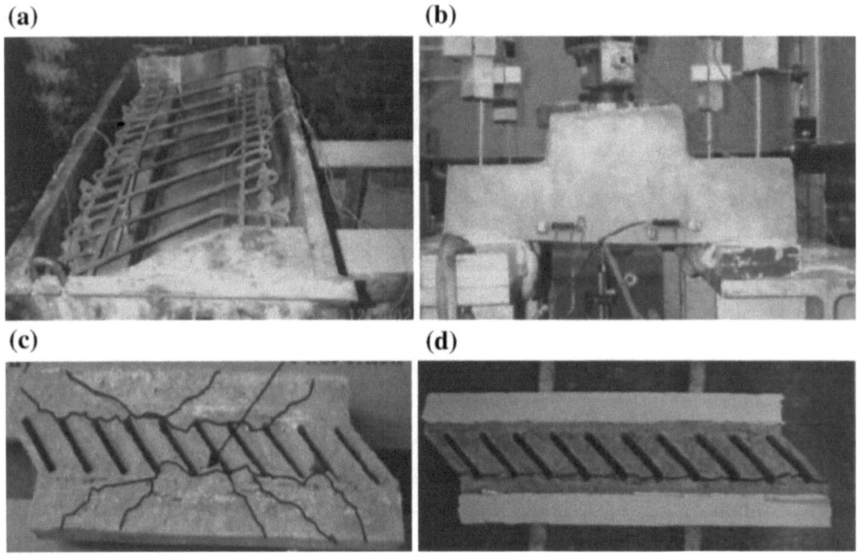

Fig. 13.24 Concrete slabs for drainage systems: **a** conventional steel reinforcement used in this tyre of slabs, **b** test setup, **c** crack pattern in the slab conventionally reinforced and **d** crack pattern of the RSFRC slabs (Tlemat 2004)

13.3 Concluding Remarks

Vegetable and recycled fibres have been used to produce construction elements for roofing, walls, permanent moulds for pre-fabricated slabs and drainage systems, as well as for the concrete reinforcement in case of beams and columns. However, the technology of extracting and processing the fibres must be significantly improved in order to constitute a more competitive reinforcement system for cement based materials. The technology for producing concrete reinforced with these types of fibres must also be improved in order to take into consideration their geometric and material properties, and therefore retrieving more efficiently their reinforcement mechanisms. Deeper research must be performed, mainly on the long term behaviour of the mechanical and durability performance, with the aim of obtaining reliable information for being used in the development of design guidelines.

References

ABNT (2002) Associação Brasileira de Normas Técnicas. NBR 14859-1. Laje pré-fabricada—Requisitos—Parte 1: Lajes unidirecionais (In Portuguese)

Agopyan V, Savastano H, John VM, Cincotto MA (2005) Developments on vegetable fibre–cement based materials in São Paulo, Brazil: an overview. Cem Concr Compos 27(5):527–536

Ali M, Liu A, Sou H, Chouw N (2012a) Mechanical and dynamic properties of coconut fiber reinforced concrete. Constr Build Mater 30:814–825

Ali M, Gultom RJ, Chouw N (2012b) Capacity of innovative interlocking blocks under monotonic loading. Constr Build Mater 37:812–821

Ali M, Briet R, Chouw N (2013) Dynamic response of mortar-free interlocking structures. Constr Build Mater 42:168–189

Bragança M, Bonato MM, Portella KF, Bronholo JL, Vieira ME, Santos JCM (2014) Desempenho de postes de concreto com adições orgânicas e fotocatalíticas. Espaço Energia 20:18–24 (in Portuguese)

Chen J, Chouuw N (2016) Nonlinear flexural behaviour of flax FRP double tube confined coconut fibre reinforced concrete. Mater Des 93:247–254

Delvasto S, Toro EF, Perdomo F, Gutiérrez RM (2010) An appropriate vacuum technology for manufacture of corrugated fique fiber reinforced cementitious sheets. Constr Build Mater 24 (2):187–192

Derr-Petrossian B (1994) Fibre-concrete roofing. J Netw Afr Countries Local Build Mater Technol 3(2):19–28

EFNARC (1996) European Federation of producers and applicators of specialist products for structures. European specification for sprayed concrete (EFNARC). Guidelines for specifiers and contractors

Ferreira SR, Silva FA, Lima PRL, Toledo Filho RD (2015) Effect of fiber treatments on the sisal fiber properties and fiber–matrix bond in cement based systems. Constr Build Mater 101: 730–740

Fisher M, McVay M (1993) Kenya: fibre-concrete roofing technology adaptation and progress. J Netw Afr Countries Local Build Mater Technol 2(3):9–20

Guimarães SS (1987) Some experiments in vegetable fiber-cement composites. In: Symposium on building materials, for low-income housing, 20–26 Jan 1987. Proceedings …, Bangkok, Thailand, pp 167–175

Hussain MW, Zakaria F (1990) Prospects for coconut fibre—reinforced thin cement sheets in the Malaysian construction industry. In: Sobral HS (ed) Proc. of 2nd Intl. Symp. of RILEM on vegetable plants and their fibres as buildings materials, Salvador, Brazil. Chapman and Hall, London, pp 77–86

John VM, Agopyan V, Derolle A. Institute of technological research, São Paulo, Brazil. In: Vegetable plants and their fibres as building materials: proceedings of the second international RILEM symposium, Routledge, 1990, p 90

Juárez C, Guevara B, Valdez O, Durán-Herrera A (2010) Mechanical properties of natural fibers reinforced sustainable masonry. Constr Build Mater 24(8):1536–1541

Kim SB, Yi NH, Kim HY, Kim JHJ, Song YC (2010) Material and structural performance evaluation of recycled PET fiber reinforced concrete. Cem Concr Compos 32(3):232–240

Lima PRL, Santos, DO, Fontes CM, Barros, JAO, Toledo Filho RD (2016a) Deflection hardening of sustainable fiber-cement composites. Green Mater 4(1):18–30

Lima PRL, Barros, JAO, Santos, DO, Fontes CM, Lima JMF, Toledo Filho RD (2016b) Experimental and numerical analysis of short sisal f iber-cement composites produced with recycled matrix. Eur J Environ Civ Eng (Under Review)

Lima PRL, Roque AB, Fontes CM, Lima JMF, Barros JAO (2016c) Potentialities of cement based recycled materials reinforced with sisal fibers as filler component of precast concrete slab floors. In: Holmer Savastano Junior Juliano Fiorelli, Sergio Francisco dos Santos (eds) Non-conventional construction materials based on inorganic bonded fiber composites, Woodhead Elsevier, Pirassununga, (Under review)

Lima PRL, Barros, JAO, Roque AB, Fontes CM, Lima JMF (2016d) Produção e modelação numérica de abobadilha em microbetão reforçado com fibras discretas de sisal. In. Encontro Nacional Betão Estrutural—BE2016, Coimbra, Novembro 2016 (in Portuguese)

Marthong C, Marthong S (2016) An experimental study on the effect of PET fibers on the behavior of exterior RC beam-column connection subjected to reversed cyclic loading. Structures 5:175–185

Micelli F, Leone M, Centonze G, Aiello MA, (2015) Go green: using waste and recycling materials. In: Lu Y (ed) Infrastructure corrosion and durability—a sustainability study,

Published by OMICS Group eBooks. Foster City, CA 94404, USA 2015. Available online http://www.esciencecentral.org/ebooks/infrastructure-corrosion-durability/go-green-using-waste-and-recycling-materials.php#64. Accessed on 30 May 2015

MC2010, CEB fib Model Code 2010 – Final Draft, 2011

Omoniyi TE, Akinyemi BA (2012) Durability based suitability of bagasse-cement composite for roofing sheets. J Civ Eng Constr Technol 3(11):280–290

Paramasivam P, Nathan GK, Das Gupta NC (1984) Coconut fibre reinforced corrugated slabs. Int J Cem Compos Lightweight Concr 6(1):19–27

Pilakoutas K, Neocleous K, Tlemat H (2004) Reuse of tyre steel fibres as concrete reinforcement. Proc ICE: Eng Sustain 157(3):131–138

RILEM TC 162-TDF (2003) Test and design methods for steel fibre reinforced concrete.1327 r-e-design method. Final Recommendation. Mater Struct 36:560–567

Roma LC, Martello LS, Savastano H (2008) Evaluation of mechanical. Physical and thermal performance of cement-based tiles reinforced with vegetable fibers. Constr Build Mater 22 (4):668–674

Savastano H, Agopyan V, Nolasco AM, Pimentel L (1999) Plant fiber reinforced cement components for roofing. Constr Build Mater 13(8):433–438

Schafer HG, Brunssen GW (1990) Sisal-fiber reinforced lost formwork for floor slabs. In: Vegetable plants and their fibers as building materials. Proceedings of the second international symposium, Proceedings …, vol 7. Routledge, pp 162–174

SKAT (1987) FCR—fiber concrete roofing. In: Gram HE, Parry JPM, Rhyner K, Schaffner B, Stulz R, Wehrle K, Wehrli H (eds) A comprehensive report on the potential of Fiber Concrete Roofing, SKAT. Swiss Center for Appropriate Technology at ILE, Institute for Latin-American Research and for Development Co-operation, University of Saint-Gall IT Intermediate Technology Publications Ltd

Skatun O, Spigerverk C (1986) Applications of wet process steel fibre reinforced shotcrete in Scandinavia, RILEM Symposium FRC 86, developments in fibre reinforced cement and concrete, vol 2, paper 9.6

Tang Z, Ali M, Chouw N (2014) Residual compressive and shear strengths of novel coconut-fibre-reinforced-concrete interlocking blocks. Constr Build Mater 66:533–540

Tlemat H (2004), Steel FMres from waste tyres to concrete; testing, modelling and design. Thesis (PhD), Faculty of Engineering of the University of Sheffield

Tonoli GHD, Santos SF, Joaquim AP, Savastano H (2010) Effect of accelerated carbonation on cementitious roofing tiles reinforced with lignocellulosic fibre. Constr Build Mater 24(2):193–201

Tonoli GHD, Santos SF, Savastano H, Delvasto S, Gutiérrez RM, Murphy MDML (2011) Effects of natural weathering on microstructure and mineral composition of cementitious roofing tiles reinforced with fique fibre. Cement Concr Compos 33(2):225–232

United Nations Centre for Human Settlements (UNCHS) (1989) Kenya: standards and specifications for fibre-concrete roofing tiles. J Netw Afr Countries Local Build Mater Technol 1(1):11–12

Wells J (1994) Kenya: koma rock housing project Nairobi. Journal of the network of African countries on local building materials and technologies 3(2):1–18

Yan L, Chouw N (2013) Experimental study of flax FRP tube encased coir fibre reinforced concrete composite column. Constr Build Mater 40:1118–1127

Yan L, Chouw N, Krishnan J (2014) Effect of column parameters on flax FRP confined coir fibre reinforced concrete. Constr Build Mater 55:299–312

Zamanzadeh Z, Lourenço LAP, Barros JAO (2015a) Recycled steel fibre reinforced concrete failing in bending and in shear. Constr Build Mater 85:195–207

Zamanzadeh Z, Lourenço LAP, Barros JAO (2015b) Thin panels of cement composites reinforced with recycled fibres for the shear strengthening of reinforced concrete elements, fibre concrete 2015, FC2015. Czech Republic, Prague

Conclusions

Liberato Ferrara, Enzo Martinelli and Joaquim A.O. Barros

In this book the research activities performed as part of the FP7 project EnCoRe, "Environmentally friendly Concrete with Recycled and natural constituents", have been reviewed, dealing with the concept, material characterization and engineering/structural applications of sustainable concrete and cement based construction materials incorporating different kinds of recycled constituents from industrial and "natural" sources. These have ranged from the "classical" use of secondary raw materials as cement substitutes, e.g. *fly ash*, to "consolidating" and emerging technologies, such as the use of recycled concrete aggregates (RAC) from construction and demolition waste, to steel fibres from tyre recycling industry and natural fibres, by-products of food and agriculture industry.

The results of the research have been analysed in the context of the existing and challengingly evolving "state of the art" of related scientific and technological knowledge and development, but also in the framework of the existing guidelines and, in case codified, design approaches, also to verify their applicability or pointing out the need for their extension. As a matter of fact, this framework unavoidably governs the possibility of "spreading" the use of these materials and technologies into the building and civil engineering sectors.

For instance, a physically consistent approach aimed at controlling the resulting compressive strength of RAC mixtures has been formulated. Far from having achieved its final calibration, the proposed formulation is a promising conceptual tool for designing RAC mixtures. A more stable calibration of the relevant parameters and relationships would require further experimental work intended at creating the wide database of results needed for a statistically consistent calibration of the proposed relationships. Moreover, since this book mainly covers the compressive strength of RAC, the quantitative definition of relationships capable of predicting their relevant properties (both in the fresh and hardened state) as a

J.A.O. Barros et al. (eds.), *Recent Advances on Green Concrete for Structural Purposes*, Research for Development, DOI 10.1007/978-3-319-56797-6 function of the mixture and constituents properties is still an open issue and, as such, it deserves significant consideration in the future development of research.

Similar considerations can be made about cementitious composites internally reinforced by either recycled or natural fibers, whose physical behaviour in the fresh state, mechanical response under multiaxial stress states as well as under long term environmental actions deserve further investigation. Crucial aspect in this regards is a reliable methodology to evaluate the reinforcement mechanisms provided by recycled steel fibres, due to the erratic nature of the geometric characteristics and the possibility of having rubber particles remaining in the fibres, which pose extra challenges in the design context, both at analytical level, as well as using advanced numerical approaches.

Experimental research has been complemented by modelling activities, also with the aim of consciously, from the engineering point of view, and coherently proposing tailored design approaches, or extending the existing ones under sound and not merely heuristic proof of their reliability. The aforementioned predictive modelling approaches must "serve" to design the intended engineering applications and to address their concept, not merely in the framework of replicating commonly existing applications, but for the replacement of one or the other constituent with a by-product or a recycled "ingredient".

Therefore, challenging paths of investigation with the objective of *product innovation* have been identified in the synergy effects brought by (or surprisingly inborn in) the simultaneous use of more than one recycled constituent. This may not only mutually temper likely "individual" drawbacks (as in the synergy between fly ashes and recycled concrete aggregates) bus also add new functionalities to the composite, such as, *e.g.*, the use of natural fibres as healing promoters if used in combination with other constituents.

In this respect, the modelling activity, such as a "classical" finite element engineering modelling, has to be intended not only to support the experimental work in assessing and deeper understanding the material signature properties, and predict, in a design-wise perspective its in-structure performance. The modelling activity, in case tools and approaches taken from engineering fields other than the civil one (such as in the prediction of the casting flow processes of fresh concrete, which has several analogies with the modelling of plastic and metal forming processes) has to serve to address and support the *process innovation*, which only can highlight the "competitiveness" of this category of advanced sustainable cement based materials. This also with the ambitious aim of promoting and consolidating the penetration capacity and impact of these materials into the construction market. Specifically, since no comprehensive models are nowadays available for predicting *long-term behaviour* and *durability performance* of the "environmentally-friendly" cementitious composites addressed in this book, theoretical models should specifically be intended at filling the gap of knowledge on these aspects. In fact, they are needed as key information in carrying out a sound sustainability evaluation of RAC based on well-established analytical techniques, such as Life-Cycle Analysis.

Moreover, last but not the least, the modelling must be always conceived in a design-wise perspective, with the aim of creating and shaping those *framework innovation* that only can be the driving engine through which the results of fundamental and applied research can serve to a true and sustainable economic and social development.

MIX
Papier aus verantwortungsvollen Quellen
Paper from responsible sources
FSC® C105338

If you have any concerns about our products,
you can contact us on
ProductSafety@springernature.com

In case Publisher is established outside the EU,
the EU authorized representative is:
Springer Nature Customer Service Center GmbH
Europaplatz 3, 69115 Heidelberg, Germany

Printed by Libri Plureos GmbH
in Hamburg, Germany